Mechanical Design of Process Systems

Volume I

Piping and Pressure Vessels

Volume 1: Piping and Pressure Vessels

1. Piping Fluid Mechanics
2. The Engineering Mechanics of Piping
3. Heat Transfer in Piping and Equipment
4. Engineering Mechanics of Pressure Vessels
Appendix A. Partial Volumes and Pressure Vessel Calculations
Appendix B. National Wind Design Standards
Appendix C. Properties of Pipe
Appendix D. Conversion Factors
Index

Volume 2: Shell-and-Tube Heat Exchangers; Rotating Equipment; Bins, Silos and Stacks

5. The Engineering Mechanics of Bins, Silos and Stacks
6. Rotating Equipment
7. The Mechanical Design of Shell-and-Tube Heat Exchangers
8. External Loadings on Shell Structures
Appendix A. Partial Volumes and Pressure Vessel Calculations
Appendix B. National Wind Design Standards
Appendix C. Properties of Pipe
Appendix D. Conversion Factors
Index

Mechanical Design of Process Systems

Volume I

Piping and Pressure Vessels

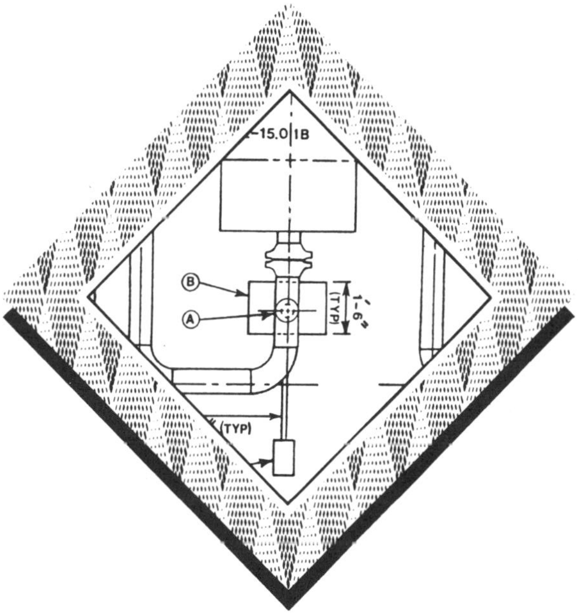

A. Keith Escoe

Gulf Publishing Company
Book Division
Houston, London, Paris, Tokyo

Mechanical Design of Process Systems

Volume 1

Piping and Pressure Vessels

Library of Congress Cataloging-in-Publication Data

Escoe, A. Keith.

Mechanical design of process systems.

Bibliography: p.

Includes index.

1. Chemical plants—Design and construction.

I. Title.

TP155.5.E83 1986 660.2′81 85-22005

ISBN 0-87201-562-9 (Vol. 1)

ISBN 0-87201-565-3 (Vol. 2)

Contents

Foreword vii
by John J. McKetta

Preface ix

Chapter 1
Piping Fluid Mechanics 1

Basic Equations, 1
Non-Newtonian Fluids, 5
Velocity Heads, 8
Pipe Flow Geometries, 22
Compressible Flow, 25
Piping Fluid Mechanics Problem Formulation, 25
Example 1-1: Friction Pressure Drop for a Hydrocarbon Gas-Steam Mixture in a Pipe, 27
Example 1-2: Frictional Pressure Drop for a Hot Oil System of a Process Tank, 33
Example 1-3: Friction Pressure Drop for a Waste Heat Recovery System, 42
Example 1-4: Pressure Drop in Relief Valve Piping System, 43
Notation, 45
References, 45

Chapter 2
The Engineering Mechanics of Piping 47

Piping Criteria, 47
Primary and Secondary Stresses, 49
Allowable Stress Range for Secondary Stresses.
Flexibility and Stiffness of Piping Systems, 52
Stiffness Method Advantages. Flexibility Method Advantages.
Stiffness Method and Large Piping, 58
Flexibility Method of Piping Mechanics. Pipe Loops.
Pipe Restraints and Anchors, 68
Pipe Lug Supports. Spring Supports. Expansion Joints. Pre-Stressed Piping.
Fluid Forces Exerted on Piping Systems, 81
Extraneous Piping Loads, 83
Example 2-1: Applying the Stiffness Method to a Modular Skid-Mounted Gas Liquefaction Facility, 88
Example 2-2: Applying the Flexibility Method to a Steam Turbine Exhaust Line, 95
Example 2-3: Flexibility Analysis for Hot Oil Piping, 96
Example 2-4: Lug Design, 98
Example 2-5: Relief Valve Piping System, 99
Example 2-6: Wind-Induced Vibrations of Piping, 100
Notation, 101
References, 101

Chapter 3
Heat Transfer in Piping and Equipment . . . 103

Jacketed Pipe versus Traced Pipe, 103
Tracing Piping Systems, 106
Traced Piping without Heat Transfer Cement. Traced Piping with Heat Transfer Cement. Condensate Return. Jacketed Pipe. Vessel and Equipment Traced Systems.
Heat Transfer in Residual Systems, 132
Heat Transfer through Cylindrical Shells. Residual Heat Transfer through Pipe Shoes.
Example 3-1: Steam Tracing Design, 136
Example 3-2: Hot Oil Tracing Design, 137
Example 3-3: Jacketed Pipe Design, 139
Example 3-4: Thermal Evaluation of a Process Tank, 140
Example 3-5: Thermal Design of a Process Tank, 142
Internal Baffle Plates Film Coefficient. Film Coefficient External to Baffles—Forced Convection. Heat Duty of Internal Vessel Plates. Outside Heat Transfer Jacket Plates. Heat Duty of Jacket Plates Clamped to Bottom Vessel Head. Total Heat Duty of Tank.

Example 3-6: Transient and Static Heat Transfer Design, 148
Static Heat Transfer Analysis. Total Heat Removal. Water Required for Cooling. Transient Heat Transfer Analysis.
Example 3-7: Heat Transfer through Vessel Skirts, 152
Example 3-8: Residual Heat Transfer, 154
Example 3-9: Heat Transfer through Pipe Shoe, 156
Notation, 156
References, 157

Chapter 4
The Engineering Mechanics of Pressure Vessels 159

Designing for Internal Pressure, 159
Designing for External Pressure, 160
Design of Horizontal Pressure Vessels, 166
Longitudinal Bending Stresses. Location of Saddle Supports. Wear Plate Design. Zick Stiffening Rings.
Steel Saddle Plate Design, 174
Saddle Bearing Plate Thickness, 180
Design of Self-Supported Vertical Vessels, 180
Minimum Shell Thickness Required for Combined Loads, 181
Support Skirt Design, 183
Anchor Bolts, 184
Base Plate Thickness Design, 186
Compression Ring and Gusset Plate Design, 189
Anchor Bolt Torque, 189
Wind Analysis of Towers, 190
Wind Design Speeds. Wind-Induced Moments. Wind-Induced Deflections of Towers. Wind-Induced Vibrations on Tall Towers. Ovaling. Criteria for Vibration Analysis.
Seismic Design of Tall Towers, 209
Vertical Distribution of Shear Forces.
Tower Shell Discontinuities and Conical Sections, 214
Example 4-1: Wear Plate Requirement Analysis, 215
Example 4-2: Mechanical Design of Process Column, 215
Section Moments of Inertial Tower Section Stress Calculations. Skirt and Base Plate Design. Section Centroids. Vortex-Induced Vibration. Equivalent Diameter Approach versus ANSI A58.1-1982.
Example 4-3: Seismic Analysis of a Vertical Tower, 237
Example 4-4: Vibration Analysis for Tower with Large Vortex-Induced Displacements, 241
Moments of Inertia. Wind Deflections.
Example 4-5: Saddle Plate Analysis of a Horizontal Vessel, 249
Saddle Plate Buckling Analysis. Horizontal Reaction Force on Saddle.
Notation, 252
References, 254

Appendix A
Partial Volumes and Pressure Vessel Calculations 255

Partial Volumes of Spherically Dished Heads, 256
Partial Volumes of Elliptical Heads, 257
Partial Volumes of Torispherical Heads, 259
Internal Pressure ASME Formulations with Outside Dimensions, 261
Internal Pressure ASME Formulations with Inside Dimensions, 262

Appendix B
National Wind Design Standards 265

Criteria for Determining Wind Speed, 265
Wind Speed Relationships, 266
ANSI A58.1-1982 Wind Categories, 267

Appendix C
Properties of Pipe 271

Insulation Weight Factors, 278
Weights of Piping Materials, 279

Appendix D
Conversion Factors 303

Alphabetical Conversion Factors, 304
Synchronous Speeds, 311
Temperature Conversion, 312
Altitude and Atmospheric Pressures, 313
Pressure Conversion Chart, 314

Index 315

Foreword

The engineer who understands the impact of process design decisions on mechanical design details is in a position to save his client or his company a lot of money.

That is because the test of any process design is in how cost-effectively it yields the desired product, and how "cost" generally translates to "equipment": How much will the process require? How long will it last? How much energy will it consume per unit of product?

In this two-volume work on *Mechanical Design of Process Systems,* A. K. Escoe has performed a monumental service for mechanical design engineers and chemical process engineers alike. It is presented in such a manner that even the neophyte engineer can grasp its full value. He has produced an in-depth review of the way in which process design specifications are interpreted into precise equipment designs. Perhaps most valuable of all are the extensive *worked examples* throughout the text, of actual designs that have been successfully executed in the field.

The piping system is the central nervous system of a fluid flow process, and the author has treated this with proper respect in two excellent chapters on fluid mechanics and the engineering mechanics of piping.

The chapter on heat transfer in vessels and piping illustrates lucidly the interrelationship between process and mechanical design. Every engineer working with industrial process systems will benefit from reading this chapter.

Although the author has made a herculean effort in covering the mechanical design of pressure vessels, heat exchangers, rotating equipment, and bins, silos and stacks, it is true that there are omissions. It is hoped that, as the author hints in his preface, a future volume might be added covering multiphase flow, specific cogeneration processes, turbines, and detailed piping dynamics.

Still, at this writing these two volumes comprise an outstanding practical reference for chemical and mechanical engineers and a detailed instructional manual for students.

I recommend these volumes highly for each design engineer's professional library.

John J. McKetta, Ph.D., P.E.
Joe C. Walter Professor of Chemical Engineering
University of Texas, Austin

Dedication

To the memory of my beloved parents, Aubrey H. Escoe and Odessa Davies Escoe; and to the dedicated engineer, Dr. Judith Arlene Resnik, U.S. astronaut aboard the ill-fated space shuttle *Challenger* (Flight 51-L).

Preface to Volume 1

This book's purpose is to show how to apply mechanical engineering concepts to process system design. Process systems are common to a wide variety of industries including petrochemical processing, food and pharmaceutical manufacturing, power generation (including cogeneration), ship building, and even the aerospace industry. The book is based on years of proven, successful practice, and almost all of the examples described are from process systems now in operation.

While practicality is probably its key asset, this first volume contains a unique collection of valuable information, such as velocity head data; comparison of the flexibility and stiffness methods of pipe stress analyses; analysis of heat transfer through pipe supports and vessel skirts; a comprehensive method on the design of horizontal vessel saddles as well as a method to determine when wear plates are required; detailed static and dynamic methods of tower design considering wind gusts, vortex-induced vibration and seismic analysis of towers; and a comparative synopsis of the various national wind codes.

Topics included in the text are considered to be those typically encountered in engineering practice. Therefore, because most mechanical systems involve single-phase flow, two-phase flow is not covered. Because of its ubiquitous coverage in the literature, flange design is also excluded in this presentation. Since all of the major pressure vessel codes thoroughly discuss and illustrate the phenomenon of external pressure, this subject is only mentioned briefly.

This book is not intended to be a substitute or a replacement of any accepted code or standard. The reader is strongly encouraged to consult and be knowledgeable of any accepted standard or code that may govern. It is felt that this book is a valuable supplement to any standard or code used.

The book is slanted toward the practices of the ASME vessel and piping codes. In one area of vessel design the British Standard is favored because it provides excellent technical information on Zick rings. The book is written to be useful regardless of which code or standard is used. The intent is not to be heavily prejudiced toward any standard, but to discuss the issue—engineering. If one feels that a certain standard or code should be mentioned, please keep in mind that there are others who may be using different standards and it is impossible to discuss all of them.

The reader's academic level is assumed to be a bachelor of science degree in mechanical engineering, but engineers with bachelor of science degrees in civil, chemical, electrical, or other engineering disciplines should have little difficulty with the book, provided, of course, that they have received adequate academic training or experience.

Junior or senior undergraduate engineering students should find the book a useful introduction to the application of mechanical engineering to process systems. Professors should find the book a helpful reference (and a source for potential exam problems), as well as a practical textbook for junior-, senior-, or graduate-level courses in the mechanical, civil, or chemical engineering fields. The book can also be used to supplement an introductory level textbook.

The French philosopher Voltaire once said, "Common sense is not very common," and unfortunately, this is sometimes the case in engineering. Common sense is often the by-product of experience, and while both are essential to sound engineering practice, neither can be learned from books alone. It is one of this book's goals to unite these three elements of "book learning," common sense, and experience to give the novice a better grasp of engineering principles and procedures, and serve as a practical design reference for the veteran engineer.

Finally, I wish to thank Dr. John J. McKetta, professor of chemical engineering at the University of Texas at Austin, who had many helpful comments, suggestions, and words of encouragement. I also wish to thank other engineering faculty members at the University of Texas at Austin for their comments. I must express thanks to Larry D. Briggs for reviewing some calculations in Chapter 4; and last, but certainly not least, I wish to express gratitude to William J. Lowe and Timothy W. Calk of Gulf Publishing Company, whose hard work and patience made this book possible.

A. Keith Escoe, P.E.

1

Piping Fluid Mechanics

The study of fluid energy in piping systems is a comprehensive subject that could in itself fill countless volumes. This chapter is primarily concerned with fluid energy dissipated as friction resulting in a head loss. Although this topic is popularly known in industry as "hydraulics," the term "piping fluid mechanics" is used here to avoid confusion.

BASIC EQUATIONS

The basic equation of fluid mechanics, originally derived by Daniel Bernoulli in 1738, evolved from the principle of conservation of energy:

$$-\frac{dP}{\rho} = \frac{d\,(V^2)}{2g_c} + \frac{g}{g_c}\,dY + dF + dH_A + dH_E \qquad (1\text{-}1)$$

where ρ = density, lb_m/ft^3 or g_m/cm^3
P = pressure, lb_f/ft^2 or kg_f/cm^2
V = velocity, ft/sec or cm/sec
g_c = conversion constant, 32.17 ($ft\text{-}lb_m/sec^2 lb_f$)
g = gravitational acceleration = 32.2 ft/sec^2, cm/sec^2; g/g_c = 1.0
Y = height above datum, ft, cm
dY = differential between height above datum and reference point, ft, cm
F = head loss, friction loss, or frictional pressure drop, $ft\text{-}lb_f/lb_m$, $cm\text{-}kg_f/g_m$
H_A = energy added by mechanical devices, e.g. pumps, $ft\text{-}lb_f/lb_m$, $cm\text{-}kg_f/g_m$
H_E = energy extracted by mechanical devices, e.g. turbines, $ft\text{-}lb_f/lb_m$, $cm\text{-}kg_f/g_m$

Rewriting Equation 1-1 along a fluid streamline between points 1 and 2 with steady, incompressible flow and no mechanical energy added or extracted results in

$$\frac{P_1 - P_2}{\rho} = \frac{V_1^2 - V_2^2}{2g_c} + (Y_1 - Y_2)\frac{g}{g_c} + F \qquad (1\text{-}2)$$

where subscripts 1 and 2 refer to flow upstream (after the flow process) and downstream (before the flow process), respectively, and

$$\frac{P_1 - P_2}{\rho} = \text{change in pressure head}$$

$$\frac{V_1 - V_2}{2g_c} = \text{change in velocity head (kinetic energy)}$$

$$= dZ$$

$$(Y_1 - Y_2)\frac{g}{g_c} = \text{change in static head (potential energy)}$$

$$F = \text{friction loss in } \frac{ft\,(lb_f)}{lb_m}, \frac{cm\,(kg_f)}{g_m}$$

The following are expressions of the Bernoulli equation when applied to various incompressible and compressible flow conditions:

Incompressible flow—

$$\frac{P_1 - P_2}{\rho} = \frac{V_1^2 - V_2^2}{2g_c} + (Z_1 - Z_2)\frac{g}{g_c} + F + H_A + H_E$$

Compressible-isothermal flow—

$$\left(\frac{P_1}{\rho_1}\right) \ln\left(\frac{P_1}{P_2}\right) = \frac{V_2^2}{2g}\left[1 - \left(\frac{A_2}{A_1}\right)^2\left(\frac{P_2}{P_1}\right)^2\right] + (Z_2 - Z_1)$$
$$+ F + H_A + H_E$$

Compressible-adiabatic flow—

$$\left(\frac{k}{k-1}\right)\left(\frac{P_1}{\rho_1}\right)\left[1-\left(\frac{P_2}{P_1}\right)^{(k-1)/k}\right] = \frac{V_2^2}{2g}\left[1-\left(\frac{P_2}{P_1}\right)^{2/k}\left(\frac{A_2}{A_1}\right)^2\right] + (Z_2 - Z_1) + F + H_A + H_E$$

where $\left(\frac{\rho}{\rho_1}\right)^k = \left(\frac{P}{P_1}\right)$ = general gas law

k = specific heat ratio (adiabatic coefficient), C_p/C_v
C_p = specific heat at constant pressure, Btu/lb_m-°F
C_v = specific heat at constant volume, Btu/lb_m-°F

Equation 1-2 is the analytical expression that states a pressure loss is caused by a change in velocity head, static head, and friction head. The most common units are "feet of head." lb_m and lb_f do not cancel out and the expression is exactly "energy (ft-lb_f) per pound of mass."

In most industrial fluid problems, Equation 1-2 is cumbersome to use, because the friction loss is the parameter most often desired. The friction loss is the work done by the fluid in overcoming viscous resistance. This friction loss can only rarely be analytically derived and is determined by empirical data developed through experimental testing.

Forcing a fluid through a pipe component requires energy. This energy is expended by shear forces that develop between the pipe wall and the fluid, and to a lesser extent among the fluid elements themselves. These shear forces are opposed to fluid flow and require excess energy to overcome. Figure 1-1 shows a simple version of this phenomenon and illustrates how shear stresses increase in the radial direction away from the pipe center line and are maximum within the boundary layer next to the wall. Friction energy loss is a result of these shear stresses next to the pipe wall. Excess loss in energy occurs because of local turbulence and changes in the direction and speed of flow. As a fluid changes direction, energy is expended because of a change in momentum. The methods used to determine energy loss caused by wall friction are essentially the same, where the pipe component is treated as a straight piece of pipe. However, the methods used to determine energy loss caused by change in momentum differ, and a couple are described as follows.

Equivalent Length

In this approach to determining energy loss caused by a change in fluid momentum, a piping component is extended a theoretical length that would yield the same energy loss as the actual component. This length is called the "equivalent length" because it is that length required to obtain the same amount of friction pressure drop as the piping component alone. The major problem with

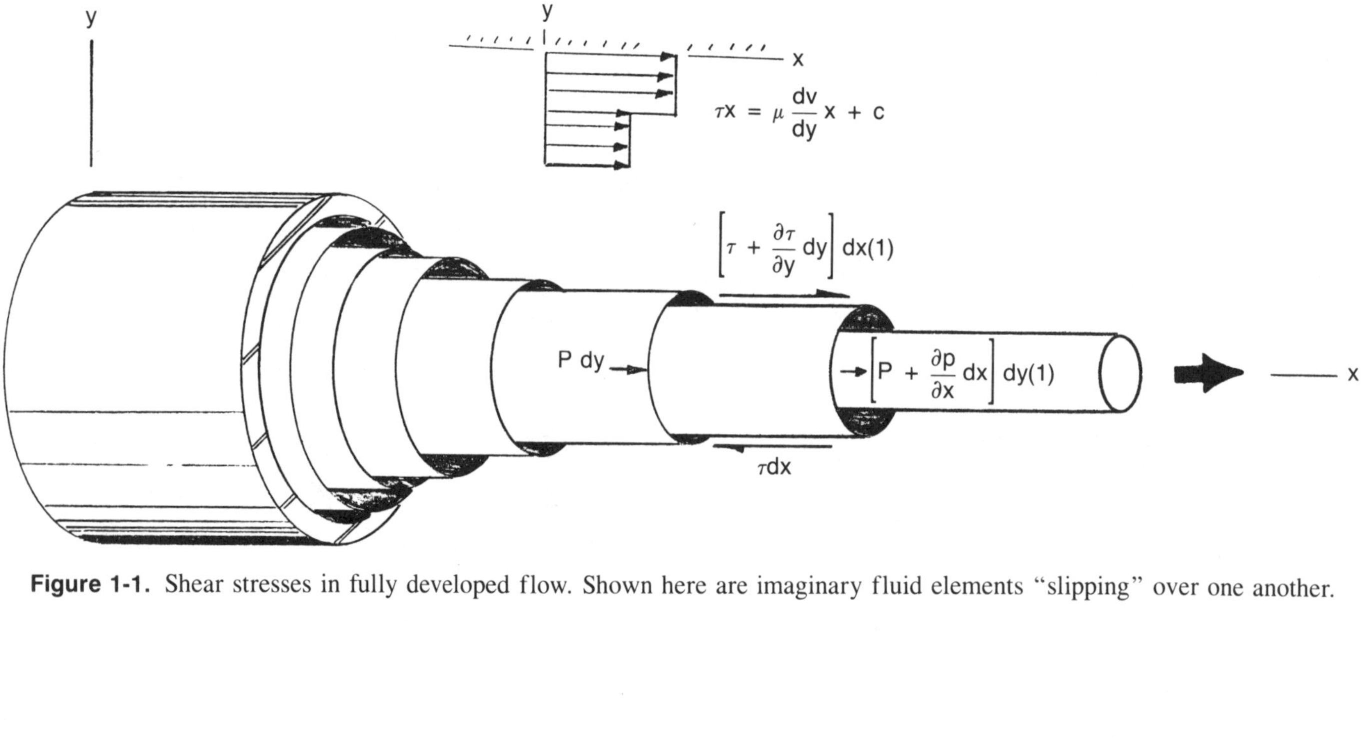

Figure 1-1. Shear stresses in fully developed flow. Shown here are imaginary fluid elements "slipping" over one another.

this method is that the equivalent length for a pipe component varies with the Reynolds number, roughness, size, and geometry of the pipe. All these parameters must be analyzed in using this method.

Velocity Head Method

Since the excess head loss is mostly attributed to fluid turbulence, the velocity head method is widely accepted and is replacing the equivalent length method in fluid calculations. Throughout this book, the velocity head approach will be used.

The velocity head is the amount of kinetic energy in a fluid, $V^2/2g_c$. This quantity may be represented by the amount of potential energy required to accelerate a fluid to a given velocity. Consider a tank holding a fluid with a pipe entrance shown in Figure 1-2. We draw a streamline from point 1 of the fluid surface to point 2 at the pipe entrance. Applying Equation 1-2 at point 1 we obtain the following:

$$\frac{P_1}{\rho} = Y_1 \frac{g}{g_c}$$

And applying Equation 1-2 at point 2 we have

$$\frac{P_1 - P_2}{\rho} = \frac{P_1}{\rho} - \frac{V_2^2}{2g_c}$$

in which the change in fluid pressure between points 1 and 2 is $V_2^2/2g_c$ or one velocity head. A pressure gauge mounted on the pipe entrance would record the difference of pressure of one velocity head. This term is accounted for in Equation 1-2 by $V_1^2 - V_2^2/2g_c$.

Analyzing a simple conversion from potential to kinetic energy is an elementary procedure, as demonstrated. After the fluid passes through the pipe entrance into the piping system, the factor F in Equation 1-2 becomes the desired parameter. This friction loss is the work done by the fluid in overcoming viscous resistance and loss attributed to turbulence. The parameter F is composed of two components, pipe wall friction and losses for the various pipe fittings, pipe entrances, pipe exits, and fluid obstructions that contribute to a loss in fluid energy. These latter losses are described in terms of velocity heads, K_i. In solving for F in Equation 1-2, we first obtain pressure loss attributed to pipe wall friction, represented by

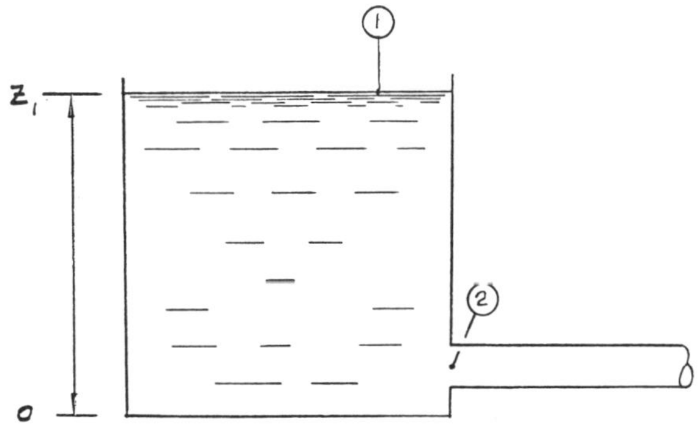

Figure 1-2. Storage tank.

$$-\Delta P_F = F = \frac{\rho V^2}{2g_c}\left(\frac{fL}{d}\right) \quad (1\text{-}3)$$

By adding values of velocity head losses to Equation 1-3, we obtain the following for any piping system:

$$-\Delta P_f = \left(\frac{fL}{d} + \sum K_i\right)\frac{\rho V^2}{2g_c} \quad (1\text{-}4)$$

where fL/d is the dependent pipe friction of the pipe of diameter d over the length L, and ΣK_i the summation of velocity head losses. Equation 1-4 provides the friction pressure drop in a pipe for a steady-state incompressible fluid of fully developed flow with a flat velocity profile. Examples of this equation are given after the terms in Equation 1-4 are further explained.

The term $(fL/d)(\rho V^2/2g_c)$ expresses the amount of energy loss attributed to shear forces at the pipe wall and is based on experimental evidence. It is a function of the pipe component length and diameter and the velocity of the fluid. Writing the relationship for friction pressure drop as a result of pipe wall friction results in

$$F_{P_f} = \frac{fL}{144d}\frac{\rho V^2}{2g_c} \quad (1\text{-}5)$$

where F_{P_f} = friction loss, psi
L = length of pipe, in.
d = corroded inside diameter, in.

The other terms are explained with Equation 1-1. Equation 1-5 may be expressed in various forms. To express flow rate in gpm (w) and d in inches use

$$F_{P_f} = 0.000217\ fLW/d^5 \quad (1\text{-}5a)$$

Equation 1-5 is the most commonly used relationship and is known as the Fanning equation. Dividing the equation by $\rho/144$ yields feet of friction loss rather than psi.

The reader is cautioned in applying the friction factor f, because it is not always defined as above and some authors use $4f_1$ in place of f. If such factors are used, particular attention should be paid to the specific friction factor chart used.

The friction factor f is dependent upon the dimensionless term expressing the roughness of the pipe (E/D, where E is the depth of the pipe) and the dimensionless Reynolds number $N_{Re} = d\rho V/M$, where μ is the absolute viscosity of the fluid, in-lb_f-sec/ft^2. The Reynolds number is the single most important parameter in fluid mechanics because it establishes flow regimes and dynamic similarity. The relationship between the friction factor f, the pipe roughness, and the Reynolds number is shown in the classic relationship given by Moody in Figure 1-3.

Figure 1-3 may be presented in a more convenient form as shown in Figure 1-4, where the relative roughness of the pipe is based on a single value of roughness. This value of roughness must be an average value estimated to simplify the problem. The figures presented herein are the best available until more reliable friction factor data can be obtained and better understood through use of new methods for measuring roughness.

Figure 1-3 is broken into three flow regimes—laminar, transition from laminar to turbulent, and turbulent. The Reynolds numbers establishing these zones are 2,100 for laminar, 2,100 to 3,000 for transition zone, and 3,000 or more for turbulent.

The basis for Figure 1-3 is the classic Colebrook equation

$$\frac{1}{(f)^{0.5}} = -2 \log_{10}\left(\frac{\epsilon/d}{3.7} + \frac{2.51}{N_{Re}(f)^{0.5}}\right) \qquad (1\text{-}6a)$$

$$\text{for } (3{,}000 \text{ to } 4{,}000) \leqslant N_{Re} \leqslant 10^8$$

For laminar flow the friction factor is determined by the simple expression

$$f = \frac{64}{N_{Re}} \qquad (1\text{-}6b)$$

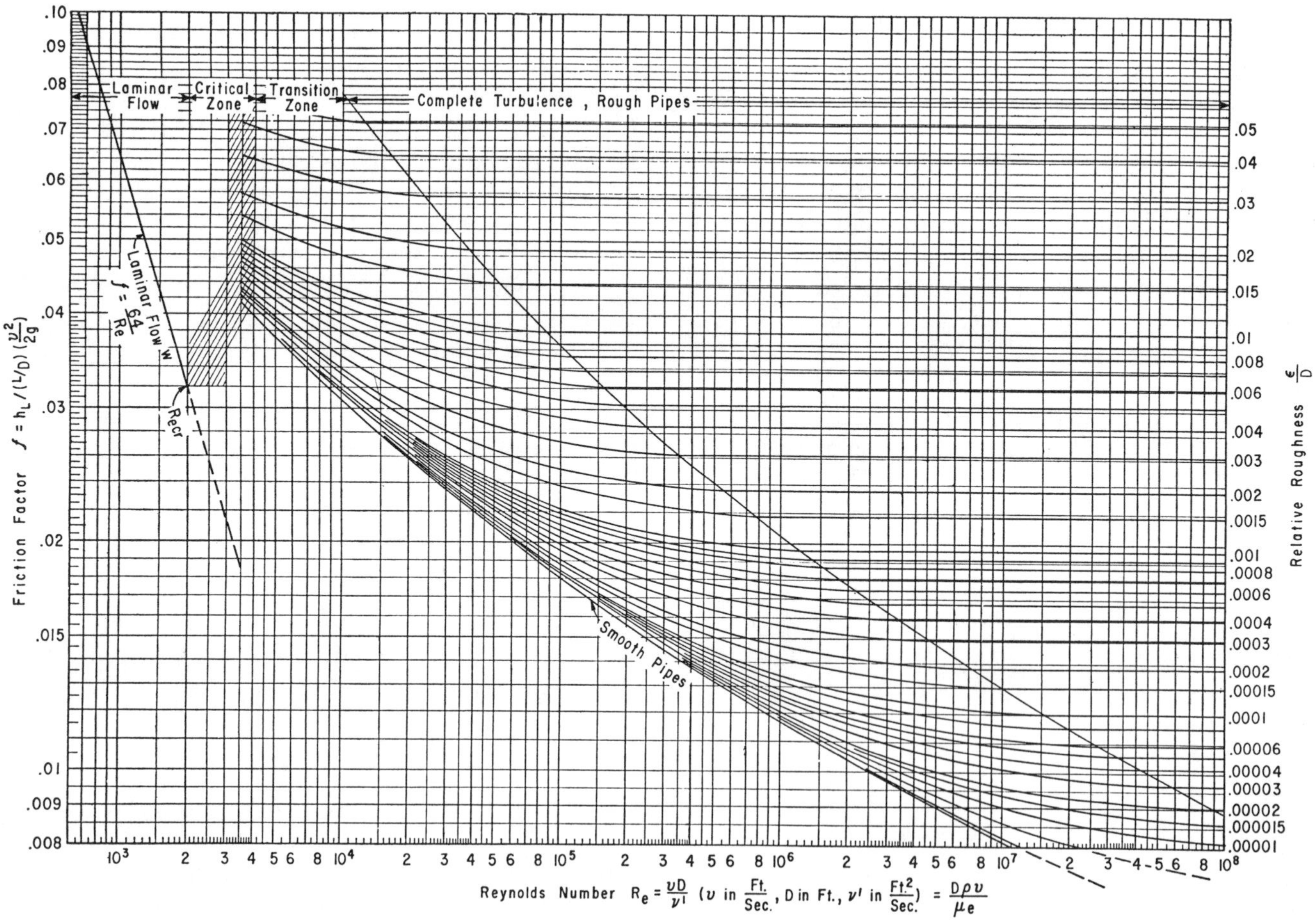

Figure 1-3. Moody friction factors. (Reprinted from *Pipe Friction Manual,* © 1954 by Hydraulic Institute. Data from L. F. Moody, *Friction Factors for Pipe Flow,* permission of ASME.)

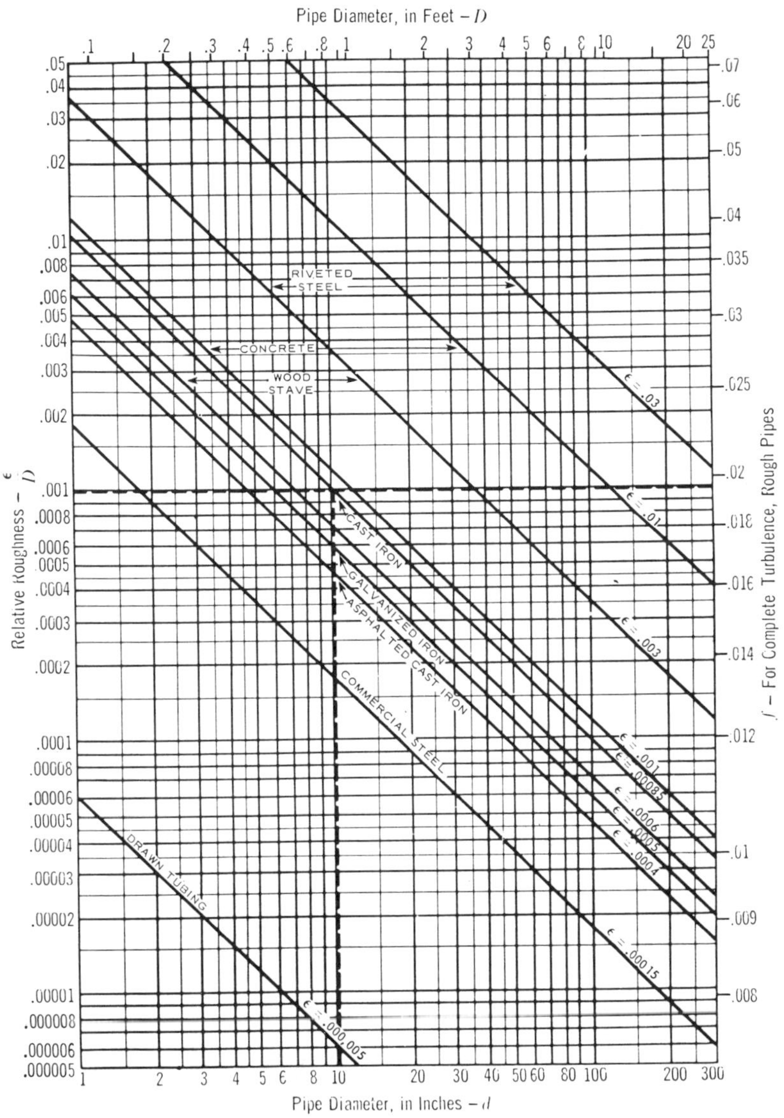

Figure 1-4. Relative roughness of pipe materials and friction factors for complete turbulence. (Courtesy of Crane Company [5]. Data from L. F. Moody, *Friction Factors for Pipe Flow,* permission of ASME.)

Equation 1-6a, which describes the friction factor for turbulent flow in pipe of any roughness, is a simple addition of the Prandtl solution for smooth pipe and the von Karman solution for rough pipe. The relationship holds for the transition between rough and smooth pipe.

To solve Equation 1-6a for the friction factor f an iterative analysis is required because the function is nonhomogeneous and inseparable. There are several empirical relations of f expressed as an independent separate function of f (ϵ/d, N_{Re}), but with today's micro-computers Equation 1-6 can be solved more accurately and expediently with iteration.

Dimensional forms of Equation 1-4 are presented in Table 1-1 [1], where the equation is conveniently shown in various units that are used to solve fluid pressure loss problems.

NON-NEWTONIAN FLUIDS

The Colebrook equation holds for fluids whose flow properties are dependent on the fluid viscosity. These fluids consist of all gases, liquids, and solutions of low molecular weight and are known as Newtonian fluids. In

Table 1-1
Dimensional Forms Used With Equation 1-4 [1]

Pressure Drop, Reynolds Number, and Velocity				
Flow rate	$-\Delta P_f$ or ρH_f	N_{Re}	V	H_V
W	$\left(\frac{bfL}{D} + \Sigma K_i\right)\frac{aW^2}{\rho D^4}$	$\frac{cW}{\mu D}$	$\frac{dW}{\rho D^2}$	$\frac{\rho V^2}{e}$
Q	$\left(\frac{bfL}{D} + \Sigma K_i\right)\frac{a\rho Q^2}{D^4}$	$\frac{cQ\rho}{\mu D}$	$\frac{dQ}{D^2}$	$\frac{\rho V^2}{e}$

Friction Factor

If $N_{Re} < 2{,}000$, $f = 64/N_{Re}$

If $N_{Re} \geqslant 2{,}000$, $f = [2 \log_{10} (0.27\, \epsilon/D + (7/N_{Re}{}^{0.9})]^{-2*}$

Units and constants

	Conventional units				**[Gas]**	**Metric units**			
$-\Delta P_f(H_f)$	psi	psi	(ft)	(ft)	$inH_2O[60°F]$	bar	bar	Pa	(m)
W(Q)	lb/h	(gpm)	lb/h	(gpm)	(acfm)	kg/s	(L/s)	kg/s	(m^3/s)
D	in.	in.	in.	in.	in.	mm	mm	m	m
ϵ	in.	in.	in.	in.	in.	mm	mm	m	m
L	ft	ft	ft	ft	ft	m	m	m	m
ρ	lb/ft^3	lb/ft^3	lb/ft^3	lb/ft^3	lb/ft^3	kg/m^3	kg/m^3	kg/m^3	kg/m^3
μ	cp	cp	cp	cp	cp	mPa-s(cp)	mPa-s(cp)	Pa-s	Pa-s
V	ft/s	ft/s	ft/s	ft/s	ft/min	m/s	m/s	m/s	m/s
H_V	psi	psi	ft	ft	in. H_2O	bar	bar	Pa	m
a	2.799×10^{-7}	1.801×10^{-5}	4.031×10^{-5}	2.593×10^{-3}	0.02792	8.106×10^6	8.106	0.8106	0.08265
b	12	12	12	12	12	1,000	1,000	1	1
c	6.316	50.66	6.316	50.66	379.0	1.273×10^6	1.273	1.273	1.273
d	0.05093	0.4085	0.05093	0.4085	183.3	1.273×10^6	1.273	1.273	1.273
e	9,266	9,266	$64.35 \times \rho$	$64.35 \times \rho$	1.204×10^6	2×10^5	2×10^5	2	$19.61 \times \rho$

a,b,c,d,e = constants
D = pipe diameter
f = Weisbach friction factor
H_f = frictional head loss
K = number of velocity heads
L = pipe length
$-\Delta P_f$ = frictional pressure drop
Q = volumetric flowrate
N_{Re} = Reynolds number
V = velocity
H_V = velocity head
W = mass flow rate
ϵ = pipe roughness
μ = fluid viscosity
ρ = fluid density

Newtonian fluids the viscosity alone defines the rheological behavior.

Non-Newtonian fluids are those in which the viscosity alone does not define their rheological behavior. Such fluids are solutions composed of solid particles that expand. Clay and very dense slurries are examples of non-Newtonian fluids. The flow properties of such fluids are a function of the particle characteristics, e.g., size and flexibility and thermal expansion.

Purely viscous non-Newtonian fluids are classified into three categories: time-dependent and time-independent and viscoelastic. A time-dependent fluid displays slow changes in rheological properties, such as thixotropic fluids that exhibit reversible structural changes. Several types of crude oil fit into this category. Another type of time-dependent non-Newtonian fluid is rheopectic fluids. Under constant sustained shear, these fluids' rate of structural deformation exceeds the rate of structural decay. One such category of fluids is polyester. Rheopectic fluids are less common than thixotropic fluids.

Time-independent fluids that are purely viscous are classified as pseudoplastic, dilatant, Bingham, and yield-pseudoplastic fluids. In *pseudoplastic fluids* an infinitesimal shear stress will initiate motion and the ratio of shear stress with velocity decreases with increasing velocity gradient. This type of fluid is encountered in solutions or suspensions of fine particles that form loosely bounded aggregates that can break down or reform with an increase or decrease in shear rate. Such solutions are aqueous dispersions of polyvinyl acetate and of an acrylic copolymer; aqueous solutions of sodium carboxymethyl cellulose, and of ammonium polymethacrylate; and an aqueous suspension of limestone.

In dilatant fluids an infinitesimal shear stress will start motion and the ratio of shear stress to velocity increases as the velocity is increased. A dilatant fluid is characterized by an increase in volume of a fixed amount of dispersion, such as wet sand, when subjected to a deformation that alters the interparticle distances of its constituents from their minimum-size configuration. Such fluids are titanium dioxide particles in water or sucrose solution. Dilatant fluids are much rarer than pseudoplastic fluids.

In *Bingham fluids* a finite shearing stress is required to initiate motion and there is a linear relationship between the shearing stress—after motion impends—and the velocity gradient. Such fluids include thickened hydrocarbon greases, certain asphalts, water suspensions of clay, fly ash, finely divided minerals, quartz, sewage sludge, and point systems.

Yield-pseudoplastic fluids are similar to Bingham fluids, but the relationship between the excess shearing stress after motion impends and velocity gradient is nonlinear. Fluids in this category are defined by their rheograms, where relationships between the shear stress and rate of shear exhibit a geometric convexity to the shear stress axis. Such fluids are many clay-water and similar suspensions and aqueous solutions of carboxypolymethylene (carbopol).

Viscoelastic fluids make up the last category of non-Newtonian fluids. The term "viscoelastic fluid" is applied to the most general of fluids—those that exhibit the characteristic of partial elastic recovery of the fluid structure. Whenever a viscoelastic fluid is subjected to a rapid change in deformation, elastic recoil or stress relaxation occurs. Many solutions exhibit viscoelastic properties under appropriate conditions—molten polymers, which are highly elastic; and solutions of long-charged molecules, such as polyethylene oxide and polyacrylamides. Processes such as coagulation, oil-well fracturing, and high-capacity pipelines rely on polymeric additives to cause pressure drops. Viscoelastic fluids exhibit the "Weissenberg effect," which is caused by normal stresses and produces unusual phenomena, such as the tendency of the fluid to climb up a shaft rotating in the fluid.

For any time-independent non-Newtonian fluid, Metzer and Reed [2] have developed the following generalized Reynolds number fraction:

$$N_{Re} = \frac{D^n U^{2-n} \rho}{\gamma} \tag{1-7}$$

where D = pipe ID, ft
U = average bulk velocity, ft/sec
ρ = density, lb_m/ft^3
γ = generalized viscosity coefficient, lb_m/ft sec = $g_c\ c\ 8^{n-1}$ (see Table 1-1)
C = experimentally determined flow constant, μ/g_c for a Newtonian fluid
n = empirical constant that is a function of non-Newtonian behavior (flow behavior index), 1.0 for Newtonian fluids

For $n = 1.0$ and $C = \mu/g_c$, Equation 1-7 reduces to $N_{Re} = Du\,\rho/\mu$ for Newtonian fluids. For $2{,}100 \leqslant N_{Re} \leqslant 100{,}000$ the following empirical relations can be used for determining the friction factor:

$$f = \frac{a_n}{(N'_{Re})\ b_n}$$

where $b_n = 0.0019498\ (n)^{-4.511}$
$n = (7.8958 \times 10^{-7})\ (a_n)\ 182.1321$

Typical values for γ and n are given in Table 1-2 [3]. Values for γ and n not available in literature must be determined by viscosimeter measurements.

Figure 1-5 shows the rheological classification of non-Newtonian mixtures that behave as single-phase flow. The reader is urged to refer to Govier [4] for further information on non-Newtonian fluid or other complex mixtures. Usually, the mechanical design of process systems does not involve non-Newtonian fluids, but knowledge of them and their peculiarities is a must if the need arises.

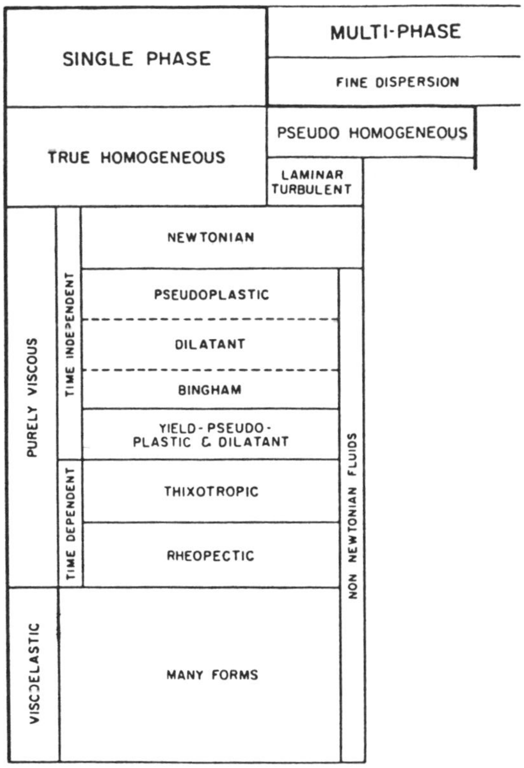

Figure 1-5. Rheological classification of complex mixtures that behave as single phase fluids [4].

Table 1-2
Rheological Constants for Some Typical Non-Newtonian Fluids* [3]

Composition of Fluid	Rheological Constants n	γ	Composition of Fluid	Rheological Constants n	γ
23.3% Illinois yellow clay in water	0.229	0.863	18.6% solids, Mississippi clay in water	0.022	0.105
0.67% carboxy-methyl-cellulose (CMC) in water	0.716	0.121	14.3% clay in water	0.350	0.0344
1.5% CMC in water	0.554	0.920	21.2% clay in water	0.335	0.0855
3.0% CMC in water	0.566	2.80	25.0% clay in water	0.185	0.204
33% lime water	0.171	0.983	31.9% clay in water	0.251	0.414
10% napalm in kerosene	0.520	1.18	36.8% clay in water	0.176	1.07
4% paper pulp in water	0.575	6.13	40.4% clay in water	0.132	2.30
54.3% cement rock in water	0.153	0.331	23% lime in water	0.178	1.04

* Reproduced by permission: A. B. Metzner and J. C. Reed, *AICHE Journal*, 1, 434 (1955).

VELOCITY HEADS

Returning to Equation 1-4, let's focus on the term ΣK_i. This term represents the *excess* velocity heads lost in fluid motion due to fluid turbulence caused by local turbulence at the pipe wall and change in flow direction. The latter is the greatest contributor to the ΣK_i term. When a fluid strikes a surface and changes flow direction, it loses momentum and, therefore, energy. Considering the 90° elbow in Figure 1-6, we see that the fluid changes direction from the x to the y direction and imparts reactions F_x and F_y, each a function of the pressure and velocity of the fluid. End conditions of the elbow determine some of the velocity head loss, that is, where the fitting is a "smooth elbow" or a "screwed elbow." A smooth elbow is one that is either flanged or welded to the pipe such that a smooth internal surface is encountered by the flow. In a screwed elbow there are abrupt changes in the wall causing local turbulence and hence increased velocity head loss.

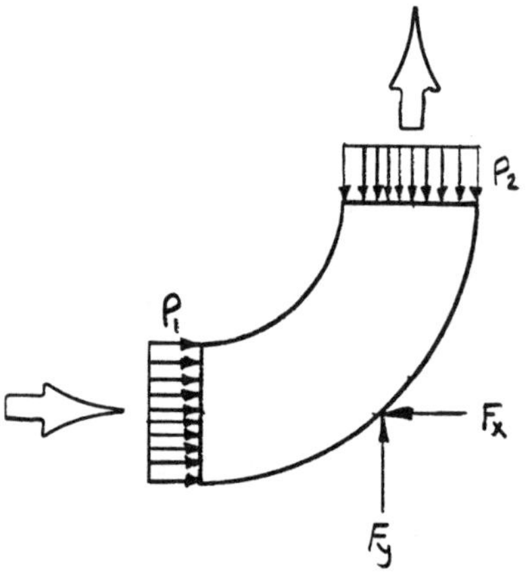

Figure 1-6. Reactions on an elbow induced by a change of flow.

Analytical determination of velocity heads can only be accomplished in a few simplified cases. The values for velocity heads must be determined and verified empirically. Comprehensive listings of such velocity head (K) values are given in Figures 1-7 [5], 1-8 [5], 1-9 [6], and 1-10. Using these values in Equation 1-4, you can analyze most cases of friction pressure drop for pipe under 24 inches in diameter. For pipe with diameter greater than 24 inches, an additional analysis must be made in solving for the velocity head term. This method, presented by Hooper [7] is called the "two-K method."

TWO-K METHOD

As explained previously, the value of K does not depend on the roughness of the fitting or the fitting size, but rather on the Reynolds number and the geometry of the fitting. The published data for single K values apply to fully-developed turbulent flow, and K is independent of N_{Re} when N_{Re} is well into the turbulent zone. As N_{Re} approaches 1,000, the value of K increases. When $N_{Re} < 1{,}000$, the value of K becomes inversely proportional to N_{Re}. In large diameter pipe (> 24 in.) the value of N_{Re} must be carefully considered if values of 1,000 or less are encountered. The two-K method accounts for this dependency in the following equation:

$$K = K_1/N_{Re} + K_\infty (1 + 1/d) \tag{1-8}$$

where K_1 = K for the fitting of $N_{Re} = 1$
K_∞ = K for a large fitting of $N_{Re} = \infty$
d = internal diameter of attached pipe, in.

Text continued page 22

Representative Resistance Coefficients (K) for Valves and Fittings

PIPE FRICTION DATA FOR CLEAN COMMERCIAL STEEL PIPE WITH FLOW IN ZONE OF COMPLETE TURBULENCE

Nominal Size	½″	¾″	1″	1¼″	1½″	2″	2½, 3″	4″	5″	6″	8-10″	12-16″	18-24″
Friction Factor (f_T)	.027	.025	.023	.022	.021	.019	.018	.017	.016	.015	.014	.013	.012

FORMULAS FOR CALCULATING "K" FACTORS* FOR VALVES AND FITTINGS WITH REDUCED PORT

• **Formula 1**

$$K_2 = \frac{0.8\left(\sin\frac{\theta}{2}\right)(1 - \beta^2)}{\beta^4} = \frac{K_1}{\beta^4}$$

• **Formula 2**

$$K_2 = \frac{0.5\,(1 - \beta^2)\sqrt{\sin\frac{\theta}{2}}}{\beta^4} = \frac{K_1}{\beta^4}$$

• **Formula 3**

$$K_2 = \frac{2.6\left(\sin\frac{\theta}{2}\right)(1 - \beta^2)^2}{\beta^4} = \frac{K_1}{\beta^4}$$

• **Formula 4**

$$K_2 = \frac{(1 - \beta^2)^2}{\beta^4} = \frac{K_1}{\beta^4}$$

• **Formula 5**

$$K_2 = \frac{K_1}{\beta^4} + \text{Formula 1} + \text{Formula 3}$$

$$K_2 = \frac{K_1 + \sin\frac{\theta}{2}\left[0.8\,(1 - \beta^2) + 2.6\,(1 - \beta^2)^2\right]}{\beta^4}$$

• **Formula 6**

$$K_2 = \frac{K_1}{\beta^4} + \text{Formula 2} + \text{Formula 4}$$

$$K_2 = \frac{K_1 + 0.5\sqrt{\sin\frac{\theta}{2}}\,(1 - \beta^2) + (1 - \beta^2)^2}{\beta^4}$$

• **Formula 7**

$$K_2 = \frac{K_1}{\beta^4} + \beta\,(\text{Formula 2} + \text{Formula 4}) \text{ when } \theta = 180°$$

$$K_2 = \frac{K_1 + \beta\left[0.5\,(1 - \beta^2) + (1 - \beta^2)^2\right]}{\beta^4}$$

$$\beta = \frac{d_1}{d_2}$$

$$\beta^2 = \left(\frac{d_1}{d_2}\right)^2 = \frac{a_1}{a_2}$$

Subscript 1 defines dimensions and coefficients with reference to the smaller diameter.

Subscript 2 refers to the larger diameter.

SUDDEN AND GRADUAL CONTRACTION

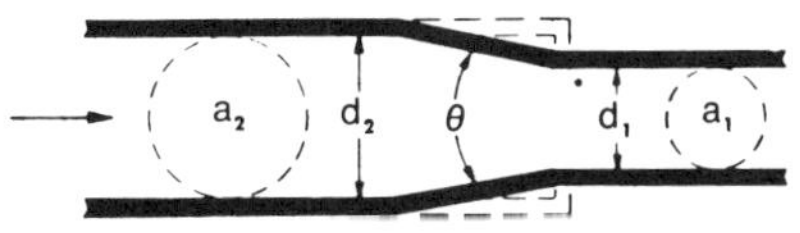

If: $\theta \leq 45°$ . . . . K_2 = Formula 1

$45° < \theta \leq 180°$ . . . K_2 = Formula 2

SUDDEN AND GRADUAL ENLARGEMENT

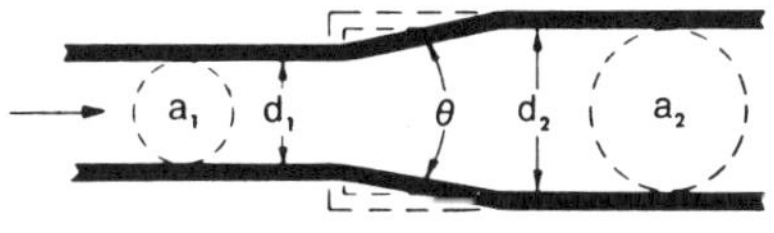

If: $\theta \leq 45°$ . . . . K_2 = Formula 3

$45° < \theta \leq 180°$ . . . K_2 = Formula 4

Figure 1-7A. Selected Crane Company velocity head values. (Courtesy Crane Company [5].)

GATE VALVES
Wedge Disc, Double Disc, or Plug Type

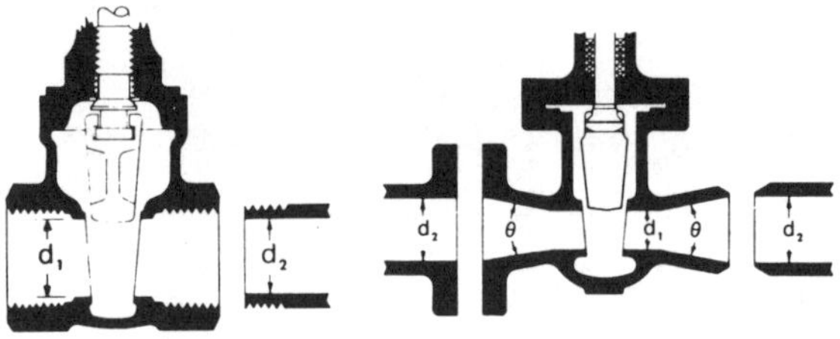

If: $\beta = 1, \theta = 0$ . . . . $K_1 = 8 f_T$

$\beta < 1$ and $\theta \leqq 45°$ . . . . K_2 = Formula 5

$\beta < 1$ and $45° < \theta \leqq 180°$ . . . K_2 = Formula 6

SWING CHECK VALVES

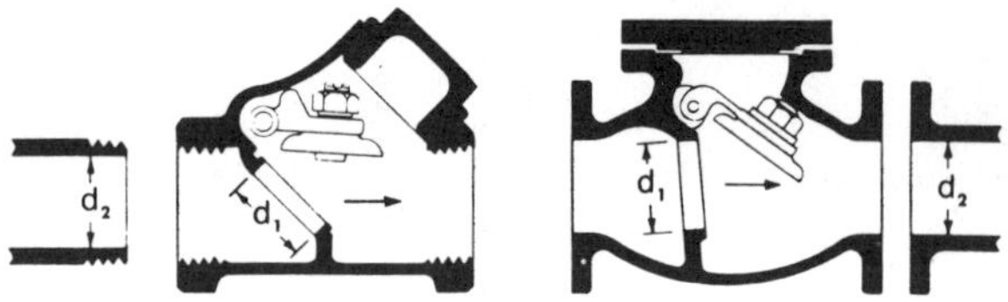

$K = 100 f_T$

Minimum pipe velocity (fps) for full disc lift $= 35 \sqrt{\bar{V}}$

$K = 50 f_T$

Minimum pipe velocity (fps) for full disc lift $= 48 \sqrt{\bar{V}}$

GLOBE AND ANGLE VALVES

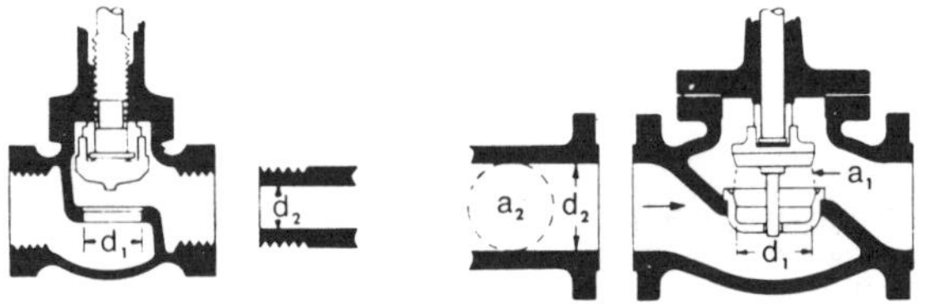

If: $\beta = 1 \ldots K_1 = 340 f_T$

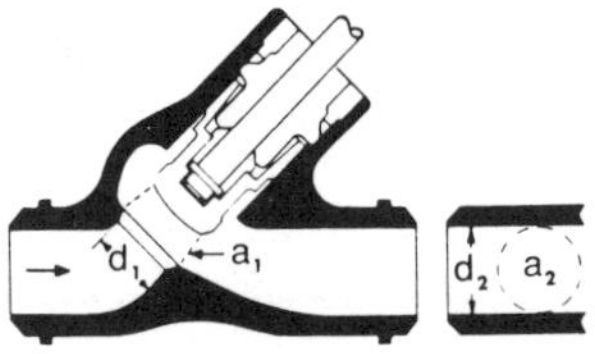

If: $\beta = 1 \ldots K_1 = 55 f_T$

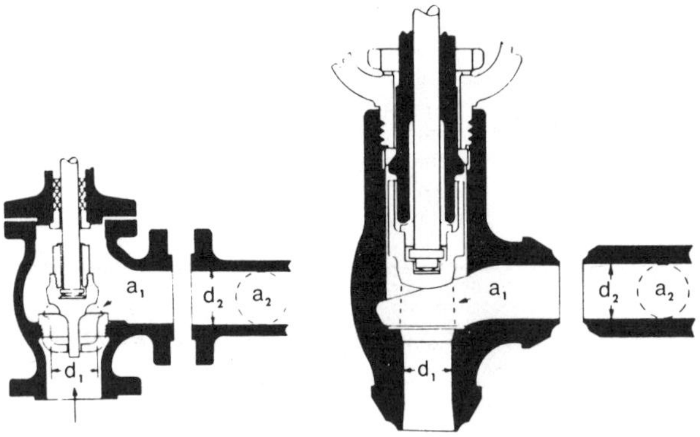

If: $\beta = 1 \ldots K_1 = 150 f_T$ If: $\beta = 1 \ldots K_1 = 55 f_T$

All globe and angle valves, whether reduced seat or throttled,
If: $\beta < 1 \ldots K_2$ = Formula 7

LIFT CHECK VALVES

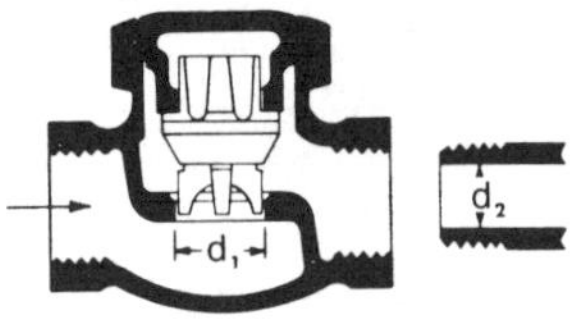

If: $\beta = 1 \ldots K_1 = 600 f_T$

$\beta < 1 \ldots K_2$ = Formula 7

Minimum pipe velocity (fps) for full disc lift $= 40\ \beta^2 \sqrt{\bar{V}}$

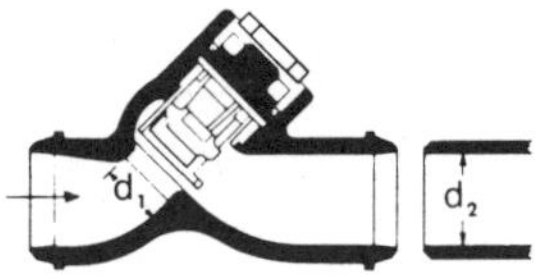

If: $\beta = 1 \ldots K_1 = 55 f_T$

$\beta < 1 \ldots K_2$ = Formula 7

Minimum pipe velocity (fps) for full disc lift $= 140\ \beta^2 \sqrt{\bar{V}}$

TILTING DISC CHECK VALVES

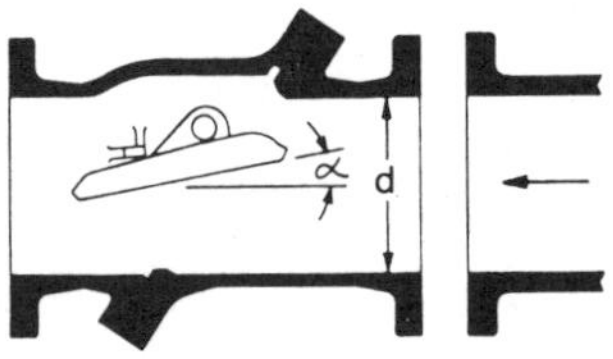

	$\alpha = 5°$	$\alpha = 15°$
Sizes 2 to 8″. . . K =	$40 f_T$	$120 f_T$
Sizes 10 to 14″. . . K =	$30 f_T$	$90 f_T$
Sizes 16 to 48″. . . K =	$20 f_T$	$60 f_T$
Minimum pipe velocity (fps) for full disc lift =	$80 \sqrt{\bar{V}}$	$30 \sqrt{\bar{V}}$

Figure 1-7B. Selected Crane Company velocity head values. (Courtesy Crane Company [5].)

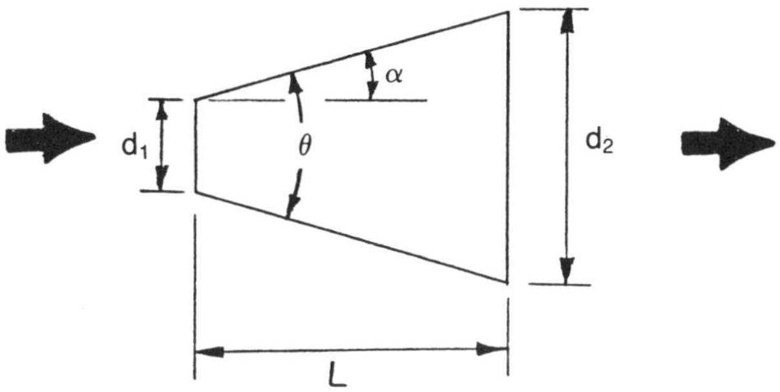

If $\theta \leqslant 45°$ or $\alpha \leqslant 22.5°$

$$K_1 = 2.6(1 - \beta^2)^2 \sin \alpha$$

If $45° < \theta \leqslant 180°$ or $22.5° < \alpha \leqslant 90°$

$$K_1 = (1 - \beta^2)^2$$

$$\alpha = \arcsin\left(\frac{d_2 - d_1}{2L}\right) = \frac{\theta}{2}$$

Figure 1-8. Calculated Crane K-values for concentric conical diffusers are tabulated in Table 1-6.

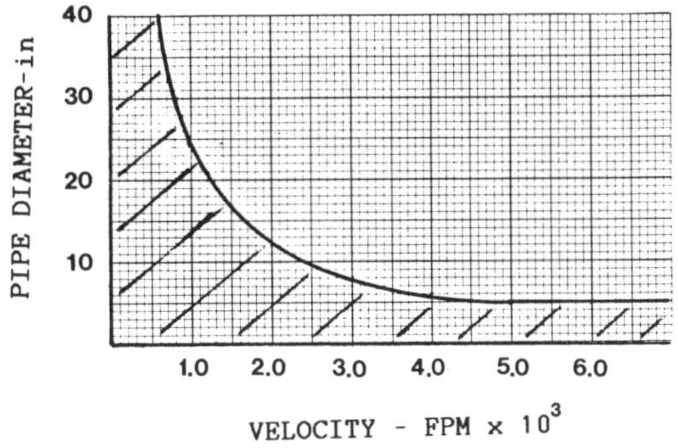

SHADED AREA UNDER CURVE IS CRITERIA FOR MULTIPLYING CORRECTION FACTOR, a, BY THE VALUE OF K FOR EACH FITTING AND PIPE COMPONENT.

CORRECTION FACTOR TABLE

$N \times 10^{-4}$	1	2	3	4	6	8	10	≥14
a	1.40	1.26	1.19	1.14	1.09	1.06	1.04	1.0

Figure 1-9. Correction factor.

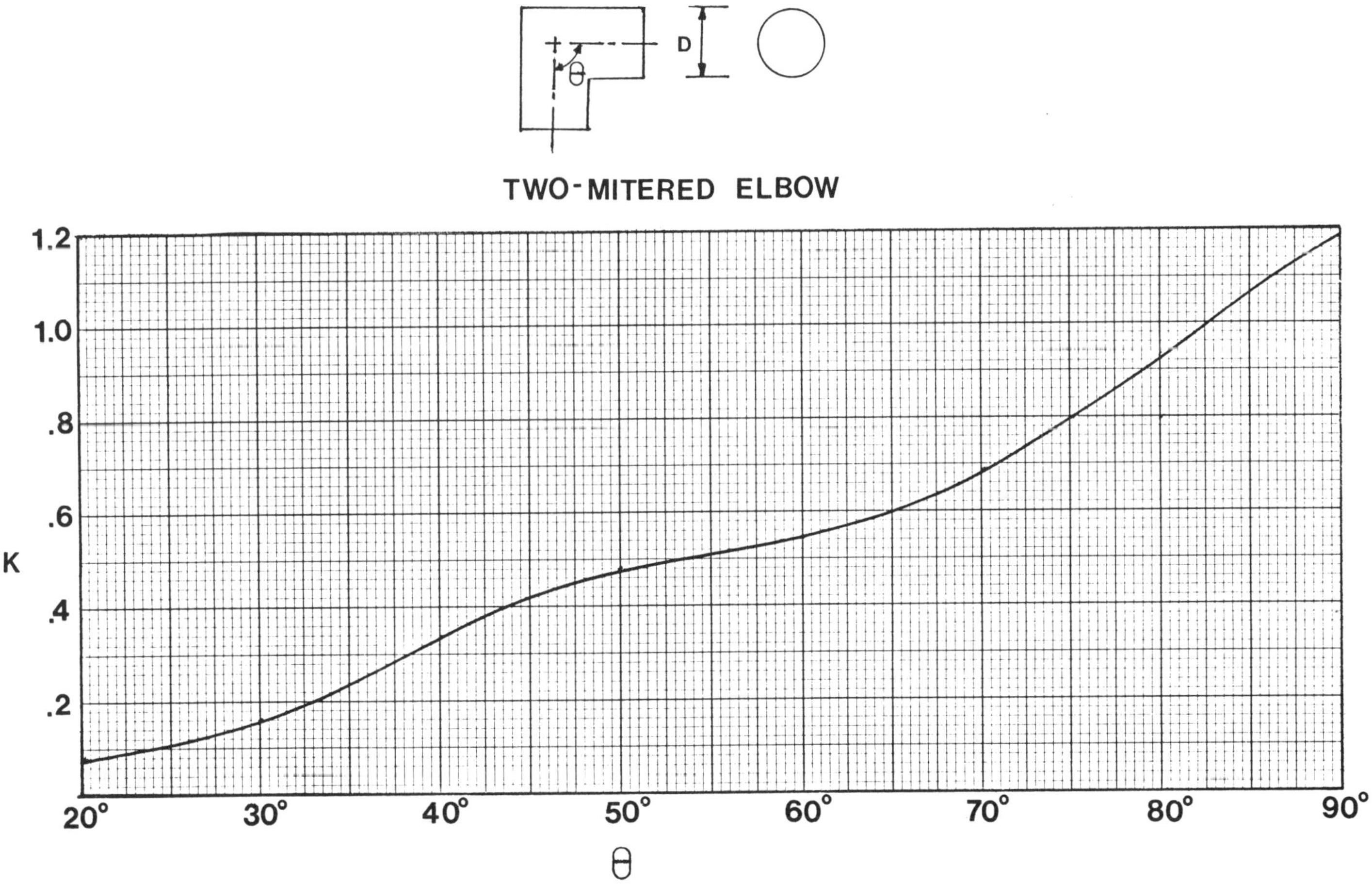

Figure 1-10A. Velocity heads for change of flow [6].

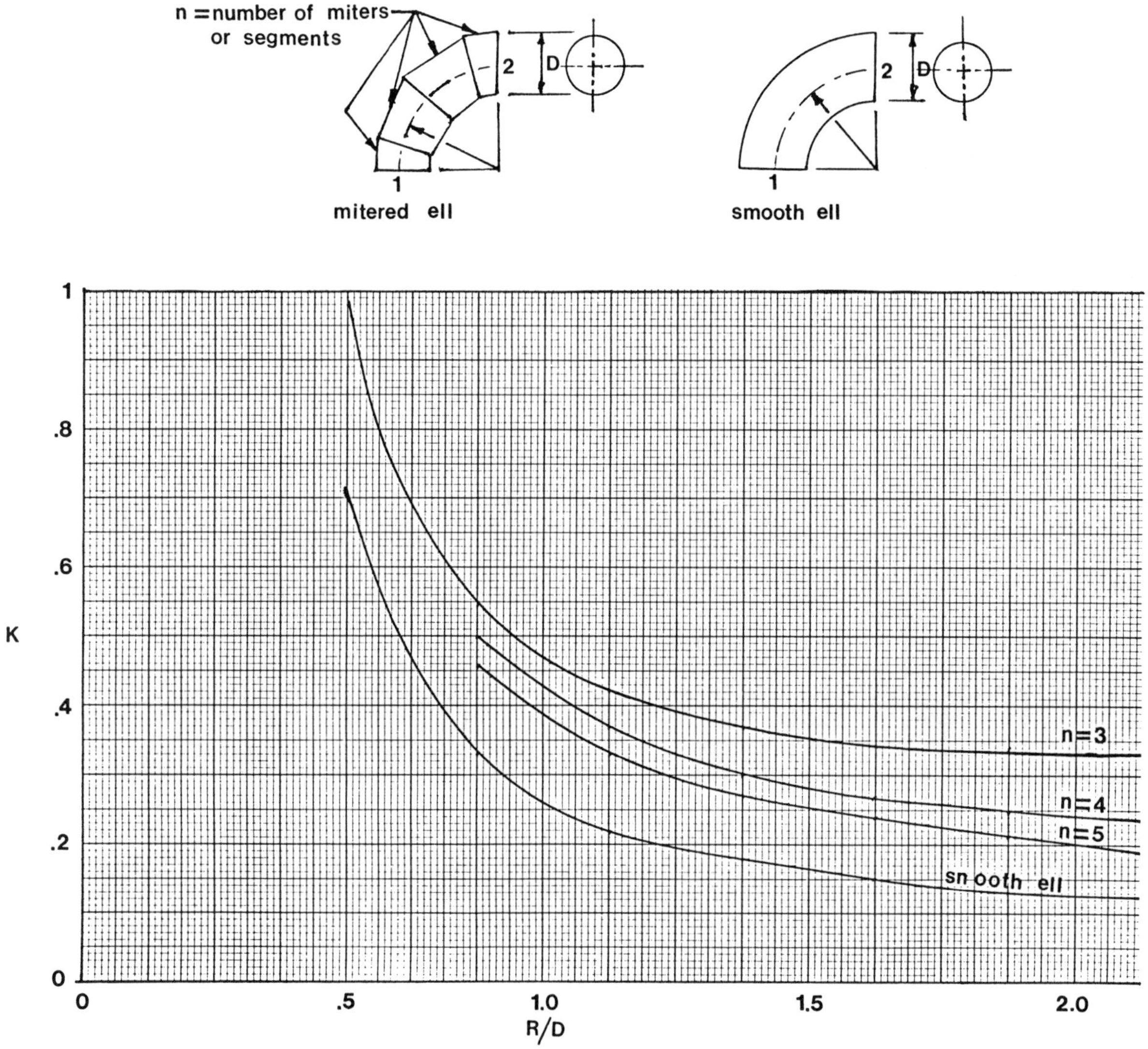

Figure 1-10B. Velocity heads for change of flow [6].

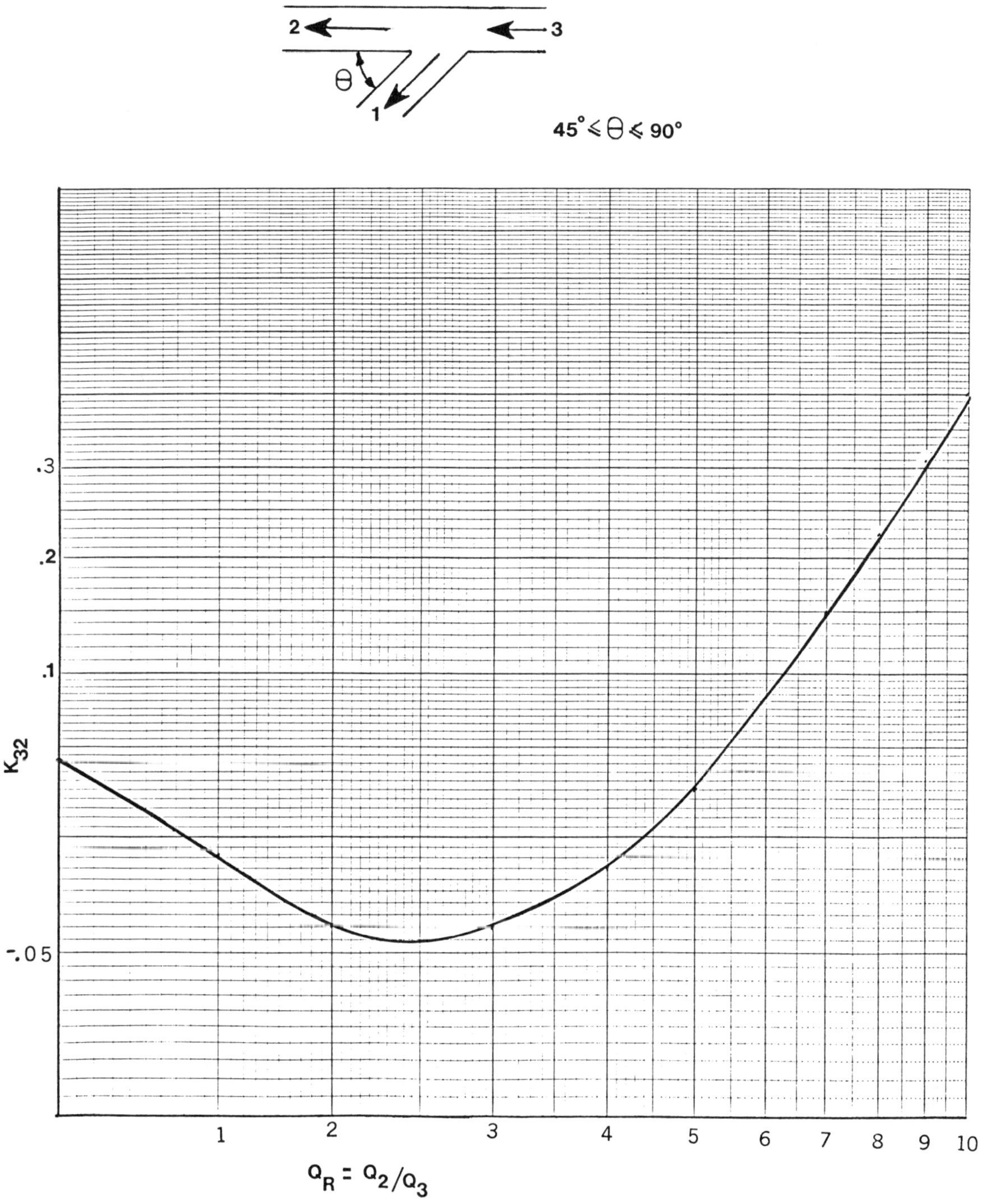

Figure 1-10C. Velocity heads for change of flow [6].

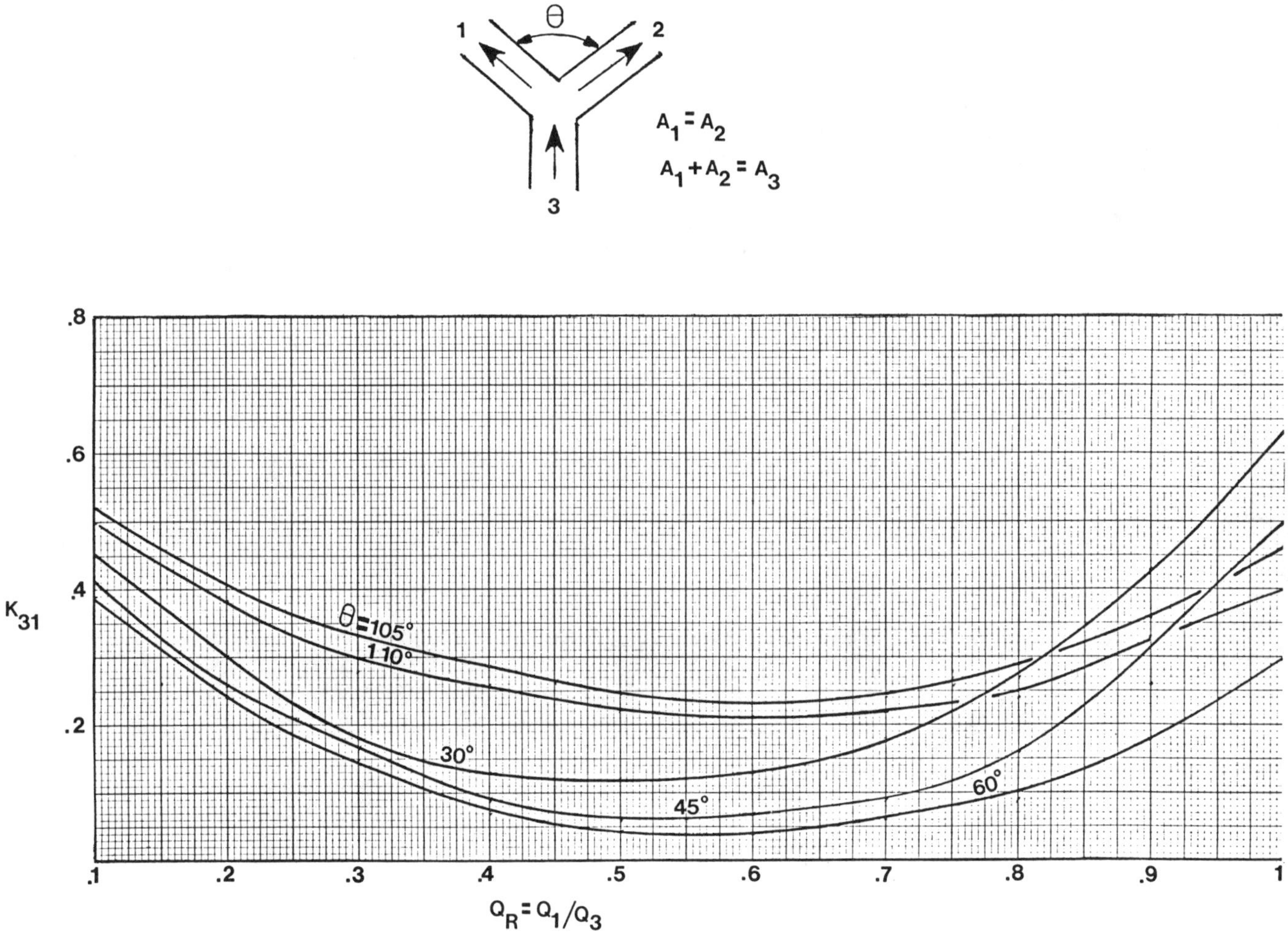

Figure 1-10D. Velocity heads for change of flow [6].

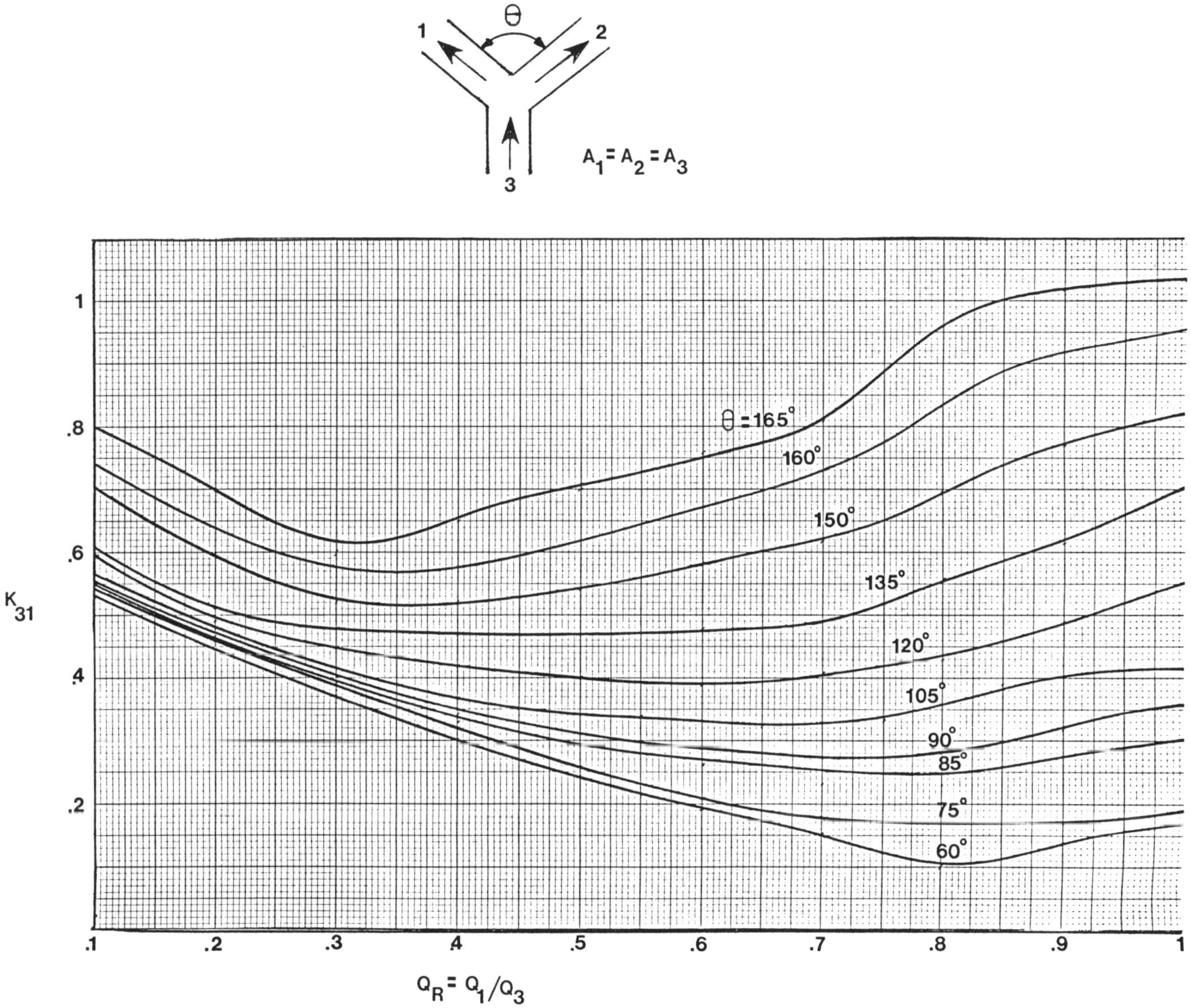

Figure 1-10E. Velocity heads for change of flow [6].

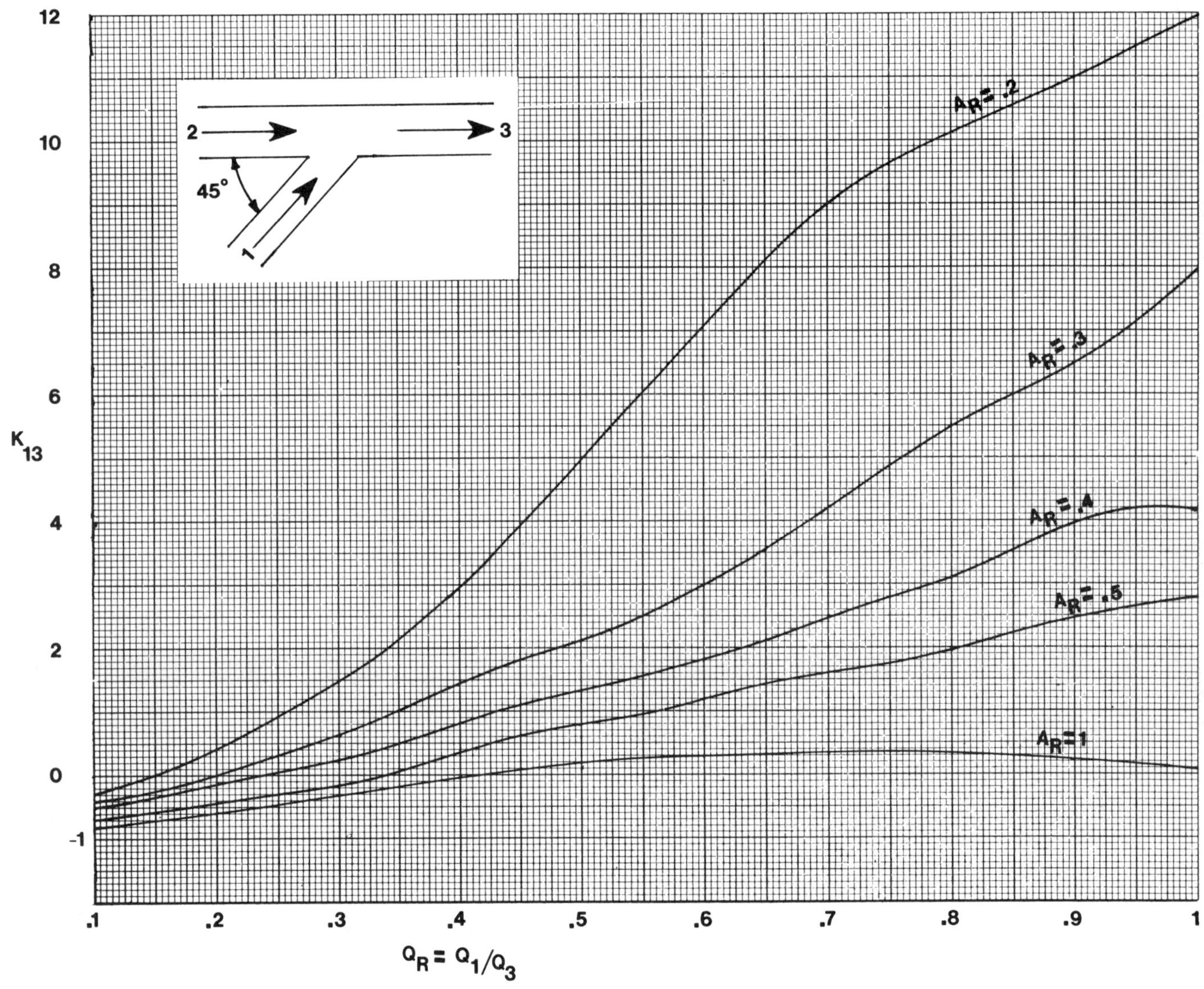

Figure 1-10F. Velocity heads for change of flow [6].

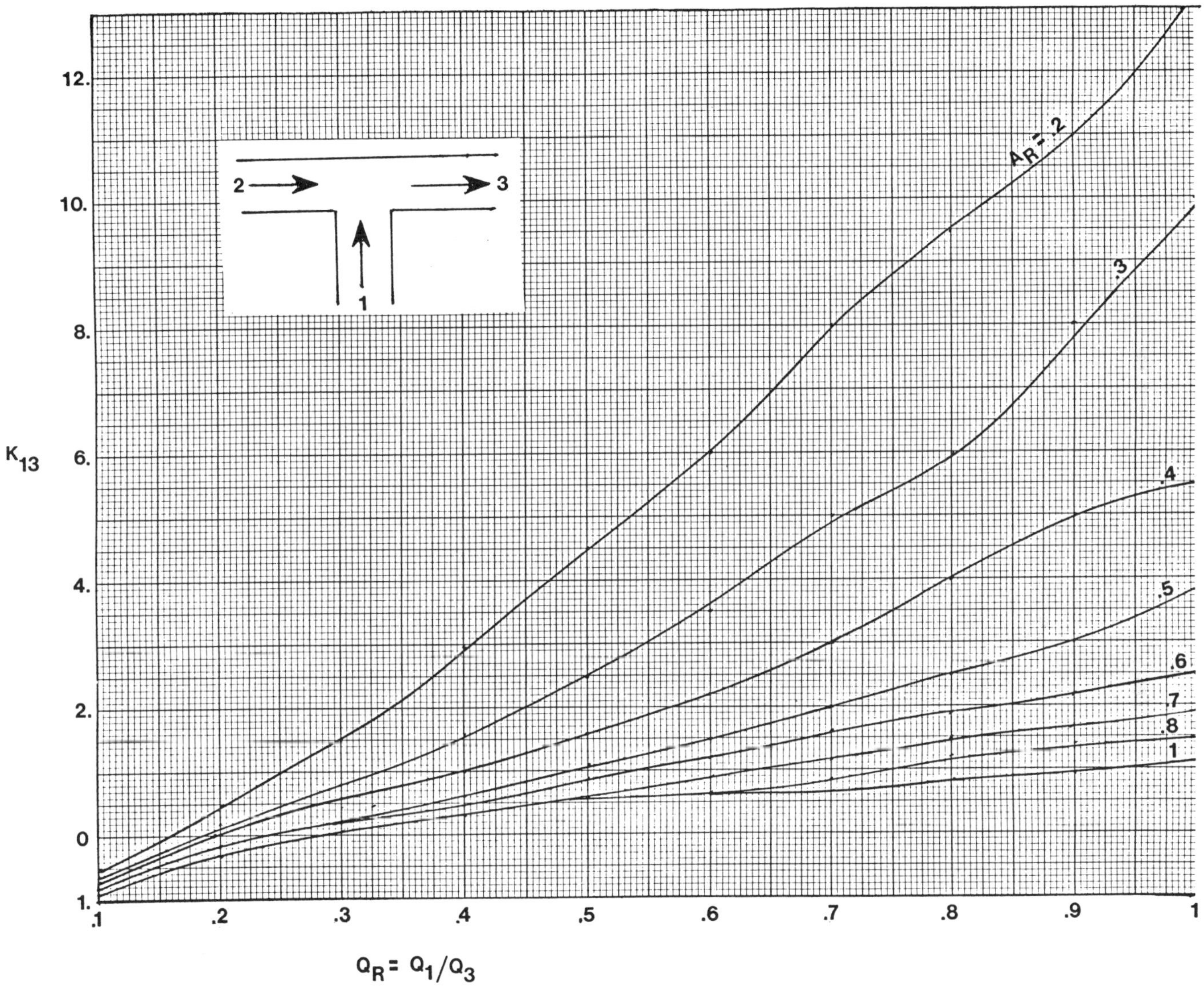

Figure 1-10G. Velocity heads for change of flow [6].

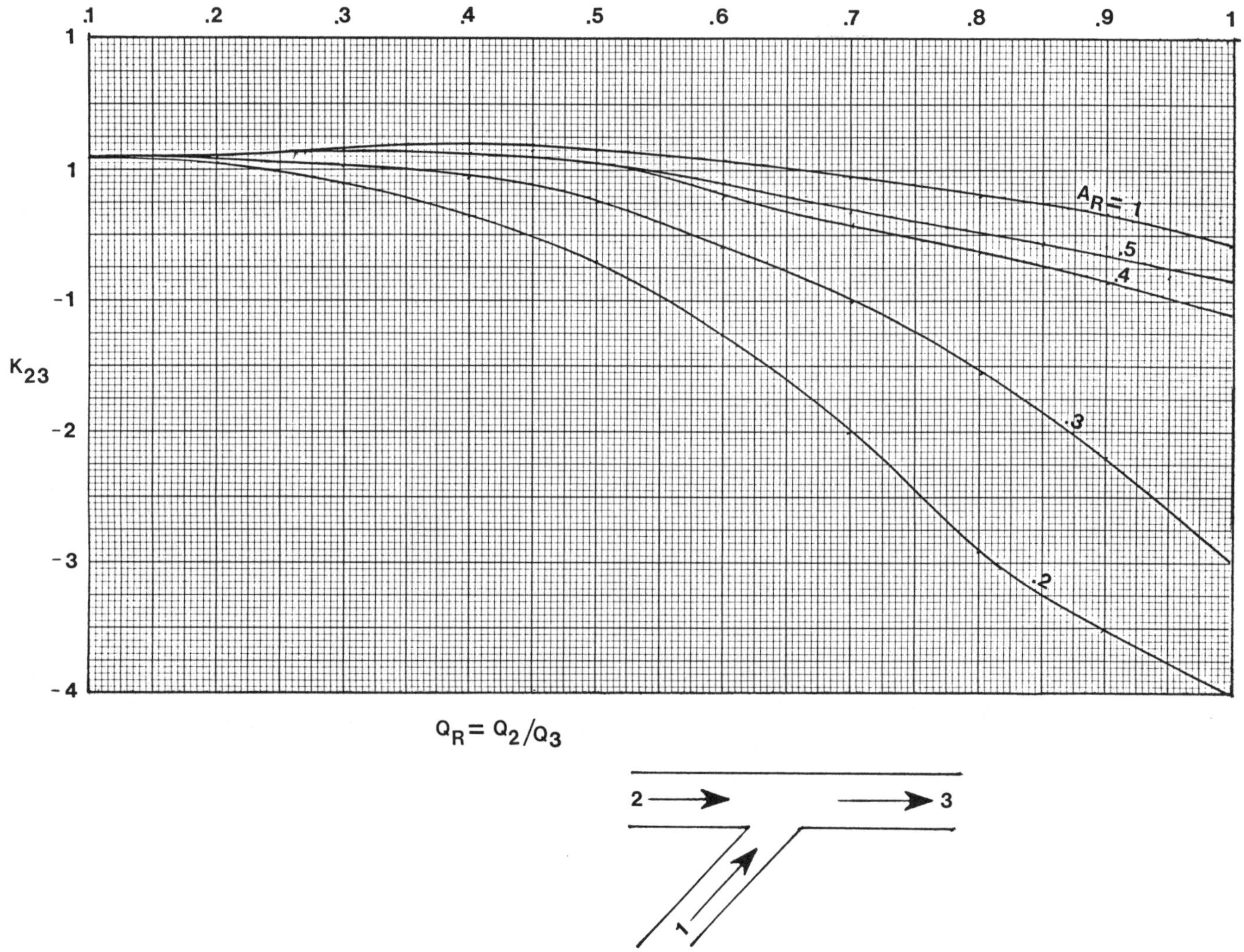

Figure 1-10H. Velocity heads for change of flow [6].

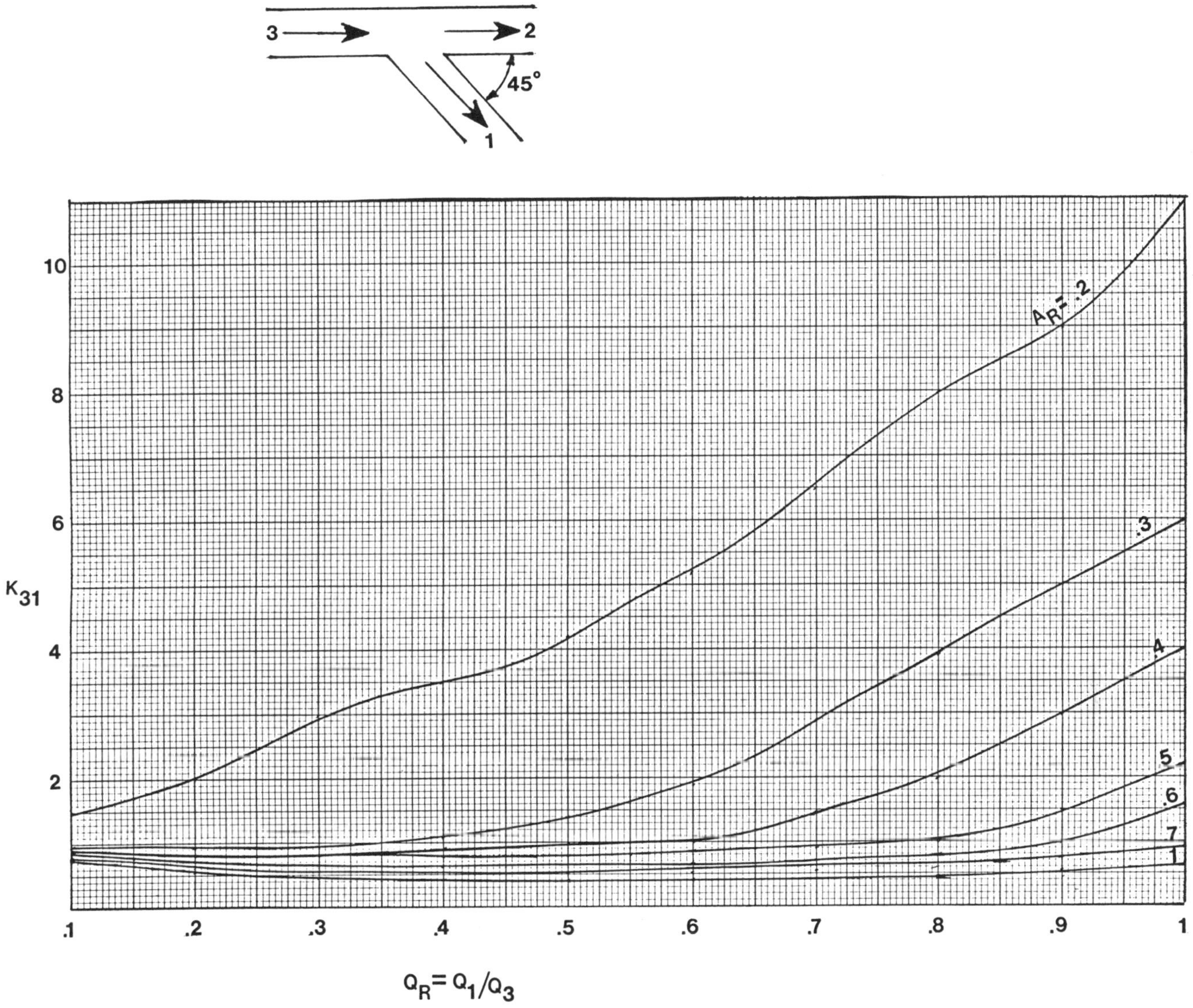

Figure 1-10l. Velocity heads for change of flow [6].

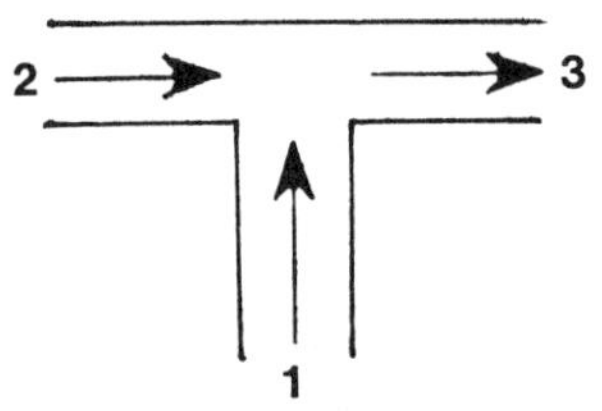

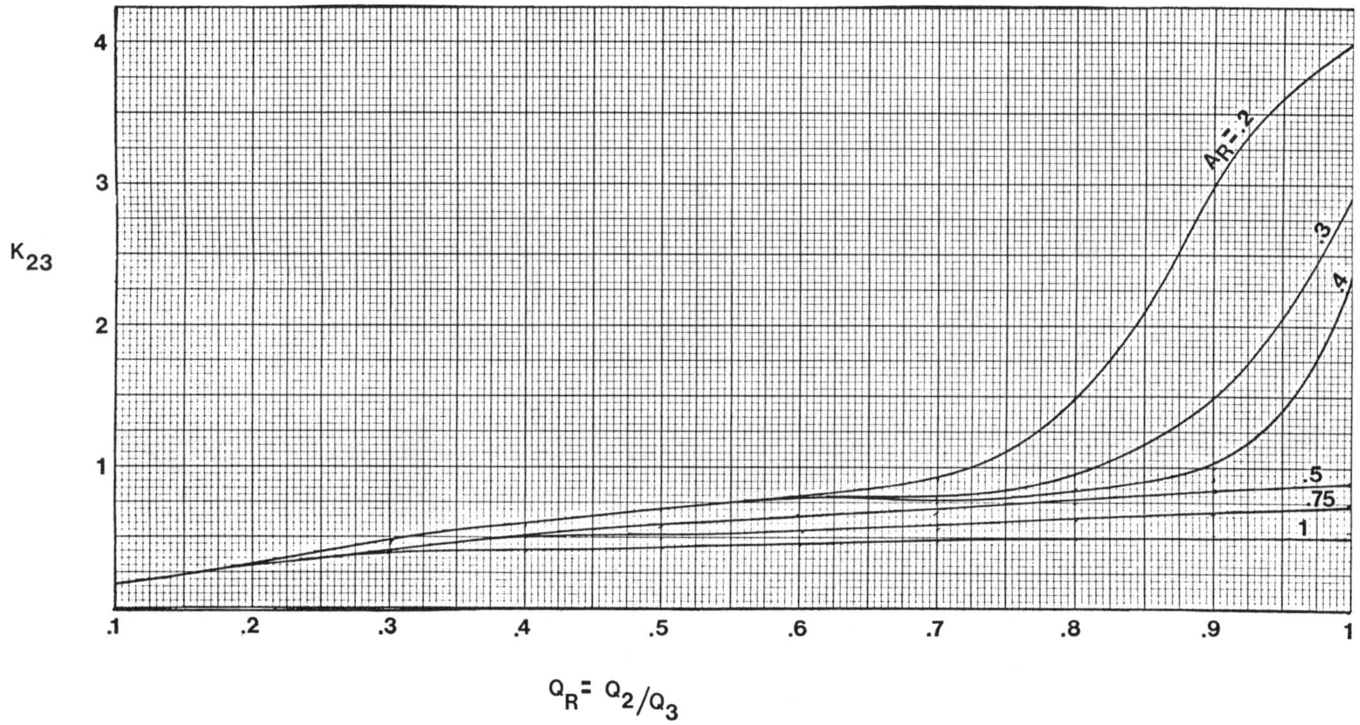

Figure 1-10J. Velocity heads for change of flow [6].

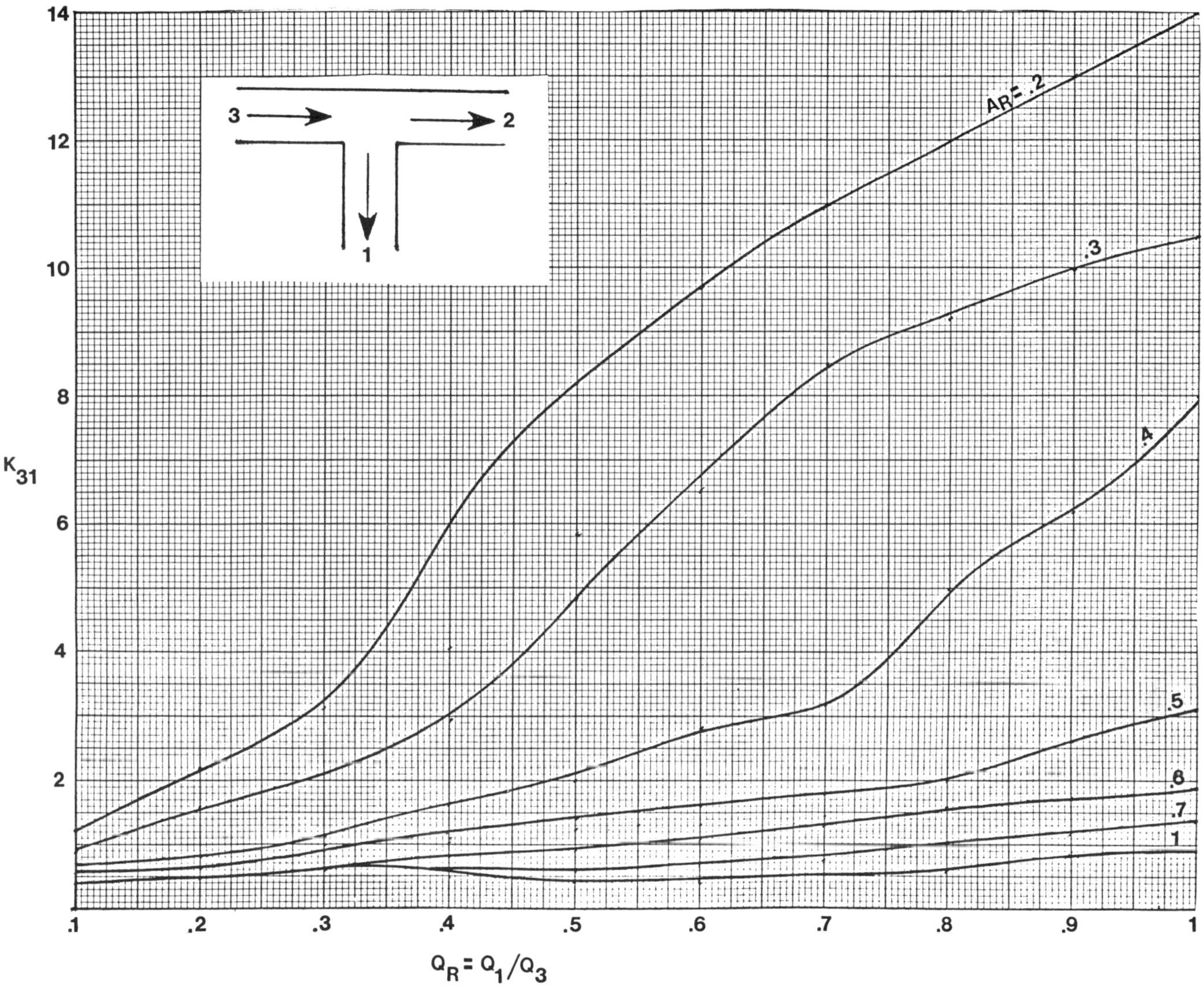

Figure 1-10K. Velocity heads for change of flow [6].

Small pipe fittings have more surface roughness and abrupt changes in cross sections, making K_1 insignificant at values of $N_{Re} > 10{,}000$. For this reason, the new Crane method is recommended for pipe diameters 24 in. and less. Comparison of the methods for elbows is depicted in Figure 1-11. Table 1-3 lists K_1 and K_∞ values.

The two-K method is preferred over the equivalent length method because in large, multi-alloy systems the equivalent length method could predict losses 300% too high, resulting in oversized pumps and equipment. With laminar flow, the equivalent length method predicts head losses too low. Also, in the equivalent length method, every equivalent length has a specific friction factor associated with it, because the equivalent lengths are derived from the expression $L_e = K\,D/f$.

The Hydraulic Institute's widely used K-factors are good for 1-in. to 8-in. pipe, but result in errors in larger piping.

The disadvantage of the two-K method is it is limited to the number of values of K_1 and K_∞ available, shown in Table 1-3. For other fittings, approximations must be made from data in Table 1-3.

Table 1-3
Constants for the Two-K Method [7]

		Fitting Type	K_1	K_∞
Elbows	90°	Standard (R/D = 1), screwed	800	0.40
		Standard (R/D = 1), flanged/welded	800	0.25
		Long-radius (R/D = 1.5), all types	800	0.20
		Mitered elbows (R/D = 1.5): 1-Weld (90°)	1,000	1.15
		2-Weld (45°)	800	0.35
		3-Weld (30°)	800	0.30
		4-Weld (22½°)	800	0.27
		5-Weld (18°)	800	0.25
	45°	Standard (R/D = 1), all types	500	0.20
		Long-radius (R/D = 1.5), all types	500	0.15
		Mitered, 1 weld, 45°	500	0.25
		Mitered, 2 weld, 22½°	500	0.15
	180°	Standard (R/D = 1), screwed	1,000	0.60
		Standard (R/D = 1), flanged/welded	1,000	0.35
		Long radius (R/D = 1.5), all types	1,000	0.30
Tees	Used as elbow	Standard, screwed	500	0.70
		Long-radius, screwed	800	0.40
		Standard, flanged or welded	800	0.80
		Stub-in-type branch	1,000	1.00
	Run-through tee	Screwed	200	0.10
		Flanged or welded	150	0.50
		Stub-in-type branch	100	0.00
Valves	Gate, ball, plug	Full line size, $\beta = 1.0$	300	0.10
		Reduced trim, $\beta = 0.9$	500	0.15
		Reduced trim, $\beta = 0.8$	1,000	0.25
	Globe, standard		1,500	4.00
	Globe, angle or Y-type		1,000	2.00
	Diaphragm, dam type		1,000	2.00
	Butterfly		800	0.25
	Check	Lift	2,000	10.00
		Swing	1,500	1.50
		Tilting-disk	1,000	0.50

Note: Use R/D = 1.5 values for R/D = 5 pipe bends, 45° to 180°. Use appropriate tee values for flow through crosses.

PIPE FLOW GEOMETRIES

In using Equation 1-4 the geometry of the flow area must be considered if the area is noncircular. In calculating the Reynolds number and the diameter for a noncircular cross section, the hydraulic radius is applied:

$$R_H = \text{hydraulic radius} = \frac{\text{cross-sectional flow area}}{\text{wetted perimeter}}$$

This relationship applies to noncircular cross sections flowing full or partially full, oval, rectangular, etc., but not to extremely narrow shapes, such as annular or elongated openings, where the width is small relative to the length. In such cases the value of R_H is approximately one-half the width of the passage.

The value of $4R_H$ is substituted for d in Equation 1-4.

$$N_{Re} = \frac{V(4R)}{\nu}$$

Table 1-4 provides hydraulic radii for various cross sections.

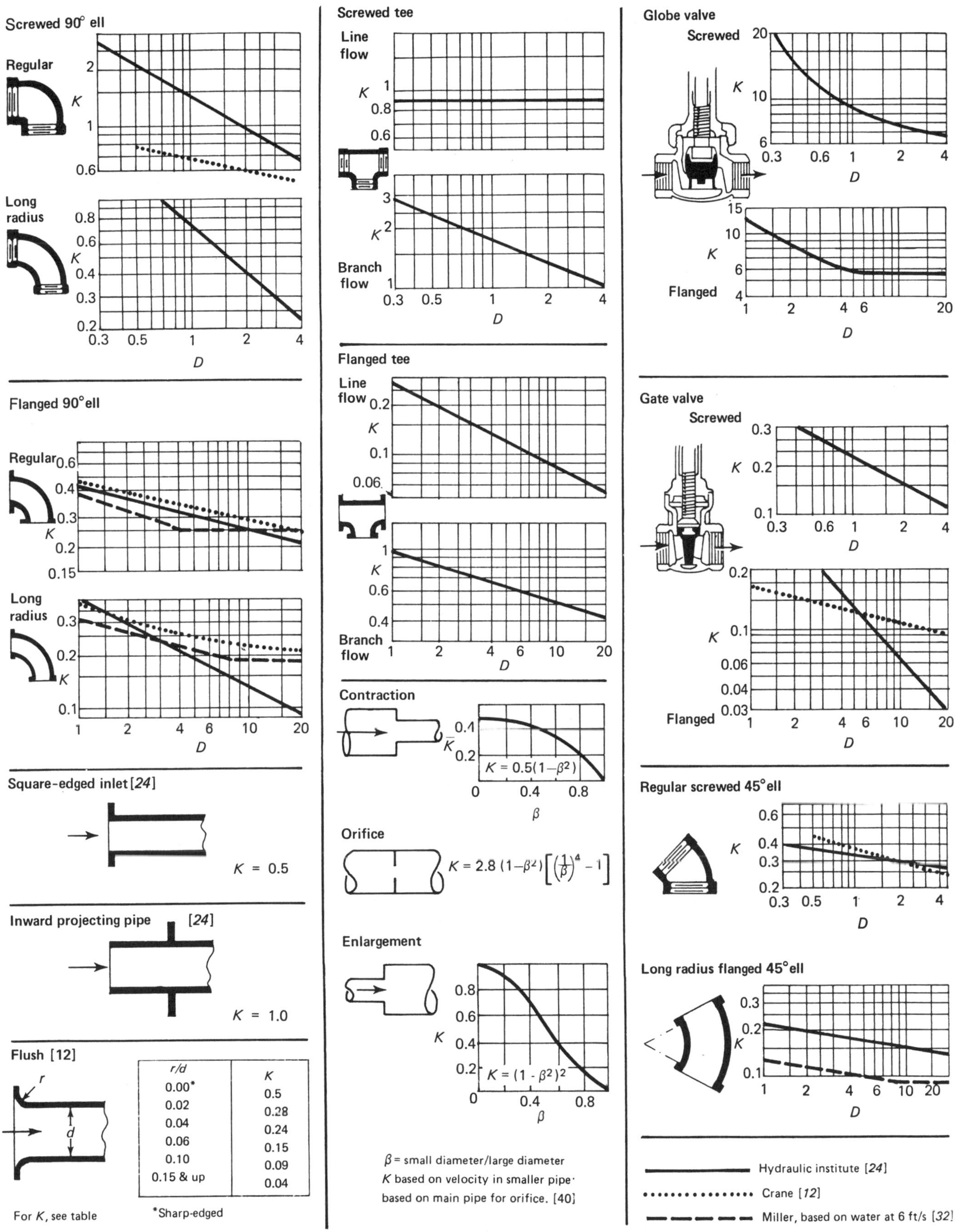

r/d	K
0.00*	0.5
0.02	0.28
0.04	0.24
0.06	0.15
0.10	0.09
0.15 & up	0.04

Figure 1-11. Velocity head values for common piping components [1]. (Reprinted by special permission from *Chemical Engineering*, © 1978, by McGraw-Hill, Inc., New York.)

Swing check valve

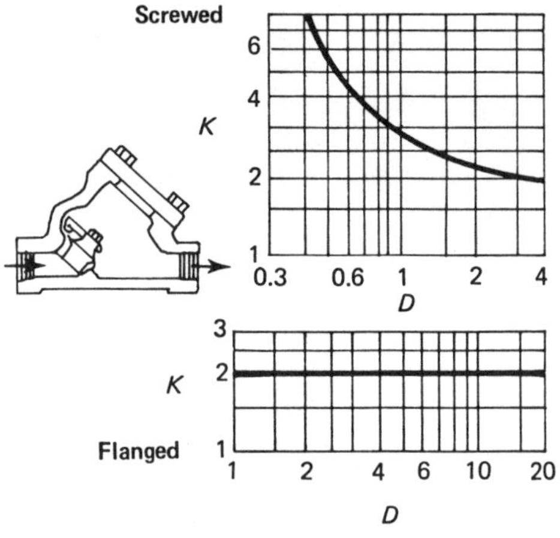

Angle valve

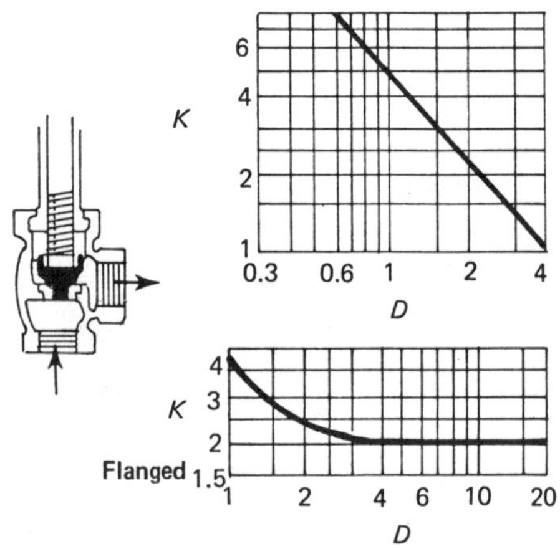

Screwed return bend

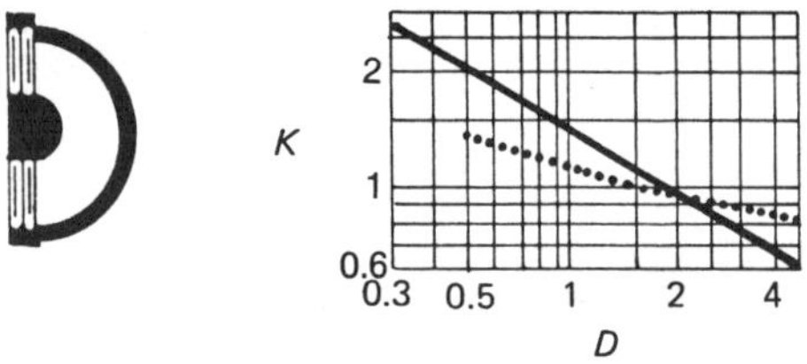

Flanged return bend

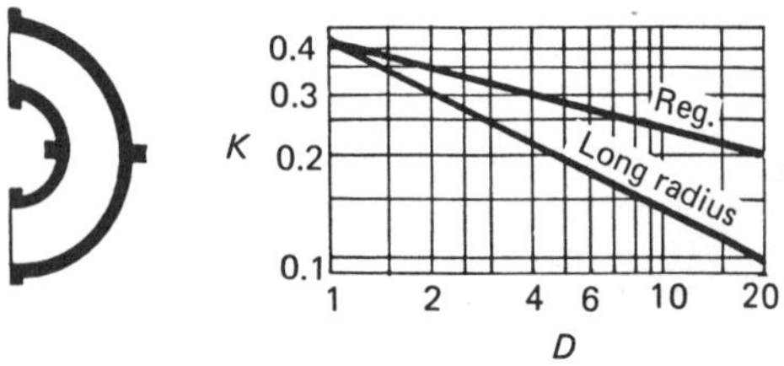

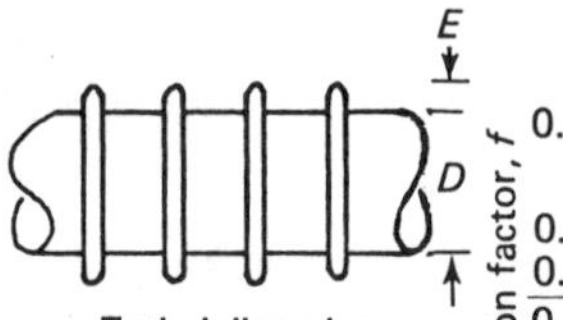

Typical dimensions

Nominal size (in.)	I.D. (in.)	E/D
1/2	0.520	0.250
3/4	0.750	0.275
1	1.000	0.187
1-1/2	1.500	0.200
2	2.000	0.161
3	2.962	0.143
4	4.000	0.081

Fanning friction factor for flexible metal hose [14]

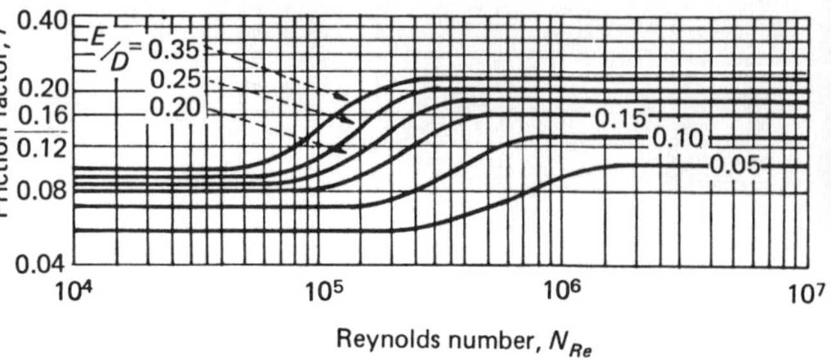

Head loss in conical diffusers

[19]

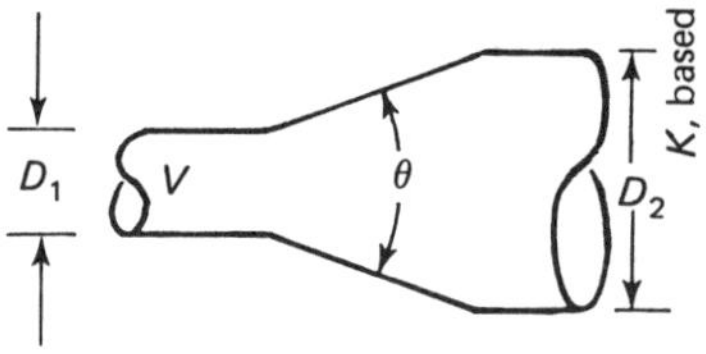

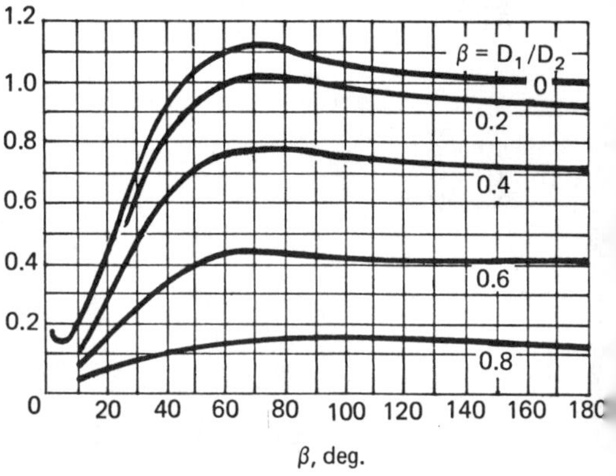

Head loss in circular miters

[47]

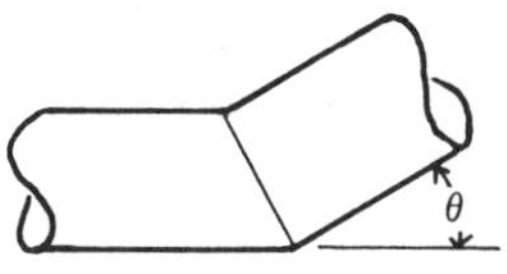

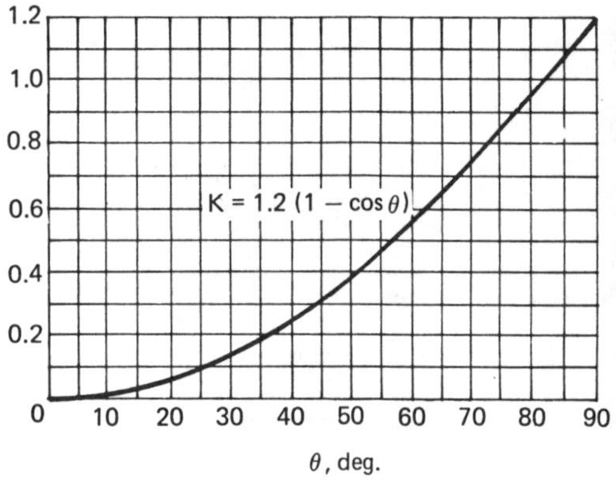

Plug cock valve [19]

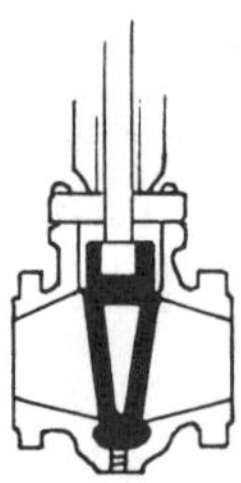

θ	K
5°	0.05
10°	0.29
20°	1.56
40°	17.3
60°	206.0

θ is angle between pipe axis and plug cock axis

Butterfly valve [19]

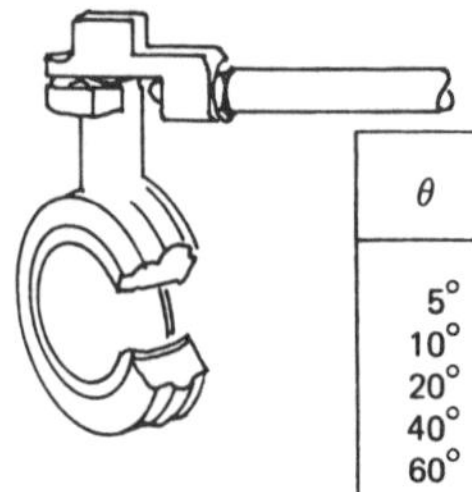

θ	K
5°	0.24
10°	0.52
20°	1.54
40°	10.8
60°	118.0

θ is angle between pipe axis and flapper plate

Figure 1-11. Continued.

Table 1-4
Values of hydraulic radius (R_H) for Various Cross Sections

Cross Section	R_H
CIRCULAR 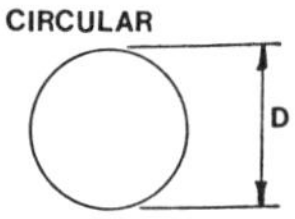	D/4
ANNULUS 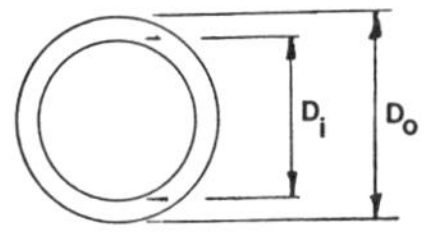	$(D_o - D_i)/4$
180° ANNULUS 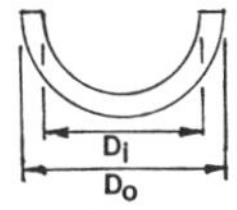	$0.153\ D_i$
SQUARE 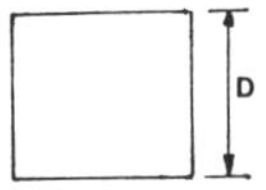	D/4
RECTANGLE 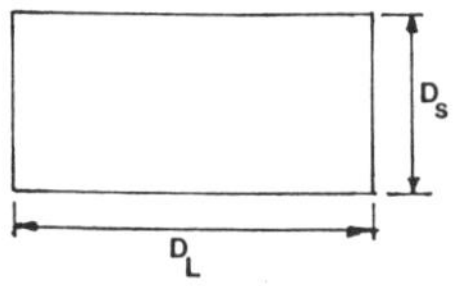	$\frac{D_L D_S}{2(D_L + D_S)}$

COMPRESSIBLE FLOW

The preceding analysis assumes steady and incompressible fluid flow. This is a reasonable assumption since most liquids are steady flow, but frequently the assumption is valid for gases. Because some liquids and all gases are compressible, a criterion is needed to determine what percent of error is incurred assuming constant density.

An estimate of the error can easily be made for a barotropic fluid—a fluid whose density is a function of pressure. Sabersky and Acusta [8] have shown that for a fluid stream of velocity V_1, static pressure P, and density ρ_1 that

$$V_1^2 = \frac{2}{\rho_1}(P_2 - P_1)\left(1 - \frac{V_1^2}{4a_1^2}\right) \qquad (1\text{-}9)$$

where $V_1^2/4a_1^2 = 0$ for incompressible flow since $a_1^2 \rightarrow \infty$, the term a_1^2 multiplied by the fluid density ρ_1 is the bulk compressibility modulus of the fluid and gives the pressure change for the fractional change in density. Values of the bulk compressibility modulus for various substances are given in Appendix A. The term a_1 is the velocity of sound waves propagated in a compressible medium.

As Equation 1-9 shows, the velocity of the fluid is compared to the fluid velocity of sound in the term $V_1^2/4a_1^2$. If this ratio is small, compressibility effects can be ignored because the error is the difference between this term and unity. This analysis is valid only for barotropic fluids, which are typical of most industrial applications involving flow of gas through a nozzle and the flow of water in conduits or over obstacles.

Compressibility effects of a fluid are small when the fluid velocity is small, compared to the fluid sonic velocity. If V_1/a_1 is equal to 0.3, the error in the velocity is less than 1% when using the incompressible assumption. For ambient air, this limitation corresponds to a velocity of 300 ft/sec without causing significant error.

The phenomenon of nonsteady flow is somewhat more complex than that of steady flow. The acceleration or deceleration of liquid particles immersed in a two-phase solution is one such example. The time required for the nonsteady phenomenon to occur is compared to the time required for a sound wave to traverse the flow in which substantial differences in velocity occur. If the time differences are small, then the incompressible Bernoulli equation (Equation 1-2) may be applied.

PIPING FLUID MECHANICS PROBLEM FORMULATION

To solve piping fluid problems a firm understanding of basic equations and units is essential. The units should be carefully defined and used throughout the calculations. Table 1-5 presents reasonable velocities for various services used in mechanical systems. These velocities are only guides intended to give the reader foresight for trial values and are only for mechanical systems; for such values of chemical processes the reader should consult chemical engineering sources.

Viscosity

Widely misunderstood and often improperly applied, viscosity is perhaps the most recondite of all the properties associated with fluid mechanics. However, a clear conception of this physical property is critical to the suc-

Table 1-5
Reasonable Velocities

Liquids	Feet per Second
Service water mains	2 to 5
General service water piping	4 to 10
Boiler feed water piping	6 to 13
Heat medium oils	2 to 6
Lubricating oils	2 to 6
Gases	
Low-pressure steam heating and process piping	15 to 70
Low-pressure steam mains	70 to 165
High-pressure steam mains	165 to 400
Steam engine and pump piping	100 to 150
Steam turbine piping	150 to 330
Natural gas	100
Air, 0 to 30 psig	70
Forced draft ducts	40 to 60
Induced-draft flues	30 to 50
Chimneys and stacks	35
Ventilating ducts	7

cessful design of hydraulic systems and rotating equipment that transport fluids (see Chapter 6).

Viscosity is the property of a fluid to resist flow. Consider how much more freely and easily gasoline pours from a container than does black strap molasses. In fluid mechanics terminology, the heavier, bulkier nature of molasses is caused by the fluid's high shear stresses. These high shear stresses make the molasses very resistant to flow.

The fundamental measurement of flow resistance is the dynamic or absolute viscosity. In the cgs (centimeter-gram-second) system of units the basic unit of viscosity is the poise, which is equal to one hundred centipoises, (For a detailed explanation of how absolute viscosities of fluids are determined, the reader should consult a basic text on fluid mechanics.) The centipoise (cp) is now the standard unit of absolute viscosity, but because other units are still used, as illustrated in some of the examples scattered throughout this book, methods for converting to and from centipoises are provided. With the centipoise, one must be careful in using the English system of units when converting to the kinematic viscosity. Illustrating this conversion we have the following:

μ = absolute viscosity, centipoise
$= 1.0\ \text{lb}_f\text{-sec/ft}^2 = 478.7\ \text{poise} = 4.787\ \text{centipoise}$
ν = kinematic viscosity, centistoke

$$= \frac{\mu\, g_c}{w}, \text{ for the English system of units}$$

where $g_c = 32.17\ \dfrac{\text{ft-lb}_m}{\text{lb}_f\text{-sec}^2}$

w = mass density of the fluid, lb_m/ft^3

Thus, if one has a fluid such as a fuel oil (see Example 6-1), which for a given temperature has an absolute viscosity of 139.53 centipoise, we calculate the kinematic viscosity, ν, in the English system of units as follows:

$\mu = 139.53$ cp at 90°F
$w = 54.725\ \text{lb}_m/\text{ft}^3$

$$\nu = \frac{(139.53)\text{cp}\left(\dfrac{0.0000209}{1\ \text{cp}}\right)\dfrac{\text{lb}_f\text{-sec}}{\text{ft}^2}(32.17)\dfrac{\text{ft-lb}_m}{\text{lb}_f\text{-sec}^2}}{(54.725)\dfrac{\text{lb}_m}{\text{ft}^3}}$$

$$\nu = 0.0017\ \frac{\text{ft}^2}{\text{sec}}$$

$$\nu = \frac{0.0017\ \dfrac{\text{ft}^2}{\text{sec}}}{0.000010769\ \dfrac{\text{ft}^2}{\text{sec}}}\ \text{centistoke} = 159.261\ \text{centistokes}$$

Since the kinematic viscosity is a function of the fluid density, the above value is only valid at the specified temperature of 90°F. In the metric system the kinematic viscosity can be obtained by dividing the absolute viscosity by the specific gravity. *This is only for the metric system of units.* It is a common mistake in using the English system of units to compute the kinematic viscosity by dividing the absolute viscosity by the specific gravity of the fluid. Equipment manufacturers often use other units of viscosity. One of the most widely used units is the Seconds Saybolt Universal (SSU). This unit represents the number of seconds required for sixty cubic centimeters of liquid at a constant temperature to flow through a calibrated orifice. For liquids of high viscosity a larger orifice is used and the unit applied is termed Seconds Saybolt Furol (SSF). It is customary to specify these units of viscosity at standard temperatures.

The following are formulas for converting SSU's and SSF's to centistokes. Below the value of 32 the SSU is undefined and below the value of 25 the SSF is undefined. Throughout this book, the centipoise and the centistoke are the standard units of absolute and kinematic viscosity, respectively. Where the need arises, the centistoke is converted to SSU.

SSU to centistokes

t = Seconds Saybolt Universal
ν = centistokes

For $32 \leqslant t \leqslant 99$, $\quad 0.226t - \dfrac{195}{t} = \nu$

For $t \geqslant 100$, $\quad 0.220t - \dfrac{135}{t} = \nu$

SSF to centistokes

t = Seconds Saybolt Furol

For $25 \leqslant t \leqslant 39$, $\quad 2.24t - \dfrac{184}{t} = \nu$

For $t \geqslant 40$, $\quad 2.16t - \dfrac{60}{t} = \nu$

EXAMPLE 1-1: FRICTION PRESSURE DROP FOR A HYDROCARBON GAS-STEAM MIXTURE IN A PIPE

An amine still reboiler boils off a hydrocarbon gas-steam mixture that flows in a 304 ss line connecting the reboiler with an amine still tower. It is desired to determine the maximum pressure drop in the line as shown in Figure 1-12. The gas is a small tract of amine immersed with CO_2 in steam. The properties of the mixture are as follows:

μ = 0.01322 cp $\qquad$ t = 245°
ρ = 0.085 lb/ft^3 $\qquad$ p = 10 psig
ϵ = 0.0015 in.

The velocity head approach is used in determining the friction pressure drop. The line shown in Figure 1-12 is coming off two nozzles on top of the reboiler and merging at a tee before entering the amine still tower. The dimensions shown are identical with both sides coming off the reboiler nozzles (exemplified by the word "TYP," meaning for both sides). To solve this problem, we must apply Equation 1-4. To use this equation, we divide the connecting pipe into three components (see Figures 1-13—1-15)—an 18-in. ϕ portion with W = 25,291 lb/hr; a 24-in. ϕ portion with W = 25,291 lb/hr; and a 24-in. ϕ with W = 50,582 lb/hr, Equation 1-4 is applied to each portion and the pressure drop for each is added to

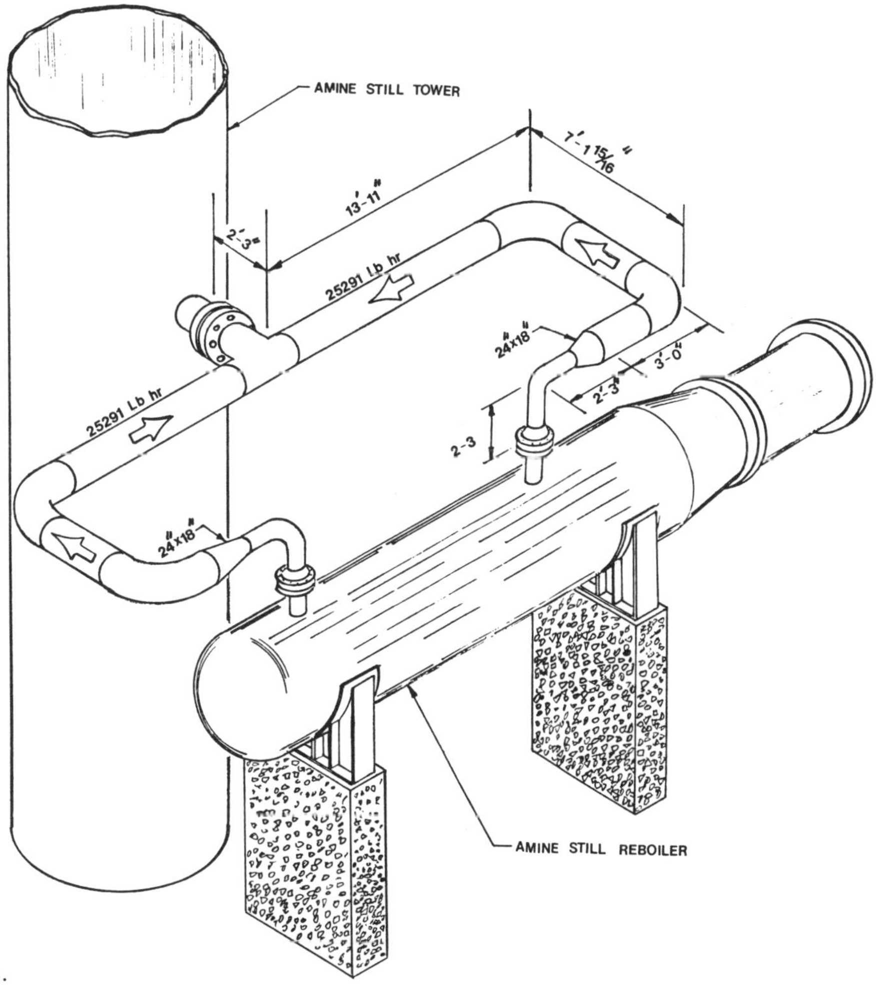

Figure 1-12. Amine still reboiler hydraulics.

FLUID ANALYSIS FOR SINGLE PHASE FLOW - GAS OR LIQUID

CONFIGURATION　　LINE NUMBER *18"ϕ PORTION - 18"ϕ SCHEDULE 10S*　　PAGE ___ OF ___

$$L_1 = (2'-3") + (2'-3") + (1'-8") = 6'-2" = 6.167\ FT$$

$$V = \frac{(25291)\frac{LB}{HR}\left(\frac{1\ HR}{3600\ SEC}\right)}{(0.085)\frac{LB}{FT^3}(240.5)\ IN^2\left(\frac{1\ FT^2}{144\ IN^2}\right)} = 49.487\ \frac{FT}{SEC}$$

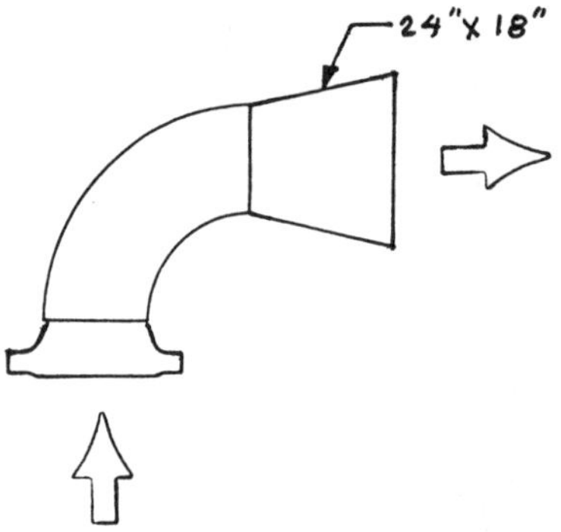

SERVICE ____________　　REYNOLDS NO = $DV\rho/\mu$

REASONABLE VELOCITY FOR SERVICE = ________　　K_i = NO OF VELOCITY HEADS; $K = DV^2/2$

DEPENDENT PIPE FRICTION = fL/D

K VALUES [L_D = 50K, D = INSIDE DIAMETER (ft)]

FOR COMPONENTS:

$\sum K$:

PIPE ENTRANCE =	0.780
24"x 18" DIFFUSER (CENTRIC) =	0.079
18" LR 90° ELL =	0.036
	$\sum K = 1.219$

μ = 0.01322 cp

ρ = 0.085 Lb/cu ft

L = 6.167 ft

ϵ = 0.0015 in.

γ = ________

D = 17.50 in.

ΔP = 0.029 psi

V = 49.487 ft/sec

N_{RE} = 690,491

f = 0.014

Q = ________ gpm

W = 25291 Lb/hr

Figure 1-13. Fluid analysis for single phase flow—gas or liquid.

FLUID ANALYSIS FOR SINGLE PHASE FLOW - GAS OR LIQUID

CONFIGURATION

LINE NUMBER 24"Ø PORTION - SCHEDULE 10S PAGE ___ OF ___

$$L_2 = (3'\text{-}0") + (7'\text{-}1\ ^{15}/_{16}") + (13'\text{-}11") = 24'\text{-}0\ ^{15}/_{16}" = 24.078\ ft$$

$$V = \frac{(25291)\frac{LB}{HR}\left(\frac{1\ HR}{3600\ SEC}\right)}{(0.085)\frac{LB}{ft^3}(433.7)\,in^2\left(\frac{1\ ft^2}{144\ in^2}\right)} = 27.442\ \frac{ft}{sec}$$

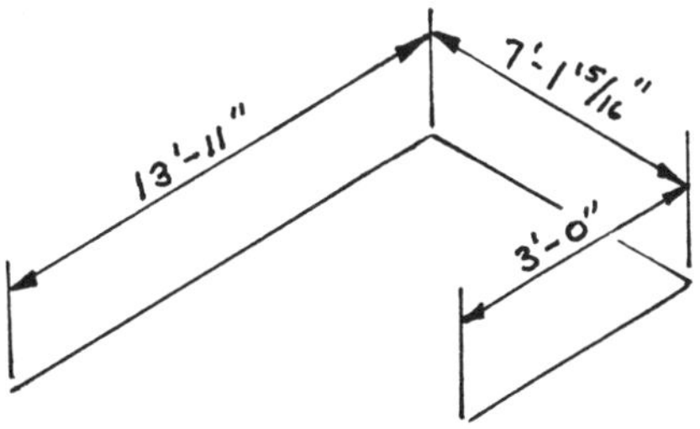

SERVICE ______ REYNOLDS NO = $DV\rho/\mu$

REASONABLE VELOCITY FOR SERVICE = ______

K_i = NO OF VELOCITY HEADS; $K = DV^2/2$

DEPENDENT PIPE FRICTION = fL/D

K VALUES [L_D = 50K, D = INSIDE DIAMETER (ft)]

FOR COMPONENTS:

$\sum K$:

TWO 90° LR ELLS = K = 0.720

μ = 0.01322 cp

ρ = 0.085 Lb/cu ft

L = 24.078 ft

ϵ = 0.0015 in.

γ = ______

D = 23.5 in.

ΔP = 0.029 psi

V = 27.442 ft/sec

N_{RE} = 514,177.125

f = 0.014

Q = ______ gpm

W = 25291 Lb/hr

Figure 1-14. Fluid analysis for single phase flow—gas or liquid.

FLUID ANALYSIS FOR SINGLE PHASE FLOW – GAS OR LIQUID

CONFIGURATION LINE NUMBER 24"ϕ PORTION - 24"ϕ SCHEDULE 10S PAGE ___ OF ___

$L_3 = 2'\text{-}11'' = 2.917$ ft

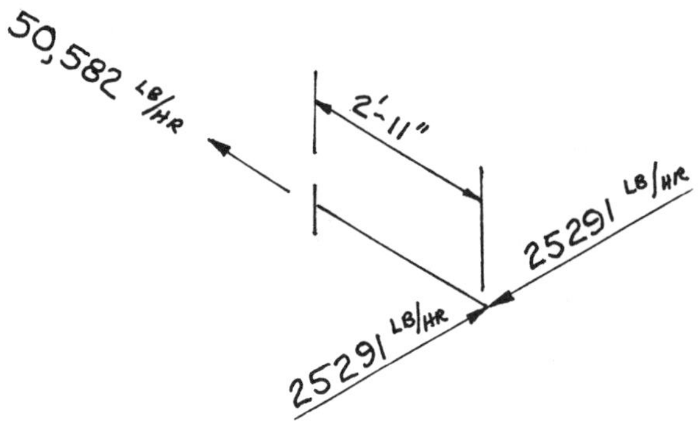

$$V = \frac{(50582)\frac{LB}{HR}\left(\frac{1}{3600\ SEC}\right)}{(0.085)\frac{LB}{FT^3}(433.7)IN^2\left(\frac{1\ FT^2}{144\ IN^2}\right)} = 54.884\ \frac{ft}{sec}$$

SERVICE ___ REYNOLDS NO = $DV\rho/\mu$

REASONABLE VELOCITY FOR SERVICE = ___ K_i = NO OF VELOCITY HEADS; $K = DV^2/2$

DEPENDENT PIPE FRICTION = fL/D

K VALUES [L_D = 50K, D = INSIDE DIAMETER (ft)]

FOR COMPONENTS:

$\sum K$:

COMBINING FLOW TEE = 1.2

PIPE EXIT = 1.0

$\sum K = 2.2$

μ = 0.01322 cp

ρ = 0.085 Lb/cu ft

L = 2.917 ft

ϵ = 0.0015 in

γ = ___

D = 23.5 in.

ΔP = 0.061 psi

V = 54.884 ft/sec

N_{RE} = 1,028,354.250

f = 0.013

Q = ___ gpm

W = 50582 Lb/hr

Figure 1-15. Fluid analysis for single phase flow—gas or liquid.

the other portions to give the total frictional pressure drop for the line (For velocity head value of concentric conical diffusers, the reader is referred to Table 1-6). The calculations are as follows.

$L_1 = 6.167 \text{ ft}$

$$\mu = (6.72 \times 10^{-4})(0.01322) = (8.884 \times 10^{-6}) \frac{\text{lb}_m}{\text{ft-sec}}$$

$$N_{Re} = \frac{DV\rho}{\mu} = \frac{\left(\frac{17.50}{12}\right)\text{ft } (49.487) \frac{\text{ft}}{\text{sec}} (0.085) \frac{\text{lb}_m}{\text{ft}^3}}{(8.884 \times 10^{-6}) \frac{\text{lb}_m}{\text{ft-sec}}}$$

$$= 690{,}491.450$$

$$(f)^{-0.5} = -2 \log_{10} \left[\frac{\epsilon/d}{3.7} + \frac{2.51}{N_{Re}(f)^{0.5}}\right] \quad (1\text{-}6a)$$

Let $f = 0.014$

$$(0.014)^{-0.5} = 8.452 = -2 \log_{10} [(2.317 \times 10^{-5}) + (3.072 \times 10^{-5})] = -8.537$$

$$f \approx 0.014$$

$$\Delta P_f = \left(\frac{fL}{d} + \sum K_i\right) \frac{\rho V^2}{2g_c} \quad (1\text{-}4)$$

$$\Delta P_{f_1} = \left[\frac{(0.014)(6.167)(12)}{(17.50)} + 1.219\right] \times \frac{(0.085) \frac{\text{lb}_m}{\text{ft}^3} (49.487)^2 \frac{\text{ft}^2}{\text{sec}^2} \left(\frac{1 \text{ ft}^2}{144 \text{ in.}^2}\right)}{2(32.2) \frac{\text{ft lb}_m}{\text{sec}^2 \text{ lb}_f}}$$

$\Delta P_{f_1} = 0.029 \text{ psi}$

$L_2 = 24.078 \text{ ft}$

Similarly,

$N_{Re} = 514{,}177.125$ and $f = 0.014$

$$\Delta P_f = \left[\frac{(0.014)(24.078)}{(23.50)} + 0.720\right] \times \frac{(0.085) \frac{\text{lb}_m}{\text{ft}^3} (27.442)^2 \frac{\text{ft}^2}{\text{sec}^2} \left(\frac{1 \text{ ft}^2}{144 \text{ in.}^2}\right)}{2(32.2) \frac{\text{ft lb}_m}{\text{sec}^2 \text{ lb}_f}}$$

Table 1-6
K-Values for Concentric Conical Diffusers

Size		d_2(in.)	d_1(in.)	L(in.)	α(deg)	K_1
Sch 80 1/2 ×	1/4	0.546	0.302	—	—	—
	3/8	0.546	0.423	—	—	—
Sch 80 3/4 ×	3/8	0.742	0.423	1.500	6.104	0.126
	1/2	0.742	0.546	1.500	3.746	0.036
Sch 80 1 ×	3/8	0.957	0.423	2.000	7.672	0.225
	1/2	0.957	0.546	2.000	5.898	0.121
	3/4	0.957	0.742	2.000	3.081	0.022
Sch 80 1 1/4 ×	1/2	1.278	0.546	2.000	10.545	0.318
	3/4	1.278	0.742	2.000	7.701	0.153
	1	1.278	0.957	2.000	4.603	0.040
Sch 80 1 1/2 ×	1/2	1.500	0.546	2.500	10.999	0.373
	3/4	1.500	0.742	2.500	8.720	0.225
	1	1.500	0.957	2.500	6.235	0.099
	1 1/4	1.500	1.278	2.500	2.545	0.009
Sch 40 2 ×	3/4	2.067	0.742	3.000	12.758	0.436
	1	2.067	0.957	3.000	10.661	0.297
	1 1/4	2.067	1.278	3.000	7.556	0.131
	1 1/2	2.067	1.500	3.000	5.423	0.055
Sch 40 2 1/2 ×	1	2.469	0.957	3.500	12.474	0.406
	1 1/4	2.469	1.278	3.500	9.796	0.237
	1 1/2	2.469	1.500	3.500	7.957	0.143
	2	2.469	2.067	3.500	3.292	0.013
Sch 40 3 ×	1	3.068	0.957	—	—	—
	1 1/4	3.068	1.278	3.500	14.816	0.454
	1 1/2	3.068	1.500	3.500	12.944	0.337
	2	3.068	2.067	3.500	8.221	0.111
	2 1/2	3.068	2.467	3.500	4.909	0.028
Sch 40 3 1/2 ×	1 1/4	3.548	1.278	4.000	16.484	0.559
	1 1/2	3.548	1.500	4.000	14.833	0.449
	2	3.548	2.067	4.000	10.668	0.210
	2 1/2	3.548	2.469	4.000	7.751	0.093
	3	3.548	3.068	4.000	3.440	0.010

Table 1-6 continued

Size		d_2(in.)	d_1(in.)	L(in.)	α(deg)	K_1
Sch 40	1½	1.500	4.026	4.000	18.406	0.609
	2	2.067	4.026	4.000	14.174	0.345
4 ×	2½	2.469	4.026	4.000	11.223	0.197
	3	3.068	4.026	4.000	6.878	0.055
	3½	3.548	4.026	4.000	3.425	0.088
Sch 40	2	2.067	5.047	5.000	17.338	0.537
	2½	2.469	5.047	5.000	14.940	0.388
5 ×	3	3.068	5.047	5.000	11.414	0.205
	3½	3.548	5.047	5.000	8.621	0.100
	4	4.026	5.047	5.000	5.860	0.035
Sch 40	2½	2.469	6.065	5.500	19.081	0.592
	3	3.068	6.065	5.500	15.810	0.392
6 ×	3½	3.548	6.065	5.500	13.228	0.257
	4	4.026	6.065	5.500	10.682	0.151
	5	5.047	6.065	5.500	5.310	0.023
Sch 40	3	3.068	7.981	6.000	24.168	0.726
	3½	3.548	7.981	6.000	21.680	0.618
8 ×	4	4.026	7.981	6.000	19.243	0.476
	5	5.047	7.981	6.000	14.152	0.229
	6	6.065	7.981	6.000	9.188	0.074
Sch 40	4	4.026	10.020	7.000	25.350	0.703
	5	5.047	10.020	7.000	20.807	0.514
10 ×	6	6.065	10.020	7.000	16.409	0.295
	8	7.981	10.020	7.000	8.375	0.051
Sch 40	5	5.047	11.938	8.000	25.511	0.674
	6	6.065	11.938	8.000	21.535	0.525
12 ×	8	7.981	11.938	8.000	14.319	0.197
	10	10.020	11.938	8.000	6.885	0.027
Sch 40	6	6.065	13.250	13.000	16.042	0.449
	8	7.981	13.250	13.000	11.692	0.214
14 ×	10	10.020	13.250	13.000	7.136	0.059
	12	11.938	13.250	13.000	2.892	0.005
Std Wt	6	6.065	15.250	14.000	19.150	0.604
	8	7.981	15.250	14.000	15.047	0.356
16 ×	10	10.020	15.250	14.000	10.765	0.157
	12	11.938	15.250	14.000	6.793	0.046
	14	13.250	15.250	14.000	4.096	0.011

Table 1-6 continued

Size		d_2(in.)	d_1(in.)	L(in.)	α(deg)	K_1
Std Wt	8	7.981	17.250	15.000	17.997	0.496
	10	10.020	17.250	15.000	13.946	0.275
18 ×	12	11.938	17.250	15.000	10.199	0.125
	14	13.250	17.250	15.000	7.662	0.058
	16	15.250	17.250	15.000	3.823	0.008
Std Wt	8	7.981	19.250	—	—	—
	10	10.020	19.250	—	—	—
	12	11.938	19.250	20.000	10.533	0.180
20 ×	14	13.250	19.250	20.000	8.627	0.108
	16	15.250	19.250	20.000	5.739	0.036
	18	17.250	19.250	20.000	2.866	0.005
Std Wt	10	10.020	21.250	—	—	—
	12	11.938	21.250	—	—	—
	14	13.250	21.250	20.000	11.537	0.194
22 ×	16	15.250	21.250	20.000	8.627	0.092
	18	17.250	21.250	20.000	5.739	0.030
	20	19.250	21.250	20.000	2.866	0.004
Std Wt	10	10.020	23.250	—	—	—
	12	11.938	23.250	—	—	—
	14	13.250	23.250	—	—	—
24 ×	16	15.250	23.250	20.000	11.537	0.169
	18	17.250	23.250	20.000	8.627	0.079
	20	19.250	23.250	20.000	5.739	0.026
Std Wt	12	11.938	25.250	—	—	—
	14	13.250	25.250	—	—	—
	16	15.250	25.250	—	—	—
26 ×	18	17.250	25.250	24.000	9.594	0.123
	20	19.250	25.250	24.000	7.181	0.057
	22	21.250	25.250	24.000	4.780	0.018
	24	23.250	25.250	24.000	2.388	0.003
Std Wt	14	—	—	—	—	—
	16	—	—	—	—	—
	18	—	—	—	—	—
30 ×	20	19.250	29.250	24.000	12.025	0.174
	22	—	—	—	—	—
	24	23.250	29.250	24.000	7.181	0.044
	26	25.250	29.250	24.000	4.780	0.014
	28	27.250	29.250	24.000	2.388	0.002

Table 1-6 continued

Size		d_2(in.)	d_1(in.)	L(in.)	α(deg)	K_1
Std Wt	16	—	—	—	—	—
	18	—	—	—	—	—
	20	—	—	—	—	—
	22	—	—	—	—	—
34 ×	24	23.250	33.250	24.000	12.025	0.141
	26	25.250	33.250	24.000	9.594	0.078
	28	—	—	—	—	—
	30	29.250	33.250	24.000	4.780	0.011
	32	31.250	33.250	24.000	2.388	0.001
Std Wt	16	—	—	—	—	—
	18	—	—	—	—	—
	20	—	—	—	—	—
	22	—	—	—	—	—
36 ×	24	23.250	35.250	24.000	14.478	0.207
	26	25.250	35.250	24.000	12.025	0.128
	28	—	—	—	—	—
	30	29.250	35.250	24.000	7.181	0.032
	32	31.250	35.250	24.000	4.780	0.010
	34	33.250	35.250	24.000	2.388	0.001
Std Wt	24	23.250	41.250	24.000	22.024	0.454
	26	25.250	41.250	24.000	19.471	0.339
42 ×	30	29.250	41.250	24.000	14.478	0.161
	32	31.250	41.250	24.000	12.025	0.098
	34	33.250	41.250	24.000	9.594	0.053
	36	35.250	41.250	24.000	7.181	0.024

$\Delta P_{f_2} = 0.005$ psi

$L_3 = 2.917$ ft

$N_{Re} = 1{,}028{,}354.250$ and $f = 0.013$

$$\Delta P_{f_3} = \left[\frac{(0.013)(2.917)}{(23.50)} + 2.20\right] \times \frac{(0.085)\ \dfrac{lb_m}{ft^3}\ (54.884)^2\ \dfrac{ft^2}{sec^2}\left(\dfrac{1\ ft^2}{144\ in.^2}\right)}{2(32.2)\ \dfrac{ft\ lb_m}{sec^2\ lb_f}}$$

$\Delta P_{f_3} = 0.061$ psi

Total Friction Pressure Drop for Line = ΔP

$\Delta P_f = \Delta P_{f_1} + \Delta P_{f_2} + \Delta P_{f_3} = (0.029 + 0.005 + 0.061)$ psi

$\Delta P_f = 0.095$ psi

EXAMPLE 1-2: FRICTIONAL PRESSURE DROP FOR A HOT OIL SYSTEM OF A PROCESS TANK

A pressure vessel storage tank contains 6,000 gallons of filler coating that must be maintained at 370°F to be used in the manufacture of roofing products. To maintain the coating mixture at the required temperature, external jacket coils are placed on the outside shell and bottom head as well as four internal coils inside the tank with an agitator. The tank is depicted in Figure 1-16 and the hot

Figure 1-16. Process surge tank. (Courtesy of Tranter, Inc.)

oil system in Figures 1-17, 1-18, and 1-19. It is desired to determine how much frictional pressure drop will be incurred for the entire tank so that pump sizes may be selected.

The tank is divided into two systems—the hot oil supply system and the hot oil return system. Each system connects to the three components—the four internal coils inside the tank, the outside shell jacket coils, and the jacket coils connected to the bottom head—and each of the three components must be analyzed separately.

Bottom Head System

A 2-in. pipe header supplies hot oil to the six inlet jacket nozzles and returns hot oil from six outlet jacket nozzles. The supply nozzles are designated by an S and the return nozzles by an R. We will analyze the supply system. The piping system is divided up into "stations," which are points designating flow change due to separating fluid. Each line following a station must be analyzed separately because the flow rate decreases after the flow separates in the tee. We will consider the pressure drop from point A to B, since that path involves more stations and the maximum amount of pressure drop.

Bottom Head Hot Oil Supply

☐ *Hot Oil Entrance from the 2-in. Header and Flow Through Station 1.* (Figure 1-17):

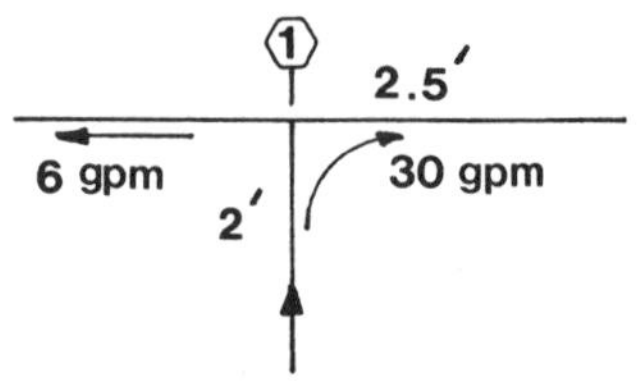

For $Q = 36$ gpm
$L = 26.5$ in., $\mu = 0.15$ cp, $\epsilon = 0.0018$
$d = 2.067$, $\rho = 58.7$ lb/ft^3
$\mu = (6.72 \times 10^{-4})(0.150) = 1.008 \times 10^{-5}$ lb$_m$/ft-sec

$$V = \frac{(36)\ \dfrac{\text{gal}}{\text{min}}\left(\dfrac{1\ \text{ft}^3}{7.479\ \text{gal}}\right)\left(\dfrac{1\ \text{min}}{60\ \text{sec}}\right)}{(3.356)\text{in.}^2\left(\dfrac{1\ \text{ft}^2}{144\ \text{in.}^2}\right)} = 3.442\ \text{ft/sec}$$

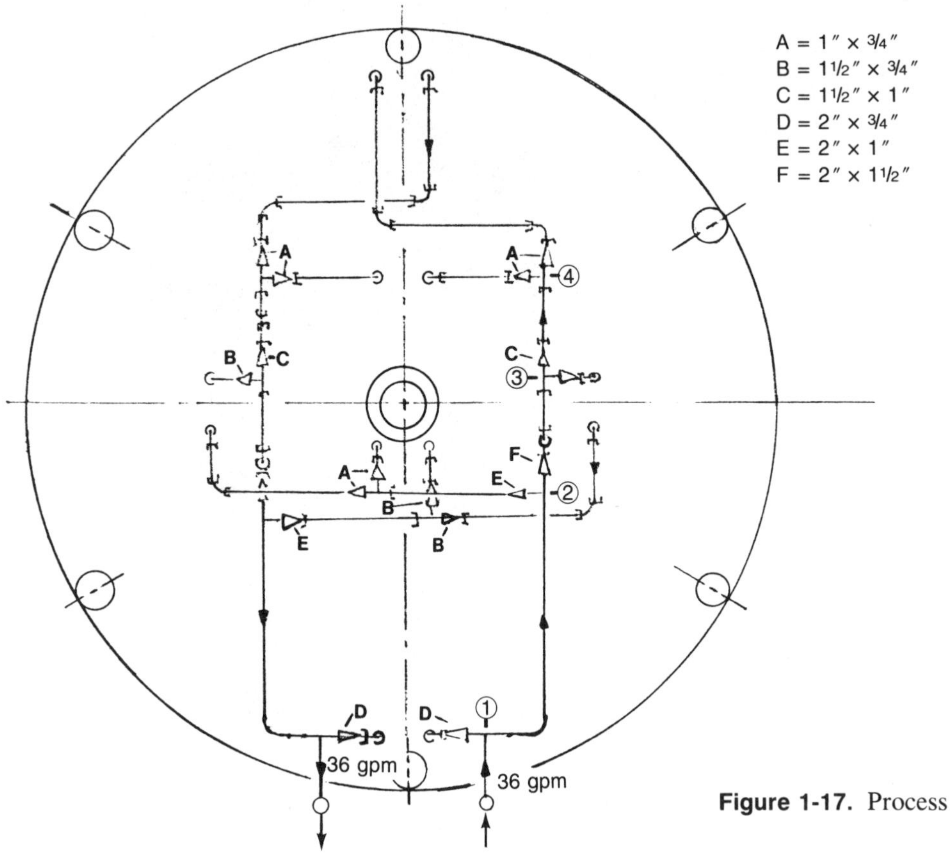

Figure 1-17. Process surge tank bottom head coils.

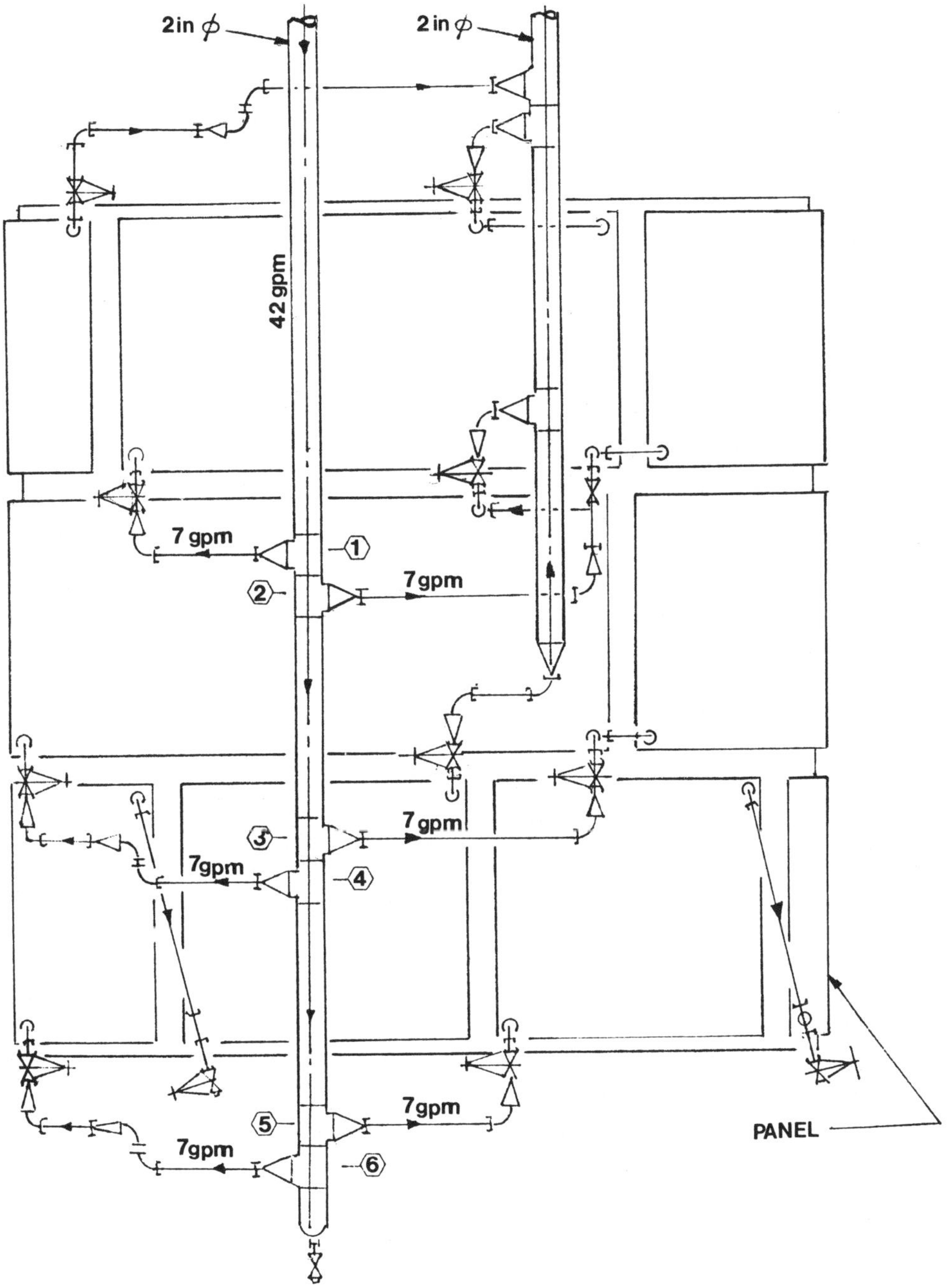

Figure 1-18. Process surge tank—shell coils (looking south).

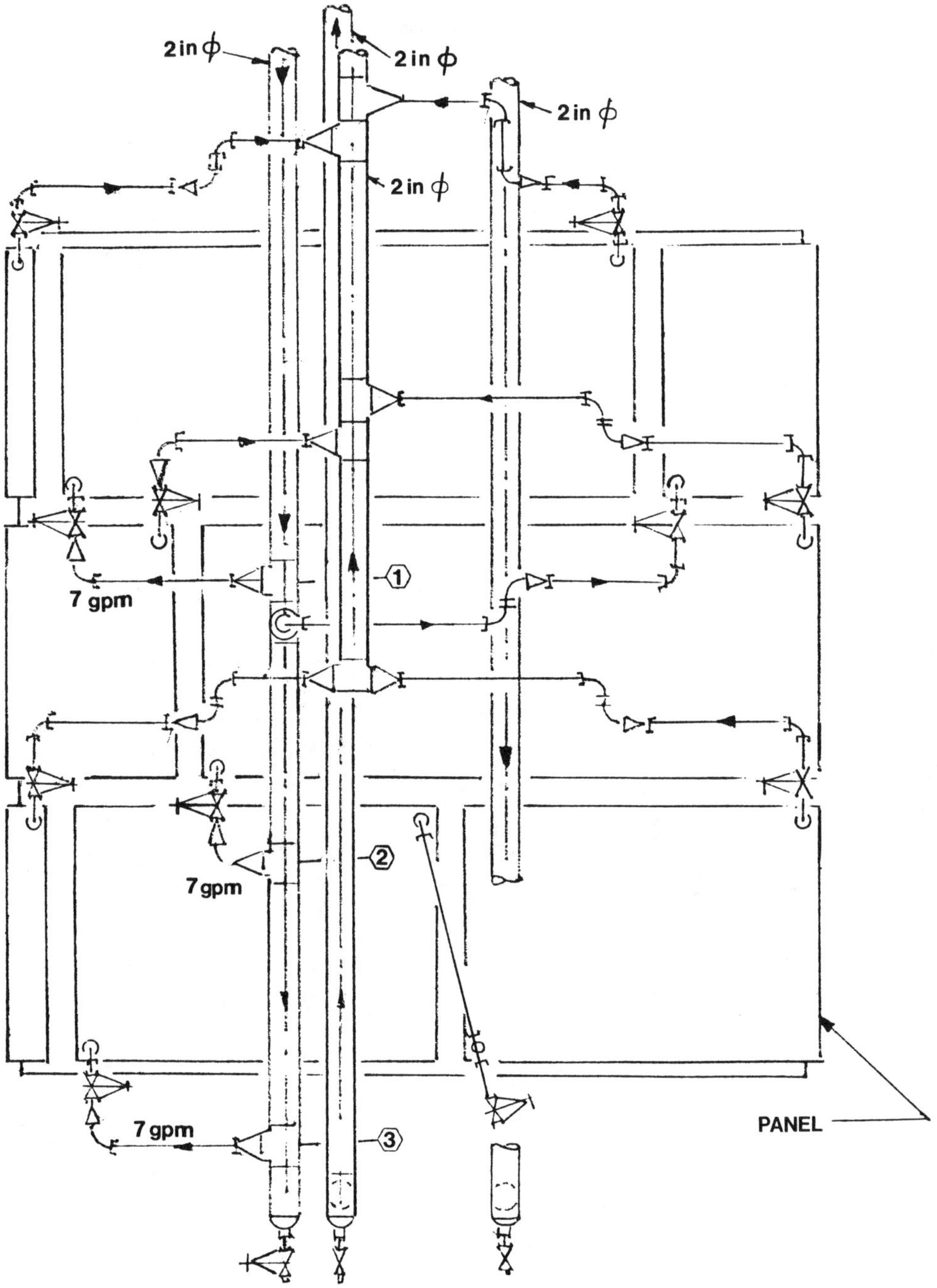

Figure 1-19. Process surge tank—shell coils (looking north).

$$N_{Re} = \frac{DV\rho}{\mu} = \frac{\left(\frac{2.067}{12}\right)\text{ft }(3.442)\ \frac{\text{ft}}{\text{sec}}\ (58.7)\ \frac{\text{lb}}{\text{ft}^3}}{(1.008 \times 10^{-5})\ \frac{\text{lb}_m}{\text{ft-sec}}}$$

$$= 3{,}452{,}910$$

$$\frac{1}{(f)^{0.5}} = -2\log_{10}\left[\frac{\left(\frac{0.0018}{0.172}\right)}{3.7} + \frac{2.51}{(3{,}452{,}910)(f)^{0.5}}\right] \Rightarrow f$$

$$= 0.040$$

K-factors

$$\text{Pipe entrance} = K = 0.78$$

$$\text{Branch flow tee} = K = 0.46 + (1.41)\left(\frac{30}{36}\right) + (-0.78)\left(\frac{30}{36}\right)^3 = 1.184$$

$$\sum K = 0.78 + 1.184 = 1.964$$

$$\Delta P_f = \left(\frac{fL}{d} + \sum K_i\right)\frac{\rho V^2}{2g_c}$$

$$= \left[\frac{(0.040)(2.208)(12)}{(2.067)} + 1.964\right] \times \frac{(58.7)\ \frac{\text{lb}}{\text{ft}^3}\ (3.442)^2\ \frac{\text{ft}^2}{\text{sec}^2}\left(\frac{1\text{ ft}^2}{144\text{ in.}^2}\right)}{2(32.2)\ \frac{\text{ft lb}_m}{\text{sec}^2\text{ lb}_f}}$$

$$\Delta P_{f_1} = 0.186 \text{ psi}$$

□ *Hot Oil Flow from Station 1 to Station 2:*

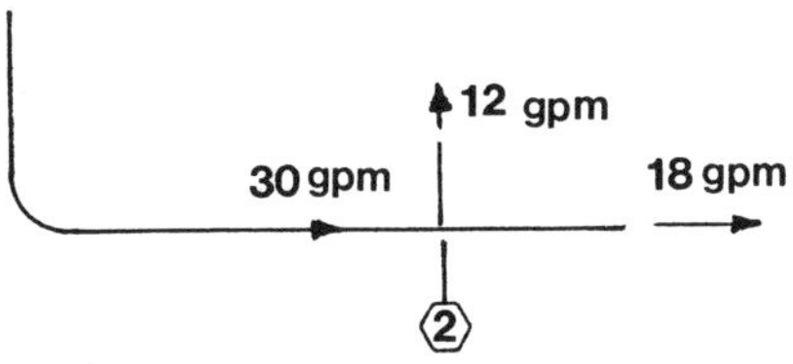

$$Q_R = \frac{Q_1}{Q_3} = \frac{18}{30} = 0.6$$

$$L = 4.0 \text{ ft}$$

$$V = \frac{(30)\ \frac{\text{gal}}{\text{min}}\left(\frac{1\text{ ft}^3}{7.479\text{ gal}}\right)\left(\frac{1\text{ min}}{60\text{ sec}}\right)}{(3.356)\text{in.}^2\left(\frac{1\text{ ft}^2}{144\text{ in.}^2}\right)} = 2.869 \text{ ft/sec}$$

$$N_{Re} = \frac{DV\rho}{\mu} = \frac{\left(\frac{2.067}{12}\right)\text{ft }(2.869)\ \frac{\text{ft}}{\text{sec}}\ (58.7)\ \frac{\text{lb}}{\text{ft}^3}}{(1.008 \times 10^{-4})\ \frac{\text{lb}_m}{\text{ft-sec}}}$$

$$= 287{,}784$$

$$\frac{1}{(f)^{0.5}} = -2\log_{10}\left[\frac{\left(\frac{0.0018}{0.172}\right)}{3.7} + \frac{2.51}{(287{,}784)(f)^{0.5}}\right] \Rightarrow f$$

$$= 0.040$$

K-factors

90° LR ell	= K =	0.570
Flow-thru tee	= K =	0.910
		1.480

$$\Delta P_{f_2} = \left[\frac{(0.040)(4.0)(12)}{(2.067)} + 1.480\right] \times \frac{(58.7)\ \frac{\text{lb}}{\text{ft}^3}\ (2.869)^2\ \frac{\text{ft}^2}{\text{sec}^2}\left(\frac{1\text{ ft}^2}{144\text{ in.}^2}\right)}{2(32.2)\ \frac{\text{ft lb}_m}{\text{sec}^2\text{ lb}_f}}$$

$$\Delta P_{f_2} = 0.126 \text{ psi}$$

□ *Hot Oil Flow from Station 2 to Station 3:*

$$L = 2 \text{ ft}$$

$$V = \frac{(18)\ \frac{\text{gal}}{\text{min}}\left(\frac{1\text{ ft}^3}{7.479\text{ gal}}\right)\left(\frac{1\text{ min}}{60\text{ sec}}\right)}{(2.036)\text{in.}^2\left(\frac{1\text{ ft}^2}{144\text{ in.}^2}\right)} = 2.837 \text{ ft/sec}$$

$$N_{Re} = \frac{\left(\frac{1.610}{12}\right)\text{ft }(2.837)\ \frac{\text{ft}}{\text{sec}}\ (58.7)\ \frac{\text{lb}}{\text{ft}^3}}{(1.008 \times 10^{-4})\ \frac{\text{lb}_m}{\text{ft-sec}}}$$

$$= 221{,}657$$

$$\frac{1}{(f)^{0.5}} = -2\log_{10}\left[\frac{\left(\frac{0.0018}{0.134}\right)}{3.7} + \frac{2.51}{(221{,}657)(f)^{0.5}}\right] \Rightarrow f$$

$$= 0.042$$

K-values

- 2-in. × 1½-in. LR ell

$$K = \frac{0.8\sin\left[\left(\frac{\theta}{2}\right)(1-\beta^2)\right]}{\beta^4};\ \beta^2 = \left(\frac{d_1}{d_2}\right)^2$$

$$= \left(\frac{1.610}{2.067}\right)^2 = 0.607$$

$$\Rightarrow \beta^4 = 0.368$$

$$K = \frac{0.8\sin[(5.423)(1-0.607)]}{(0.368)} = 0.081$$

- Flow-thru tee

$$Q_R = \frac{12}{18} = 0.667 \Rightarrow K = 0.53$$

$$A_R = \frac{A_2}{A_3} = 1.0$$

$$\sum K = 0.081 + 0.53 = 0.611$$

$$\Delta P_{f_3} = \left[\frac{(0.042)(2.0)(12)}{(1.610)} + 0.611\right]\frac{(58.7)\frac{lb}{ft^3}(2.837)^2\frac{ft^2}{sec^2}\left(\frac{1\ ft^2}{144\ in.^2}\right)}{2(32.2)\frac{ft\ lb_m}{sec^2\ lb_f}}$$

$\Delta P_{f_3} = 0.063$ psi

☐ *Hot Oil Flow from Station* 3 *to Station* 4*:*

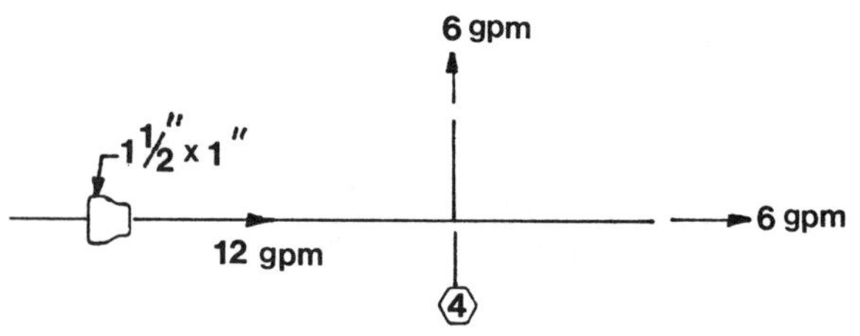

$$L = 1\ ft$$

$$V = \frac{(12)\frac{gal}{min}\left(\frac{1\ ft^3}{7.479\ gal}\right)\left(\frac{1\ min}{60\ sec}\right)}{(0.864)in.^2\left(\frac{1\ ft^2}{144\ in.^2}\right)} = 4.457\ ft/sec$$

$$N_{Re} = \frac{\left(\frac{1.049}{12}\right)ft\ (4.457)\frac{ft}{sec}(58.7)\frac{lb}{ft^3}}{(1.008\times10^{-4})\frac{lb_m}{ft\text{-}sec}}$$

$$= 226{,}889.525$$

$$\frac{1}{(f)^{0.5}} = -2\log_{10}\left[\frac{\left(\frac{0.0018}{0.087}\right)}{3.7} + \frac{2.51}{(226{,}890)(f)^{0.5}}\right] \Rightarrow f$$

$$= 0.051$$

K-values

- 1½-in. × 1-in. reducer

$$K = \frac{0.8\sin[\alpha(1-\beta^2)]}{\beta^4} = \frac{0.8\sin[(6.235)(0.593)]}{0.166}$$

$$= 0.311$$

- Flow-thru tee

$$Q_R = \frac{Q_1}{Q_3} = \frac{6}{12} = 0.5;\ A_R = \frac{A_1}{A_3} = 1.0 \Rightarrow K = 0.87$$

$$\sum K = 0.311 + 0.87 = 1.181$$

$$\Delta P_{f_4} = \left[\frac{(0.051)(1.0)(12)}{(1.049)} + 1.181\right]$$

$$\times\frac{(58.7)\frac{lb}{ft^3}(4.457)^2\frac{ft^2}{sec^2}\left(\frac{1\ ft^2}{144\ in.^2}\right)}{2(32.2)\frac{ft\text{-}lb_m}{sec^2\text{-}lb_f}} = 0.222\ psi$$

□ *Hot Oil Flow from Station 4 to Exit B:*

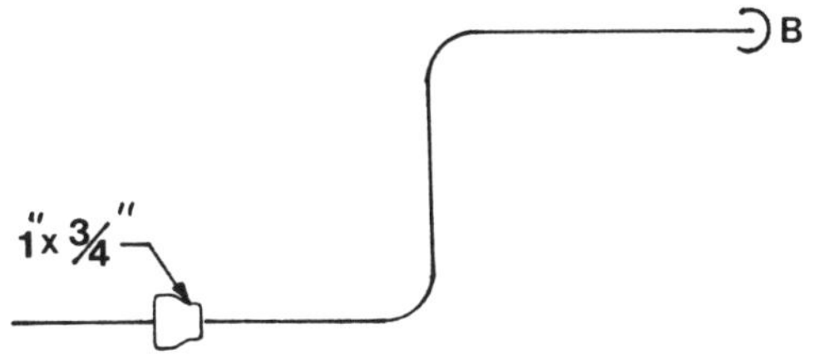

$$L = 2 \text{ ft}$$

$$V = \frac{(6)\dfrac{\text{gal}}{\text{min}}\left(\dfrac{1 \text{ ft}^3}{7.479 \text{ gal}}\right)\left(\dfrac{1 \text{ min}}{60 \text{ sec}}\right)}{(0.533)\text{in.}^2\left(\dfrac{1 \text{ ft}^2}{144 \text{ in.}^2}\right)} = 3.612 \text{ ft/sec}$$

$$N_{Re} = \frac{\left(\dfrac{0.824}{12}\right)\text{ft } (3.612)\,\dfrac{\text{ft}}{\text{sec}}\,(58.7)\,\dfrac{\text{lb}}{\text{ft}^3}}{(1.008 \times 10^{-4})\,\dfrac{\text{lb}_m}{\text{ft-sec}}} = 144{,}449$$

$$\frac{1}{(f)^{0.5}} = -2\log_{10}\left[\frac{\left(\dfrac{0.0018}{0.069}\right)}{3.7} + \frac{2.51}{(144{,}449)(f)^{0.5}}\right] \Rightarrow f$$

$$= 0.055$$

K-values

- 1-in. × 3/4-in. reducer

$$K = \frac{0.8 \sin[\alpha(1 - \beta^2)]}{\beta^4} = \frac{0.8 \sin[(3.081)(1 - 0.601)]}{(0.361)}$$

$$= 0.048$$

- 3–90° LR ells

$$K = 3(0.025)(30) = 2.250$$

Pipe exit ⇒ $K = 1.0$

$$\sum K = 0.048 + 2.250 + 1.000 = 3.298$$

$$\Delta P_{f_5} = \left[\frac{(0.055)(2.0)(12)}{(0.824)} + 3.298\right]$$

$$\times \frac{(58.7)\,\dfrac{\text{lb}}{\text{ft}^3}\,(3.612)^2\,\dfrac{\text{ft}^2}{\text{sec}^2}\left(\dfrac{1 \text{ ft}^2}{144 \text{ in.}^2}\right)}{2(32.2)\,\dfrac{\text{ft-lb}_m}{\text{sec}^2\text{-lb}_f}}$$

$$\Delta P_{f_5} = 0.405 \text{ psi}$$

Total friction pressure drop from entrance A to exit B:

Path (Figure 1-17)	ΔP_f psi
① Entrance A thru branch flow tee	0.186 psi
② Flow from station 1 to station 2	0.126 psi
③ Flow from station 2 to station 3	0.063 psi
④ Flow from station 3 to station 4	0.222 psi
⑤ Flow from station 4 to exit B	0.405 psi
Total	1.002 psi

$\sum \Delta P_f = 1.002$ psi = Total frictional pressure drop from entrance A to exit B

□ *Shell coils—South side* (Figure 1-18)

Station 1—

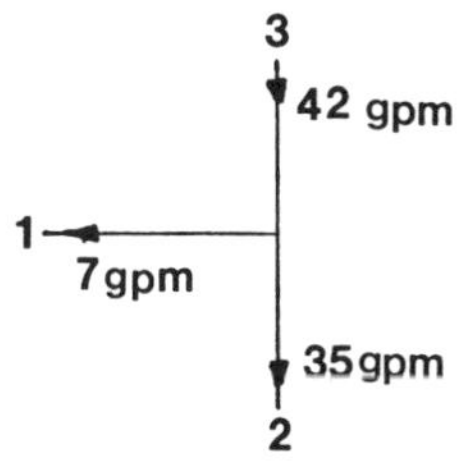

$$Q_R = \frac{Q_1}{Q_3} = \frac{7}{42} = 0.167$$

$K_{32} = -0.032$
Header entrance = K = 0.78
station 1 = K = −0.03
K = 0.75

$Q_3 = 42$ gpm
$L = 10$ ft; $\mu = 0.15$ cp; $d = 2.067$ in.
$\rho = 58.7$ lb/ft³; $\epsilon = 0.0018$
$V = 4.016$ ft/sec; $f = 0.020$; $N_{Re} = 402{,}829$
$\Delta P_1 = 0.195$ psi

Station 2—

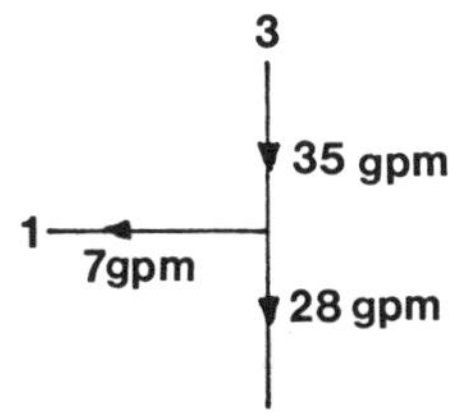

$$Q_R = \frac{Q_1}{Q_3} = \frac{7}{35} = 0.200$$

$K_{32} = -0.03$
L = 5 in. = 0.417 ft; d = 2.067 in.
Q_3 = 35 gpm
V = 3.346 ft/sec; f = 0.020; N_{Re} = 335,691
ΔP_2 = 0.001 psi

Station 3—

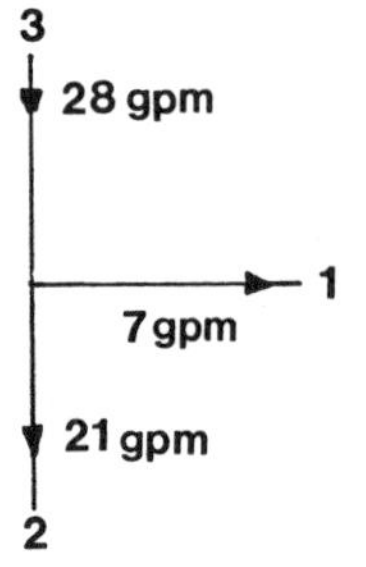

Q_3 = 28 gpm

$$\frac{Q_1}{Q_3} = \frac{7}{28} = 0.25$$

L = 10 ft
d = 2.067 in.
v = 2.677 ft/sec; $K_{32} = -0.036$
N_{Re} = 268,553
f = 0.020
ΔP_3 = 0.052 psi

Station 4—

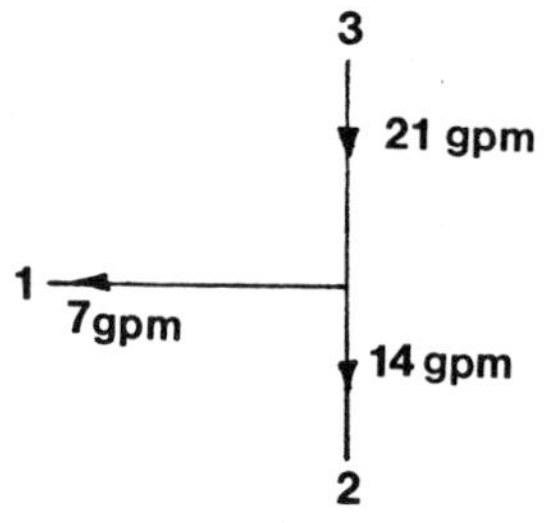

$$\frac{Q_1}{Q_3} = \frac{7}{21} = 0.333$$

$K_{32} = -0.030$
L = 5 in. = 0.417 ft; d = 2.067 in.
Q_3 = 21 gpm
V = 2.008 ft/sec; f = 0.021
N_{Re} = 201,415
ΔP_4 = 0.001 psi

Station 5—

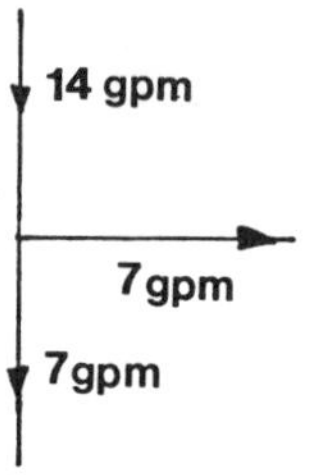

$$\frac{Q_1}{Q_3} = \frac{7}{14} = 0.500$$

L = 10 ft
K_{32} = 0.015
V = 1.339 ft/sec; f = 0.021
N_{Re} = 134,276
ΔP_5 = 0.014 psi

Friction pressure drop from station 6 to coil entrance

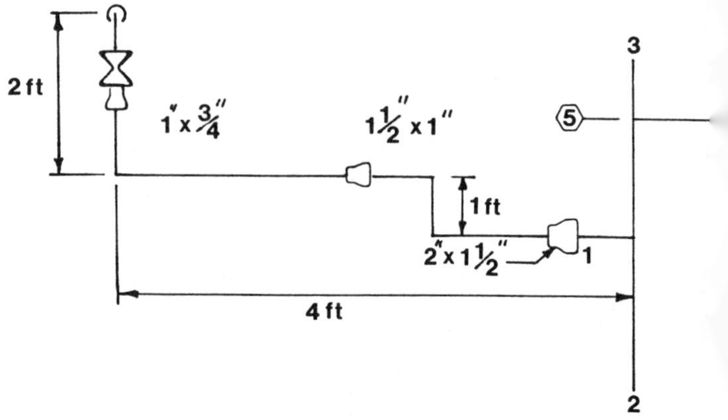

$$Q_R = \frac{Q_1}{Q_3} = 1.0$$

K_{31} = 1.28
For 2-in. × 1½-in. reducer, K = 0.129
1½-in. × 1-in. reducer, K = 0.311
1-in. × ¾-in. reducer, K = 0.048
¾-in. plug valve, K = 18(0.025) = 0.450
2-1½-in. 90° LR ells, K = 2(30)(0.021) = 1.260
1-1-in. 90° LR ell, K = (30)(0.023) = 0.690
1-¾-in. 90° LR ell, K = (30)(0.025) = 0.750
Exit into coil, K = 1.0
Q = 7 gpm; L = 7.25 ft

$$\sum K = 5.168$$

- For 2-in. ϕ pipe, d = 2.067 in.
  L = 7 in. = 0.583 ft
  K = 1.049
  V = 0.669 ft/sec

$N_{Re} = 67{,}138$
$f = 0.023$
$\Delta P = 0.004$ psi

- For $1^1/_2$-in. ϕ pipe, $d = 1.610$ in.
  $L = 3$ ft
  $K = 1.571$
  $V = 1.103$ ft/sec; $N_{Re} = 86{,}195$;
  $f = 0.023$
  $\Delta P = 0.016$ psi

- For 1-in. ϕ pipe, $d = 1.049$ in.
  $L = 2$ ft
  $K = 0.738$
  $V = 2.599$ ft/sec; $N_{Re} = 132{,}292$;
  $f = 0.024$
  $\Delta P = 0.055$ psi

- For $^3/_4$-in. ϕ pipe, $d = 0.824$ in.
  $L = 2$ ft
  $K = 2.20$
  $V = 4.211$ ft/sec; $N_{Re} = 168{,}416$;
  $f = 0.025$
  $\Delta P = 0.330$ psi

Total frictional pressure drop from station 1 to bottom shell coil is

$$\sum \Delta P = \underbrace{0.195 \text{ psi}}_{\text{station 1}} + \underbrace{0.001 \text{ psi}}_{\text{station 2}} + \underbrace{0.052 \text{ psi}}_{\text{station 3}} + \underbrace{0.001 \text{ psi}}_{\text{station 4}}$$

$$+ \underbrace{0.014 \text{ psi}}_{\text{station 5}} + \underbrace{(0.004 + 0.016 + 0.055 + 0.330) \text{ psi}}_{\text{station 6}}$$

$\sum \Delta P = 0.668$ psi – Total Drop for Shell Coil on South Side

□ *Shell coils—North side* (Figure 1-18)

Station 1—

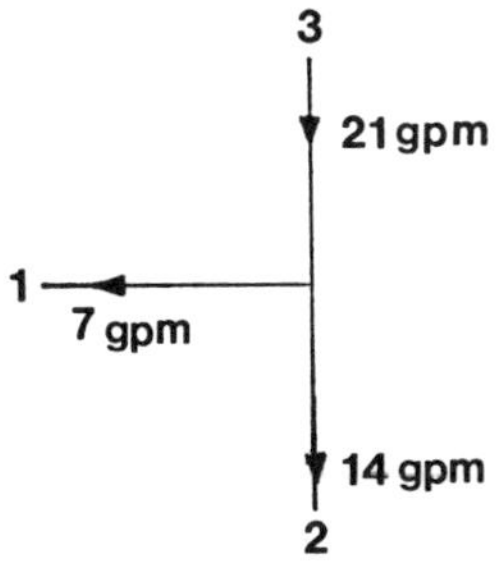

$Q_R = \dfrac{Q_1}{Q_3} = 0.333$
$K_{32} = -0.030$
$L = 10$ ft
$Q_3 = 21$ gpm
$\rho = 58.7$ lb/ft^3; $\mu = 0.015$ cp; $\epsilon = 0.0018$;
$d = 2.067$ in.
$V = 2.008$ ft/sec; $N_{Re} = 201{,}415$; $f = 0.021$
$\Delta P_1 = 0.030$ psi

Station 2—

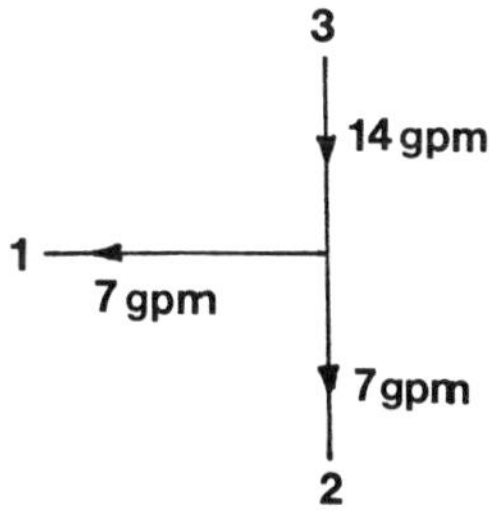

$Q_R = \dfrac{Q_1}{Q_3} = 0.500$; $L = 10$ ft
$K_{32} = 0.015$
$Q_3 = 14$ gpm
$V = 1.339$ ft/sec; $N_{Re} = 134{,}276$; $f = 0.021$
$\Delta P = 0.014$ psi

Friction pressure drop from station 2 to coil entrance:

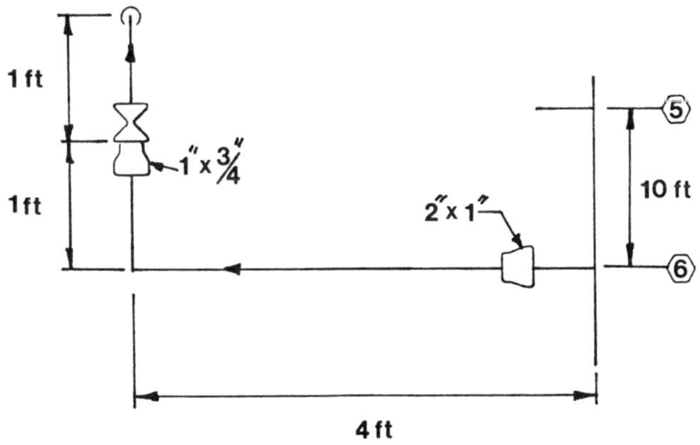

$Q_R = \dfrac{Q_1}{Q_3} = 1.0$; $K_{31} = 1.28$
For 2-in. × 1-in. reducer, $K = 2.538$
1-in. × $^3/_4$-in. reducer, $K = 0.048$
$^3/_4$-in. plug valve, $K = 0.450$
Exit into coil, $K = 1.0$
For 1-in. 90° LR ell, $K = 0.690$
For 1-in. ϕ pipe, $d = 1.049$ in.

$$Q = 7 \text{ gpm}, L = 4 \text{ ft}, K = 0.738$$
$$V = 2.599 \text{ ft/sec}; f = 0.024$$
$$N_{Re} = 132{,}292$$

$$\underline{\Delta P = 0.079}$$

For 2-in. ϕ pipe: $Q = 7$ gpm, $L = 11$ ft

$$K = 2.538$$
$$V = 0.669 \text{ ft/sec}$$
$$N_{Re} = 67{,}138.184$$
$$f = 0.023$$
$$\underline{\Delta P = 0.011 \text{ psi}}$$

For 3/4-in. ϕ pipe; $d = 0.824$; $L = 2$ ft; $K = 1.450$

$$V = 4.211 \text{ ft/sec}$$
$$N_{Re} = 168{,}416$$
$$f = 0.025$$
$$\underline{\Delta P = 0.245 \text{ psi}}$$

Total frictional pressure drop from station 1 to bottom shell coil is

$$\sum \Delta P = \underbrace{0.030 \text{ psi}}_{\text{station 1}} + \underbrace{0.014 \text{ psi}}_{\text{station 2}} + \underbrace{(0.011 + 0.079 + 0.245)\text{psi}}_{\text{station 3}}$$

$$\sum \Delta P = 0.379 \text{ psi} = \text{Total Drop for Shell Coil on North Side}$$

Maximum friction pressure drop in supply system is incurred at bottom head coil line with $\Delta P = 1.002$ psi.

EXAMPLE 1-3: FRICTION PRESSURE DROP FOR A WASTE HEAT RECOVERY SYSTEM

A gas turbine manufacturer specifies that the maximum back pressure on the unit used in this system be 10 in. of water pressure, therefore, the waste heat recovery system should be designed so that the frictional pressure drop does not exceed 10 in. of water. The system is shown in Figure 1-20.

□ *Turbine exhaust data for outside air at 60°F*

Temp. of exhaust gas = 795°F; V = 131 ft/sec

$$\mu = 0.0759 \frac{\text{lb}}{\text{ft-hr}} (0.4132) \frac{\text{cp}}{\text{lb/ft-hr}} = 0.031 \text{ cp}$$

$$L = 120 \text{ ft}; \frac{\epsilon}{D} = 0.000042 \text{ (commercial steel)}$$

$$\epsilon = 0.00015$$

$$N_{Re} = \frac{VD\rho}{\mu}; \mu = 0.0759 \frac{\text{lb}}{\text{ft-hr}} \left(\frac{1 \text{ hr}}{3{,}600 \text{ sec}}\right)$$

$$= 2.108 \times 10^{-5} \frac{\text{lb}}{\text{ft-sec}}$$

$$N_{Re} = \frac{(131.0) \frac{\text{ft}}{\text{sec}} (41.25)\text{in.} \left(\frac{1 \text{ ft}}{12 \text{ in.}}\right) (0.031) \frac{\text{lb}}{\text{ft}^3}}{(2.108 \times 10^{-5}) \frac{\text{lb}}{\text{ft-sec}}}$$

$$= 662{,}224$$

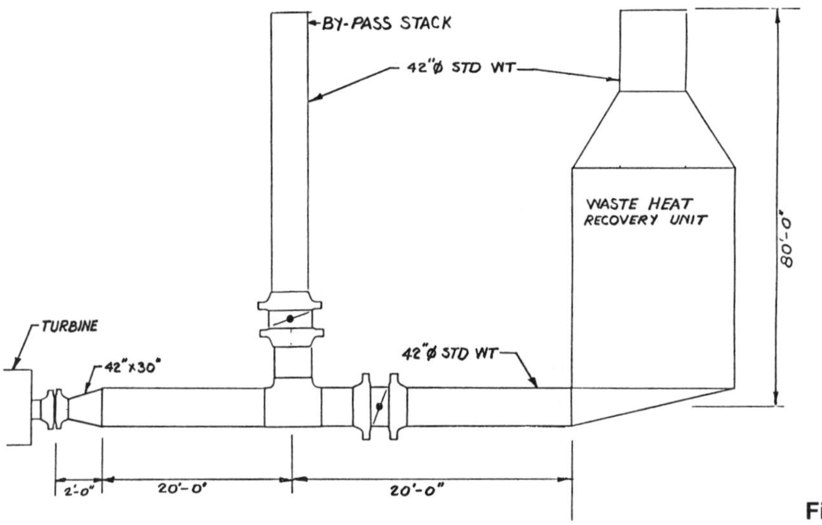

Figure 1-20. Waste heat recovery system.

From Equation 1-6a,

$$\frac{1}{(f)^{0.5}} = -2 \log_{10}\left(0.000011 + \frac{0.0000038}{(f)^{0.5}}\right)$$

$$f = 0.0130$$

$$D = 41.25 \text{ in.} = 3.438 \text{ ft}$$

$$K = \frac{fL}{D} = \frac{(0.0130)(120)}{3.438} = 0.454$$

From Equation 1-8,

$$K = K_1/N_{Re} + K_\infty\,(1 + {}^1/d)$$

For straight pipe, 42-in. ϕ section

$$K = \frac{fL}{d} = \frac{(0.0130)(120 - 30)}{3.438} = 0.340$$

K-values

Valves and Fittings	n	K_1	nK_1	K_∞	nK_∞
Butterfly valve	1	800	800	0.25	0.25
Run-thru tee	1	150	150	0.50	0.50
			950		0.75

$$K = \frac{950}{662{,}224} + (0.75)\left(1 + \frac{1}{41.25}\right) = 0.770$$

Other K-values

Pipe entrance at turbine nozzle	1.000
42-in. ϕ × 30-in. ϕ	0.161
Waste heat recovery unit entry duct	1.000
Stack exit	1.000
	3.161

For a rectangular duct,

$$R_H = ab/2(a + b)$$

For round pipe,

$$R_H = D/4$$

$$\Rightarrow D = \frac{2ab}{a + b} = \frac{2(10)\text{ft}\,(4)\text{ft}}{10 \text{ ft} + 4 \text{ ft}} = 5.714 \text{ ft}$$

or

D = 68.571 in. = Equivalent circular diameter

For 42-in. ϕ portion,

$$K_{total} = 0.770 + 3.161 + 0.340 = 4.271$$

From equation 1-4,

$$\Delta P = \left(\frac{fL}{d} + \sum K_i\right)\frac{\rho V^2}{2g_c}$$

$$\Delta P = \left[\frac{(0.0130)(120)}{(3.44)} + 4.271\right]\frac{(0.031)\,\frac{\text{lb}}{\text{ft}^3}\,(131.00)^2\,\frac{\text{ft}^2}{\text{sec}^2}}{2(32.2)\,\frac{\text{ft}}{\text{sec}^2}\,(144)\,\frac{\text{ft}^2}{\text{in.}^2}}$$

$$\Delta P = 0.271 \text{ psi}$$

For 10 ft × 4 ft × 42-in. ϕ transition piece,

D = 68.571 in.; L = 4.0 ft; K = 0.615

$$\Delta P = \left[\frac{(0.0130)(4.0)}{5.714} + 0.615\right]\frac{(0.031)\,\frac{\text{lb}}{\text{ft}^3}\,(47.458)^2\,\frac{\text{ft}^2}{\text{sec}^2}}{2(32.2)\,\frac{\text{ft}}{\text{sec}^2}\,(144)\,\frac{\text{ft}^2}{\text{in.}^2}}$$

$$\Delta P = 0.005 \text{ psi}$$

$$\Delta P = 0.005 \text{ psi } (27.912)\,\frac{\text{in. } H_2O \text{ at } 62°F}{1 \text{ psi}}$$

$$= 0.140 \text{ in. } H_2O$$

ΔP thru heating coils = 2 in. H_2O

Total ΔP = 7.564 in. + 0.140 in. + 2.000 in.

or

ΔP = 9.704 in. H_2O < 10 in. allowed

EXAMPLE 1-4 PRESSURE DROP IN RELIEF VALVE PIPING SYSTEM

Relief valve piping systems are designed to have minimum pressure drop. In this application the plant rules stipulate that the pressure drop will not exceed 3% of the valve set pressure. The system is to have two valves, shown in Figure 1-21.

The relieving fluid is Freon 114 and the flow rate is W = 243,755 lb/hr. First we compute the velocity heads, or K-values.

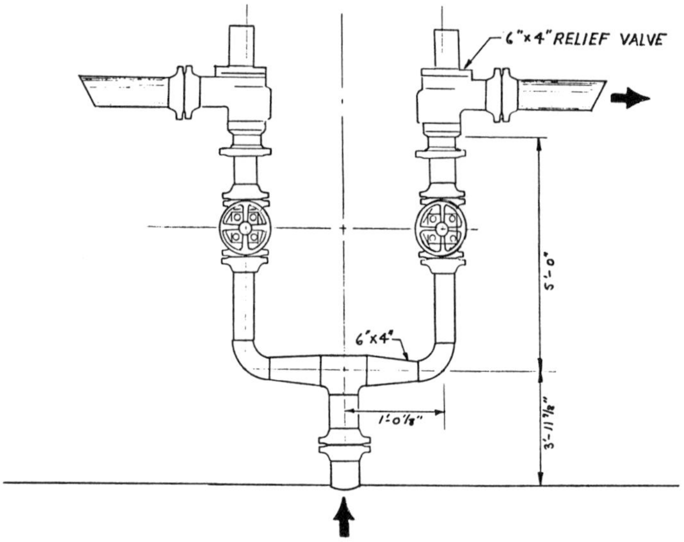

Figure 1-21. Relief valve piping system.

6-in. tee

$$K = 60\ f_t = 60(0.015) = 0.900$$

6-in. × 4-in. swage nipple ⇒ $\theta = 60°$

$$\beta = \frac{d_1}{d_2} = \frac{4.026}{6.065} = 0.664$$

From Figure 1-7,

$$K = \frac{0.5(1 - 0.441)(\sin 30°)^{0.5}}{0.194} = 1.019$$

4-in. ϕ 90° LR elbow,
$k = 30\ f_t = 30(0.015) = 0.450$

4-in. gate valve,

$$\beta = \frac{d_1}{d_2} = 1.0 \Rightarrow K = 8\ f_t = 8(0.015) = 0.120$$

Entrance, K = 0.78

6-in. ϕ line from entrance thru swage, $L_6 = 5$ ft

$\sum K = 0.78 + 0.90 + 1.019 = 2.699$
$\Delta P = 3.869$ psi; $N_{Re} = 14{,}931{,}929$
$V = 37.002$ ft/sec; $f = 0.01741$

4-in. line from swage to relief valve, $L_4 = 5$ ft

$\sum K = 0.450 + 0.120 = 0.570$
$\Delta P = 5.935$ psi; $N_{Re} = 22{,}494{,}325$
$V = 83.973$ ft/sec; $f = 0.01913$

The total pressure drop for 6-in. and 4-in. lines $= P_T$

$P_T = 3.869$ psi $+ 5.935$ psi $= 9.804$ psi
Set pressure = 205 psi

$$\%\ \Delta P = \frac{9.804}{205} = 4.8\% > 3\%$$

Consider moving 6-in. × 4-in. swages above gate valves and making 90° LR and gate valve 6-in., as shown in Figure 1-22. Recomputing the K-values we have

6-in. 90° LR ell, $K = 30\ f_t = 30(0.015) = 0.450$
6-in. gate valve, $K = 8\ f_t = 8(0.015) = 0.120$
Entrance, $K = 0.780$
Tee, $K = 0.900$
6-in. 90° LR ell, $K = 0.450$
6-in. gate valve, $K = 0.120$
6-in. × 4-in. swage, $K = 0.019$

$$\sum K = 3.269$$

For 6-in. ϕ line from entrance through swage, $L_6 = 10$ ft

$\Delta P = 4.869$ psi; $N_{Re} = 14{,}931{,}929$
$V = 37.002$ ft/sec; $f = 0.01741$

$$\%\ \Delta P = \frac{4.869}{205} = 0.024 = 2.4\% < 3\%$$

The pressure drop in the system in Figure 1-22 does not exceed 3% up to the relief valve as the plant rules require, thus, Figure 1-22 is the final configuration. Later, in example 2-5, we will examine the structural integrity of the system.

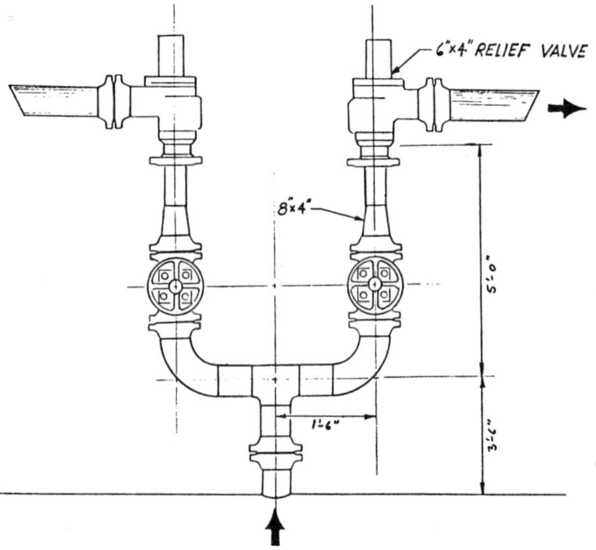

Figure 1-22. Relief valve piping system.

NOTATION

a_1 = sonic velocity of sound waves in compressible medium, ft/sec
a_n = rheological variable, dimensionless
A_R = ratio of branch area to header area, dimensionless
b_n = rheological variable, dimensionless
c = experimentally determined flow constant where $c = \mu/g_c$ for a Newtonian fluid
d = inside diameter (ID) of pipe, in.
D = inside diameter (ID) of pipe, ft
f = friction factor, dimensionless
F = head loss, friction loss or frictional pressure drop, $\frac{ft(lb_f)}{lb_m}$, $\frac{cm(kg_f)}{g_m}$
g = gravitational acceleration constant, 32.2 ft/sec^2; cm/sec^2
g_c = English system conversion factor, 32.17 $\frac{(lb_m)(ft/sec^2)}{lb_f}$
H_A = energy added by mechanical devices, e.g. pumps, $ft(lb_f)/lb_m$, $cm(kg_f)/g_m$
H_E = energy extracted by mechanical devices, e.g. turbines, $ft(lb_f)/lb_m$, $cm(kg_f)/g_m$
k = specific heat ratio (adiabatic coefficient), C_p/C_v
K = velocity head, $(ft)(lb_f)/lb_m$
K_∞ = velocity head for a large fitting at $N_{Re} = \infty$
L = length of pipe or piping component, in.
n′ = rheological variable, dimensionless
N_{Re} = Reynolds number, dimensionless
P = pressure, lb_f/ft^2, kg_f/cm^2
R_H = hydraulic radius, ft, in.
u = average bulk velocity, ft/sec
v = velocity, lb/ft^2, kg_f/cm^2
Y = height above datum, ft, cm

Greek Symbols

θ = angle, degrees
β = ratio of smaller diameter of pipe fitting to larger diameter
γ = generalized viscosity coefficient $lb_m/(ft)(sec)$
ϵ = absolute roughness or effective height of pipe wall irregularities, ft
μ = absolute (dynamic) viscosity, centipoise
ν = kinematic viscosity, centistokes
α = angle, degrees

REFERENCES

1. Simpson, L. L. and Weirick, M. L., "Designing Plant Piping," *Chem. Eng.*, April 3, 1978.
2. Metzer, A. B. and Reed, N. C. *A.I.Ch.E. Journal,* vol. 1, no. 434, A.S.M.E., New York, 1955.
3. Rase, H. F., *Piping Design for Process Plants,* John Wiley, New York, 1963.
4. Govier, G. W. and Aziz, K., *The Flow of Complex Mixtures in Pipes,* Robert E. Krieger Publishing Co., Huntington, New York, 1977.
5. Crane Co., *Technical Paper No. 410 Flow of Fluids,* Crane Co., New York, 1981.
6. SMACNA, *HVAC Duct System Design,* SMACNA, Vienna, Virginia 1981.
7. Hooper, B., "The Two-K Method Predicts Head Losses in Pipe Fittings," *Chem. Eng.*, Aug. 24, 1981.
8. Sabersky, R. H. and Acosta, A. J. *Fluid Flow—A First Course in Fluid Mechanics,* The MacMillan Company, New York, 1964.

2

The Engineering Mechanics of Piping

Static and dynamic analyses require clear and precise definition of terms—their misuse can often lead to misunderstandings, a problem the engineer greatly appreciates. The application of engineering mechanics to piping is mainly referred to in industry as "pipe stress analysis." However, the term is not comprehensive enough because engineers are usually more concerned about forces and moments exerted on equipment than stress. Certainly, stress is a concern and is discussed along with other phenomena in the chapter.

Another popular term used in industry is "piping flexibility analysis." The word flexibility can pose a problem because in the stiffness method of analysis it is actually the structural stiffness of pipe supports, rather than flexibility, that is important. For this reason the term "piping flexibility analysis" is avoided.

A piping component is any constituent part of a piping system, of any finite length of pipe—a valve, flange, elbow, pump, or anything else within the piping system. Piping is supported for various reasons—an obvious one being to counteract the force of gravity—and to begin to understand the applications we must start with some basic concepts.

Consider a piping component as shown in Figure 2-1. Here we have a three-dimensional axis system with the component—a short length of straight pipe—subjected to forces and moments about each axis. The forces and moments are considered as vector quantities and often expressed in terms of resultant vectors. For convenience we will express resultant vectors in terms of a resultant vector operator defined as follows:

$$R(x,y,z) = \sqrt{x^2 + y^2 + z^2} \qquad (2\text{-}1)$$

The resultant force and moments change in magnitude and direction along the length of the piping system. These forces and moments are controlled by structural supports attached to the piping using pipe supports to control forces and moments in the pipe and attaching components bring up two fundamental concepts—stiffness and flexibility—which are discussed later in this chapter.

PIPING CRITERIA

In analyzing piping mechanics, the following parameters must be considered:

1. The appropriate code that applies to the system.
2. The design pressure and temperature.
3. The type of material.
4. The pipe size and wall thickness of each pipe component.
5. The piping geometry including movements of anchors and restraints.
6. The allowable stresses for the design conditions set by the appropriate code.
7. Limitations of forces and moments on equipment nozzles set by API, NEMA, or the equipment manufacturers.
8. Metallurgical considerations, such as protecting material from critical temperatures, like carbon steel below its transition temperature.

For any piping system, these criteria must be considered and satisfied. While it is sufficient to analyze a piping system, it is not always necessary. For example, a system having only two terminal points and pipe of uniform size does not require a formal analysis if the following approximate criterion is satisfied:

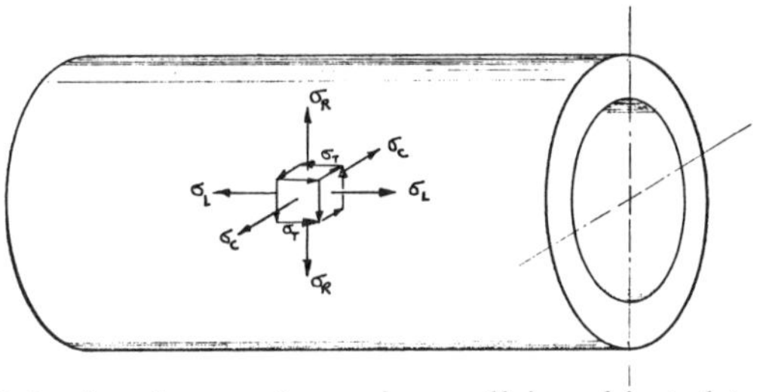

Figure 2-1. An element in a pipe wall is subjected to four stresses.

$$\frac{D_y}{(L - U)^2} \leqslant C \qquad (2\text{-}2)$$

where D_y = outside diameter (OD) of pipe, in. (mm)
y = resultant of total displacement strains to be absorbed by the piping system, in. (mm)
L = developed length of line axis between anchors, ft (m)
U = anchor distance (length of straight line joining anchors), ft (m)
C = 0.03 for U.S. units
= 208.3 for SI units, in parentheses

Usually, however, the piping system has either more than two terminal points or not all of the previous criteria are met, and a formal analysis is required.

After the first five criteria are considered the next and foremost factor to consider is Step 6—the allowable stress of the pipe. To determine this, one must refer to the appropriate code that governs the piping system. The following are codes applicable to industrial piping in the United States:

ASME B31.1—*Power Piping*—governs piping in the power industries (e.g., high-pressure steam lines)
ASME B31.3—*Chemical Plant and Petroleum Refinery Piping*—governs piping systems used in the chemical and petroleum industry
ASME B31.4—*Liquid Petroleum Transportation Piping Systems*
ASME B31.5—*Refrigeration Piping*
ASME B31.8—*Gas Transmission and Distribution Piping Systems*
ASME Section III—*Nuclear Piping.*

Most foreign codes are similar to the ASME (American Society of Mechanical Engineers) codes, particularly as far as the theoretical basis is concerned. The only code that is different from the ASME codes is the German DIN code, where the basis of yield is different. The code basis and theories of yield are discussed later.

Regardless of what ASME codes are used, the user is cautioned that the codes are written by ASME to be guidelines and not design handbooks. The intent of the codes is merely to set minimum rules and procedures for *design*. This does not include operation of plants. Operational problems are not intended to be governed by ASME codes. Such problems as bowing of the pipe and geysering are considered operational and are not considered as design phenomena.

Piping codes are not the only ones with which the design engineer should be familiar. It would be expedient and helpful if he or she is familiar with ASME Section VIII Div. I and II. Also, the AISC (American Institute of Steel Construction) *Manual of Steel Construction* is mandatory in the design of structural supports—a requirement that will be obvious later.

The reader will notice a stark contrast between the ASME and AISC philosophies of codes. The AISC *Manual of Steel Construction* is intended to be a design handbook and is considered as such. AISC, unlike ASME, covers all industries of steel construction, from the building of tall office buildings to major chemical plants. Unlike ASME, the AISC codes give a commentary on what bases are used in formulating the code and why these bases were used. It cannot be emphasized too much that engineering mechanics crosses mechanical and civil engineering disciplines as known in the United States. A knowledge of some of both is necessary to understand the overall perspective of piping mechanics.

In satisfying Step 6 in the list of criteria, once the appropriate code is selected, the system must be analyzed to determine if any portion of the system exceeds the allowable stress given by the code. The allowable stress given by the code is, in the ASME and most foreign codes, based on the maximum shear stress theory of failure. This theory is based on the fact that a material yields when the maximum shear stress equals the yield stress. This theory is in good agreement with experimental data under steady state and fatigue stress conditions and for this reason has been adopted.

Since knowledge of the different theories of yield is not directly pertinent to industrial applications of piping mechanics, the reader is referred to Faires [1] for further discussion. It is pertinent to note what stresses are required by the codes in analyzing piping systems.

An element of pipe wall subjected to four stresses is shown in Figure 2-1. The pipe is under internal pressure and the four stresses are as follows:

σ_L = longitudinal stress
σ_C = circumferential or hoop stress

σ_R = radial stress
σ_T = shear or torsional stress

The longitudinal stress is the sum of the following three components:

1. Bending stress induced by thermal expansion.
   For straight pipe:

$$\sigma_B = \frac{M}{Z_M} \tag{2-3}$$

For curved pipe:

$$\sigma_B = \frac{M}{Z_M}\, i \tag{2-4}$$

2. Bending stress induced by the weight of the pipe. (This stress should not be a consideration if the piping is properly supported and will not be considered in this analysis.)
3. Longitudinal stress induced by internal pressure.

$$\sigma_P = P\,\frac{Ai}{A_m} \tag{2-5}$$

Because both longitudinal stress caused by internal pressure and bending stress act in the same direction,

$$\sigma_L = \sigma_{BL} + \sigma_P \tag{2-6}$$

The circumferential or hoop stress is caused primarily by internal pressure. Thus,

$$\sigma_C = \frac{P(D - 2Py)}{2tE} \tag{2-7}$$

For thin-walled cylinders σ_R is negligible. However, for thick-walled pipe, the following relationship may be used for determining the radial stress:

$$\sigma_r = \frac{r_i^2 P_i - r_0^2 P_0}{r_0^2 - r_i^2} - \frac{r_0^2 r_i^2 (P_i - P_0)}{(r_0^2 - r_i^2) r} \tag{2-8}$$

and where external pressure, $P_0 - 0$, we have

$$\sigma_r - \frac{r_i^2 P_i}{r_0^2 - r_i^2} - \frac{r_0^2 r_i^2 P_i}{(r_0^2 - 4_i^2) r} \tag{2-9}$$

Torsional or shear stress is

$$T = \frac{T}{2Zm} \tag{2-10}$$

where torsion is generated in a multiplane system.

Direct shear stress is negligible and is not considered when caused by the piping temperature, because local yielding or "creep" reduces the stress at piping components. Local strain hardening restricts the local yielding and prevents the material from rupturing. This phenomenon of local yielding reducing stress is termed "self-springing," and has the same or similar effect as cold or hot springing. The operating stress ("operating" is used because it can be either hot or cold) diminishes with time. This change in stress is compensated for by the allowable stress range, which is the sum of the operating and down condition stresses and remains practically constant for one cycle. This sum is obtained as follows:

$$\sigma_A = f\,(1.25\,\sigma_o + 0.25\,\sigma_D) \tag{2-11}$$

f = stress range reduction factor for cyclic condition

Total no. of full temp. cycles over expected life
≤ 7,000
≤ 14,000
≤ 22,000
≤ 45,000
≤ 100,000

Expansion stress, caused by thermal expansion, must not exceed the allowable stress range, σ_A, and is defined:

$$\sigma_E = [(\sigma_B)^2 + 4(\sigma_T)^2] \tag{2-12}$$

The piping codes further state that the sum of the longitudinal stresses caused by pressure, weight, and other sustained loadings shall not exceed σ_D. This also includes the longitudinal stress caused by internal pressure, σ_P, defined above.

When torsional stress becomes significant, as in many multiplane systems, the resultant fiber stress, or combined stress, is determined by the following:

$$\sigma = \frac{1}{2}\,[\sigma_L + \sigma_D + [4(\sigma_T)^2 + (\sigma_L - \sigma_D)^2]^{0.5}] \tag{2-13}$$

PRIMARY AND SECONDARY STRESSES

These two concepts are very important in analyzing piping mechanics problems. A more detailed discussion of the various types of primary stresses is given in Chapter 4. The reader is encouraged to review Chapter 4 for an understanding of pressure vessels, as well as this chapter for help in solving piping mechanics problems.

Secondary stresses are called self-limiting or self-equilibrating because as they increase in magnitude, local yielding causes local deformation which in turn reduces the stresses. Self-springing is an example of this phenomenon.

Primary stresses are not self-limiting because as they increase, local yielding does not reduce them. One example of primary stress is internal pressure. Under sufficient pressure a pipe will undergo local yielding and deform, but the stress will not diminish and the pipe wall deformations will be excessive and unacceptable. For this reason, it is necessary to assign lower allowable stress limits to primary stresses than to secondary stresses. This fact is extremely important, as primary and secondary stresses are evaluated differently, and have different allowable limits. It must be remembered that piping and vessel codes give allowable stresses *only* for primary stresses. Secondary stresses must be assigned allowable limits as shown in the following discussion.

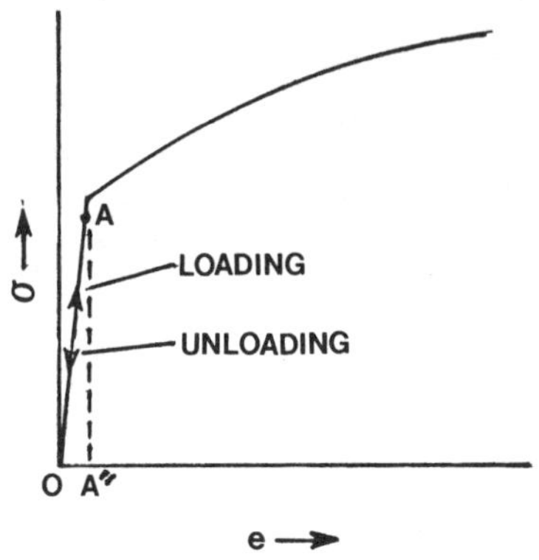

Figure 2-2. Stress-strain curve.

ALLOWABLE STRESS RANGE FOR SECONDARY STRESSES

The most important secondary stresses are those induced by thermal expansion (or contraction) and surface discontinuities, the latter being more relevant to vessels. The most widely used approach in designing equipment, vessels, and piping is to keep the induced stresses in the elastic range. In the case of ductile materials, the elastic range is well defined by the minimum yield point. Ductile materials are often used in piping systems subjected to loads that induce secondary stresses. Materials that do not have a well defined minimum yield point are designed on the basis of their ultimate yield strength, which is the maximum tensile load divided by the original cross-sectional area of the specimen. The minimum yield point is the tensile load required to develop permanent deformation in the material. Materials that do not have a well defined minimum yield point are generally not used in piping systems subjected to extreme temperatures and pressures. Thus, this discussion applies to those materials with minimum yield points.

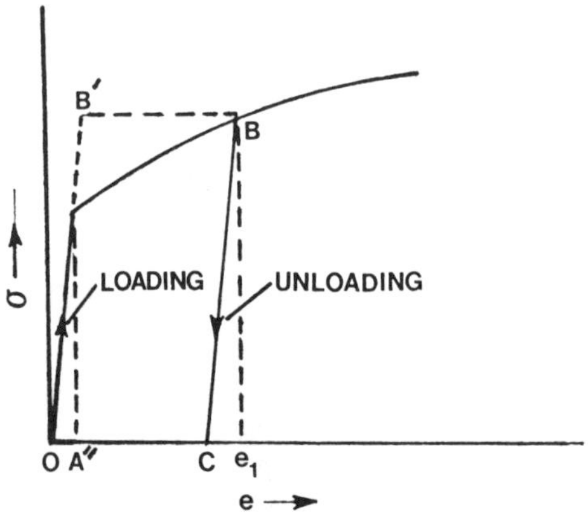

Figure 2-3. Stress-strain curve.

Consider the stress-strain curve shown in Figure 2-2. Here the metal specimen is loaded to point A and then unloaded. Because point A is the minimum yield point, no deformation occurs because the material is still in the elastic range. Now, consider Figure 2-3 where the material is loaded beyond point A. Because the minimum yield point is exceeded, plastic deformation sets in that permanently deforms the material to point B. When the specimen is unloaded, e_1 is the amount of permanent deformation, denoted by point C. Point B′ is the theoretical stress point if the material had not deformed to point B. Figure 2-4 shows a case where a specimen is loaded

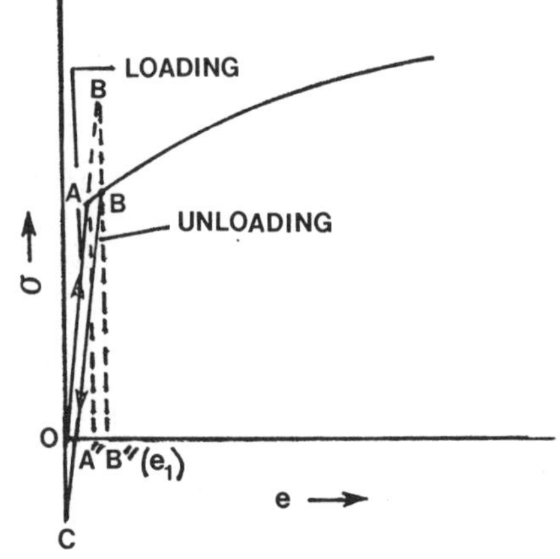

Figure 2-4. Stress-strain curve.

into the plastic region. For complete plastic deformation to occur, the entire area of the pipe wall must exceed the minimum yield point. This would not be acceptable in practice because of permanent deformation and the possibility of rupture.

There are acceptable cases where the loads will fall between Figure 2-2 and Figure 2-3. This condition is shown on Figure 2-4, where part of the pipe wall is in the elastic range and the other part is on the plastic region. For cases where the portion in the plastic range is small compared with the portion in the elastic range, the amount of permanent plastic deformation is imperceptible. For this reason, the distance between points A and B in Figure 2-4 is small compared to Figure 2-3 because the portion of material in the elastic range limits the amount of permanent deformation. Thus, when the specimen is unloaded, residual stresses are developed that cause reverse yielding when the material exceeds the compressive yield point. This is shown graphically in Figure 2-5. The specimen is loaded to point A and an excessive load deforms it to point B. At point B, part of the material is in the plastic range and the other portion is in the elastic range. When the specimen is unloaded, the stresses in the material go into compression shown at point C. Residual stresses caused by the combination of material in the elastic and plastic regions make part of the material exceed the compressive yield point and the specimen deforms from point C to point D. Upon application of the same initial tensile load, the material is loaded to point E. Point E is larger in value and, thus, to the right of point A, because the initial loading of part of the specimen into the plastic range causes strain harden-

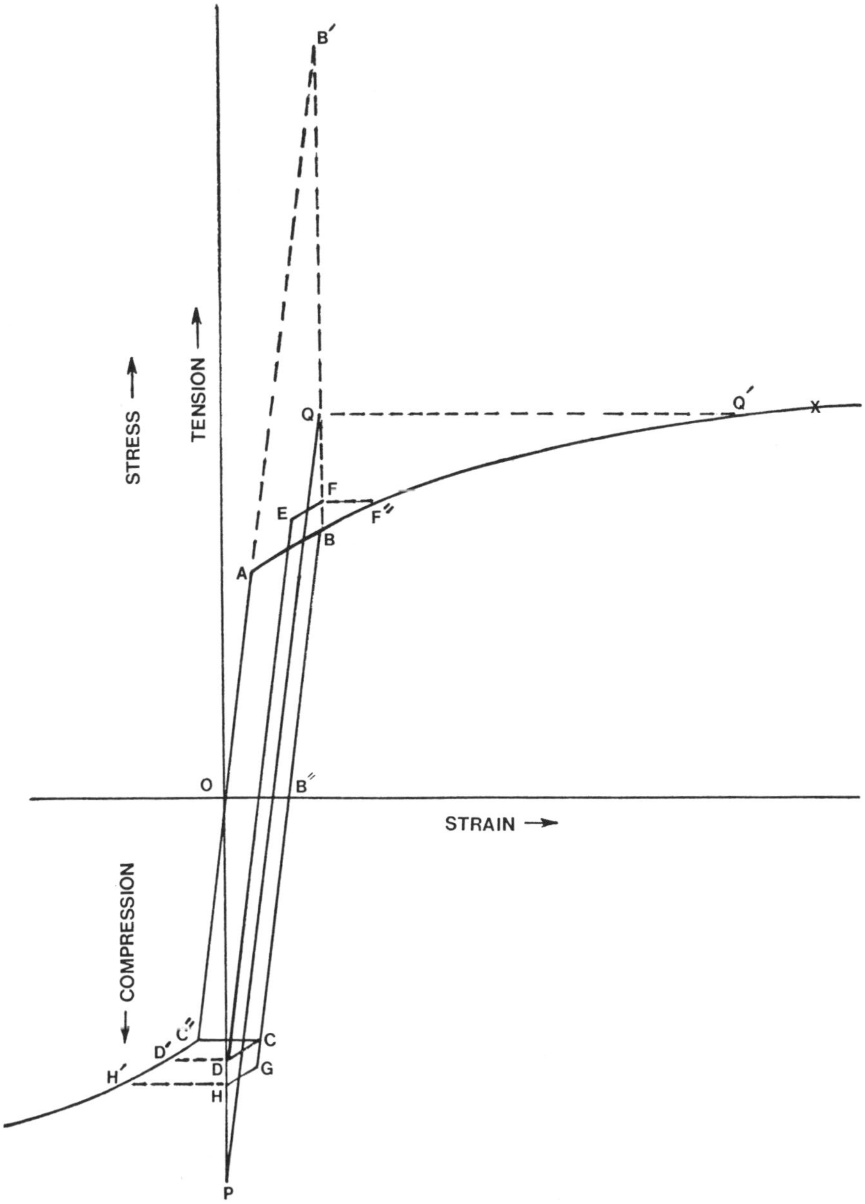

Figure 2-5. Stress-strain curve.

ing and, thus, increases the minimum yield point of the material. As excessive loads are applied, the minimum yield point E is exceeded and the material deforms to point F. As the material is unloaded again the initial process repeats itself and the stresses in the material move to point G and then to point H as the compressive yield point is exceeded.

Point Q represents the stress in the loaded condition after several loading cycles, and point P represents the stress in the unloaded condition. It is possible that no significant plastic deformation will occur after many load cycles. However, should stress values of Q and P exceed the fatigue limit of the material, small cracks will propagate throughout the strain-hardened material. After the small cracks appear, further cyclic loading will result in brittle fracture failure. The stress magnitude P results from the specimen being unloaded when the load condition, point Q, is reached. Thus, since Q is the tensile stress opposite to the compressive stress P in the parallelogram OB′QP, the sides OB′ and QP are equal in length. Therefore, $Q = 0.5\,B'$. Fracture by strain hardening will not occur if the theoretical tensile stress B′ does not exceed twice the minimum yield stress of point A, and the magnitude of Q does not exceed the ultimate yield strength of point A.

When a ductile material, that is a material with a defined minimum yield point, is subjected to repeated loading, a certain behavior occurs. When a component, such as a nozzle on a header pipe, is repeatedly loaded and unloaded, the strain hardening makes the material stronger from load cycle to load cycle. As the material becomes harder, it is better able to resist yield. However, the maximum point at which this repeated loading cycle can occur is $2\sigma_{YP}$. The stress $\sigma = 2\sigma_{YP}$ is the limit of the maximum stress range. This process is called elastic shakedown, that is, the material "shakes down" to an elastic response, and undergoes deformations or strains induced by loads beyond the minimum yield point of the material.

It must be noted that at elevated temperatures the value of $2\sigma_{YP}$ can be altered by hydrogen embrittlement. Carbon steel exposed to hydrogen at elevated temperatures can fail during elastic shakedown because the hydrogen combines with the carbon causing embrittlement.

The relationship between the maximum stress range and the initial yield point can be expressed as

$$\sigma_{MR} \leq 2\sigma_{YP} \tag{2-14}$$

where MR = maximum local stress range not producing fatigue failure, psi
YP = initial yield point of the material at the operating temperature, psi

This analysis indicates that the allowable stress should be based on the yield point rather than ultimate strength.

The material's ability to revert into compression and limit itself to the amount of permanent plastic deformation is termed "shake down." The material "shaking down" limits the amount of deformation and, thus, has an elastic response.

From this discussion, we see that there is a range of allowable stresses available. Direct membrane stresses are limited by σ_Y, bending stress is limited by $1.5\sigma_Y$, and a limited, one-time permanent deformation from A to B occurring from secondary stresses is limited by $2\sigma_Y$.

Table 2-1 gives recommended values for design allowable stresses. As shown in ASME Section VIII, Division I, paragraph UA-5e, different stress levels for different stress categories are acceptable.

FLEXIBILITY AND STIFFNESS OF PIPING SYSTEMS

There are two basic approaches to piping mechanics—flexibility and stiffness. The former approach is more common and easier to understand. Piping mechanics (more popularly known as "pipe stress") is often referred to as "flexibility analysis," but it will become obvious in the following discussion that such a term is not complete.

In the flexibility approach, the piping configuration is made more flexible by using loops that allow the pipe to

Table 2-1
Allowable Stresses*

Pressure Component Design Conditions	
1. Internal pressure	σ_A
2. Internal pressure plus thermal loading	$1.25\,(\sigma_A + \sigma_D)$
3. Temporary mechanical overload	$1.33\sigma_A < \sigma_Y$
4. Hydrotest	$\sigma_D \times$ hydrotest factor
Non-pressure Components Design Conditions	
1. Pipe supports and connections other than bolts	$1.33\sigma_A$
2. Bolting	Per AISC Manual of Steel Construction considerable savings in material can be incurred if high strength bolts are utilized, such as SA-193-B7. Following AISC guidelines in non pressure components will result in prudent economical design.

* Courtesy of American Society of Mechanical Engineers

displace itself, resulting in lower stresses, forces, and moments in the system. This method is often the most desirable when relatively inexpensive piping material is used (pipe elbows can be very expensive in alloy piping) and space is available for the loop(s).

However, the stiffness method becomes quite important when the flexibility method is neither practical nor economical. When limited space reduces piping flexibility or makes it impossible or undesirable to use flexibility loops, restraining the piping using the stiffness of pipe supports becomes the alternative. This approach is gaining popularity with the increased use of modular designs of petrochemical plants, offshore platforms, and other industrial facilities.

The following is a summary of the advantages of both methods:

Stiffness Method Advantages

1. Requires less pipe fittings and is thus more economical than flexibility method, because pipe restraints required are far less expensive than the number of fittings they replace. In alloy piping these savings are enormous.
2. Requires far less space for piping, such as in modular skid-mounted plants, offshore platforms, and ships.
3. Method is safer because in case of a failure, such as a leak in a weld crack, the pipe restraints can (and have) kept systems from blowing apart.
4. Piping and system is more resistant to dynamic loads, such as vibration and seismic shock loads.

Flexibility Method Advantages

1. Utilizes simpler pipe supports, and requires less piping engineering skill.
2. Is more desirable in noncritical systems, e.g. exhaust and flare lines.
3. Many solutions do not require a computer. The problems can be solved manually.

To better understand these two methods of piping mechanics, it is necessary to examine some basics of structural analysis.

Stiffness is the amount of force or moment required to produce unit displacement, either translational or angular movement. The simplest concept of stiffness is to imagine using X pounds to compress a spring one inch. Thus, the spring stiffness is in terms of pounds per inch. This simple example illustrates translational stiffness. Rotational stiffness can be thought of in a similar manner as a spring that resists rotational movement, foot-pounds per unit degree of movement.

A piping element has six degrees of freedom, three in translation and three in rotation, as shown in Figure 2-6. The amount of force or moment required to produce unit displacement in each degree of freedom at points all along the piping element is described mathematically as the stiffness matrix, K, which is defined as

$$P = KU$$

where we havc an elastic element subjected to a set of n forces and moments

$$P = \{P_1, P_2, \ldots, P_i, \ldots, P_n\} \tag{2-15}$$

the corresponding displacement of each P_i is described by the matrix

$$U = \{U_1, U_2, \ldots, U_i, \ldots, U_n\} \tag{2-16}$$

Therefore, the stiffness matrix can be expressed as

$$K = \frac{P}{U} \tag{2-17}$$

which can be in pounds per inch or foot pounds per degree. The relationship

$$C = \frac{U}{P} \tag{2-18}$$

is defined as the compliance or flexibility matrix and can be in inches per pounds or degrees per foot-pounds. Thus, the stiffness K of a system is the inverse of the system compliance or flexibility, C, that is, the piping system becomes more flexible, or less stiff than its initial configuration.

The system stiffness matrix, K, is made up of elements that are either direct stiffness or indirect stiffness components. The direct stiffness component K_{ij} is the value of stiffness at the point i when the displacement U_j is produced by a force or moment P acting in the direction of U_j. The indirect stiffness K_{ij} is the value of stiffness at the point j, with the displacement U_j acting in the direction of j, due to a force or moment at another point i in the direction of i. The indirect stiffness can also be thought of as relative stiffness—those stiffness values induced by forces and moments in the system other than the point in consideration. It is the combined grouping of the complete direct and indirect stiffness values that form what is called the "stiffness matrix." Each direct and indirect stiffness is considered in the matrix when all other matrix components are zero. Such as the system described in the following:

$$P = \underbrace{\begin{bmatrix} K_{11} & K_{12} & K_{13} & K_{14} & K_{15} & K_{16} \\ K_{21} & K_{22} & K_{23} & K_{24} & K_{25} & K_{26} \\ K_{31} & K_{32} & K_{33} & K_{34} & K_{35} & K_{36} \\ K_{41} & K_{42} & K_{43} & K_{44} & K_{45} & K_{46} \\ K_{51} & K_{52} & K_{53} & K_{54} & K_{55} & K_{56} \\ K_{61} & K_{62} & K_{63} & K_{64} & K_{65} & K_{66} \end{bmatrix} \begin{bmatrix} U_1 \\ U_2 \\ U_3 \\ U_4 \\ U_5 \\ U_6 \end{bmatrix}}_{KU} \quad (2\text{-}19)$$

where the values K_{11}, K_{22}, K_{33}, K_{44}, K_{55}, and K_{66} are known as direct stiffness values and all the other components are known as indirect stiffness values. Each value of U represents a unit displacement. The components of the stiffness matrix are determined by the nature (axial force, bending moment, shear force) of the force or moment inducing unit displacement U at or away from the point in question. To effectively see how these stiffness components are utilized in practical applications, we will consider each type of force or moment inducing displacements, that is, each component of the P matrix corresponding to each value of the U matrix. Table 2-2 lists the direct values of stiffness induced by direct and indirect loadings shown in Figure 2-6. For analytic derivations, the reader is referred to Przemieniecki [2].

To illustrate how these concepts apply to piping mechanics, let us consider both a 4-in. schedule 40 pipe and a 10-in. schedule 40 pipe shown in Figure 2-7. Here we are considering two pipe spool pieces subjected to a force F shown. Referring to Table 2-2, we see that the translational stiffness for a beam element fixed on one end and pinned at the other end is

$$K_{ij} = \frac{3EI}{L^3}$$

For the 4-in. pipe,

$$K_4 = \frac{3(29 \times 10^6)\ \dfrac{\text{lb}}{\text{in.}^2}(7.23)\ \text{in.}^4}{(48)^3\ \text{in.}^3} = 5{,}687.66\ \frac{\text{lb}}{\text{in.}}$$

For the 10-in. pipe,

$$K_{10} = \frac{3(29 \times 10)\ \dfrac{\text{lb}}{\text{in.}^2}(160.8)\ \text{in.}^4}{(48)^3\ \text{in.}^3} = 126{,}497.40\ \frac{\text{lb}}{\text{in.}}$$

The force required to move the 4-in. pipe 1/4 in. is

$$F = (5{,}687.66)\ \frac{\text{lb}}{\text{in.}}\ (0.25)\ \text{in.} = 1{,}421.92\ \text{lb}$$

To generate the same amount of force in a 10-in. pipe the same length would have to move

$$\Delta = \frac{1{,}421.92\ \text{lb}}{126{,}497.40\ \dfrac{\text{lb}}{\text{in.}}} = 0.011\ \text{in.}$$

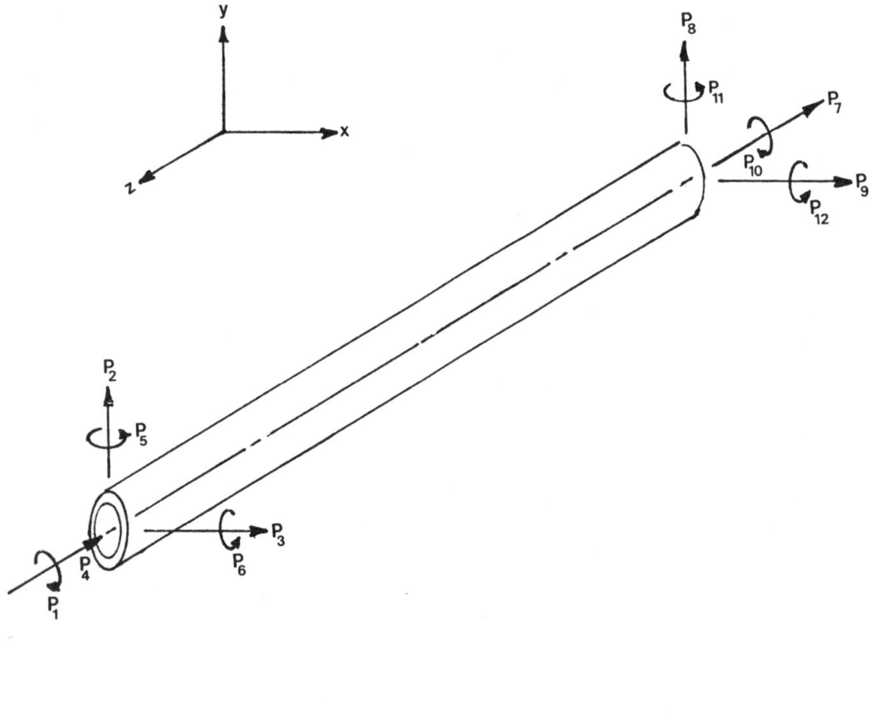

Figure 2-6. Pipe element.

Table 2-2
Stiffness Properties of Piping Elements

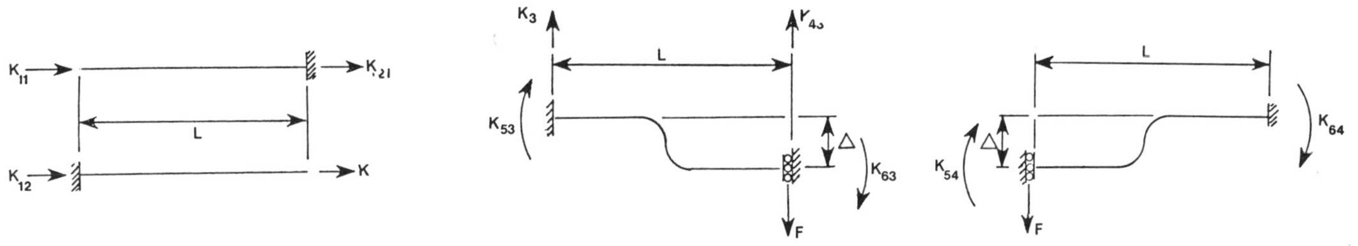

$K_{11} = K_{22} = \dfrac{AE}{L}$	$K_{13} = K_{23} = 0$	$K_{14} = K_{24} = 0$
$K_{21} = K_{12} = \dfrac{AE}{L}$	$K_{33} = \dfrac{12EI}{(1 + \Phi)\, L^3}$	$K_{44} = \dfrac{12EI}{(1 + \Phi)\, L^3}$
$K_{31} = K_{41} = K_{51} = K_{61} = 0$	$K_{43} = \dfrac{-12EI}{(1 + \Phi)\, L^3}$	$K_{34} = \dfrac{-12EI}{(1 + \Phi)\, L^3}$
$K_{32} = K_{42} = K_{52} = K_{62} = 0$	$K_{53} = K_{63} = \dfrac{-6EI}{(1 + \Phi)\, L^2}$	$K_{54} = K_{64} = \dfrac{6EI}{(1 + \Phi)\, L^3}$

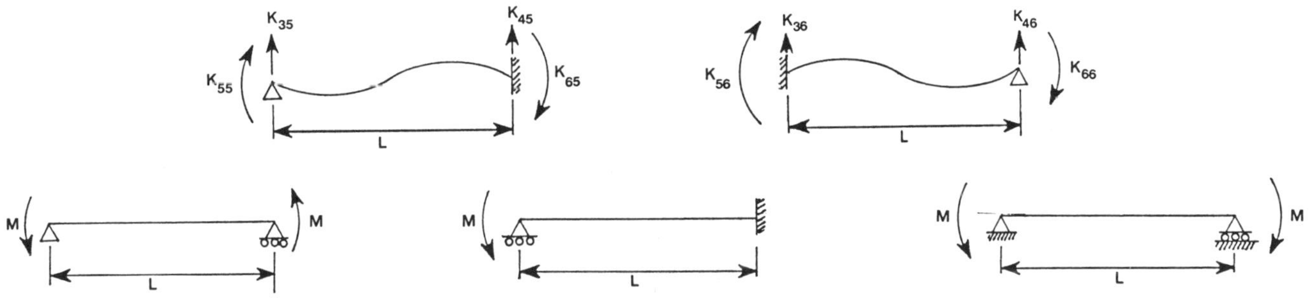

$K_{45} = K_{46} = \dfrac{6EI}{(1 + \Phi)\, L^2}$	$K_{55} = K_{66} = \dfrac{(4 + \Phi)EI}{L(1 + \Phi)}$	$K_{65} = K_{56} = \dfrac{(2 - \Phi)EI}{L(1 + \Phi)}$
$K_{35} = K_{36} = \dfrac{-6EI}{(1 + \Phi)L^2}$	Note: In all cases $\Phi = \dfrac{12EI}{GAL^2}$ and $K = \dfrac{GJ}{L}$ = torsional stiffness	

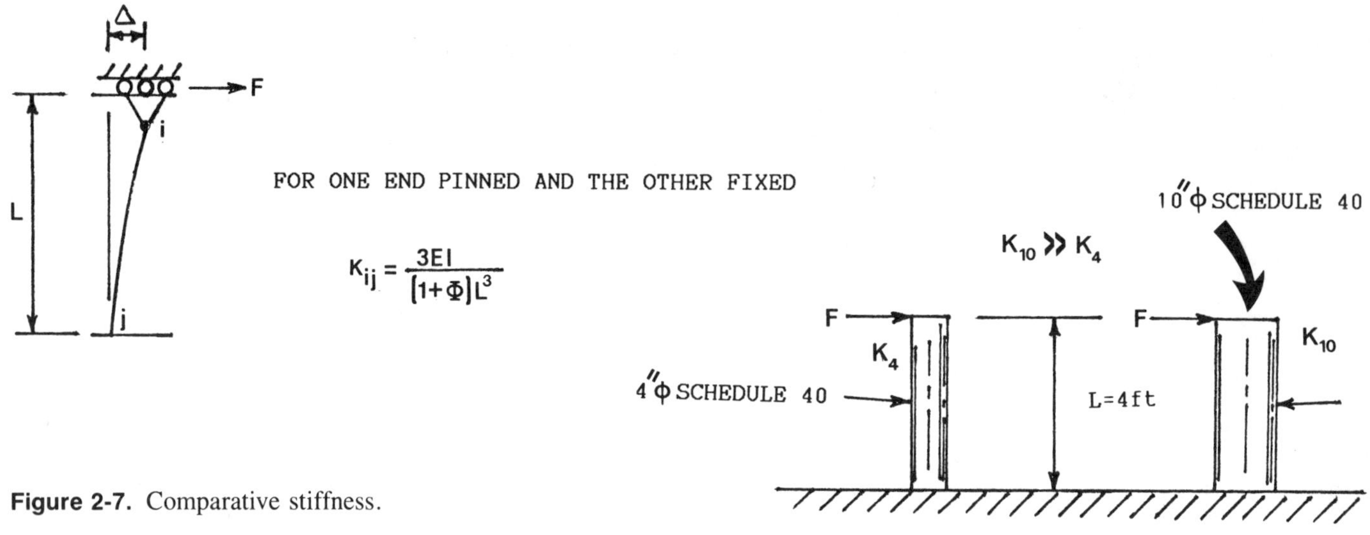

Figure 2-7. Comparative stiffness.

In other words, if the pipe itself moved because of thermal expansion and there was a restraint of a given spring constant restraining the movement, the 10-in. pipe would only have to move 0.011 in. to exert the same force as the 4-in. pipe moving 1/4 in. Thus, the 10-in. pipe is more than 22 times stiffer than the 4-in. pipe, which is a significant point because it indicates that the larger the piping, the less it must move to exert excessive forces and moments on nozzle connections and pipe supports. From this example it is obvious that the larger the piping, the greater the stiffness. Piping designers often fail to realize that larger piping does not have to move very much to generate great loads. This basic fact is important in the design of pipe supports, particularly using the stiffness approach.

Carrying the analysis further, consider the two piping configurations shown in Figure 2-8. This situation is similar to Figure 2-7 in that one end is fixed and the other pinned, i.e., both systems have the same boundary conditions. The segment B-C is flexible enough to bend with enough rotational flexibility to consider that end as a pinned joint. If the temperature of the piping is −100° f, the segment B-C moves

$$\Delta = (-1.75)\,\frac{M}{100\text{ ft}}\,(4)\text{ ft} = -0.070\text{ in.}$$

The force required to move a 4-in. schedule 40 pipe 0.070 in. is

$$F_4 = (5{,}687.66)\,\frac{\text{lb}}{\text{in.}}\,(0.070)\text{ in.} = 398.14\text{ lb.}$$

$K = \frac{3EI}{L^3}$

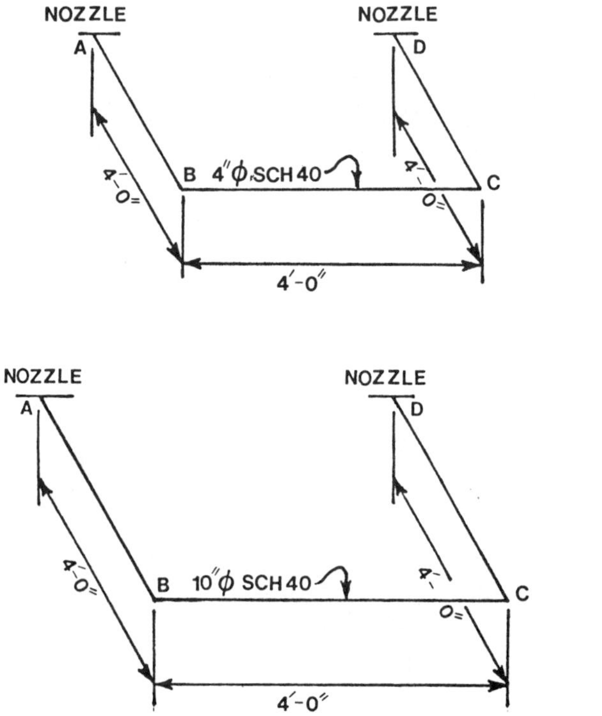

Figure 2-8. Pipe size makes a significant difference in nozzle loadings.

The force required to move a 10-in. schedule 40 pipe 0.070 in. is

$$F_{10} = (126{,}497.40)\ \frac{\text{lb}}{\text{in.}}\ (0.070)\ \text{in.} = 8{,}854.82\ \text{lb}$$

yielding a moment of

$$M_{10} = (8{,}854.82)(4) = 35{,}419.27\text{ft-lb}/2 = 17{,}709.64\text{ft-lb}$$

at the nozzles A and B.
The 4-in. force of 398.14 lb would produce a moment of

$$M_4 = (398.14)\ \frac{(4)}{2} = 796.28\ \text{ft-lb}$$

at nozzles A and B.

It is clear that the 10-in. pipe would exert moments well above the allowable moments for most rotating and stationary equipment. To reduce the loading at the nozzle, the engineer is faced with two options—make the piping configuration more flexible or restrain the piping. To fabricate the piping configuration to within a tolerance of 0.070 in. would be well beyond the practical range of any fabricating shop.

First, we will analyze a case where space is premium and there is not enough room to make the piping more flexible. This requires using piping restraints to transfer loads from the pipe to structural steel or concrete. Consider the piping system in Figure 2-9, where two aluminum heat exchangers are piped parallel to one another. Here we use the fewest 90° elbows needed to give the system enough flexibility to stay within the maximum allowable stress range for the material at the given temperature. Piping restraints are then placed close to the heat exchangers to transfer loads from the pipe to the steel instead of the nozzle of the exchanger.

Now, we analyze the component that makes the system work—the pipe restraint at the equipment nozzle. The restraint's function is to transfer forces and moments exerted by the pipe to the structural steel below, simultaneously allowing the equipment to move freely. This requires a more careful design of the piping restraint, as we are expecting it to do more.

In this example the piping restraints must allow the exchangers to move upward as shown in Figure 2-9. A restraint that resists moments by transferring the moments from the pipe to the steel is termed a moment restraining support (MRS). Different types of MRS supports are shown in Figure 2-10. An MRS can vary from a boiled plate connection shown in Figure 2-10A to a sophisticated type in Figure 2-10C. MRS restraints' sophistication is a function of how much rotation is resisted and how much translational movement is allowed. The most

Aluminum exchanger flange

Figure 2-9. An MRS support-restraint designed to reduce forces and moments on an aluminum plate-fin heat exchanger.

simple MRS restraint is the anchor, where the pipe itself or a pipe attachment is welded down to structural steel or immersed in concrete. In that case, it is resisting three degrees of freedom in translation and three degrees of freedom in rotation. In most applications, the moments at nozzle connections can become excessive, and it is often desirable to resist rotation in one, two, or three axes while allowing translational movement. Resisting rota-

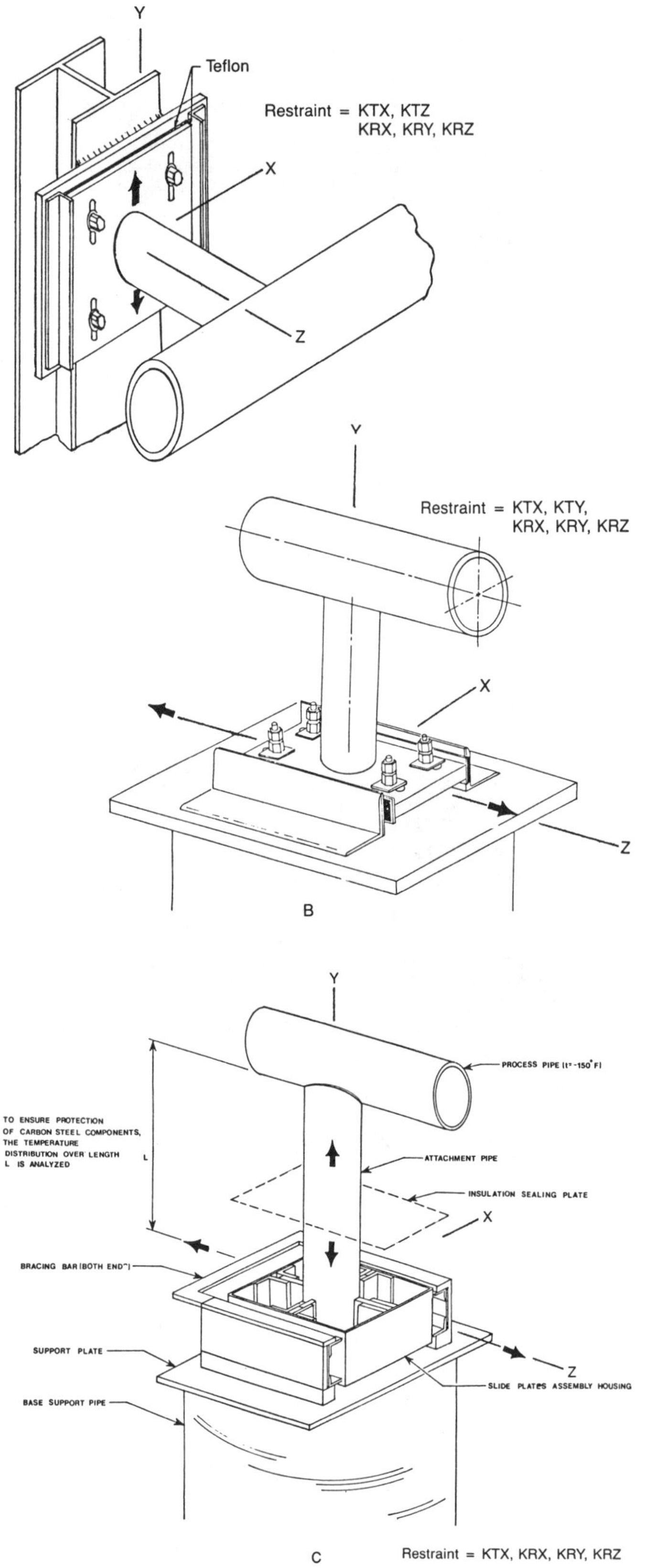

Figure 2-10. Various designs of moment restraint supports (MRS)—arrows indicate direction of allowed movement.

tion along three axes is, if not impossible, wholly impractical. An MRS allowing two degrees of freedom in translation and resisting three degrees of rotation is quite complicated, although practical, very useful, and economical when the situation warrants. In designing such restraints Teflon and other materials with very low friction coefficients are desirable. Care must be made in assuring that such material selected can withstand the forces and moments being resisted. If the material used is not resistant to shear, cold flow will result, leading to uneven surfaces and an improperly functioning restraint.

In the engineering of MRS restraints, the principles discussed previously must continuously be applied. No support or restraint can be expected to be infinitely rigid along the degrees of freedom that are being restrained. Placing MRS devices in front of equipment nozzles will not stop all loading exerted by the piping, because all restraints have a corresponding stiffness value for each deg of freedom, either lbs/in. for translation or ft-lbs/deg for rotation. The engineer must also understand what assumptions are being made by the piping stress program being applied. Almost all computerized pipe stress packages consider an anchor as six springs, three resisting translational forces of 10^9 lbs/in. and three resisting rotational forces of 10^9 ft-lbs/deg. There is no infinitely rigid anchor in nature, but 10^9 lbs/in. is sufficient to be called an anchor in almost all applications.

In modular plant design it is often desirable for the engineer to enter the actual stiffness of any anchor or restraint to obtain an accurate model of the piping system being analyzed on the computer.

STIFFNESS METHOD AND LARGE PIPING

Large piping is more difficult to restrain than small piping because of the surface to be restrained. The terms "large" and "small" are quantified in the following discussion. The most common complication of restraining large piping is the phenomenon of shear flow, which occurs longitudinally and circumferentially. As illustrated in Figure 2-11, longitudinal shear flow transfers bending moments and shear forces to the equipment nozzle.

In modular construction longitudinal shear flow does not become a problem until one starts using 10-in. pipe and larger. Shear flow can be resisted to some degree by making the attachment pipe size or structural member size close to that of the pipe, but is most often impractical. What is often desirable is to mount an MRS on opposite ends of the pipe, either top and bottom or off to both sides, depending on what space is available. In piping 30-in. and larger MRS restraints must be attached on four sides for the MRS effect to be effective. In pipe di-

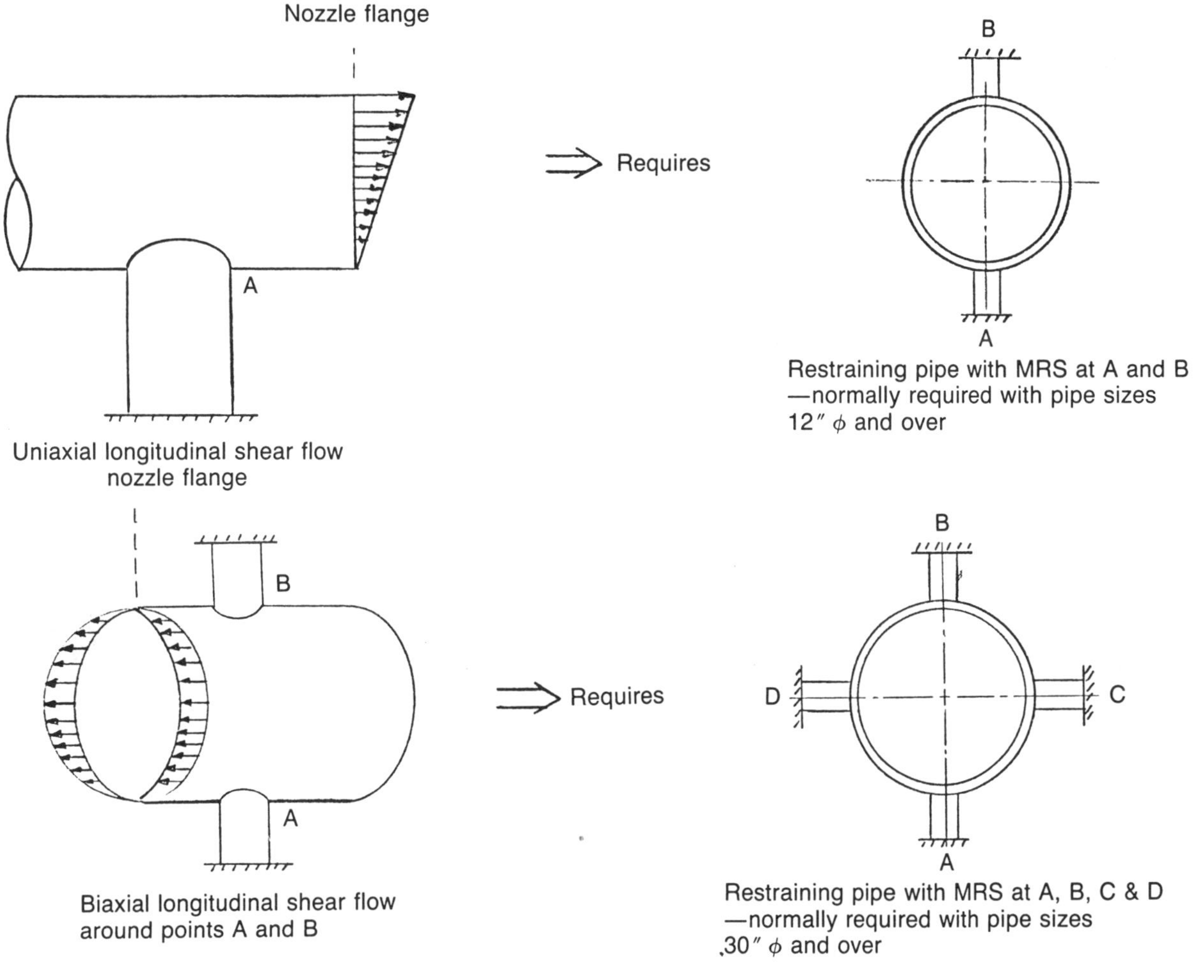

Figure 2-11. Longitudinal shear flow—a phenomenon of large pipe.

ameters 8-in. and smaller, attaching an MRS on one side is sufficient for most modular construction.

Circumferential shear flow, on the other hand, is not a factor in most installations because torsion is very effectively transferred to the structural steel by the MRS restraint.

Using piping restraints to transfer loads to structural steel or concrete to lower loads at equipment nozzles is becoming quite popular and more widespread because it is more economical in modular skid design. Also, where expensive piping materials are used, the stiffness method can help reduce the number of elbows used for flexibility and, thus, reduce the cost of the job because restraints and supports are far cheaper than piping elbows.

FLEXIBILITY METHOD OF PIPING MECHANICS

In non-modular skid construction (block-mounted plants) and areas where there is ample space to place equipment, it is often more economical and desirable to design the piping to be flexible enough to reduce loadings on supports and equipment nozzles. For pipe racks, long headers, etc. this method is the only practical approach to solving piping mechanics problems. Tools used in this approach include such well known devices and techniques as piping loops, cut short and cut long, and expansion joints.

PIPE LOOPS

The most common types of pipe loops used today are "U" shapes, "Z" shapes, and "L" shapes. Curves for these shapes showing stresses plotted against the loop dimensions are shown in Figures 2–12 and the equations are as follows:

$$F_X = A_X B \frac{I_P}{L^2} \text{ lb}, \quad I_P = \text{in.}^4$$

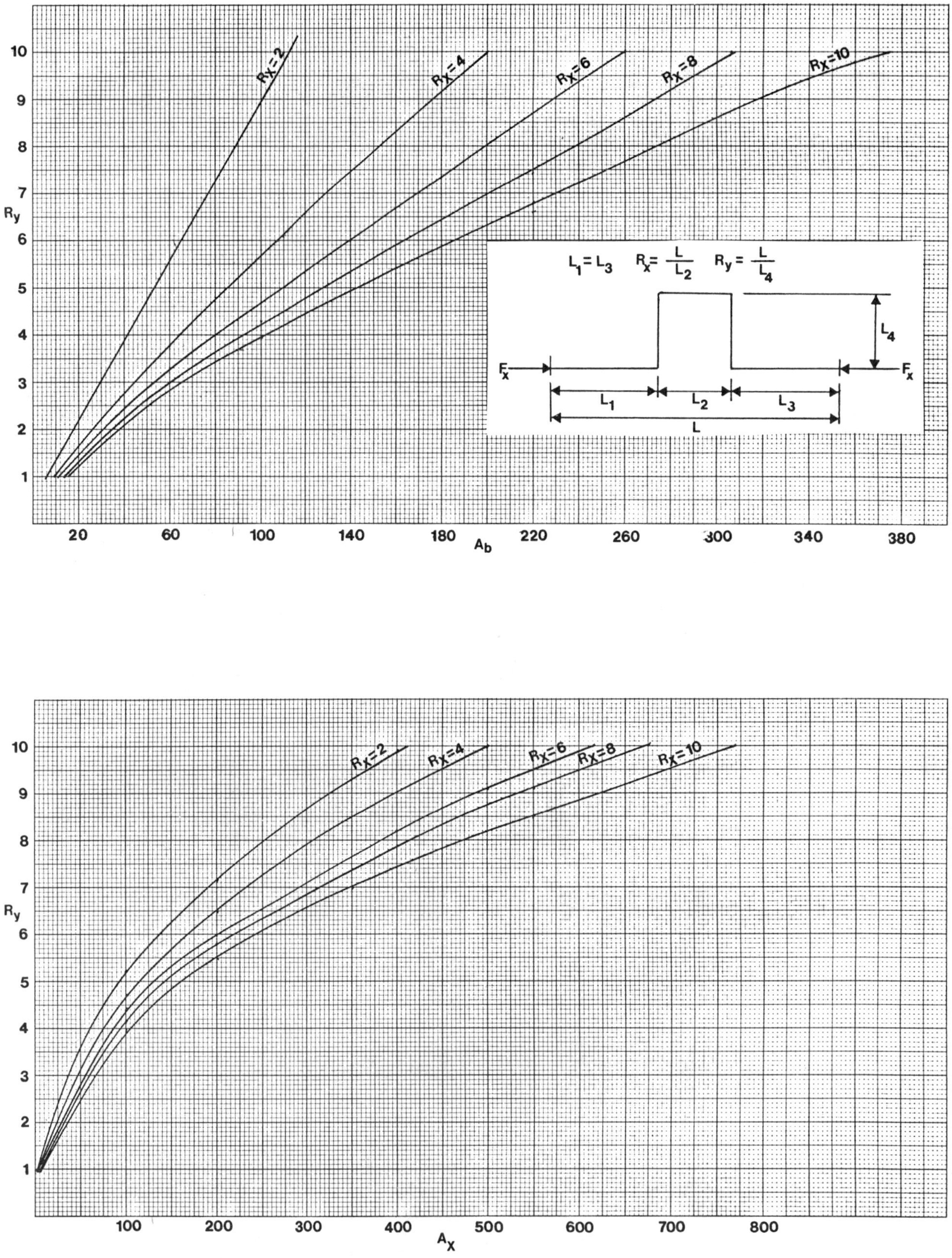

Figure 2-12A. U-loop with equal legs.

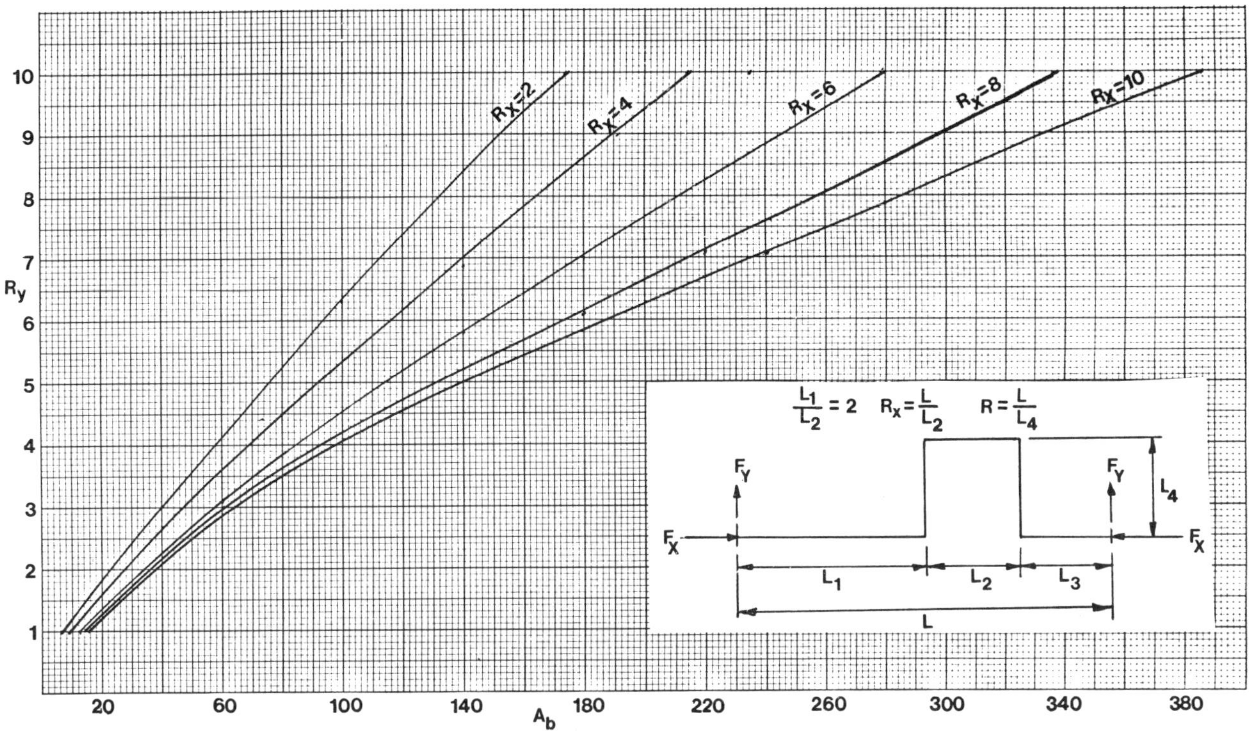

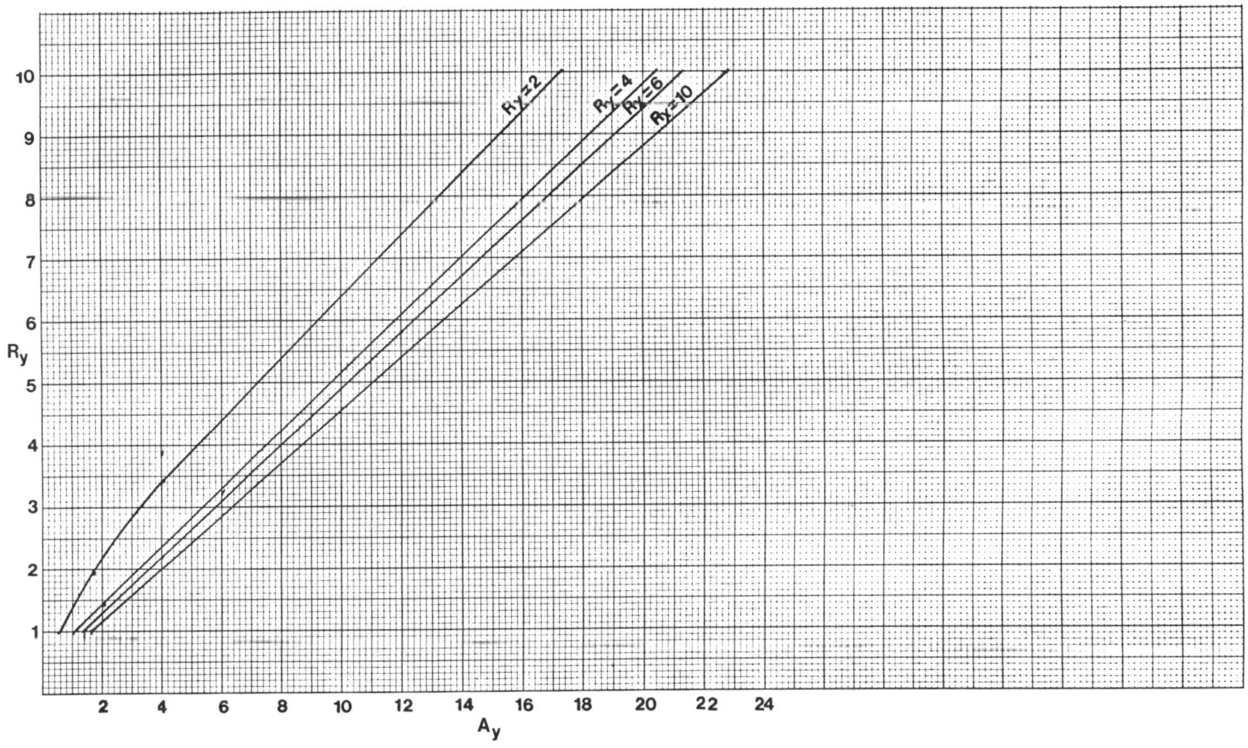

Figure 2-12B. U-loop with one leg twice the other leg.

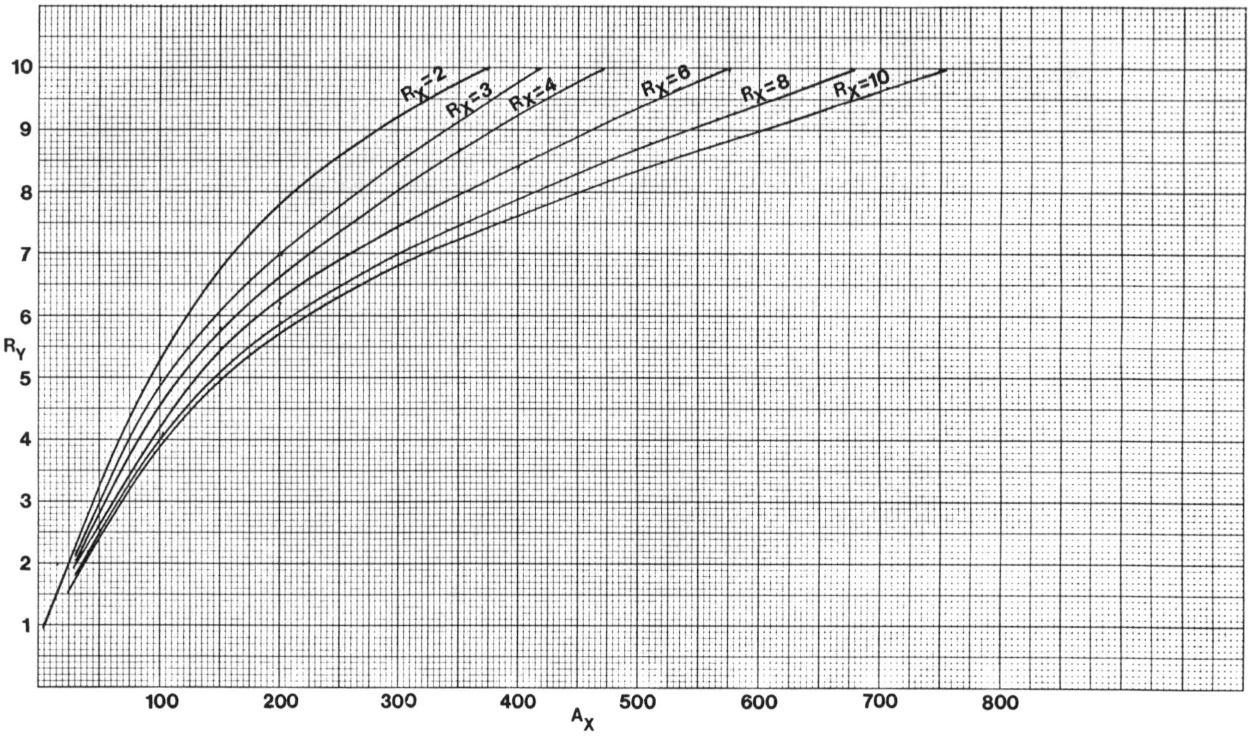

Figure 2-12B (continued). U-loop with one leg twice the other leg.

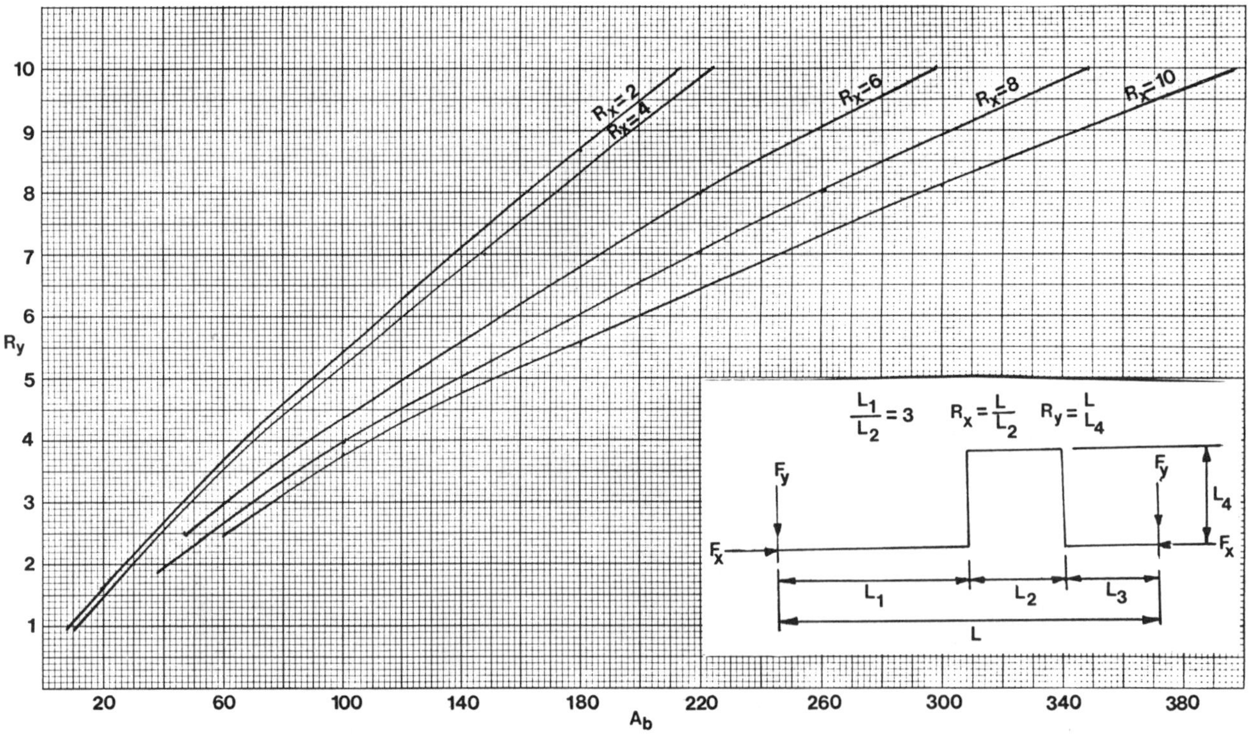

Figure 2-12C. U-loop with one leg three times the other leg.

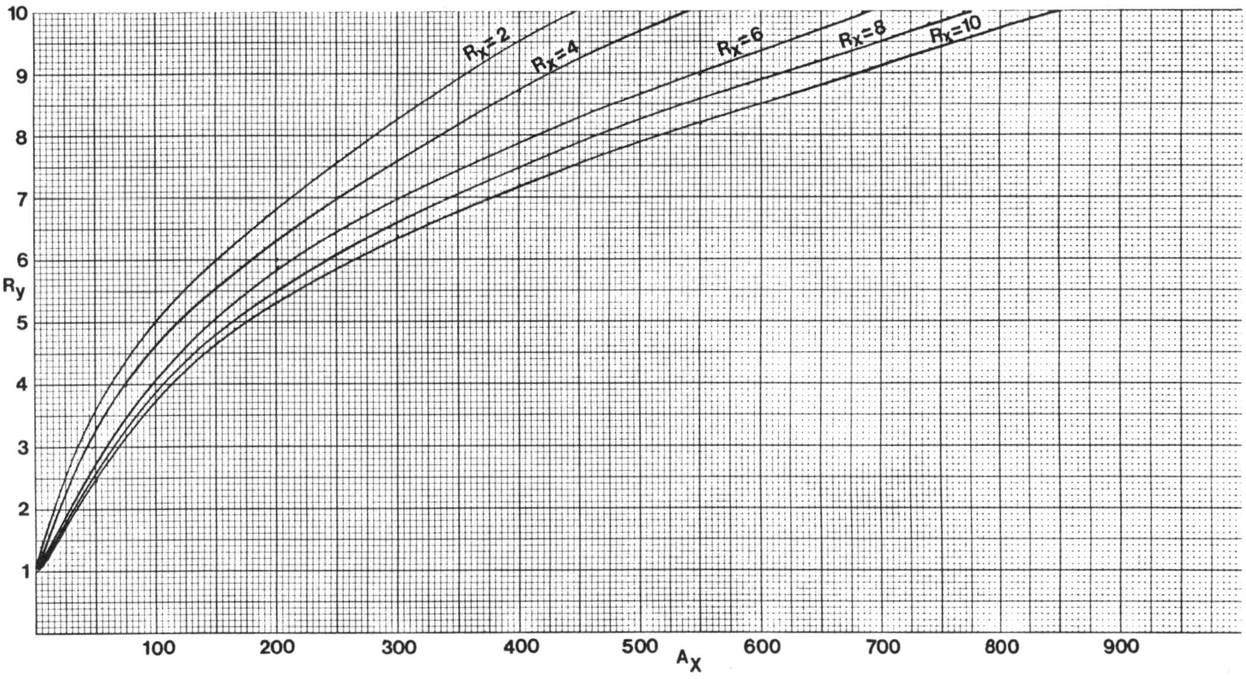

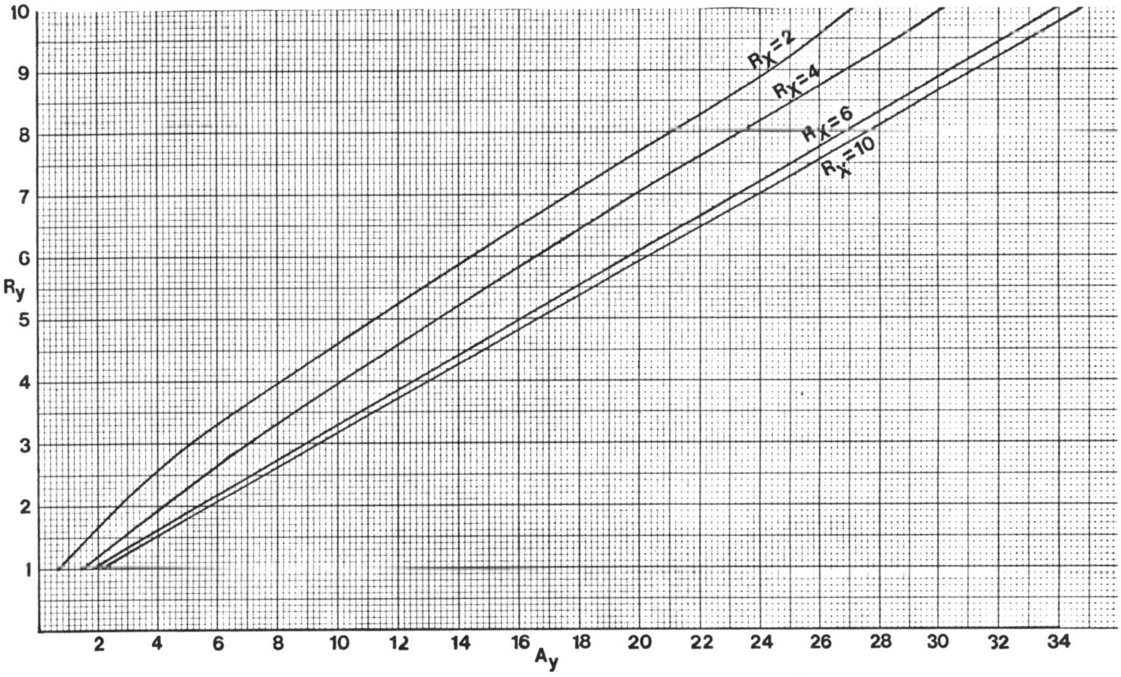

Figure 2-12C (continued). U-loop with one leg three times the other leg.

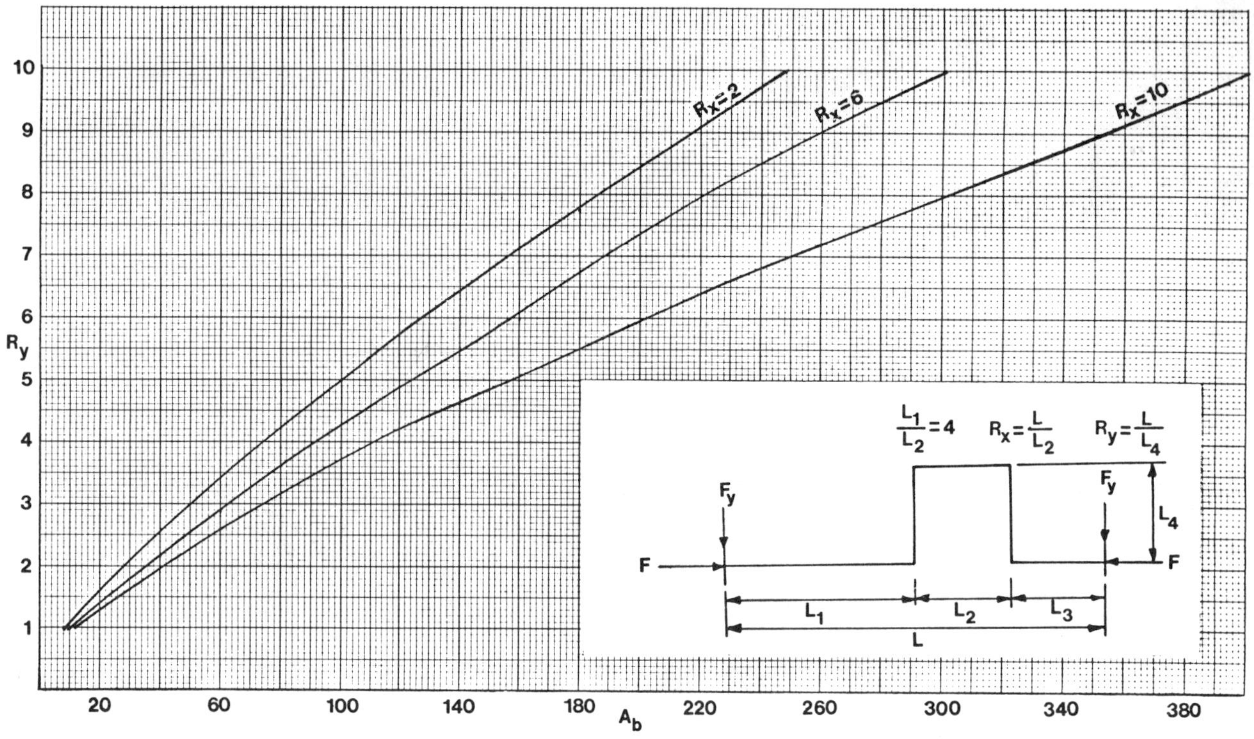

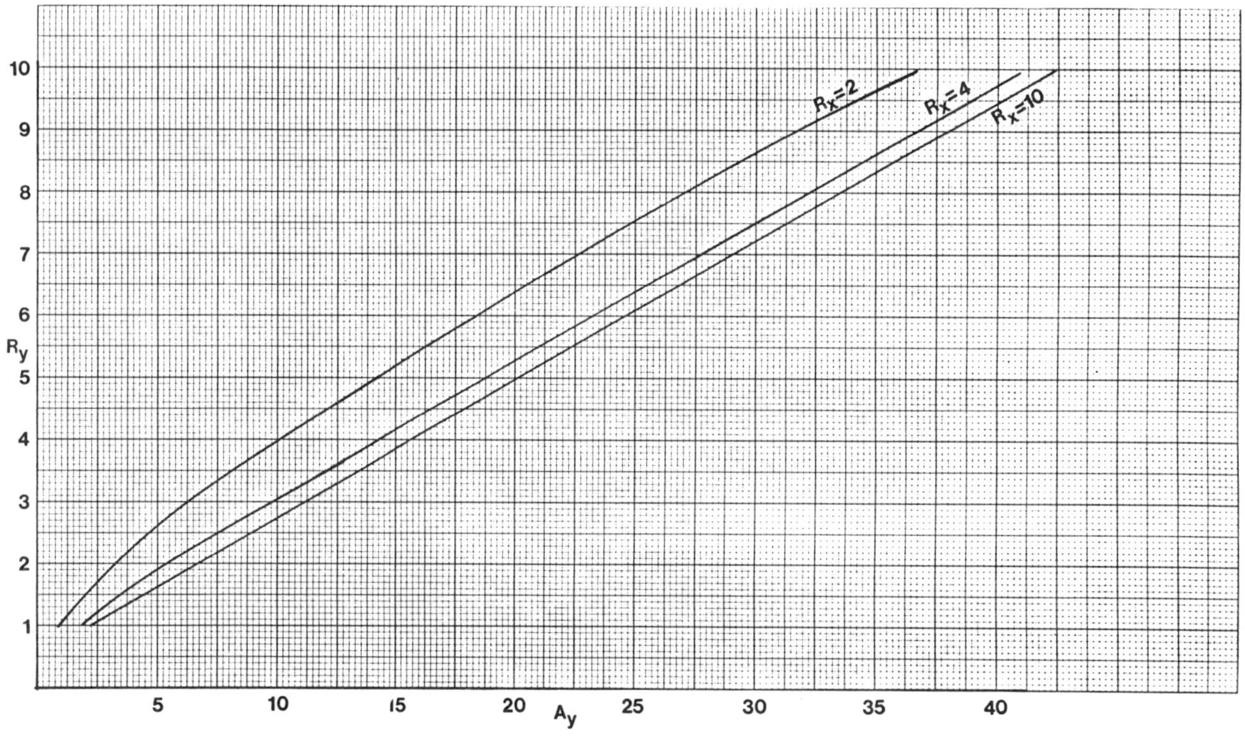

Figure 2-12D. U-loop with one leg four times the other leg.

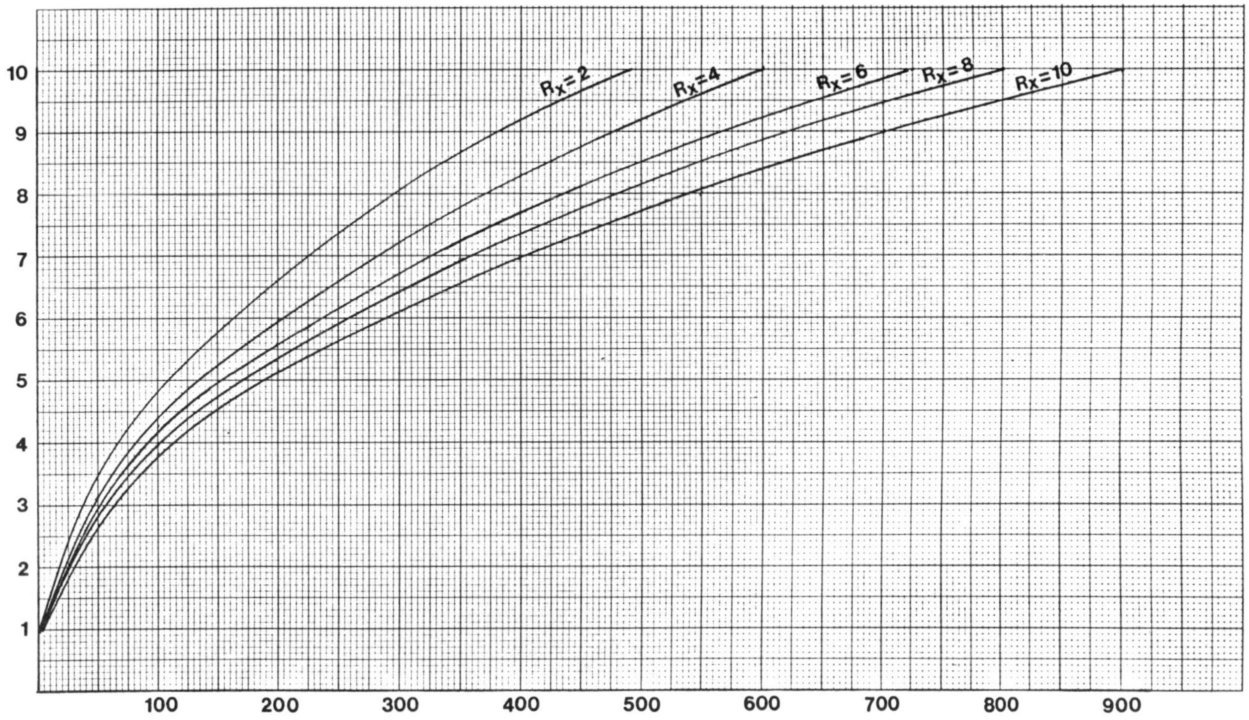

Figure 2-12D (continued). U-loop with one leg four times the other leg.

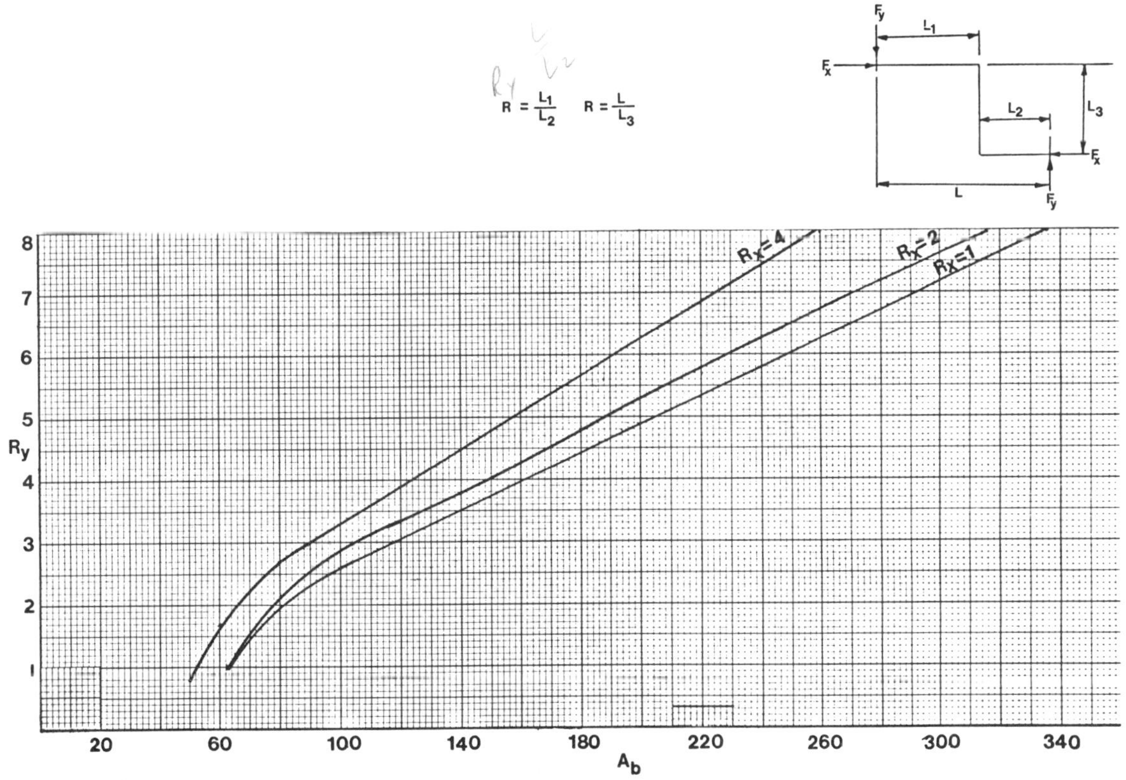

Figure 2-12E. U-loop: "Z" configuration.

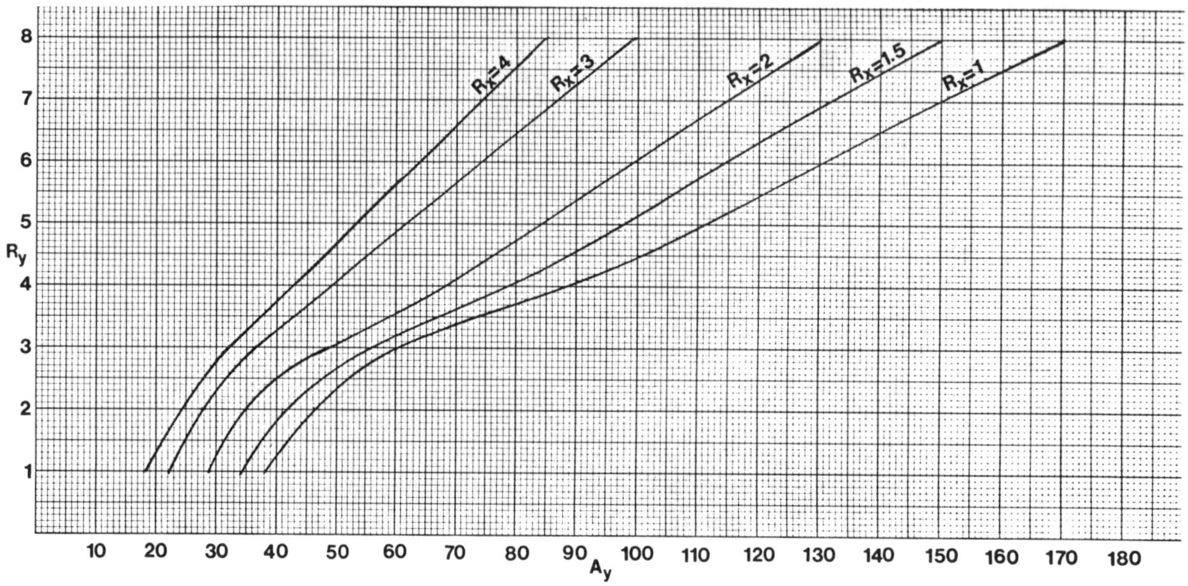

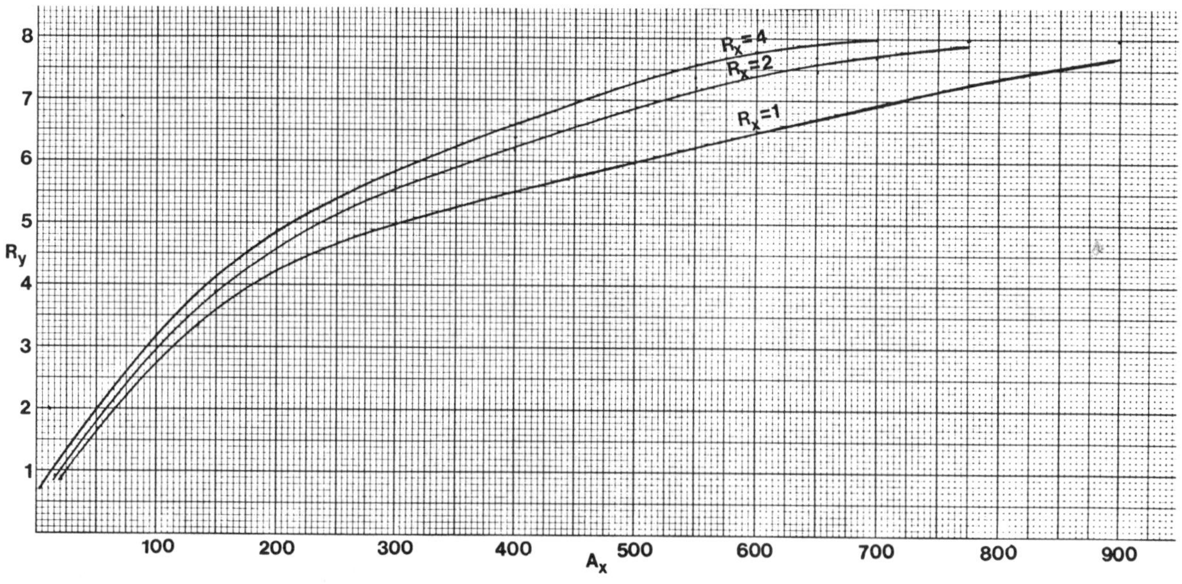

Figure 2-12E (continued). U-loop: "Z" configuration.

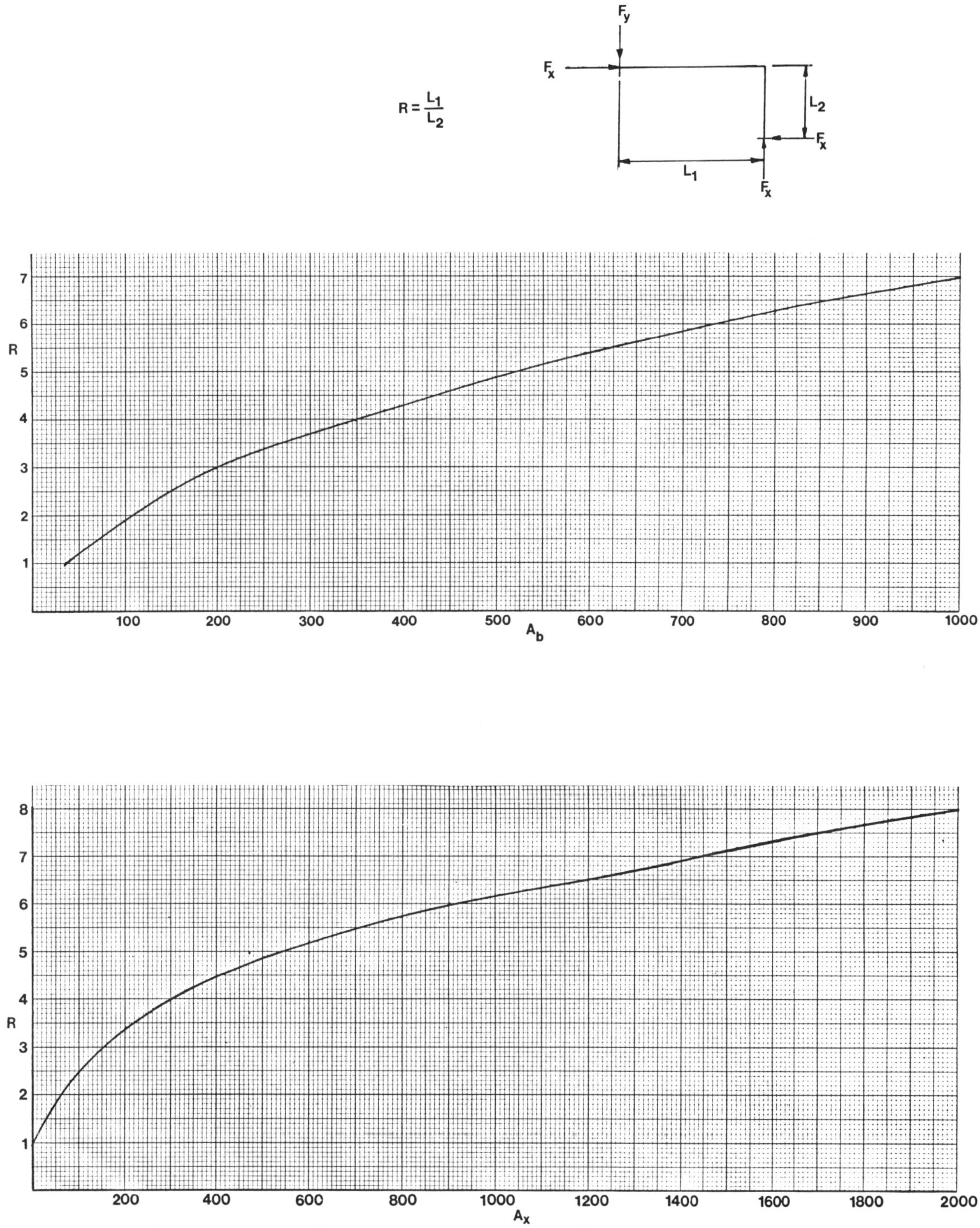

Figure 2-12F. U-loop: "L" configuration.

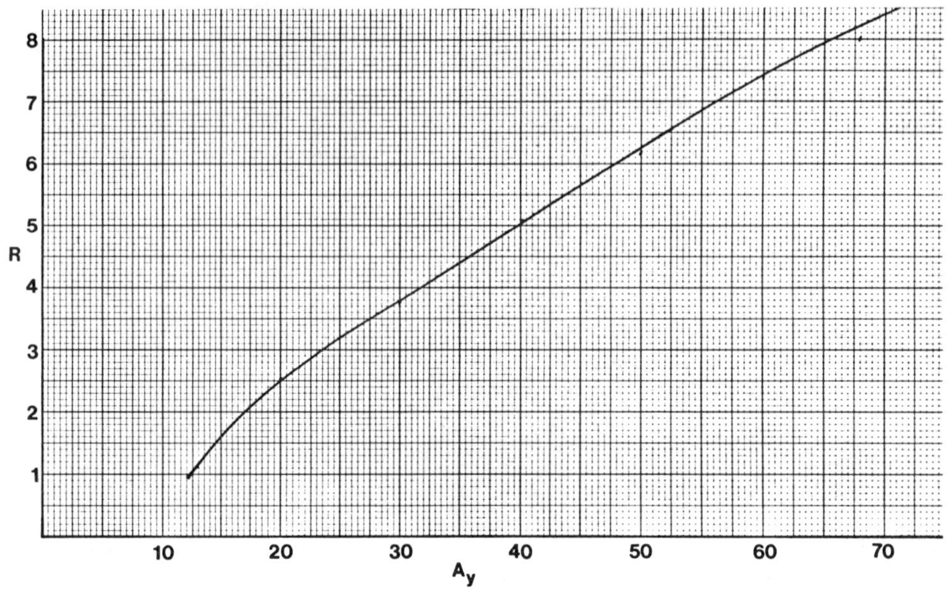

Figure 2-12F (continued). U-loop: "L" configuration.

$$F_Y = A_Y B \frac{I_P}{L^2} \text{ lb,}$$

$$\sigma_B = A_b B \frac{D}{L} \text{ psi, } L = \text{ft, } D = \text{in.}$$

$$B = \frac{\text{Thermal movement (in./100 ft)}E_D}{172{,}800}$$

SIF = 1.0 (Verified by computer stress analysis)

Loops such as circle bends, double offsets, and other geometrics involving completed circular geometry should be avoided because they are impractical, expensive, and unappealing to clients due to their complexity. If excessive looping is required, the stiffness method should be used to produce a practical, economical solution. The use of both the flexibility and stiffness approaches in areas, where applicable, can yield very attractive and appealing piping designs.

In pipe racks, the "U" shape loop is invariably the most practical shape to use because of its space efficiency. "U" loops are normally spaced together (i.e., lines running together on a pipe rack are, where practical, looped together as shown in Figure 2-13). It is desirable to guide the pipe on each side of the loop and at every other support thereafter as shown in Figure 2-14. Make sure the first guide is far enough from the loop to avoid jamming problems. Usually, this distance is twice the bend radius of an elbow of the pipe size being used. If you cannot put piping guides on the pipe coming down from the loop, then put them on the inside of the loop as shown in Figure 2-14.

Other configurations, such as "Z" and "L" shapes, are used in the normal routing of piping systems. It must be remembered that when these shapes are anchored on opposite ends, the ratio of the shortest leg to the longest should fall in the range of 1.0 to 10.0 to avoid overstressing the pipe. When analyzing the shapes by computer, any ratio can be used, but usually the aforementioned range is valid for most applications.

PIPE RESTRAINTS AND ANCHORS

Pipe restraints are used to counter forces of gravity, wind, earthquake, vibration, and other dynamic forces such as water hammer. The most common type is the gravity support, which merely restrains the force of gravity. A piping restraint can act in one or all degrees of freedom. As discussed previously, there are no restraints that are infinitely rigid—each has its own spring rate in each degree of freedom of translation and rotation. Even an "absolute" restraint has in each degree of freedom a translational stiffness of 10^9 lbs/in. and/or a rotational

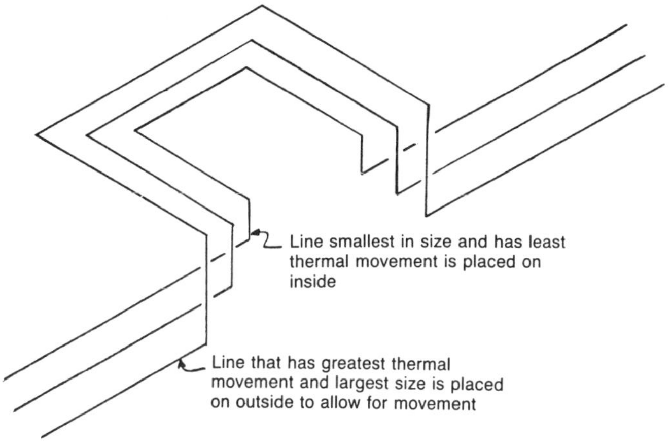

Figure 2-13. Optimum grouping of U-loops.

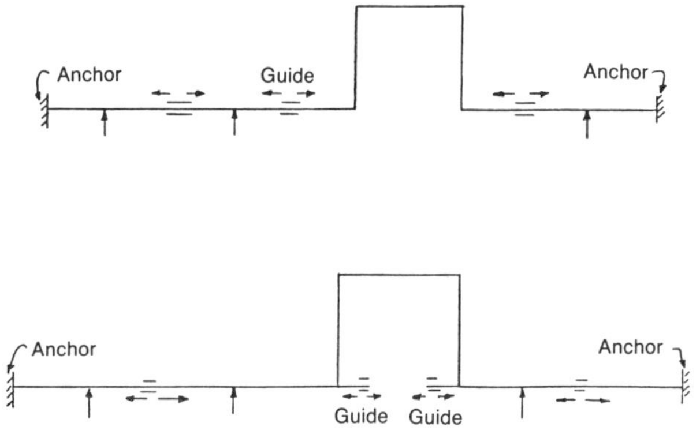

Figure 2-14. Guides are necessary for controlling movement of loops.

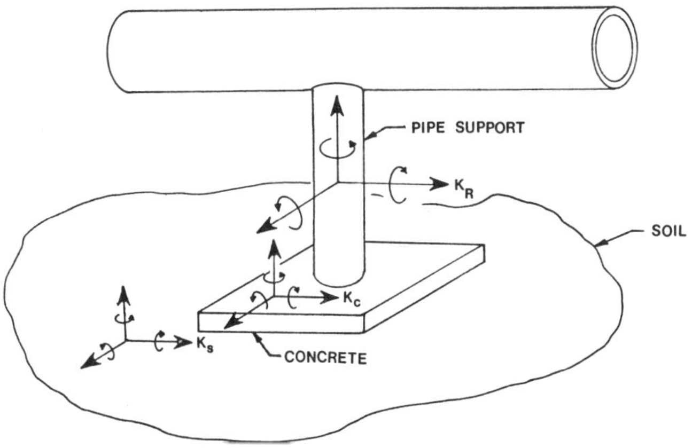

Figure 2-15. Conceptualization of system stiffness. Each component of the system—pipe, pipe supports, concrete, and soil—has translational and rotational values of stiffness (matrices) about each axis. These values can be modeled into the system as springs.

stiffness of 10^9 ft-lbs/deg. Such a restraint that restrains a pipe in all degrees of freedom is termed an anchor.

Piping guides are restraints that counter movement in one or several directions but allow total freedom of movement in one or more directions. Total freedom is defined as a stiffness value of zero. An anchor, by definition, has some value of stiffness in every degree of freedom, even though the anchor itself can move. The movement occurs while the anchor is still resisting movement at a certain stiffness in each degree of freedom. Thus, the term "sliding anchor" in place of a pipe guide is erroneous, because guides have a value of zero stiffness in one or more degrees of freedom. An anchor can restrain movement, although it may move. It is important to be cognizant of restraint terminology to avoid unnecessary confusion.

The stiffness of a support is not only a function of the restraint material, but also a function of the structural steel or concrete to which it is attached. Even though very stiff in compression, concrete is not infinitely stiff. As shown in Figure 2-15, the pipe restraint has a stiffness value K_r, the concrete a stiffness value of K_C, and the soil a value of K_S. Because $K_C >> K_S$, the concrete can sink or move in the soil if the concrete support is designed correctly or if subsidence occurs. Movements caused by soil conditions should be the responsibility of the piping engineer as well as the civil/structural engineer. The latter is responsible for limiting such movements as much as possible, and the piping engineer is responsible for entering these movements in the stress computer run or manual calculations.

It was mentioned earlier that for a pipe restraint to be considered absolute in one direction it must restrain one billion pounds per inch of translation and one billion pounds per degree of rotation. However, very few pipe restraints in nature are so rigid (an anchor being a restraint in three degrees of translation and three degrees of rotation). If the actual flexibility of the restraint is modeled into the pipe stress analysis, more realistic reactions and moments are obtained. In the case of pressure vessels much work has been done in determining realistic spring constants for nozzles. For application to rotating equipment, the reader is discouraged from using these spring constants, especially on equipment made of brittle material such as cast iron. Also, these spring constants are to be used only for ductile materials. Nozzle loadings should be based on either manufacturer recommendations or applicable standards. For further details and discussion of nozzle loadings on rotating equipment see Chapter 6.

To treat a restraint with elastic end conditions, only rotations are considered significant. Deformations induced by radial force and other translations are ignored, because their influence is insignificant.

The basic relationship for rotational deformation of nozzle ends is applying Equation 2-17 as

$$K = \frac{P}{U} = \frac{M}{\Theta} = \frac{\pi}{180}\left[\frac{EI}{D_N k_f}\right] \tag{2-17}$$

where K = KRX or KRY, ft-lb/deg
M = moment, ft-lb
Θ = angle of rotation, deg
E = modulus of elasticity of vessel metal at ambient temperature, psi
I = moment of inertia of vessel nozzle, in.^{4}
D_N = diameter of vessel nozzle, in.
k_f = flexibility factor, referred to in piping codes as "k"

The flexibility factor, k_f, is a parameter that has had several formulations over the years. One widely used variant was that proposed by the "Oak Ridge ORNL Phase 3 Report-115-3-1966." Since this document was published in 1966, several revisions have been made. The current ASME Section III Division I code gives detailed discussions on the flexibility factor. If one is designing piping for nuclear systems, then that person should only consult that code. Outside the nuclear industry the piping engineer rarely knows all the parameters that are necessary to compute the flexibility factor of Section III. Also, the piping engineer in nonnuclear work rarely knows which vendor will supply the piping components, thereby making many Section III parameters unknown.

Therefore, the more elementary "ORNL1" factors are presented here, because they produce lower values for k_f, which, in turn, produce higher, more conservative values of K. These factors are as follows:

Flexibility Factor

$$\text{Longitudinal} = K_L = \frac{C_L t_B D_B \left(\frac{D}{t}\right)^{\frac{3}{2}}}{tD}$$

$$\text{Circumferential} = K_c = \frac{C_c t_B D_B \left(\frac{D}{t}\right)^{\frac{3}{2}}}{tD}$$

Rotational Spring Rate

$$\text{Longitudinal} = R_L = \frac{EI\pi}{(2{,}160)D_B K_L}\left(\frac{\text{ft-lb}}{\text{deg}}\right)$$

$$\text{Circumferential} = R_c = \frac{EI\pi}{(2{,}160)D_B K_c}\left(\frac{\text{ft-lb}}{\text{deg}}\right)$$

Angle of Twist

$$\text{Longitudinal} = \Theta_L = \frac{MD_B K_L}{EI}\ \text{(radians)}$$

$$\text{Circumferential} = \Theta_c = \frac{MD_B K_c}{EI}\ \text{(radians)}$$

where C_L = 0.09 for in plane bending
C_c = 0.27 for out of plane bending
D = diameter of vessel or pipe header, in.
D_B = diameter of branch, in.
E = modulus of elasticity, lb/in.^{2}
I = moment of inertia of branch, in.^{4}
K_L = longitudinal flexibility factor
K_c = circumferential flexibility factor
M = applied moment, in.-lbs
Θ_L = longitudinal angle of twist, radians
Θ_c = circumferential angle of twist, radians
t = wall thickness of vessel or pipe header, in.
t_B = wall thickness of branch, in.

In-plane bending refers to *longitudinal bending* in the pipe header or vessel in the plane formed by the intersection of the branch and vessel or pipe header centerlines. *Out-of-plane* bending refers to *circumferential bending* in a plane perpendicular to the vessel or pipe header diameter. These rotational spring rates are necessary when the stiffness of an anchor must be considered in pipe stress analysis.

PIPE LUG SUPPORTS

These are about the most common pipe supports. The lug can provide a means for spring hangers or simple clevis-rod hangers. As simple as these supports are, a failure by one could result in loss of property or lives. Thus, their simplicity should not allow one to take them for granted thinking that any design will suffice.

The following method is based on the Bijlaard analysis discussed by Wichman et al. [3]. Consider a pipe subjected to a load P (lbs), as shown in Figure 2-16. The lug

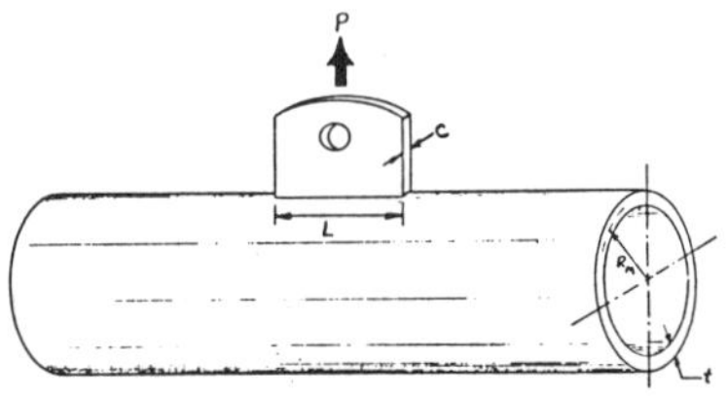

Figure 2-16. Pipe lug support for a pipe with internal pressure—primary and secondary stresses must be added.

connection is free of moments because the pin connection at the lug hole allows the pipe to twist in all directions. The usual oversight in designing a lug support is not considering the primary and secondary stresses, which must be added together and compared to the minimum tensile strength of the pipe material. First, we will discuss the Bijlaard method, which is only concerned with secondary stresses.

The pipe and lug geometries determine the attachment parameter, β, and the shell parameter, k, by

$$\beta_1 = \frac{C}{2R_M} \; ; \; \beta_2 = \frac{L}{2R_M} \; ; \; k = \frac{R_M}{t}$$

$$\text{If } \frac{\beta_1}{\beta_2} > 1 \, , \; \beta = \left[1 - \frac{1}{3}\left(\frac{\beta_1}{\beta_2} - 1\right)(1 - K_1)\right] (\beta_1\beta_2)^{0.5} \quad (2\text{-}20)$$

$$\text{If } \frac{\beta_1}{\beta_2} < 1 \, , \; \beta = \left[1 - \frac{4}{3}\left(- 1\frac{\beta_1}{\beta_2}\right)(1 - K_2)\right] (\beta_1\beta_2)^{0.5} \quad (2\text{-}21)$$

where K_1 and K_2 are determined from Table 2-3. For circumferential stress, σ_ϕ, the circumferential membrane and circumferential bending stresses are determined by

$$\frac{N_\phi}{t} = \left(\frac{N_\phi}{P/R_m}\right)\left(\frac{P}{R_m t}\right) = \text{Circumferential membrane stress}$$

$$\frac{6M}{t^2} = \left(\frac{M_\phi}{P}\right)\left(\frac{6P}{t^2}\right) = \text{Circumferential bending stress}$$

The membrane force, $N_\phi/(P/R_m)$, is determined from Figure 2-17 or Figure 2-18, and the bending moment, M_ϕ/P, is determined from Figure 2-19 or Figure 2-20.

Stress concentration factors must be accounted for in the surface discontinuity between the rectangular surface of the lug and the circular surface of the pipe. The membrane stress concentration factor for pure tension or compression is determined by

$$K_n = 1 + \left(\frac{t}{5.6w}\right)^{0.65} \quad (2\text{-}22)$$

and the concentration factor for bending stress is determined by

$$K_b = 1 + \left(\frac{2t}{9.4w}\right)^{0.8} \quad (2\text{-}23)$$

where w, the weld size, is given in Table 2-4 for various plate sizes. These values for w are only recommended

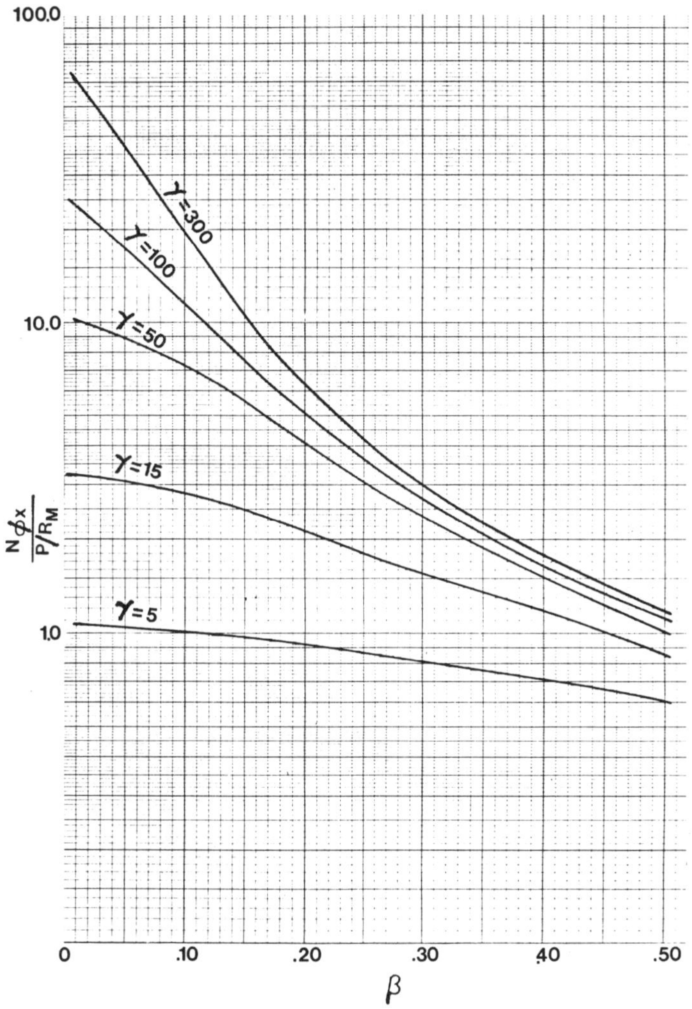

Figure 2-17. Membrane force, $N_{\phi x}/P/R_m$, induced by radial load P [3].

Table 2-3
Radial Load P

	N_ϕ	N_x	M_ϕ	M_x
K_1	0.91	1.68	1.76	1.2
K_2	1.48	1.2	0.88	1.25

Table 2-4
Recommended Minimum Weld Sizes for Plates

Thickness t of Thicker Plate Welded (in.)	Minimum Size, w, of Fillet Weld (in.)
$t < 1/2$	3/16
$1/2 < t < 3/4$	1/4
$3/4 < t < 1\,1/2$	5/16
$1\,1/2 < t < 2\,1/4$	3/8
$2\,1/4 < t \leqslant 6$	1/2
$6 < t$	5/8

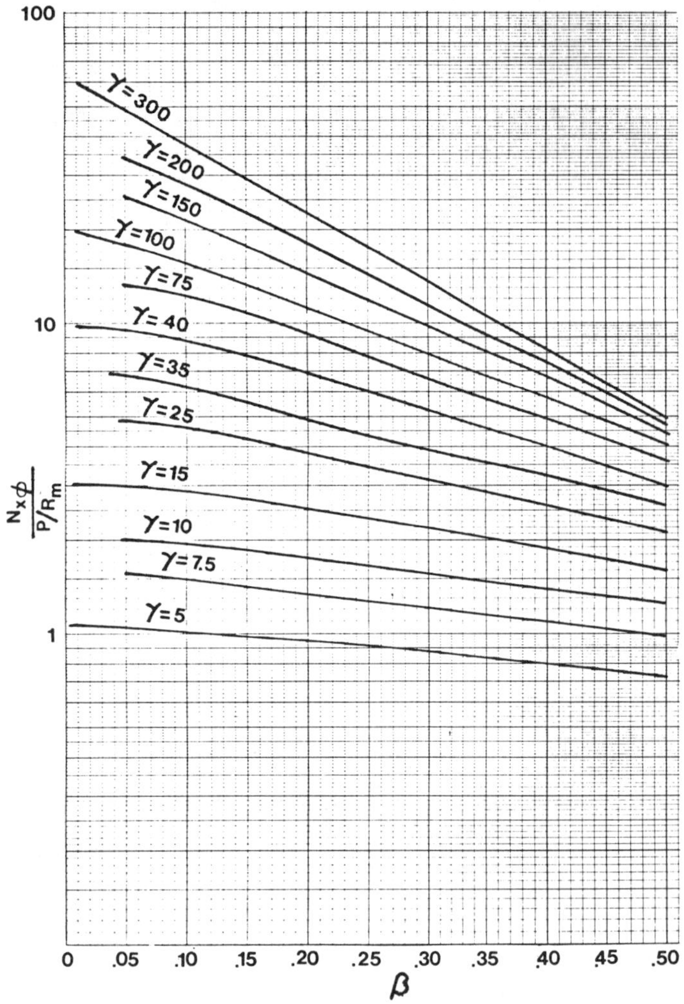

Figure 2-18. Membrane force, $N_{x\phi}/P/R_m$, induced by radial load P [3].

and the engineer should use whatever sizes are actually to be used in practice.

The total circumferential stress, σ_X, is determined by using these factors in the following equation:

$$\sigma_\phi = K_n\left(\frac{N_\phi}{t}\right) + K_b\left(\frac{6M_\phi}{t}\right) \qquad (2\text{-}24)$$

The longitudinal stress, σ_X, is determined in a similar way. The membrane force, $N_x/(P/R_m)$, and the bending moment, M_x/P, are determined from Figure 2-17 or 2-18 and Figure 2-21 or Figure 2-22, respectively.

These parameters are used to determine the longitudinal membrane and circumferential bending stresses, where

$$\frac{N_x}{t} = \left(\frac{N_x}{P/R_m}\right)\left(\frac{P}{R_m t}\right) = \text{Longitudinal membrane stress}$$

$$\frac{6M_x}{t^2} = \left(\frac{M_x}{P}\right)\left(\frac{6P}{t^2}\right) = \text{Longitudinal bending stress}$$

The total longitudinal stress is thus found by adding the two stresses,

$$\sigma_x = K_n\frac{N_x}{t} + K_b\frac{6M_x}{t^2} \qquad (2\text{-}25)$$

The longitudinal stress and circumferential stress represent the secondary stresses in the pipe wall. These stresses must be added to the primary stress which, in the case of internal pressure in the pipe, is the pressure stress. The pressure stress is determined by

$$\sigma_P = \frac{P_R(ID)}{2t}, \text{ psi} \qquad (2\text{-}26)$$

Thus, the total stress for each secondary stress is as follows:

$$\sigma_T = \sigma_\phi + \sigma_P \qquad (2\text{-}27)$$

$$\sigma_T = \sigma_X + \sigma_P \qquad (2\text{-}28)$$

where $\sigma_T < 2\sigma_A < \sigma_Y$

Often, with large piping, a simple lug will be overly stressed because of localized stresses at the lug-pipe connection. When the lug attachment dimension, c, becomes small to the pipe radius, a clamp is normally put around the entire pipe with the lug attaching to the top of the clamp. This reduces localized stresses at the pipe wall by adding extra metal. This same principle applies to vessel nozzle reinforcement, which is discussed in Chapter 8.

SPRING SUPPORTS

These supports provide loading to a pipe that has undergone displacement. Simple supports are no longer useful if the pipe raises off and loads are transferred to other supports or fragile equipment nozzles. To ensure support for the pipe while it moves, a moving support is desired. The most practical device to fill this requirement is the spring.

Springs come in two basic categories—variable springs and constant springs. The former, which is by far the most common, provides loading to a pipe at a fixed spring rate, lb/in., but the amount of force to deflect the spring varies with the amount of deflection. This force versus spring rate is a linear relationship and is the reason for a "variable" spring. The constant spring is a

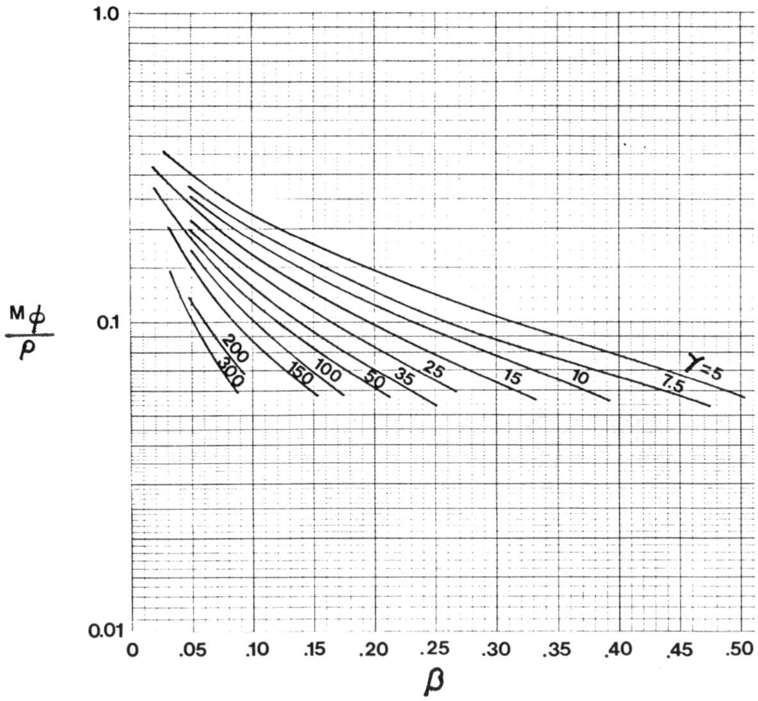

Figure 2-19. Bending moment, M_ϕ/P, induced by radial load P [3].

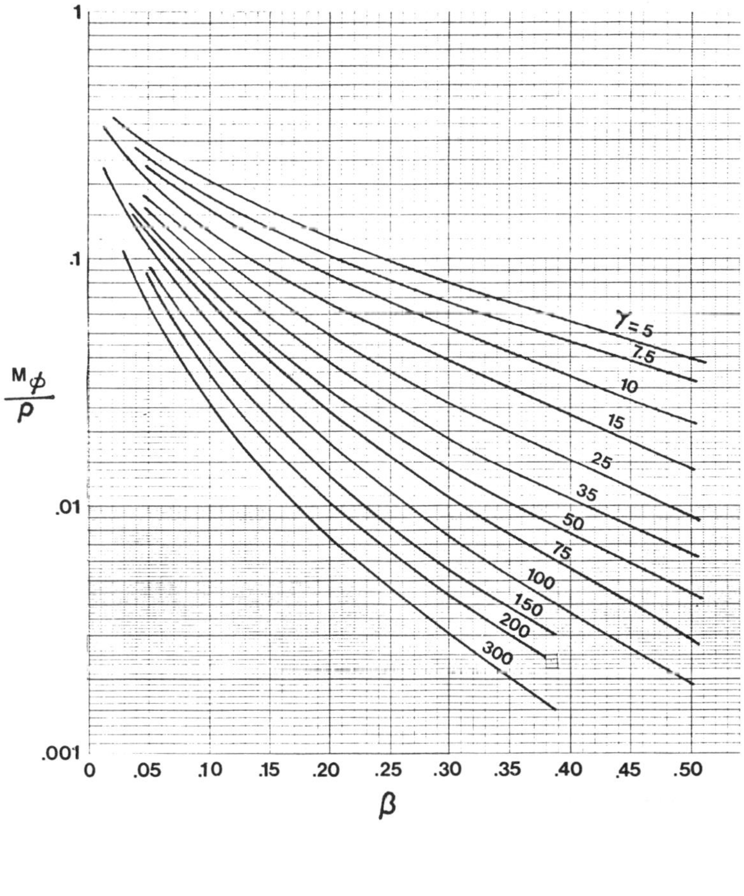

Figure 2-20. Bending moment, M_ϕ/P, induced by radial load P [3].

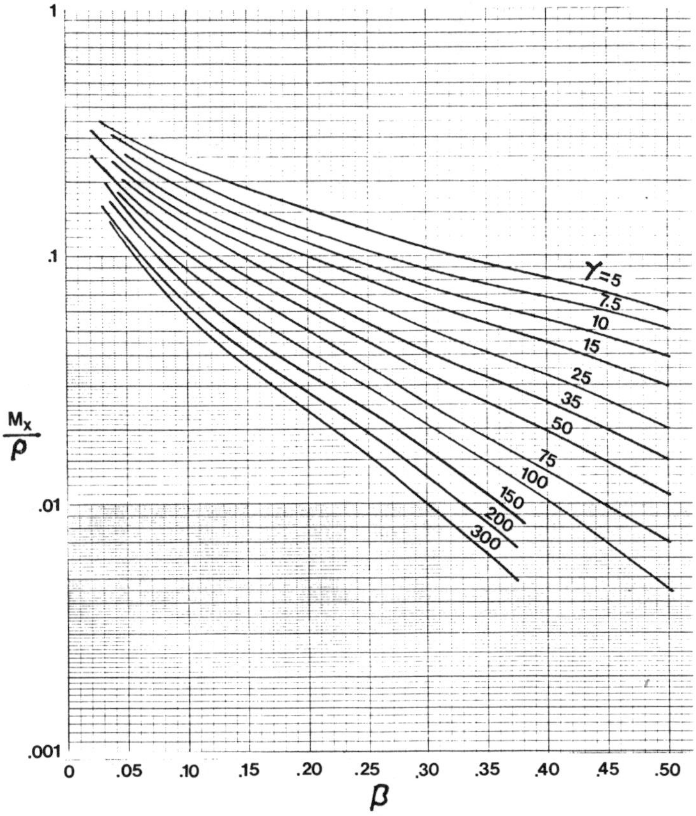

Figure 2-21. Bending moment, M_x/P, induced by radial load P [3].

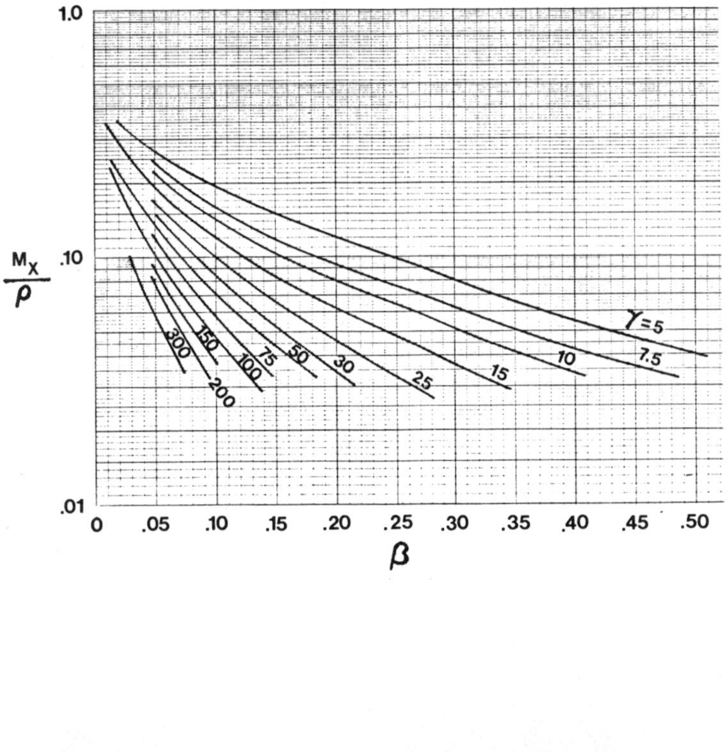

Figure 2-22. Bending moment, M_x/P, induced by radial load P [3].

spring that will provide the same spring rate for any force great enough to cause initial deflection. Constant springs are used in critical installations where forces or deflections induced on the piping system are critical. These springs are considerably more expensive than the variable types and are usually avoided by piping engineers when not needed.

Constant springs provide constant supporting force for the pipe throughout its full range of contraction and expansion. As shown in Figure 2-23, this constant support mechanism consists of a helical coil spring working in conjunction with a bell crank lever in such a manner that the spring force times its distance to the lever pivot is always equal to the pipe load times its distance to the lower pivot. Thus, the constant spring is used where it is not desirable for piping loads to be transferred to connecting equipment or other supports.

Variable springs are used where a variation in piping loads can be tolerated. As an example, consider the following example shown in Figure 2-24. The spring is above the pipe and is attached to it with a rod and clevis. This arrangement is called a spring hanger. As seen in Figure 2-24A, the spring supports the weight of the pipe and insulation. As the pipe heats up and expands it moves upward. The amount of deflection, Δ, relates to the amount of excessive force as

$$F_e = \Delta K, \text{ lb} \qquad (2\text{-}29)$$

where K = spring constant of spring, lb/in.
Δ = deflection, in.

It is common practice to calibrate the hanger in such a manner that when the piping is at its operating (hot or cold) condition, the supporting force of the spring is equal to the weight of the pipe. This means that the maximum variation in supporting force occurs when the pipe is in the down condition, when primary stresses are nonexistent because of no internal pressure. Therefore, in the cold position, the supporting force of the spring is

$$F = F_e + WP \qquad (2\text{-}30)$$

where WP = pipe and insulation weight

To reduce the amount of variability, it is desirable to use the smallest type of variable spring provided that the deflections will not exceed those of the spring range. Typical spring sizes and ranges are shown in Table 2-5.

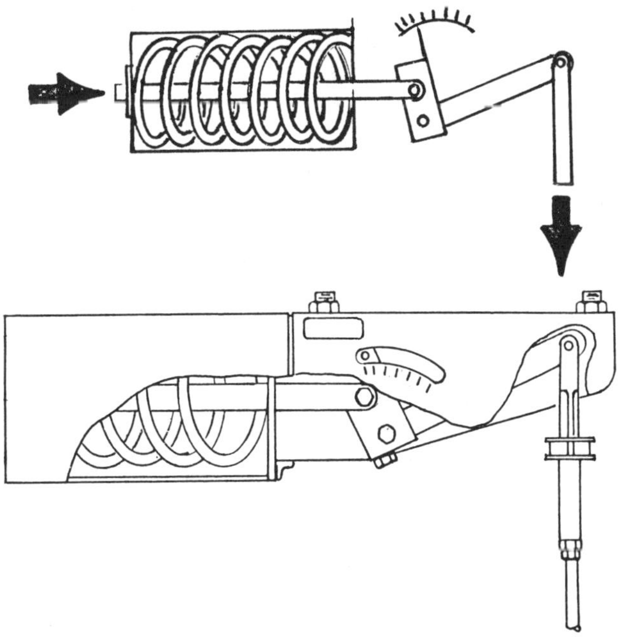

Figure 2-23. A constant load spring support provides constant support loading in critical situations.

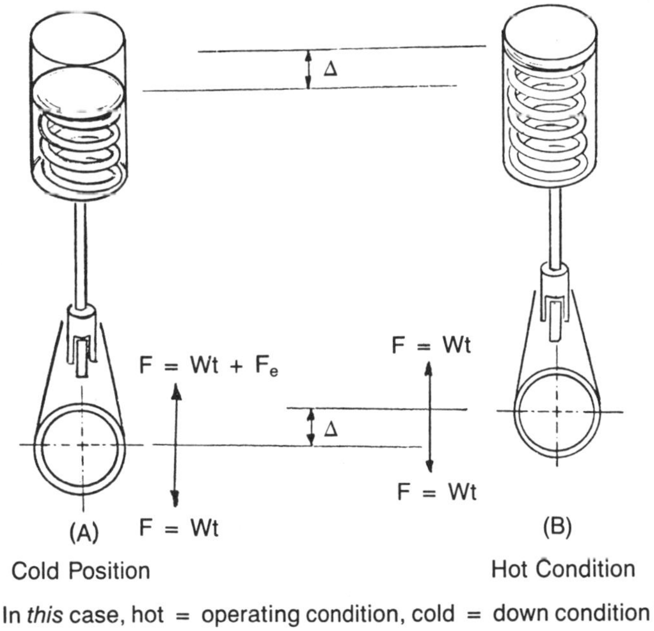

Figure 2-24. The "cold" and "hot" loading positions of a variable spring hanger.

Table 2-5
Typical Spring Sizes and Ranges*

working range, in.			hanger size																							spring deflection, in.		
fig. 98	fig. B-268	fig. 82	0	1	2	3	4	5	6	7	8	9	10	11	12	13	14	15	16	17	18	19	20	21	22	fig. 82	fig. B-268	fig. 98
			43	63	81	105	141	189	252	336	450	600	780	1020	1350	1800	2400	3240	4500	6000	7990	10610	14100	18750	25005	0	0	0
			44	66	84	109	147	197	263	350	469	625	813	1063	1406	1875	2500	3375	4688	6250	8322	11053	14688	19531	26047			
			46	68	88	114	153	206	273	364	488	650	845	1105	1463	1950	2600	3510	4875	6500	8655	11495	15275	20313	27089			
			48	71	91	118	159	213	284	378	506	675	878	1148	1519	2025	2700	3645	5063	6750	8987	11938	15863	21094	28131			
0	0	0	50	74	95	123	165	221	294	392	525	700	910	1190	1575	2100	2800	3780	5250	7000	9320	12380	16450	21875	29173	¼	½	1
			52	76	98	127	170	228	305	406	544	725	943	1233	1631	2175	2900	3915	5438	7250	9652	12823	17038	22656	30215			
			54	79	101	131	176	236	315	420	563	750	975	1275	1688	2250	3000	4050	5625	7500	9985	13265	17625	23438	31256			
			56	81	105	136	182	244	326	434	581	775	1008	1318	1744	2325	3100	4185	5813	7750	10317	13708	18213	24219	32298			
1	½	¼	58	84	108	140	188	252	336	448	600	800	1040	1360	1800	2400	3200	4320	6000	8000	10650	14150	18800	25000	33340	½	1	2
			59	87	111	144	194	260	347	462	619	825	1073	1403	1856	2475	3300	4455	6188	8250	10982	14592	19388	25781	34382			
			61	89	115	149	200	268	357	476	638	850	1105	1445	1913	2550	3400	4590	6375	8500	11315	15035	19975	26563	35424			
			63	92	118	153	206	276	368	490	656	875	1138	1488	1969	2625	3500	4725	6563	8750	11647	15477	20563	27344	36466			
2	1	½	65	95	122	158	212	284	378	504	675	900	1170	1530	2025	2700	3600	4860	6750	9000	11980	15920	21150	28125	37508	¾	1½	3
			67	97	125	162	217	291	389	518	694	925	1203	1573	2081	2775	3700	4995	6938	9250	12312	16362	21738	28906	38549			
			69	100	128	166	223	299	399	532	713	950	1235	1615	2138	2850	3800	5130	7125	9500	12645	16805	22325	29688	39591			
			71	102	132	171	229	307	410	546	731	975	1268	1658	2194	2925	3900	5265	7313	9750	12977	17247	22913	30469	40633			
3	1½	¾	73	105	135	175	235	315	420	560	750	1000	1300	1700	2250	3000	4000	5400	7500	10000	13310	17690	23500	31250	41675	1	2	4
			74	108	138	179	241	323	431	574	769	1025	1333	1743	2306	3075	4100	5535	7688	10250	13642	18132	24088	32031	42717			
			76	110	142	184	247	331	441	588	788	1050	1365	1785	2363	3150	4200	5670	7875	10500	13975	18575	24675	32813	43759			
			78	113	145	188	253	339	452	602	806	1075	1398	1828	2419	3225	4300	5805	8063	10750	14307	19017	25263	33594	44801			
4	2	1	80	116	149	193	258	347	462	616	825	1100	1430	1870	2475	3300	4400	5940	8250	11000	14640	19460	25850	34375	45843	1¼	2½	5
			82	118	152	197	264	354	473	630	844	1125	1463	1913	2531	3375	4500	6075	8438	11250	14972	19902	26438	35156	46885			
			84	121	155	201	270	362	483	644	863	1150	1495	1955	2588	3450	4600	6210	8625	11500	15305	20345	27025	35938	47926			
			86	123	159	206	276	370	494	658	881	1175	1528	1998	2644	3525	4700	6345	8813	11750	15637	20787	27613	36719	48968			
5	2½	1¼	88	126	162	210	282	378	504	672	900	1200	1560	2040	2700	3600	4800	6480	9000	12000	15970	21230	28200	37500	50010	1½	3	6
			89	129	165	214	288	386	515	686	919	1225	1593	2083	2756	3675	4900	6615	9188	12250	16302	21672	28788	38281	51052			
			91	131	169	219	294	394	525	700	938	1250	1625	2125	2813	3750	5000	6750	9375	12500	16635	22115	29375	39063	52094			
			93	134	172	223	300	402	536	714	956	1275	1658	2168	2869	3825	5100	6885	9563	12750	16967	22557	29963	39844	53136			
			95	137	176	228	306	410	546	728	975	1300	1690	2210	2925	3900	5200	7020	9750	13000	17300	23000	30550	40625	54178	1¾	3½	7
			spring scale — lb. per in.																									
		fig. 82	30	42	54	70	94	126	168	224	300	400	520	680	900	1200	1600	2160	3000	4000	5320	7080	9400	12500	16670	fig. 82		
	fig. B-268		15	21	27	35	47	63	84	112	150	200	260	340	450	600	800	1080	1500	2000	2660	3540	4700	6250	8335		fig. B-268	
fig. 98			7	10	13	17	23	31	42	56	75	100	130	170	225	300	400	540	750	1000	1330	1770	2350	3125	4167			fig. 98

* Courtesy of ITT Grinnell

It is common practice to utilize the smallest spring size possible.

In critical and large systems, spring loadings should be evaluated by computer analysis. Often, in large systems, piping movements are not intuitively obvious and errors can result because the entire system must be evaluated if a correct analysis is to be obtained. In most systems, hand calculations are far too cumbersome.

Occasionally, springs are used as moment resisting devices, as shown in Figure 2-25. In such an application, the spring preloads the pipe in a specific direction. As the pipe expands or contracts, the spring counters the force created by the movement and, thus, reduces the moment at an end connection. Such a system in normal practice usually works in the operating mode but when the system shuts down the spring overloads the piece of equipment protected in the operating condition. Thus, if such a scheme is used, care must be taken to ensure that the protected items are safe in both the operating and down conditions. These schemes can be avoided by use of MRS devices where space does not warrant piping flexibility.

Location of spring supports is of critical importance. While springs should be placed where they will be most efficient, often such locations are undesirable from the structural engineer's viewpoint. The piping engineer should always be cognizant of available structural steel or concrete and loads to be placed on structures. Most springs are supported from above at either mid spans or at elbows. Many times it is desirable to support the pipe from below. When using this type of spring, one must be cautious of pipe movement, as excessive movement (> 1/4 in.) will cause such a spring to jam, as shown in Figure 2-26A. To avoid jamming, a guided load column is used to prevent such a problem (Figure 2-26B).

Springs are often used to support equipment to reduce nozzle loadings, which are discussed in Chapter 6.

EXPANSION JOINTS

These devices accommodate movement in piping caused by temperature changes. Such items range from special slip joints that only allow movement in the axial direction to corrugated bellows joints that can be designed to accommodate movement in several directions. It is the latter type that we will concentrate on, as they are by far the most numerous and complicated of expansion joints.

Corrugated bellows expansion joints have a bad reputation with some users because of ignorance. Many bellows expansion joints have been incorrectly specified and the consequences attributed to the device itself. This is unfortunate because this device is invaluable when either re-routing the line is impossible or cold spring or other alternatives are not available. The surest way to avoid problems with bellows expansion joints is to have the piping (stress) engineer specify the unit and to procure the unit from a reputable manufacturer.

The bellows expansion joint is like the MRS device discussed earlier because the more the unit is required to accomplish the more complex is its configuration. The simplest corrugated bellows expansion joint is the single

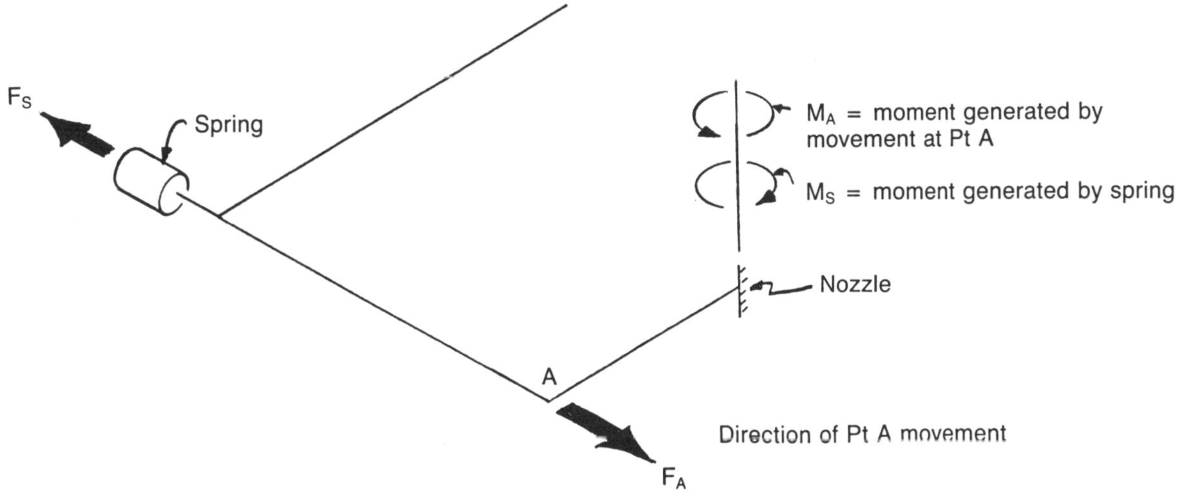

Figure 2-25. Utilizing a spring to counter a moment generated by piping is appropriate only when the spring movement, M_S, does not overload the nozzle or overstress the piping system when the latter is in the down condition and there is no movement at A. This condition is required after the operating condition is met.

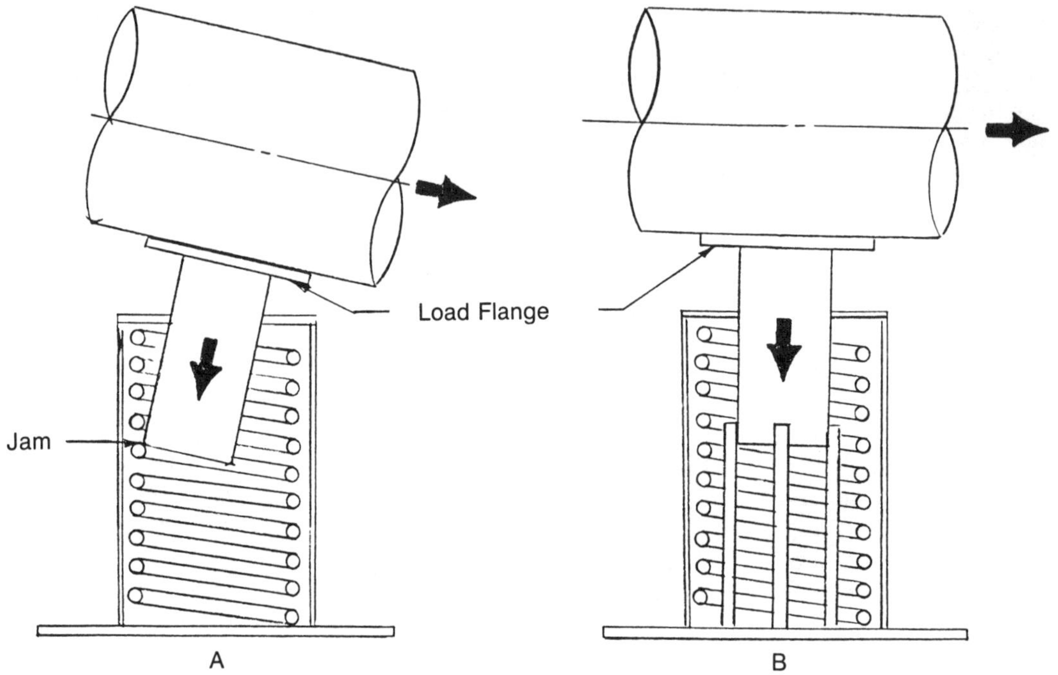

Figure 2-26. (A) Enough piping movement will cock load flange and jam spring. Note: arrows indicate direction of movement. (B) A guide load column shown here will prevent situation in (A). There are various designs for guide load columns, but for pipe movement greater than 1/4″ one should consider a column with rollers or Teflon on top of the load flange.

bellows type shown in Figure 2-27A. This specific joint is shown with flanges welded on each end, but is available from manufacturers with pipe spool pieces on each end to enable the unit to be directly welded into a line. The piping engineer should try to utilize this type of joint whenever possible because of economy and simplicity of operation. The single bellows is free to move in all degrees of freedom except about the longitudinal centerline. In fact, no bellows expansion joint can accommodate torsion and any tendency for the pipe to exert a high torsional moment could seriously damage the joint.

External restraints are placed on the joint to restrict movement in one or more degrees of freedom. Such devices are tie rods or hinges that restrict movement or pressure thrust. Figure 2-27B–E are examples of joints that are so restricted. Following the same nomenclature shown in Figure 2-10, we consider each joint in a three-dimensional axis system. KTX, KTY, and KTZ are the translational stiffness values lb/in., about the X, Y, and Z axes, respectively. KRX, KRY, and KRZ are the rotational stiffness values, ft-lb/deg, about the X, Y, and Z axes, respectively. For all bellows joints, KRZ = (1.0×10^9) ft-lb/deg, as previously stated.

In Figure 2-27A, we have finite values for KTX, KTY, KTZ, KRX, and KRY, the joint is free to translate about three axes and rotate about two axes. The bellows does not support its own weight so this joint would not be desirable where each end exceeds the maximum amount of pipe span shown, as calculated by the following equation:

$$L = 0.131 \sqrt{\frac{EI}{PA \pm K}}$$

E = modulus of elasticity of pipe material, psi
I = moment of inertia of pipe, $in.^4$
P = design pressure (psig)
A = bellows effective area, $in.^2$
K = axial bellows stiffness (KTZ in Figure 2-27)

The maximum length of unsupported pipe implies that the unit itself is within this length. Preferably, the joint is close to one support or nozzle to avoid excessive deflection.

The joint in Figure 2-27B has values of KTX and KRY and KTY = KTZ = 10^9 lbs/in. and KRX = KRZ = 10^9 lbs/deg. This means that the joint is free to translate in the X-direction and free to rotate in the y-direction and is rigid in all other directions. This type of joint is called a "hinged" joint and is self-supporting in the y-direction shown in Figure 2-27. Placing high vertical loads on a joint must be approved by the manufacturer.

The joint in Figure 2-27C has values of KRX and KRY but absolute values of KRZ, KTX, KTY, and KTZ. Normally, these joints are used in pairs to allow rotation about two axes similar to swivel or ball joints and not allowing any translation. This action is seen in Figure 2-28.

The unit in Figure 2-27D is a pressure balanced universal joint. It is free to move about all degrees of movement except KRZ and is restricted by tie rods that balance pressure thrust. This type of joint is very common in engine exhaust systems.

Figure 2-27E depicts one of the most complicated expansion designs—an in-line pressure balanced expansion joint. This joint eliminates pressure thrust, is self-supporting, and does not require a change in the piping system to install. It is desirable where structural supports are not available and a joint is needed because flexibility is required of the piping.

Pressure thrust is the amount of force generated by internal pressure and is simply internal pressure times minimum bellows radius area (PA), lbs. This force can become quite high as the pipe size and the internal pressure increase. In many applications, the piping itself is anchored and the joint is allowed to compress when the thermal compression force exceeds the pressure thrust force. As seen in Figure 2-28, when movement in the form of lateral translation is desired (KTZ and KTY), tie rods are used to restrain the joint in the axial direction (KTZ = ∞). If tie rods are being used to overcome pressure thrust, then any equipment flanged to the joint should be able to withstand the load required to overcome pressure thrust. Generally, tie rods are only used to permit lateral movement.

Bellows expansion joints can be restrained and combined in pairs or trios to perform certain tasks. It must be emphasized that just because a joint is free to move in

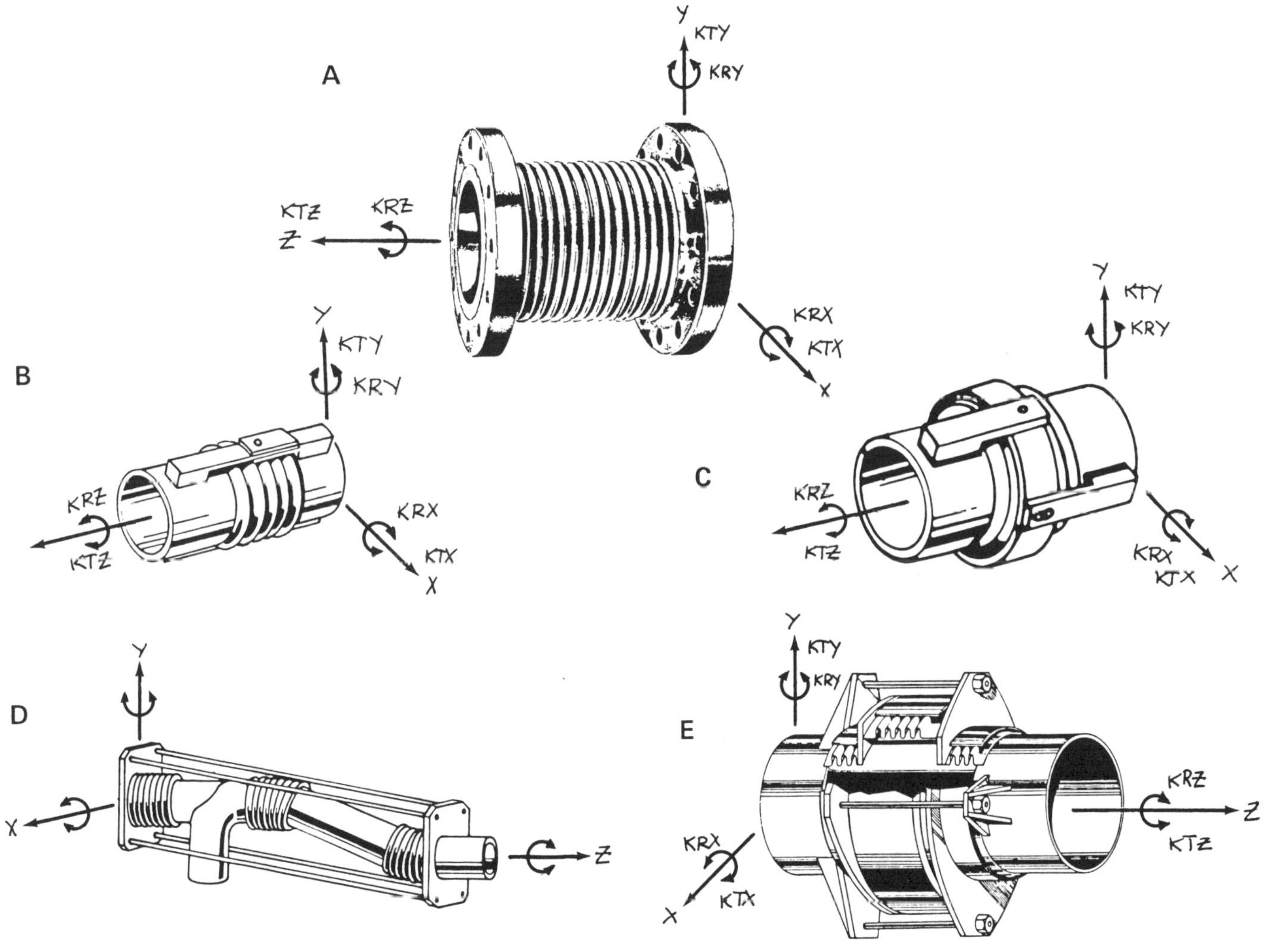

Figure 2-27. Types of bellows expansion joints: (A) flanged—flanged end simple bellows joint; (B) hinge bellows expansion joint; (C) gimbal bellows expansion joint; (D) pressure balanced bellows expansion joint; (E) "in-line" pressure balanced self-supporting bellows expansion joint. (Courtesy of Pathway Bellows, Inc.)

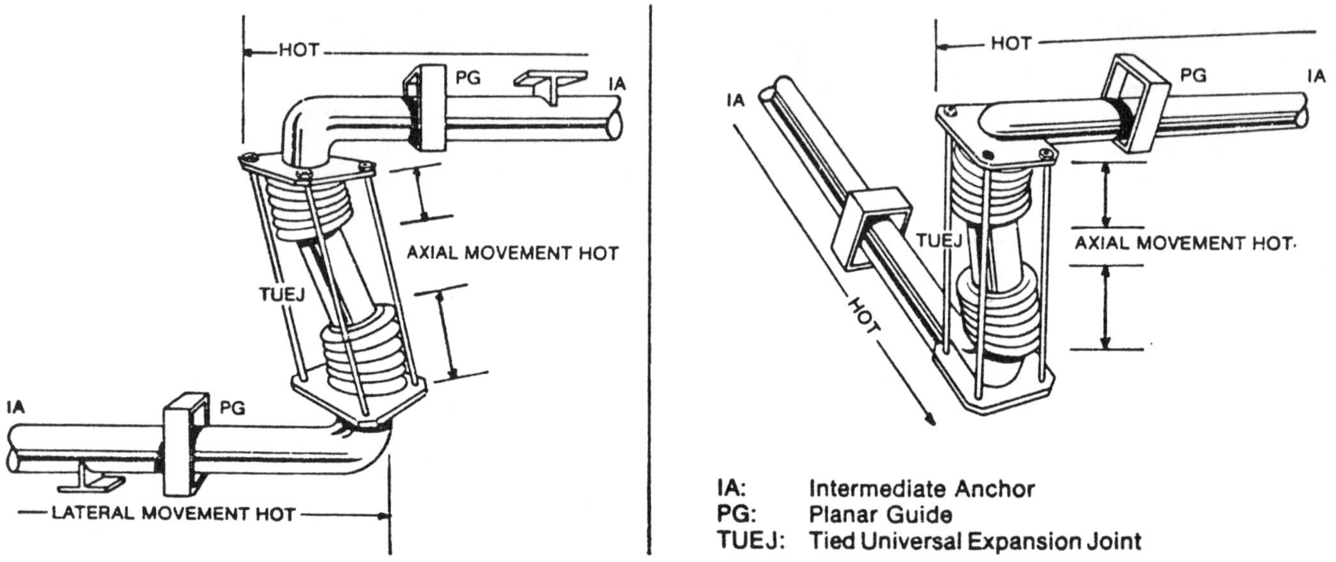

Figure 2-28. Generally the use of tie rods is to allow only lateral movement. (Courtesy of Pathway Bellows, Inc.)

directions KTX, KTY, KTZ, KRX, and KRY does not mean that the corresponding stiffness values are small. As internal pressure and pipe size increase the values of KTX, KTY, KTZ, KRX, and KRY increase, because the bellows wall thickness increases to resist increased internal pressure. The bellows can be a single wall construction (single ply) or multiple wall construction (multi-ply) and the stiffness values vary with each manufacturer.

Some people erroneously think that the purpose of using bellows expansion joints is to make the pipe stress analysis unnecessary. Such is definitely not the case, because values of stiffness in each direction must be entered in each computer stress run so that it can be *verified* that the displacement and piping loads are not excessive to the equipment nozzles. As shown in Figure 2-28 a pipe can either be properly guided or anchored, and such restrictions should be modeled into the computer stress analysis.

The piping engineer is encouraged to refer to the *Standards of the Expansion Joint Manufacturers Association* (EJMA) [4] in accessing piping layouts when using bellows expansion joints. Also, it is desirable to specify the joint such that the manufacturer is required to meet EJMA requirements. One should follow EJMA guidelines and requirements, and include modeling restraints and stiffness values in computer stress analysis to verify that attached equipment is protected. Expansion joints are not cataloged items to be bought at random but rather sophisticated pieces of equipment that *must* be engineered into the piping system. With this approach, the user should not experience any problems with bellows expansion joints.

PRESTRESSED PIPING

Piping systems are sometimes prestressed to reduce anchor and restraint forces and moments. This prestressing of the pipe is best known as cold springing, but is also called "cut short," meaning that the pipe is cut short a percentage of the amount of thermal expansion expected. The opposite is true in cold systems where the pipe contracts, so the pipe is fabricated extra long, with the extra length being a percentage of the amount of thermal contraction expected. This procedure is best known as "cut long." Some refer to cut long as "hot springing," which may cause confusion because it is not as popular as the term cold springing and to some it means hot forming, which has nothing to do with fabricating the pipe extra long.

"Credit" may not be taken for prestressing the pipe in computing the stress in the piping system. Several piping codes are specific about this and, if the piping is over the allowable stress range, one cannot cut short or long to lower the stress. However, credit may be taken for anchor and restraint reaction forces and moments.

The procedure of cutting short or long involves a percentage of thermal movement. The whole purpose of the prestressing process is to balance the forces and movements between the down and operating conditions. Thus, cutting short or long 100% (i.e., cutting short or long the exact amount of thermal movement) is normally not done. Exceeding 100% is not recommended and doesn't make good sense. Normally, the amount cut is 50% and should not exceed 66% of the thermal movement. The reactions, R_O and R_D in the operating and down conditions, respectively, are obtained from the reactions R derived from calculations based on the modulus of elasticity at ambient temperature, E_c. The relationships are as follows:

$$R_D = \left(1 - \frac{2}{3} x\right) \frac{E_O}{E_D}$$

$$R_D = X R$$

or

$$R_D = \left[1 - \frac{\sigma_O}{\sigma_E}\left(\frac{E_D}{E_O}\right)\right] R$$

whichever is greater, and with the additional condition that

$$\frac{E_D}{E_O} < 1.0$$

where X = cold (or hot) spring factor ranging from zero to one, one being 100% cold or hot spring
E = computed expansion stress, psi
E_D = modulus of elasticity in the down condition, psi
E_O = modulus of elasticity in the operating condition, psi
R = maximum reaction for full thermal movement based on E_D which is the most severe condition, lb, or in.-lb
R_D = maximum reaction in down condition, lb or in.-lb
R_O = maximum reaction in operating condition, lb or in.-lb

These formulations are not necessary nor desirable when computerized stress runs are made. All reactions that result from prestressing the pipe are much more accurately made by a computer. However, one is not always privileged to use a computer, especially at remote sites, so these formulations will yield conservative approximations to reactions.

The biggest legitimate objection to prestressing the pipe is that often it is simply not done by the pipe fabricator or construction workers. The process is often difficult, especially in large pipe, and is unpopular with fabrication personnel. When schedules get tight and people fall behind on the schedule, there is a tendency to overlook prestressing the pipe. To avoid such a problem, some large engineering companies issue cold spring reports that are signed off by inspectors. However, such reports get lost fairly easily, unless a rigid system is implemented to treat them as control documents. There is certainly nothing wrong with prestressing the pipe, except maybe a little extra paperwork.

Bellows expansion joints should be avoided if a more economical and practical method is available for providing flexibility or restraint to the pipe. In many applications, only the bellows expansion joint will suffice, e.g., movement and vibration in straight runs of pipe at elevated temperatures between different pieces of equipment can only be compensated by bellows joints. However, as the joint becomes more sophisticated and thereby more expensive, other alternatives should be considered. Such alternatives lie in either the flexibility or stiffness methods previously discussed.

FLUID FORCES EXERTED ON PIPING SYSTEMS

When fluids move in a piping system, they import energy to the system when they are forced to change direction by the pipe. In other words, it requires energy to change the direction of a moving fluid. This fundamental fact is known as the impulse-momentum principle, expressed as:

$$\left(\sum F\right) t = MV_2 - MV_1 \quad (2\text{-}31)$$

This states that the change in momentum in a system remains constant during the exchange of momentum between two or more masses of the system. Applying the equation to that of a pipe elbow shown in Figure 2-29, we apply the principle to obtain:

$$MV_{X_1} + \sum F_X \times t = MV_{X_2} \quad (2\text{-}32)$$

$$MV_{Y_1} + \sum F_Y \times t = MV_{Y_2} \quad (2\text{-}33)$$

where t = 1 for unit time
F_X = force in horizontal direction exerted by the bend on the flowing fluid, lb
F_Y = force in vertical direction exerted by the bend on the flowing fluid, lb
V_{Y_1} = horizontal velocity component at bend inlet, ft/sec
V_{y_1} = vertical velocity component at bend inlet, ft/sec
V_{X_Z} = horizontal velocity component of bend outlet, ft/sec
V_{Y_Z} = vertical velocity component of bend outlet, ft/scc
M = $W g/g_c$ = fluid mass
W = weight of fluid in bend, lb_m
g = local acceleration due to gravity, approximately 32.2 ft/sec^2
g_c = dimensional constant = 32.17 lb-ft/lb_f sec^2

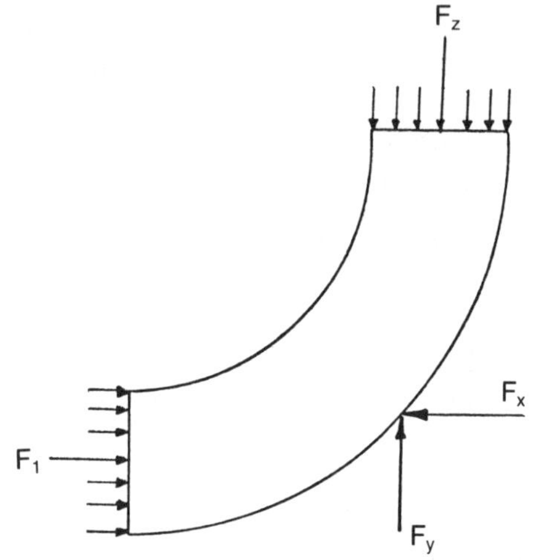

Figure 2-29. Pipe reactions induced by change of momentum of fluid flowing through elbow.

When applying Equations 2-32 and 2-33 to relief valves, the fluid dynamics of nozzles must be considered. The dynamics and thermodynamics of fluid motion through nozzles is a very involved subject and rather than investigate the various theoretical methods in this book, we will only investigate the various results and discuss their merits.

Relief valves can exert enormous forces when fluids exit the nozzles. Often, the fluid exits the nozzle at speeds exceeding Mach 1. Numerous private companies, as well as the ASME and API, have developed procedures to approximate such fluid forces. The ASME B31.1 gives a method for computing the reaction forces exerted by relief valves. The main drawback to this method is that it applies to steam only, because Code B31.1 governs only power piping. Steam is one of the most comprehensively defined substances, with all properties well known and published, but such is not the case with many chemical processes. The B31.1 method requires that the properties of the substance be well defined, to the point of being rather cumbersome to use. The ideal method would require the fewest number of physical properties, but still provide the necessary data. One such method that is very easy to use is the API formulation in API 520 Part II, paragraph 2.4, which is used for gases or vapors. This formula loses accuracy as the flow rate approaches Mach 1, so another method is desired for predicting reactions at all flow rates in processes that have poorly defined properties.

The aerospace industry has done much research in the study of fluid dynamics and thermodynamics of nozzle flow. Because relief valves operate in a closed system, the analysis of chemical rocket engines is suitable for estimating reaction forces. These calculations in such an analysis agree with those reactions computed by other methods and have been found to be slightly conservative. The method presented by Hesse [5] is desirable because of its simplicity and accuracy, and knowledge of the process fluid is limited only to the specific heat ratio, k, and the molecular weight, M. The derivation and explanation of the formulation is given by Hesse [5].

Consider the nozzle shown in Figure 2-30. The reaction force developed by a fluid exiting the nozzle is given by the following:

$$F = \lambda C_d C_v A_t P_c \left[\left(\frac{2k^2}{k-1}\right)\left(\frac{2}{k+1}\right)^{\frac{k+1}{k-1}} \times \left[1 - \left(\frac{P_e}{P_c}\right)^{\frac{k-1}{k}}\right]\right]^{0.5} + A_e(P_e - P_o) \tag{2-34}$$

where
λ = nozzle correction factor = 1/2 $(1 + \cos\alpha) \approx 1.0$ for most relief valves
C_d = nozzle discharge coefficient, which $0.97 < C_d < 1.15$, normally $C_d > 1.0$
k = specific heat ratio = C_p/C_v

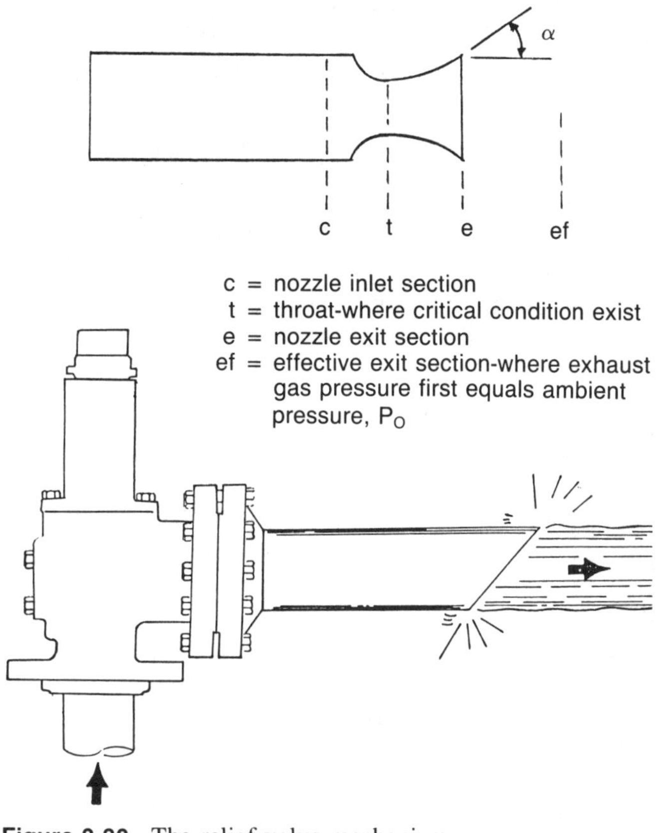

Figure 2-30. The relief valve mechanism.

$C_v = (\eta_n)^{0.5} \approx 0.95$ to 0.98
η_n = nozzle adiabatic efficiency
P_e = nozzle exit pressure, psia
P_o = ambient pressure = 14.7 psia
A_e = nozzle exit area, in.^{2}
A_t = nozzle throat area, in.^{2}

$$P_c = \text{critical pressure} = P_t\left(\frac{2}{k+1}\right)^{\frac{k}{k-1}} = \frac{\left(\frac{T_c}{M}\right)}{A_t}$$

P_c/P_e = critical pressure ratio, determined from Figure 2-31
M = molecular weight of fluid, lb/mole
G = gas flow rate, lb/sec, where

$$G = C_dA_tP_c\left[\frac{gk}{RT_c}\left(\frac{2}{k+1}\right)^{\frac{k+1}{k-1}}\right] \tag{2-34a}$$

$$T_c = \text{critical temperature} = \left(\frac{2}{k+1}\right)T_t,\ ^\circ F$$

$$\text{where } \frac{T_t}{T} = 1 + \left(\frac{k-1}{2}\right)M^2 \text{ for adiabatic process}$$

Substituting these expressions into Equation 2-34, we have

$$F = 1.028A_tP_c\left[\frac{2k^2}{k-1}\left(\frac{2}{k+1}\right)^{\frac{k+1}{k-1}}\left[1-\left(\frac{P_e}{P_c}\right)^{\frac{k-1}{k}}\right]\right]^{0.5} + A_e(P_e - P_o) \tag{2-35}$$

Equation 2-35 assumes that the flow is isentropic and in addition to relief valves, includes turbines, compressors, jet engines, rockets, injectors, ejectors, and atomizers. Most nozzles used in current applications are either convergent or convergent-divergent, also known as DeLaval or Level nozzles. Convergent-divergent nozzles are used for high pressure ratios and supersonic flow, and convergent nozzles for low pressure ratios and subsonic flow. Thus, relief valves are Level types that can handle high pressure flow.

Critical pressure of gas occurs at the point where the fluid velocity becomes Mach 1. This pressure is obtained at the minimum area of the nozzle and this minimum cross section is called the nozzle throat in the DeLaval nozzle. In the convergent nozzle, the cross section of minimum area is the exit section. The critical velocity can be expressed in terms of the inlet temperature because the process is considered adiabatic making the total temperature constant. Thus,

$$T_c = \left(\frac{2}{k+1}\right)T_t,\ ^\circ F$$

and

$$V_C = V_a = (gkRT_c)^{0.5} = \left(\frac{2gkRT_t}{k+1}\right)^{0.5} \tag{2-36}$$

Reaction forces produced by relief valves can become quite enormous and should not be overlooked. A structural failure of a relief system could well result in a catastrophe.

EXTRANEOUS PIPING LOADS

Vibration can be a real hazard in piping systems. Usually, vibration problems that occur with piping have two sources—pulsations generated by reciprocating equipment and wind. Pulsation shock phenomena on rotating equipment is briefly discussed in Chapter 6. The phenomenon of wind-induced vibrations on piping along tall towers is discussed here.

Wind-induced vibration is caused by vortex shedding on the cylindrical surface of the pipe, and becomes a problem with piping more than about thirty feet long. Vortex shedding usually occurs with piping that runs up along the height of a vertical tower.

Analyzing and solving vortex shedding vibration problems can best be handled by applying certain principles that include dimensionless parameters and experimental data. Sophisticated digital computer models are possible, and recently, vortex streets have been simulated with flow patterns around piping and structures. Such computer simulations are rigorous and expensive, so with current software they are impractical to use for all piping that may be exposed to wind.

Several proposals have been made concerning vortex-induced vibrations around cylinders, but perhaps the most straightforward is the work by Belvins [6]. He developed a dynamic model for vortex-induced vibration using random vibration theory. The theoretical basis is a representative spanwise correlation and cylinder amplitude is presented as a function of the vortex forces. When the state of resonance exists, the amplitude of the correlated lift force on the cylinder is represented as a continuous function of cylinder amplitude. Also, at resonance, the spanwise correlation of the vortex force is presented as a function of the characteristic correlation length. This model is limited to the resonance of a single mode with vortex shedding and a Reynolds number in

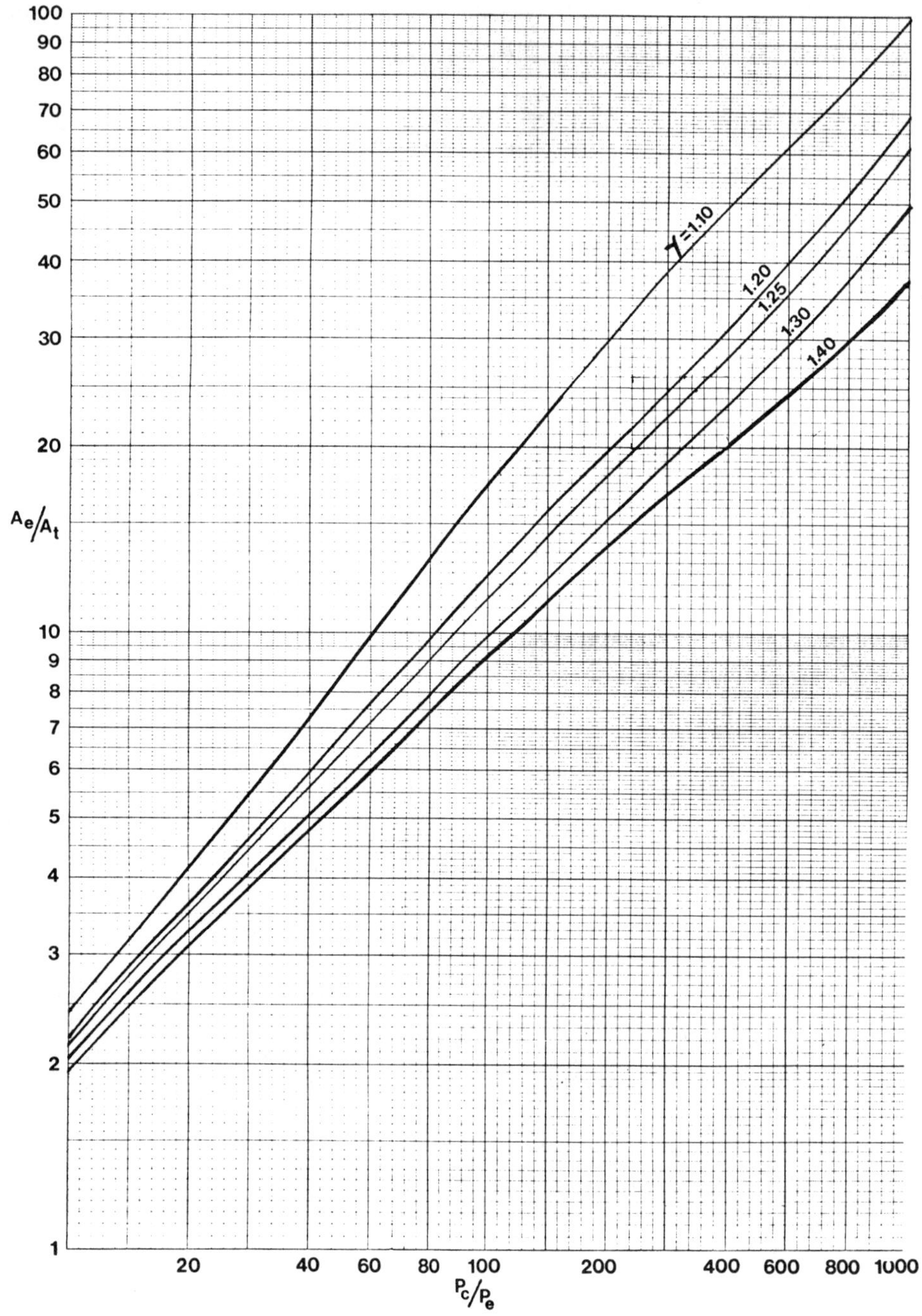

Figure 2-31. Critical pressure ratio versus area ratio for various fluid specific heat ratios (k).

the rate of 200 < N_{Re} < 200,000, where a well-formed vortex street exists.

Figure 2-32 shows flow regimes of fluid flow across stationary circular cylinders, and illustrates how the vortex streets tend to separate as the flow velocity increases. Between the range 300 N_{Re} < 300,000 the region is called subcritical because as N_{Re} approaches 300,000, the boundary layer becomes completely turbulent and the vortex shedding effect is lost. One parameter used in analyzing vortex phenomena is the Strouhal number(s), which is the numeric constant between the resonant frequency of vortex shedding (f_s) and the cylinder diameter (D), divided by the free stream velocity (V). This is analytically written as

$$S = \frac{f_s D}{V} \qquad (2\text{-}37)$$

For circular cylinders, the Strouhal versus Reynolds number is shown in Figure 2-33. In a structure, the object of design is to avoid resonance. If the inverse of the Strouhal number < 1, where f is the natural frequency of the structure, then resonances with vortex shedding from the first, second, and third harmonics are avoided. This can be accomplished by adding mass, such as insulation, and putting pipe support spacing at uneven intervals. If pipe supports are spanned evenly, periodic wave motions can form, resulting in resonance.

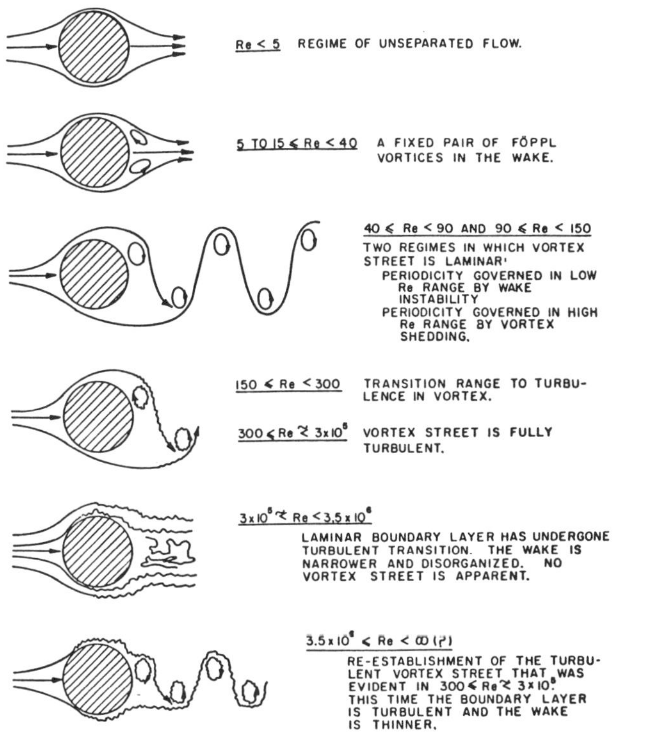

Figure 2-32. Fluid flow regimes across circular cylinder [7].

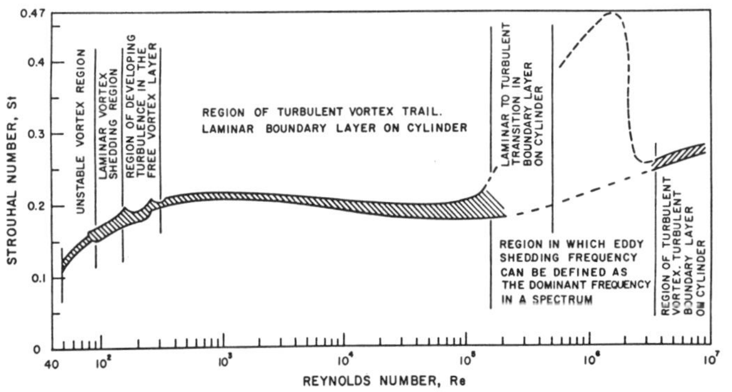

Figure 2-33. Strouhal-Reynolds number function for circular cylinders [7].

The response of a right circular cylinder at resonance with vortex shedding is a function of the following:

$$\text{Damping} = (2\pi s)^2\,\delta_r$$

$$\text{where } \delta_r = \text{reduced damping} = \frac{2m(2\pi\zeta)}{\rho D^2} \qquad (2\text{-}38)$$

and

$$\zeta = \frac{\text{energy dissipated per cycle}}{4\pi(\text{total energy of structure})}$$

Mode shape = ψ (ψ = i for a rigid cylinder)
Aspect ratio = L/D, L = length between spans

The amplitude A_Y/D can be approximated by loading the pipe with a uniform wind load and using the maximum deflection as A_Y. This can be used in Figure 2-34 to estimate the damping at resonance for a given aspect ratio. This damping is then compared with the natural frequency of the piping. The natural frequency of the pipe, especially for complex geometries, is computed by modal extraction computer analysis or any other dynamic computer software that computes the natural frequencies of piping systems. For short straight spans, the natural frequency can be determined by comparing values obtained from Table 2-6 and with the resonance damping frequency in Figure 2-34.

In practice, the greatest problem with vortex shedding occurs on tall vertical towers when pipe four inches and smaller is uninsulated and left hanging without support. It has been found that once insulation is applied to the pipe resonance vanishes.

The following simple guidelines will enable you to avoid the vast majority of wind-reduced vibrations:

1. Increase the flexural stiffness of the pipe so that its critical velocity is above the range of moderate winds.
2. Use damping devices to restrict the amplitude of vibration.
3. Reduce the effective length of the member by using intermediate struts.
4. Attach spoilers to the pipe to disrupt the flow near the tower surface; this impedes the formation of vortices and thereby eliminates the cause of vibrations.
5. Span the piping supports at uneven intervals to prevent a periodic wave function from developing.

The analysis of wind-induced vibrations on tall vertical vessels is discussed in Chapter 4.

Table 2-6
Natural Frequency of Vibration of Beam Elements

Concentrated Load on Relatively Light Beam or Spring	Uniform Load on Beam Supported at Ends	Uniform Load on Cantilever Beam
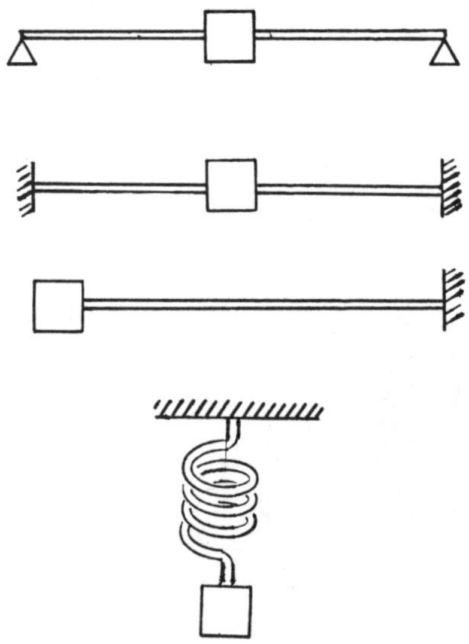	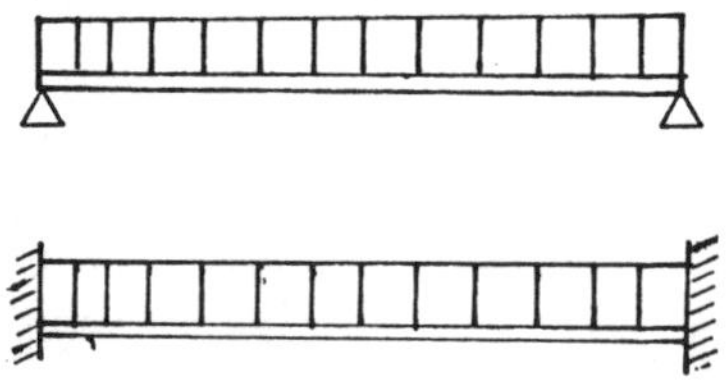	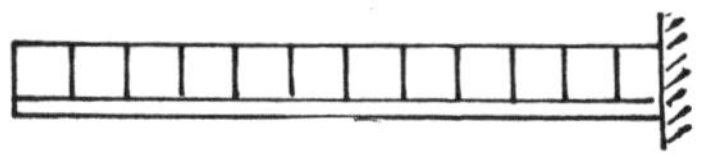
$f = \frac{1}{2\pi}\left(\frac{g}{D}\right)^{0.5}$	$f = (3.55)(D)^{-0.5}$	$f = (3.89)\ (D)^{-0.5}$

f = natural frequency of vibration, cycles per second
D = maximum static deflection of member under its own weight plus any weights that vibrate with it

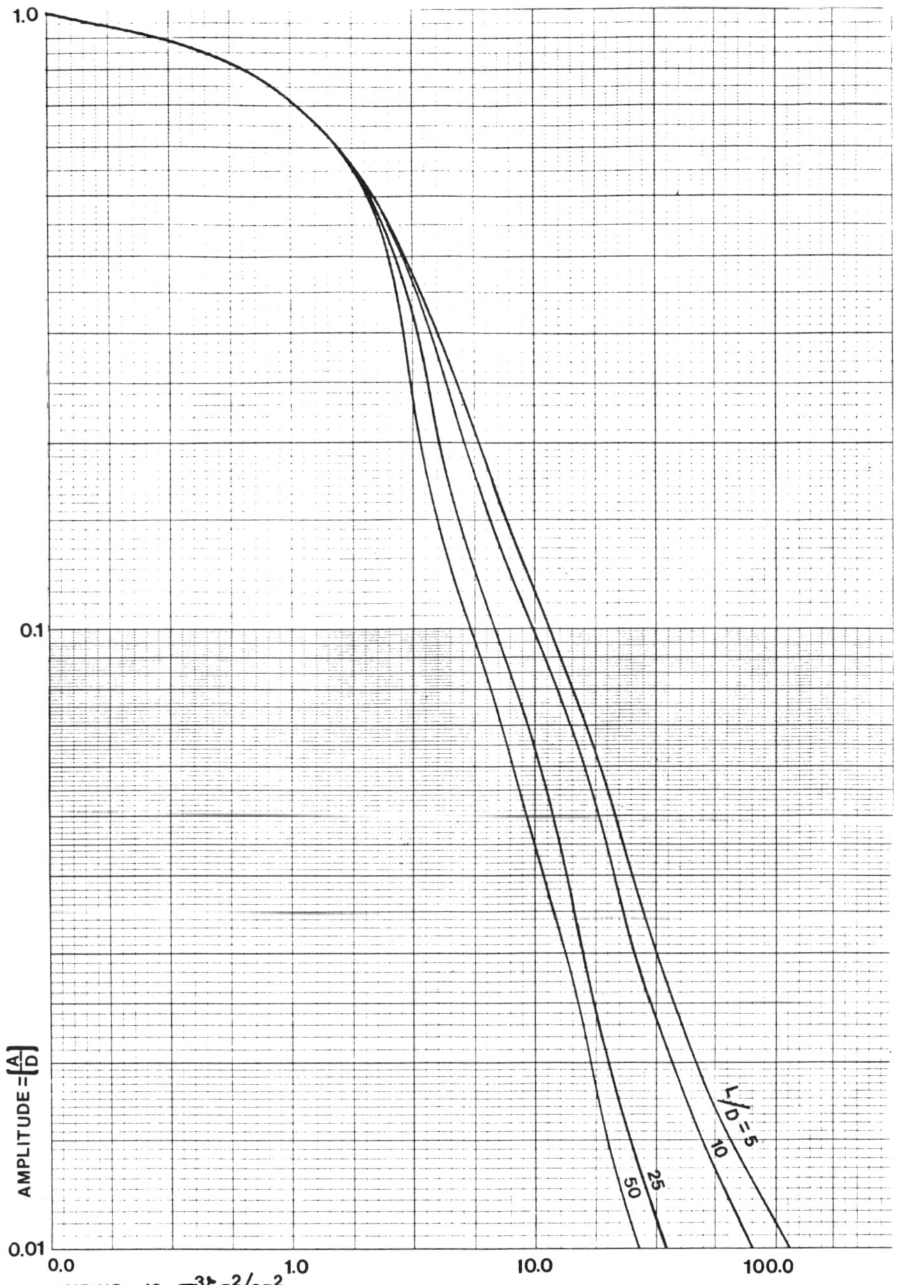

Figure 2-34. Damping, δ (dimensionless), versus amplitude, A_Y/D (dimensionless).

EXAMPLE 2-1: APPLYING THE STIFFNESS METHOD TO A MODULAR SKID-MOUNTED GAS LIQUEFACTION FACILITY

Figure 2-35 depicts the preliminary piping design of a gas liquifaction plant mounted on a skid module. Space is severely limited, as the equipment and piping are limited by the structural steel skid supports, so such devices as piping loops are unthinkable. Expansion joints are not allowed by the client, because high-pressure hydrocarbon gas is highly combustible and an expansion joint failure would mean certain gross property damage and possible loss of human lives. Therefore, the piping engineer must utilize the stiffness of pipe supports to transfer loads from the piping to the structural steel rather than to the equipment nozzles. This transfer of loads is not total, but enough to guarantee that the equipment nozzles loadings will not exceed allowable levels.

For the stiffness method to work, the piping configuration must be flexible enough for the piping itself to be within allowable stress limits set by the applicable code. This is the first significant criterion, because if the piping exceeds the allowable stress range in any part of the geometry, the system design is faulty. Conversely, the piping system can be well within the stress range and the equipment nozzles still be overloaded. Thus, the piping itself must have a certain amount of flexibility to be within code allowables. The piping supports must be stiff enough to protect equipment nozzles from excessive loads. Here our case has been stated; adding additional flexibility is not acceptable.

From computer calculations the original configuration in Figure 2-35 is found to be overstressed and the expansion stress exceeds the ASME B31.3 allowable stress range provided in Equation 2-4 for 304 SS pipe. Therefore, the piping *must* be changed to bring the maximum stress within the accepted stress range. This analysis includes the nozzle movements shown in the figure. Each nozzle is considered as an anchor.

Figure 2-36 shows the final configuration after several iterations are made to determine what configuration would best suit the structural limitations set by the module skid. This configuration is found to have a maximum allowable well within the stress range of ASME B31.3. To achieve this acceptable stress, a *limited* amount of flexibility must be added to the system. Thus, regardless which method is used—flexibility or stiffness—a certain amount of flexibility is required to make the piping system operate properly.

Once we have obtained the minimum flexible configuration required, we now focus our attention to the equipment nozzles. To consider this question, we must distinguish between the *types* of equipment. The heat exchangers HE-A and B shown in Figures 2-35 and 2-36 are aluminum plate exchangers, and the cold separator and power gas volume tank are made of reasonably thick-walled stainless steel. Thus, the critical items are the aluminum heat exchangers. The line between points 10 and 25 in Figure 2-35 must be cut because the relative Z-movement between these points overloads the nozzles at points 5 and 30, creating a very high y-moment and Z-moment, because the pipe wants to move in the −Z and +Y directions. These movements can be accommodated by using certain structural devices, such as shown in Figure 2-37. Even though flexibility has been added to the system to get the piping within the allowable stress range, the equipment nozzles are still overloaded by excessive moments above the X, Y, and Z axes—M_X, M_Y, and M_Z.

To counter the movements of the piping at the nozzles numbered 5 and 30 at HE-B and A, respectively, variable springs are placed to support the pipe while allowing the pipe to move at the same time. One spring is placed at point 20 with a simple Y support added at point 56. These additional supports help reduce the moments at nozzles 5 and 30, but not enough. So, we must add MRS restraints (see Figures 2-9 and 2-10) in pipe members 5-10 and 30-35. Each MRS is designed to allow nozzles 5 and 30 to move upward but to transfer moments M_X, M_Y, and M_Z from the pipe to the structural steel below. Also, each MRS allows pipe members 5-10 and 30-35 freedom to move along the axis so that we have the following restraints at each MRS: KTZ, KRX, KRY, and KRZ of Figure 2-38 (see Figure 2-10). Thus, we have one translational and three rotational restraints, each with a stiffness value K in lb/in. or ft-lb/deg. The pipe and exchanger are free to translate along the X and Y axes.

One can readily see that the MRS restraints must allow nozzles at points 5 and 30 to move upward, as the exchangers are bolted down to structural steel higher up on the units. Restraining the nozzles from moving upward would anchor the unit at the nozzles and at the support point causing the exchangers to rupture. Pipe members 5-10 and 30-35 must be allowed to move along the x-axis for thermal expansion.

We now have the conceptual model of what the solution looks like and the next step is to finalize the details. The MRS restraints are resisting forces and moments shown in Figure 2-38. It is necessary to design the restraints such that each has enough stiffness to transfer the loads to the steel and protect the nozzles at points 5 and 30. We will now compute the support stiffness values KTZ, KRX, KRY, and KRZ. Once these values are determined, they can be input back into the computer run and verified to be sufficient for the nozzles.

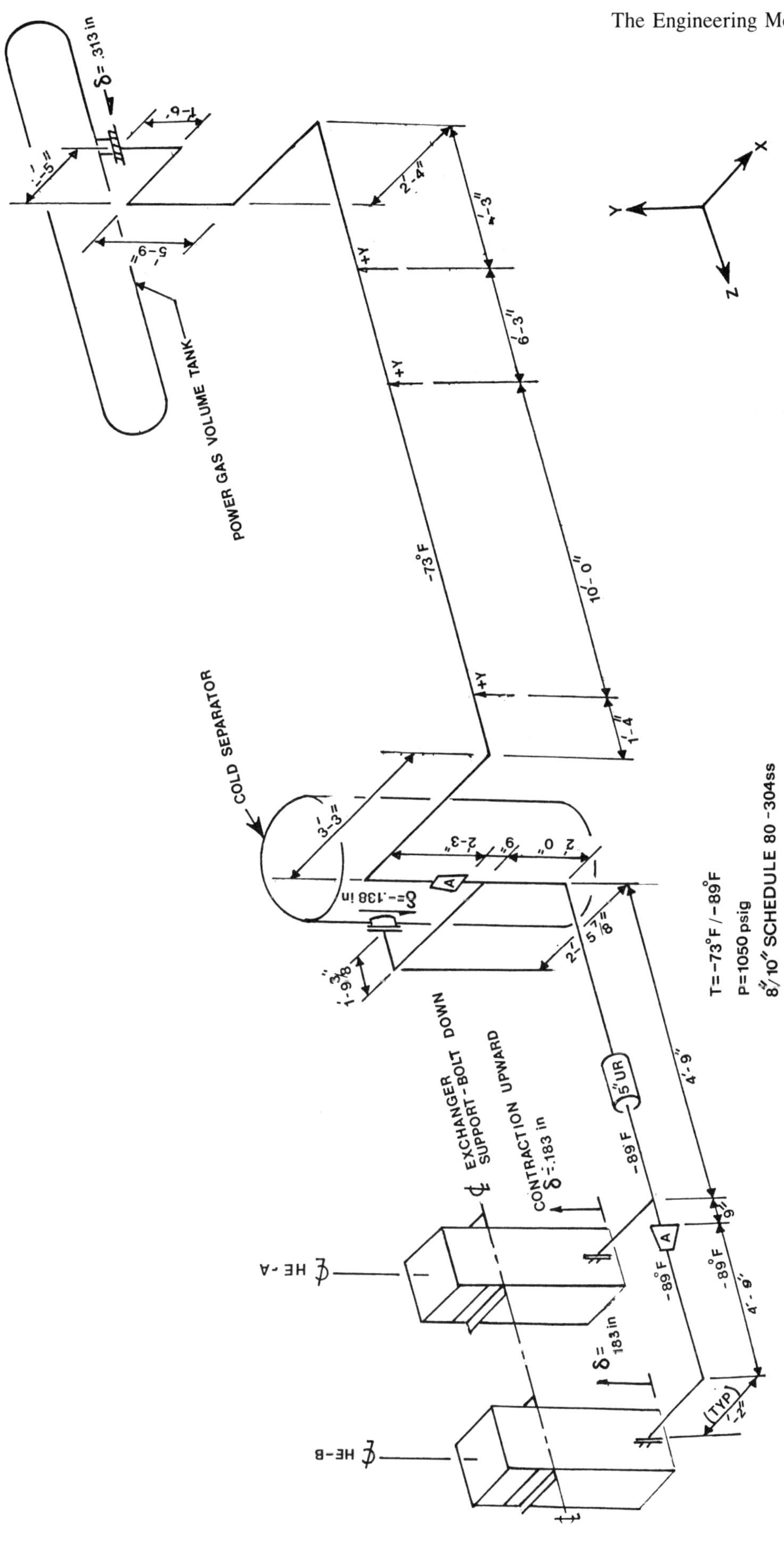

Figure 2-35. Original piping configuration.

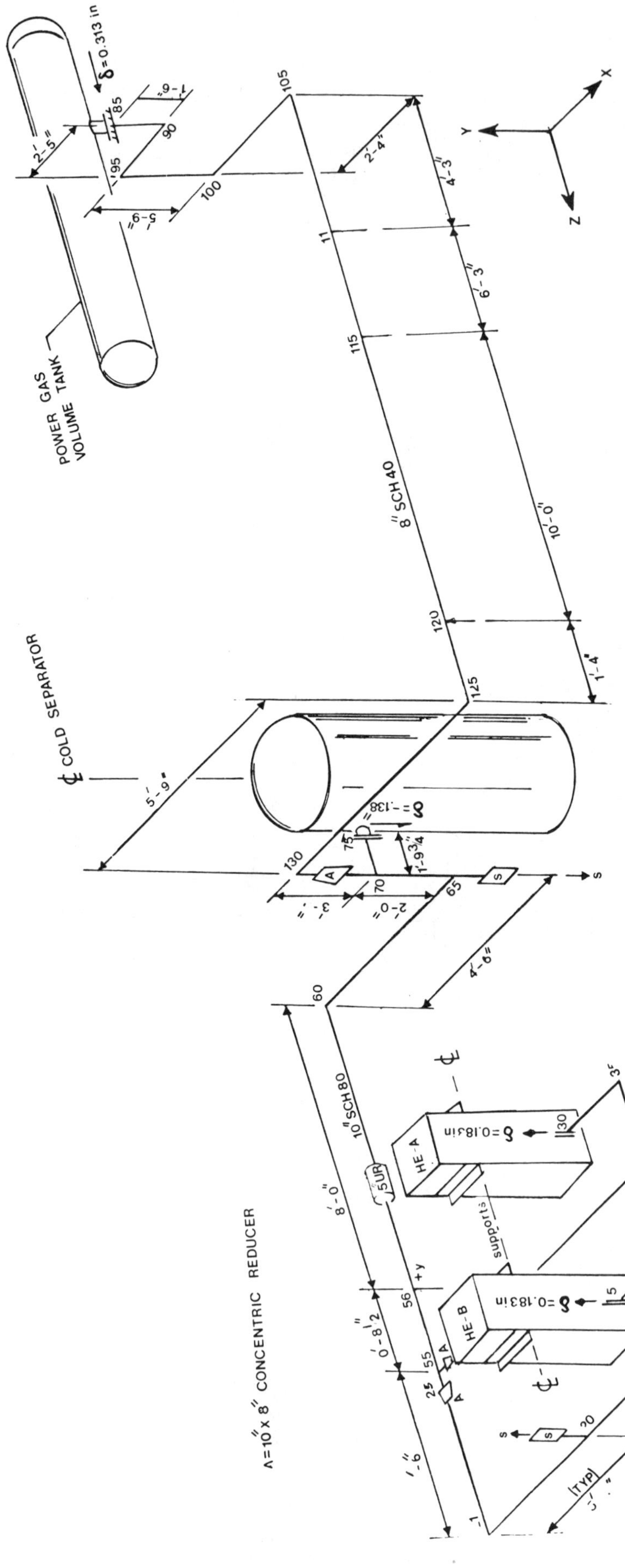

Figure 2-36. Final piping configuration.

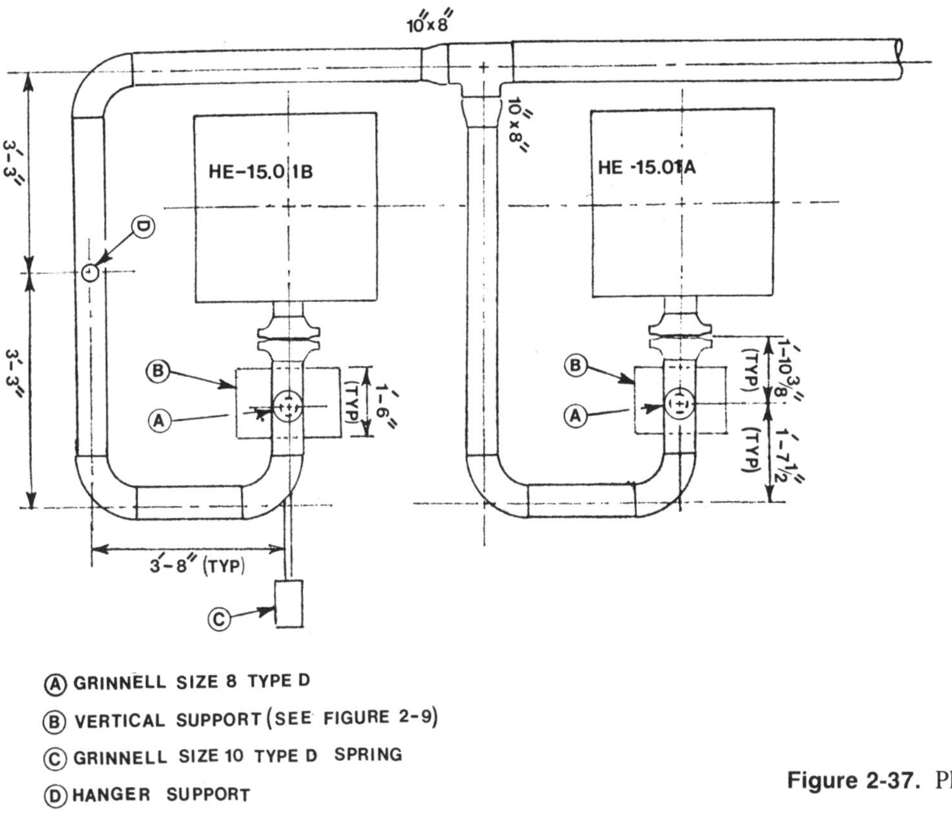

Figure 2-37. Plan view of location of MRS supports.

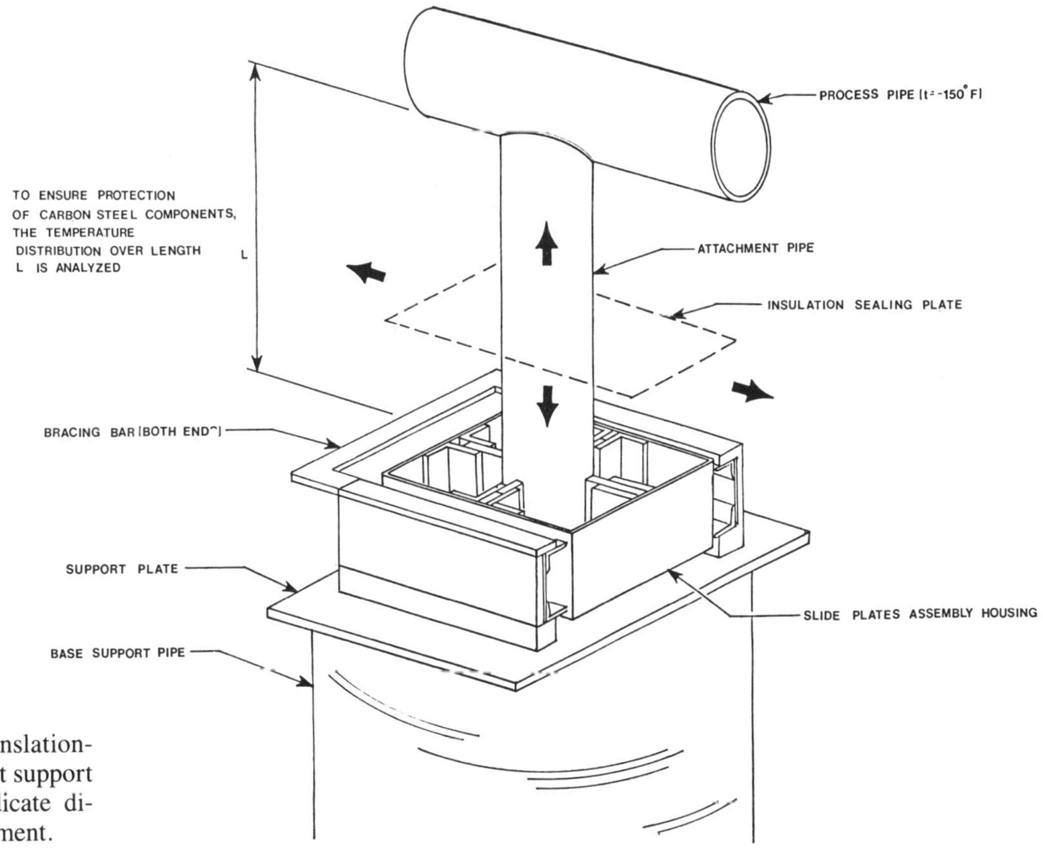

Figure 2-38. Two-axial translation-free multiple moment restraint support (BIAX-MMRS). Arrows indicate directions of freedom of movement.

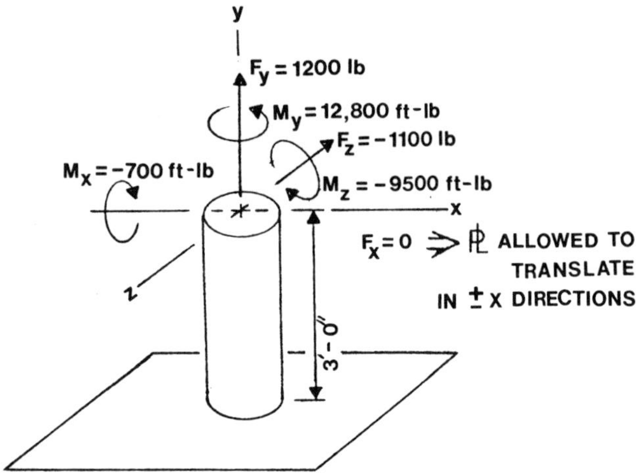

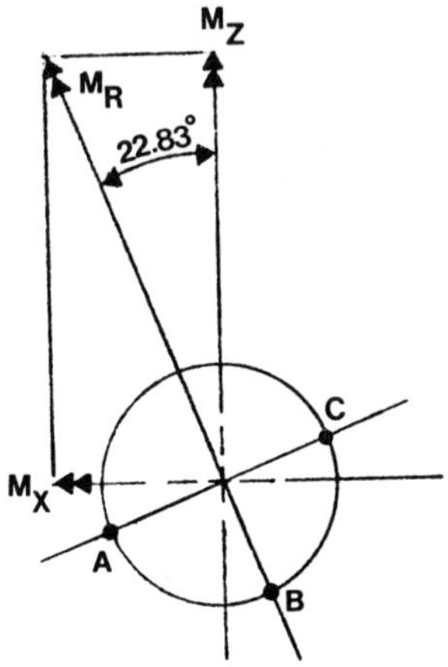

For Shear

$$c_1 = \frac{4(R^2 + Rr + r^2)}{3\pi(R + r)} \qquad \text{For 6-in. } \phi \text{ Sch 80 pipe, } R = 3.313 \text{ in., } r = 2.881 \text{ in.}$$

$$c_1 = \frac{4(10.976 + 9.545 + 8.297)}{3\pi(6.194)} = 1.975 \text{ in.}$$

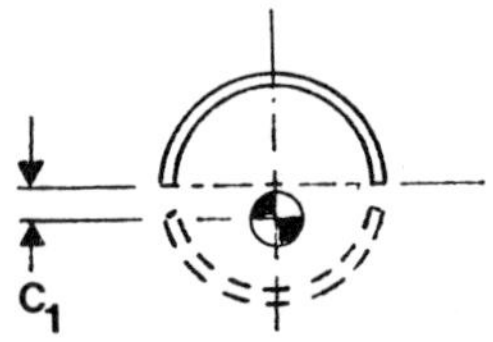

$$\tau_S = \frac{VQ}{It} = \frac{F\left(\frac{A}{2}\right)C_1}{It} = \frac{(1{,}100)\text{lb}\left(\frac{8.40}{2}\right)\text{in.}^2(1.975)\text{ in.}}{(40.49)\text{ in.}^4\,(0.432)\text{ in.}}$$

τ_S = 521.648 psi—max at neutral axis

For Bending

$$M_X = -(1{,}100 \text{ lb})(3.0) \text{ ft} - 700 \text{ ft-lb} = -4{,}000 \text{ ft-lb}$$

$$M_Z = -9{,}500 \text{ ft-lb}$$

$$M_R = R(M_X, M_Z) = [(-4{,}000)^Z + (-9{,}500)^2]^{0.5}$$

$$= 10{,}308 \text{ ft-lb}$$

$$\sigma = \frac{M}{Z} = \frac{10{,}308 \text{ ft-lb}\left(\frac{12 \text{ in.}}{1 \text{ ft}}\right)}{12.23 \text{ in.}^3} = 10{,}114 \text{ psi}$$

For Torsion

$$T = 12{,}800 \text{ ft-lb}$$

$$\tau_T = \frac{Tc}{J} = \frac{(12{,}800)\text{ft-lb}\left(\frac{12 \text{ in.}}{1 \text{ ft}}\right)\left(\frac{6.625 \text{ in.}}{2}\right)}{40.50 \text{ in.}^4}$$

$$= 12{,}563 \text{ psi}$$

For Tensile Stress

$$\sigma = \frac{P}{A} = \frac{1{,}200 \text{ lb}}{8.40 \text{ in.}^2} = 143 \text{ psi}$$

Shear Distribution

At Point A—

$$\Theta = 22.83°$$

For a circular thin-walled cross-section,

$$Q = 2R^2 t \cos\Theta$$

$$= 2\left(\frac{6.625 + 5.761}{4}\right)^2 (0.432) \cos(22.83°) = 7.635$$

$$\tau = \frac{(1{,}100)\,(7.635)}{(40.49)\,(0.432)} = 480.143 \text{ psi}$$

At Point B—

$$Q = 2(9{,}588)\,(0.432) \cos(67.166°) = 3.215$$

$$\tau = \frac{(1{,}100)\,(3.215)}{(40.49)\,(0.432)} = 202.182 \text{ psi}$$

At Point C—

$$\Theta = 22.83° \Rightarrow \tau = 480.143 \text{ psi}$$

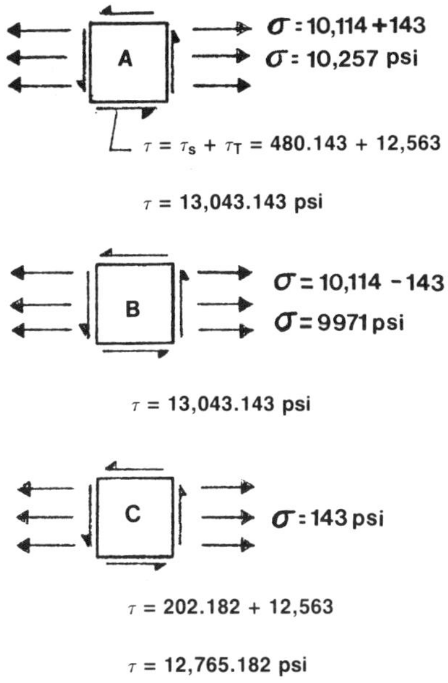

Stress Elements

Point A is the most critical point.

$$\sigma = \frac{\sigma_X - \sigma_Y}{2} \pm \left[\left(\frac{\sigma_X - \sigma_Y}{2}\right)^2 + \tau_{XY}^2\right]^{0.5} \; ; \; \sigma_Y = 0$$

$$\sigma = \frac{10{,}257}{2} \pm \left[\left(\frac{10{,}257}{2}\right)^2 + (13{,}043.143)^2\right]^{0.5}$$

$$\sigma = 19{,}143.674 \text{ psi}$$

For A312 GR TP 304SS, σ_{all} − 17,000 psi < 19,143.674 psi allowable stress, so, try 8-in. ϕ Sch 40.

Bending

$$\sigma = \frac{10{,}308 \text{ ft-lb} \left(\frac{12 \text{ in.}}{\text{ft}}\right)}{16.81 \text{ in.}^3}$$

$$\sigma = 7{,}358 \text{ psi}$$

Torsion

$$\tau_T = \frac{(12{,}800)\text{ft-lb } (12) \frac{\text{in.}}{\text{ft}} (4.313)\text{in.}}{(72.5)\text{in.}^3} = 9{,}138 \text{ psi}$$

Tensile Stress

$$\sigma = \frac{1{,}200 \text{ lb}}{8.40 \text{ in.}^2} = 143 \text{ psi}$$

Shear

$$\text{Point A, } Q = 2(18.5977)\,(0.322)\cos(22.83) = 11.039$$

$$\tau = \frac{(1{,}100)\,(11.039)}{(72.5)\,(0.322)} = 520 \text{ psi}$$

$$\sigma = 7{,}358 \text{ psi} + 143 \text{ psi} = 7{,}501 \text{ psi}$$

$$\tau = \tau_S + \tau_T = 9{,}138 \text{ psi} + 520 \text{ psi} = 9{,}658 \text{ psi}$$

$$\sigma = \frac{7{,}501}{2} + \left[\left(\frac{7{,}501}{2}\right)^2 + (9{,}658)^2\right]^{0.5}$$

$$\sigma = 14{,}111.157 \text{ psi} < 17{,}000 \text{ psi allowable stress}$$

Therefore, use 8-in. Sch 40 A-312 TP304 SS pipe for the 3-ft pipe spool piece.

From these calculations, we see that the minimum pipe size for the MRS is an 8-in. Sch 40, 3.0 feet in height. The stiffness values for an 8-in. Sch 40 pipe are as follows:

From Table 2-1, we have

$$KTX = \frac{12EI_z}{(1 + \Phi)\,L^3}$$

$$r = \frac{8.625}{2} = 4.313 \text{ in.}$$

$$\frac{r}{L} = \frac{4.313}{36.0} = 0.120 << 1 \Rightarrow \Phi = 0$$

$$KTX = \frac{12(29.0 \times 10^6)\,\frac{\text{lb}}{\text{in.}^2}\,(72.5)\text{ in.}^4}{(36.0)^3 \text{ in.}^3}$$

$$= 540{,}766.5 \text{ lb/in.}$$

$$KRX = KRZ = \frac{4(29.0 \times 10^6)\,\frac{\text{lb}}{\text{in.}^2}\,(72.5)\text{ in.}^4}{(36.0)\text{ in.}}$$

$$= 233{,}611{,}111.1 \text{ in.-lb/deg}$$

$$\text{or } KRX = KRZ = 19{,}467{,}592.6\,\frac{\text{ft-lb}}{\text{deg}}$$

$$KRY = \frac{2(29.0 \times 10^6)\,\frac{\text{lb}}{\text{in.}^2}\,(72.5)\text{ in.}^4}{(36.0)\text{ in.}}$$

$$= 116{,}805{,}55.6 \text{ in.-lb/deg}$$

$$\text{or } KRY = 9{,}733{,}796.3 \text{ ft-lb/deg}$$

Entering these stiffness values into the computer run, we see that the nozzle loads fall very sharply at points 5 and 30. Further reduction in loads can be obtained by adding springs above the MRS restraints to counter a negative moment above the Z-axis. Using springs above these supports is not always necessary, but in this case they are required because of the large vertical movement of points 5 and 30. A weight run should be made to verify that the springs do not overload the nozzles during shut-down.

The MRS restraints vary in design and are conceptually shown in Figure 2-10 and Figure 2-39. These supports are made of interlocking sliding plates with each sliding surface coated with high-strength Teflon. The precise details of such supports vary and are customized for each application.

Looking to other parts of the piping system, we notice that nozzle 75 on the cold separator has a high moment about the negative x-axis. This moment is attributed to the aluminum exchangers (HE-A and B) moving upward and the cold separator shrinking downward. Because space is premium and we are "locked-in" and can't add any more flexible piping, we add a spring at elbow 65 pulling downward to counter the excessive negative x-moment at nozzle 75. The spring is sized to be acceptable for operating and shut-down modes.

Table 2-7 lists the forces and moments at each equipment nozzle.

Upon reviewing Table 2-7, you will notice the disparity in nozzle loadings. The aluminum heat exchangers, HE-A and B, have lower loads, especially moments, than does the cold separator or power gas volume tank. This is because each has acceptable loadings that are different. The cold separator is made of 2³/₈-in. plate stainless steel, which makes the loads shown easily acceptable. (The method of determining whether such loads are acceptable on pressure vessels is discussed in Chapter 8.) Such loads would be very unacceptable for the aluminum heat exchangers because aluminum cannot withstand nearly as great a load as steel and is not very elastic, therefore making such a unit sensitive to external loads. Always be careful when subjecting rotating equipment or vessels made of light material to excessive nozzle loads.

In the final analysis the pipe loadings transferred by the MRS to the steel must be considered by the structural engineer, who must design the foundation accordingly. Sometimes it is necessary to model the stiffness of the steel foundation members when nozzle loadings become critical.

Figure 2-39. The BIAX-MMRS installed and in operation at plant facility.

Table 2-7
Equipment Nozzle Forces and Moments

Item	F_X(lbs)	F_Y(lbs)	F_Z(lbs)	F_R(lbs)	M_X(ft-lb)	M_Y(ft-lb)	M_Z(ft-lb)	M_R(ft-lb)
Heat Exchanger A	144.7	−255.9	0	293.9	0	0	−210.5	210.5
Heat Exchanger B	279.0	−624.6	0	684.1	0	0	−553.4	553.4
Process Vessel A	126.2	299.2	854.4	914.0	−6175.0	3021.7	4501.9	8217.6
Process Vessel B	−38.5	−2437.8	94.6	2440.0	877.7	−399.6	3163.0	3306.8

EXAMPLE 2-2: APPLYING THE FLEXIBILITY METHOD TO A STEAM TURBINE EXHAUST LINE

A client has added a steam turbine to a chemical plant and has piped up the turbine with make-shift parts and existing pipe, plus a newly purchased bellows expansion joint. When the turbine technicians determine they cannot cold align the turbine with the exhaust piping, the client decides that the piping must be rerouted, but requests an evaluation of the system, which is shown in Figure 2-40.

The system is modeled with a computer software package, and the results indicate that a moment about the y-axis in the magnitude of 31,000 ft/lbs is exerted on the turbine exhaust nozzle under operating conditions. Such a load is well above any turbine allowable. The reactions along the other axes are moderate and the problem of alignment must be solved. The extremely high y-moment is caused by the thermal expansion of the pipe member extending along the z-axis from point 95 to point 145 almost 7/8 in. With this expansion along the positive z-axis, the pipe rotates about the positive y-axis from point 20 through the expansion joint at point 45 to the elbow at point 75. This torsion is transmitted to the turbine nozzle at point 5. Thus, the adjustable base elbow support at point 31 is entirely useless in resisting this vertical moment and the expansion joint at point 45 transmits all of the torsion motion to the turbine nozzle at point 5.

An earlier section discussed the fact that these joints are totally rigid in torsion—a moment about the axis is parallel to the longitudinal axis, which in this example is the y-axis. In fact, with the vertical moment as great as 31,000 ft-lbs the expansion joint at point 45 will either be destroyed or have a short service life because the bel-

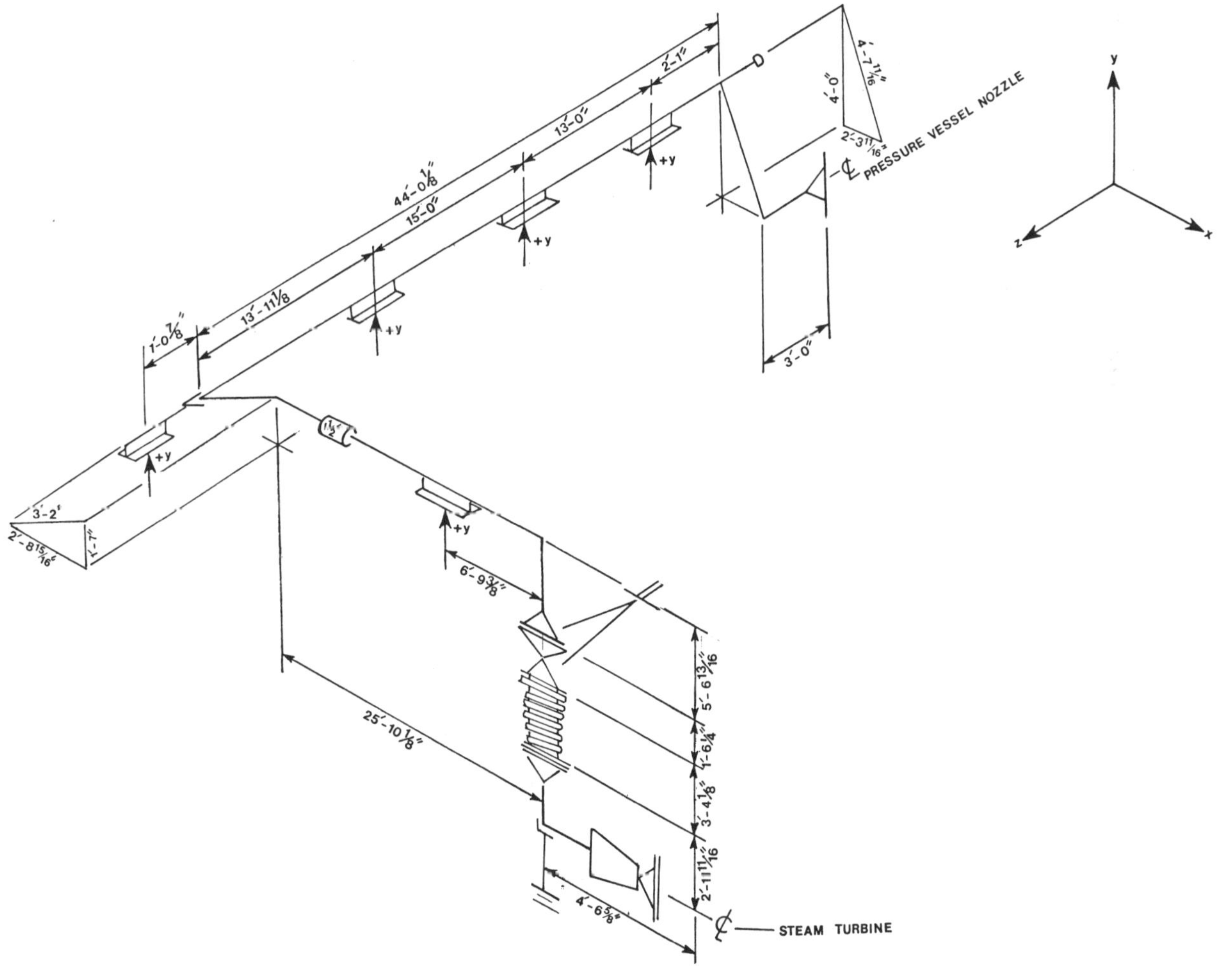

Figure 2-40. Original piping configuration of 20-in. θ steam line for turbine exhaust: temperature = 300°F, pressure = 16 Hg psia.

lows are not designed to resist such high torsional moments. Thus, the diagnosis is to avoid the high torsion and stop the .8-in. movement at point 135. To do this economically with minimum alteration to the piping, a bellows expansion joint is added at point 123 and the shoe on the dummy leg is stopped in the +z direction (i.e., movement in the +z direction is stopped, and the vessel nozzle at point 85 is protected by the joint at point 123. An expansion joint is sized based on the manufacturer's standard dimensions for a 20-in. pipe and the joint stiffness values are as follows:

KTX = KTY − 1,500 lb/in.
KRX = KRY = 200 in.-lb/deg
KTZ = 120 lb/in.

These values are provided by the joint manufacturer. The problem of turbine alignment is directly related to the inability of the turbine technicians to adjust the pipe because of the pipe's inflexibility, which is caused by the guided base elbow at point 31. The base elbow support is replaced by a spring depicted in Figure 2-41 and modeled into the computer stress program. This run is made with the added expansion joint at point 123 and the spring at point 3. The following results were obtained from the computer run:

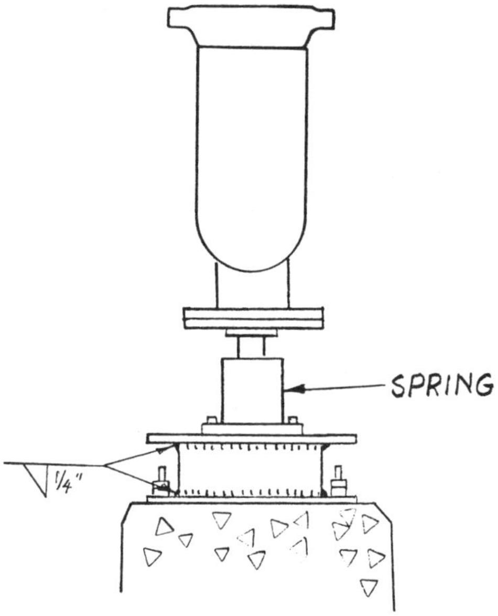

Figure 2-41. Sketch of spring that replaces base elbow support: installed load = 713 lb, operating load = 703 lb, spring = 300 lb/in.

Turbine nozzle (point 5)—

F_X = 46.5 lbs, F_Y = −530.8 lb, F_Z = 343.9 lb, F_R = 634.1 lb
M_X = 1,198.4 ft-lb, M_Y = 1,978.2 ft-lb, M_Z = 745.3 ft-lb, M_R = 2,430.0 ft-lb

Vessel nozzle (point 85)—

F_X = −46.4 lb, F_Y = −3,311.8 lb, F_Z = −3,348.5 lb, F_R = 4,709.9 lb
M_X = 5,968.7 ft-lb, M_Y = 9,742.0 ft-lb, M_Z = 5,076.0 ft-lb, M_R = 12,501.9 ft-lb

The loadings at the turbine nozzle are acceptable. (The basis for concluding this is discussed in Chapter 6.) The reactions at point 85 seem excessive and would be for a steam turbine, but considering the vessel is five feet in diameter and made of 3-in. plate, these loads are not excessive. Pressure vessel nozzle loading analysis is covered in Chapter 4, but one can deduce that pressure vessel nozzles can withstand much greater loads than most types of equipment.

The system is implemented and in two days the turbine will be fired up and operating well. The concluding remarks are that the expansion joint at point 45 is accomplishing nothing and the capital expended for its purchase was wasted. In fact, it would not hurt to move the unit, but this is not necessary since the high torsional moment has been eliminated.

The expansion joint at point 123 was specified and purchased for those stiffness values previously listed. The final configuration is shown in Figure 2-42.

EXAMPLE 2-3: FLEXIBILITY ANALYSIS FOR HOT OIL PIPING

A plant in a remote area of Brazil has an emergency need for a hot oil system. The plant manager has determined that a 3-inch Schedule 40 pipe is to be used, based on plant requirements and available pipe from local sources. We are to design the piping and ensure it will not be overstressed. There are no electronic computers available anywhere near the plant and all calculations must be made without a stress program.

For a hot oil header extending over some distance the flexibility approach is the practical method in this application. The line is to operate at 550°F at 50 psig. For a 3-in. Sch. 40 pipe, d = 3.50 in. A layout is made of the system and a preliminary loop is shown in Figure 2-43. The pipe is ASTM A-53B pipe.

$$R_x = \frac{L}{L_2} = \frac{30}{3.75} = 8.00; \; Ry = \frac{L}{L_4} = \frac{30}{4} = 7.50$$

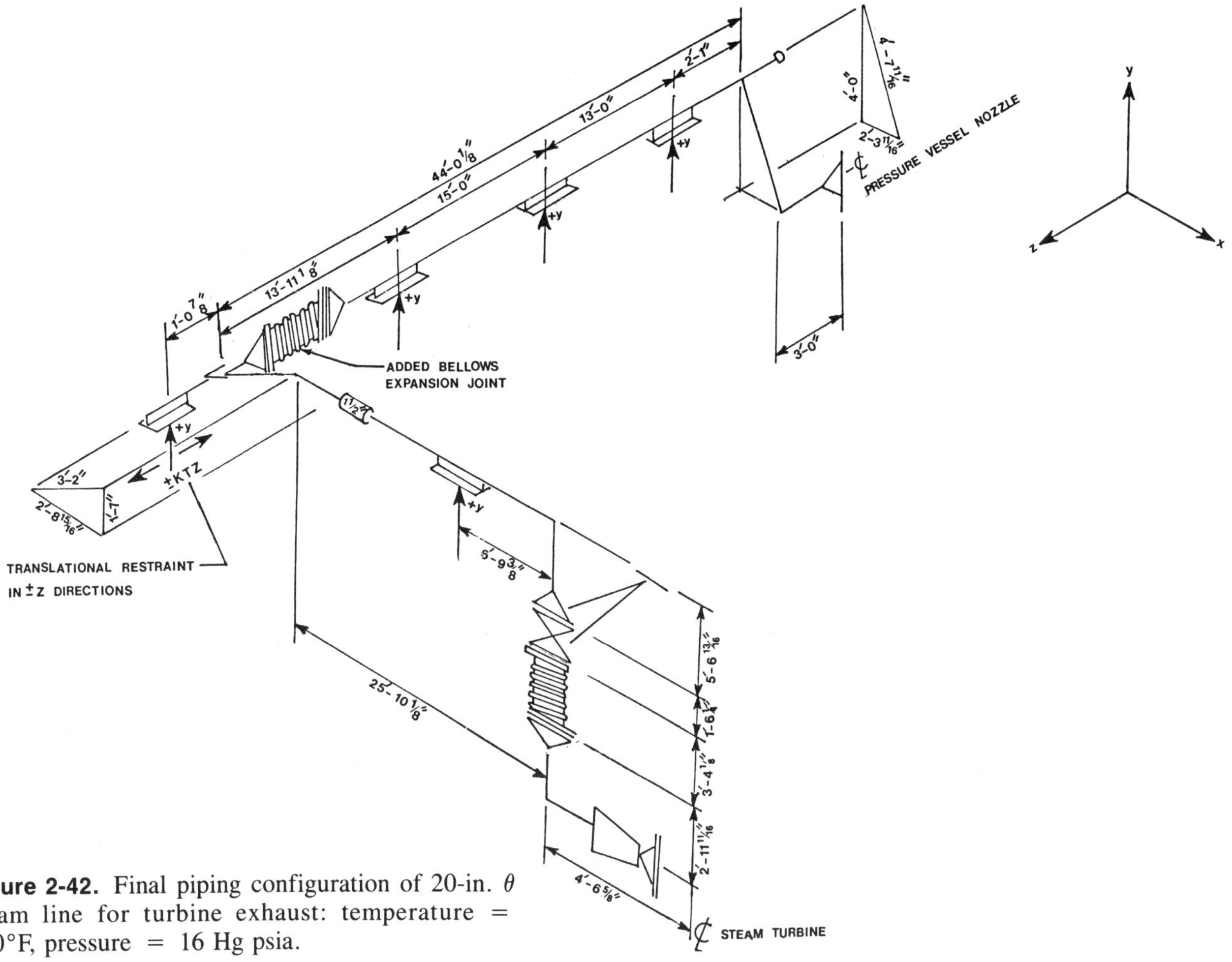

Figure 2-42. Final piping configuration of 20-in. θ steam line for turbine exhaust: temperature = 300°F, pressure = 16 Hg psia.

From Figure 2-12A, $A_b = 220$

$$B = \frac{(4.11)\,(29.0 \times 10^6)}{172{,}800} = 689.8$$

From Table 2-1 and ASME B31.3 the allowable stress is

$$\sigma_A = 1.25(20{,}000) + 0.25(18{,}100) = 29{,}525.0 \text{ psi}$$

The maximum bending stress is

$$\sigma_B = A_b B\left(\frac{D}{L}\right) = \frac{(220)\,(689.8)\,(3.50)}{30.0}$$

$$= 17{,}704.9 \text{ psi}$$

This is based on L = 30 ft between guides. Solving for L′, the distance between anchors, we have

$$\sigma_A = (17{,}704.9)\left(\frac{L'}{L}\right)$$

$$L' = \frac{(29{,}525)\,(30)}{17{,}704.9} = 50.0 \text{ ft}$$

The available steel in the plant in the area the hot oil header is to be run is spanned 4.5 m or 14.76 ft, making the anchor points spaced at 18 m or 59 ft. Thus, L′ = 59.0 ft. We change $L_4 = L_2 = 6.0 \Rightarrow R_X = R_y = 5.0$; $A_b = 98$

$$\sigma_B = \frac{(98)\,(689.8)\,(3.50)}{30} = 7{,}886.7 \text{ psi} < \sigma_A$$

Solving for total length L′,

$$L' = \frac{(29{,}525)\,(30)}{7{,}886.7} = 112.3 \text{ ft}$$

$$\Rightarrow \frac{112.3 \text{ ft}}{15 \text{ ft (between supports)}} = 7.5 \text{ supports}$$

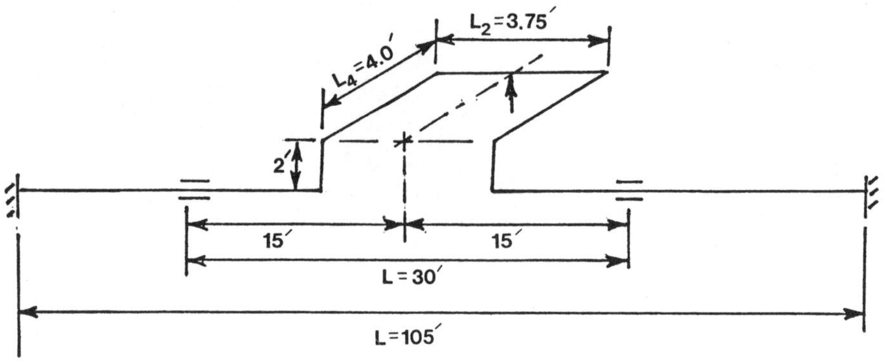

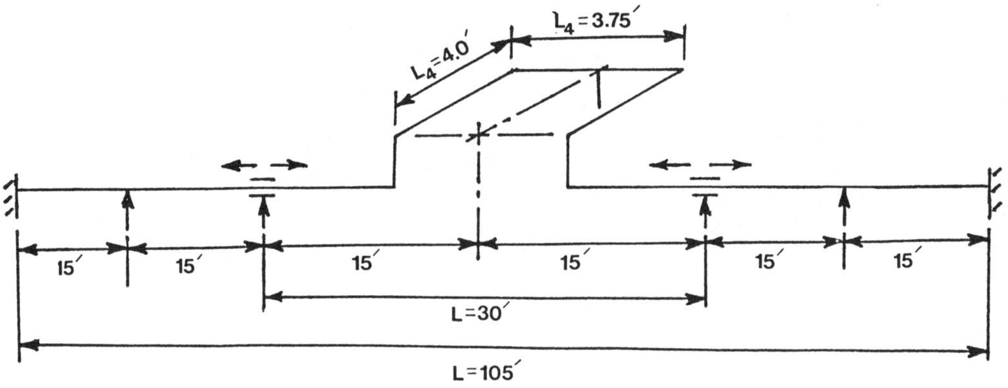

Figure 2-43. (A) Initial piping configuration; (B) final piping configuration installed and operating.

Therefore, place a loop 6 ft × 6 ft (arbitrary dimensions) every seven supports. One could increase L′ by making the loop larger (increasing L_2 and L_4), but space limitations in this application prevent it. See Figure 2-43. The stress intensification factors (SIF) in the code were made equal to one because computer stress runs have verified that the curves are conservative enough to make SIF = 1.0.

EXAMPLE 2-4: LUG DESIGN

Referring to Figure 2-16, a lug is to be designed for a pipe with the following parameters:

Pipe → 8-in. ϕ Sch. 40, C.S. SA-106B
T = 350°F, P_R = 500 psig
P = 2,000 lb
Let C = 3/8 in., L = 2½ in., and the lug has ½-in. ϕ hole
R_m = mean radius, in.
From piping properties in Appendix

$$R_m = \frac{8.625 + 7.981}{4} = 4.152 \text{ in.}$$

t = corroded pipe wall thickness, in.

$$c_1 = \frac{c}{2} = 0.188 \text{ in.}; \; c_2 = \frac{L}{2} = 1.25 \text{ in.}$$

$$\text{If } \frac{\beta_1}{\beta_2} > 1, \; \beta = \left[1 - \frac{1}{3}\left(\frac{\beta_1}{\beta_2} - 1\right)(1 - K_1)\right](\beta_1\beta_2)^{0.5}$$

$$\text{If } \frac{\beta_1}{\beta_2} < 1, \; \beta = \left[1 - \frac{4}{3}\left(1 - \frac{\beta_1}{\beta_2}\right)(1 - K_2)\right](\beta_1\beta_2)^{0.5}$$

with $\frac{\beta_1}{\beta_2} > 1$,

β = 0.19 for $N\phi$ → K_2 = 1.48
β = 0.15 for N_X → K_2 = 1.20
β = 0.11 for $M\phi$ → K_2 = 0.88
β = 0.16 for M_X → K_2 = 1.25

$\frac{N\phi}{P/R_M}$ → Figure 2-17

$\frac{M\phi}{P}$ → Figure 2-19

$$\frac{N\phi}{t} = \left[\frac{N\phi}{P/R_m}\right]\left[\frac{P}{R_m t}\right] = (1.9)\,\frac{(2{,}000)}{(4.152)\,(0.322)} = 2{,}842.30$$

$$\frac{6M\phi}{t^2} = \left[\frac{M\phi}{P}\right]\left[\frac{6P}{t^2}\right] = (0.13)\,\frac{(6)\,(2{,}000)}{(0.322)^2} = 15{,}045.72$$

Circumferential Stress, $\sigma\phi$

$$\sigma\phi = K_n\,\frac{N\phi}{t} \pm K_b\,\frac{6M\phi}{t^2}$$

$$\sigma\phi = (1.38)\,(2{,}842.30) + (1.36)\,(15{,}045.72)$$

$$= 24{,}309.65 \text{ lb}$$

Let w = 1/4-in. weld,

$$K_n = 1 + \left[\frac{1}{\frac{5.6W}{t}}\right]^{0.65} = 1 + \left[\frac{0.322}{5.6(0.25)}\right]^{0.65} = 1.38$$

$$K_b = 1 + \left[\frac{1}{\frac{9.4W}{2t}}\right]^{0.8} = 1 + \left[\frac{2(0.322)}{9.4(0.25)}\right]^{0.8} = 1.36$$

Longitudinal Stress, σ_X

$\frac{N_X}{P/R_m}$ → Figure 2-18

$\frac{M_X}{P}$ → Figure 2-21

$$\frac{N_X}{t} = \left[\frac{N_X}{P/R_m}\right]\left[\frac{P}{R_m t}\right] = (2.5)\,\frac{(2{,}000)}{(4.152)\,(0.322)}$$

$$= 3{,}739.87$$

$$\frac{6M_X}{t^2} = \left[\frac{M_X}{P}\right]\left[\frac{6P}{t^2}\right] = (0.125)\,\frac{(6)(2{,}000)}{(0.322)^2}$$

$$= 14{,}467.03$$

$$\sigma_X = K_n\,\frac{N_X}{t} \pm K_b\,\frac{6M_X}{t^2}$$

$$\sigma_X = (1.38)\,(3{,}739.87) + (1.36)\,(14{,}467.03)$$

$$= 24{,}836.19$$

Primary or pressure stress $= \sigma_P = \frac{P_R}{2t}$

$$= \frac{(500)\,(7{,}981)}{2(0.322)}$$

$$= 6{,}196.43 \text{ psi}$$

Total stress $= \sigma_T = \sigma\phi + \sigma_P = 30{,}506.08$ psi

$\sigma_T = \sigma_X + \sigma_P = 31{,}032.62$ psi

Using greatest value of σ_T,

$\sigma_T < 2\sigma_A = 2(20{,}000) = 40{,}000 \text{ psi} > 31{,}033 \text{ psi}$

Thus, a lug with these dimensions is acceptable.

Bolt shearing stress $= \tau_B$

$$\tau_B = \frac{P}{(\text{bolt or pin area})} = \frac{2{,}000}{\frac{\pi(0.5)^2}{4}}$$

$= 10{,}186 < \sigma_B$ (bolt allowable)

$\sigma_B = 25{,}000$ psi

The distance from the lug hole centerline to the lug edge is to be a minimum of AISC Table 1.16.5.1, p. 5-51.

Weld Size

$$f_l = \frac{P}{2wL} = \frac{2{,}000}{2(0.25)\,(2.50)} = 1{,}600$$

$f_w = 0.707E\sigma_{Aj}$
E = Joint efficiency
$f_w = 14{,}140.0$ psi

Weld size $= w = \frac{f_l}{f_w}$
$= 0.113 \Rightarrow$ 1/4-in. weld is acceptable.

EXAMPLE 2-5: RELIEF VALVE PIPING SYSTEM

Examine Example 1-4 relief valve system for external loadings induced by valve discharge. The gas properties are as follows:

$k = C_p/C_v = 1.451$ $\quad c_d = 1.15$

$W = 243{,}755$ lb/hr $\quad T_c = 294°F = 754°R$

$A_t = 28.89$ in.^{2} $\quad M = 170.9$

$$C_G = \frac{0.1443\, C_d \left[k\left(\frac{2}{k+1}\right)^{\frac{k+1}{k-1}}\right]^{0.5}}{\left[\frac{T_c}{M}\right]^{0.5}}$$

$$C_G = \frac{0.1443\,(1.15)\left[(1.451)\left(\frac{2}{2.451}\right)^{\frac{2.451}{0.451}}\right]^{0.5}}{\left(\frac{754}{170.9}\right)^{0.5}}$$

$$= 0.055$$

From Figure 2-31 or from the following we determine P_c and P_e as

$$P_c = \frac{G}{C_G A_t} = \frac{67.710}{(0.055)\,(28.89)} = 42.613$$

$$P_e = 14$$

$$F = (0.99)(1.15)(0.98)(28.89)(42.613)$$

$$\left[\frac{2(1.451)^2}{0.451}\left(\frac{2}{2.451}\right)^{\frac{2.451}{0.451}}\left[1 - \left(\frac{14.00}{42.613}\right)^{\frac{0.451}{1.451}}\right]\right]^{0.5}$$

$$+(28.89)\,(4)$$

$$F = 2{,}385.879 \text{ lb}$$

Reaction moment at the vessel nozzle is

$$M_R = (2{,}385.879) \text{ lb } (8.5) \text{ ft} = 20{,}279.972 \text{ ft-lb}$$

The reactions at the vessel nozzle are discussed in Chapter 4, along with external loadings on vessels.

EXAMPLE 2-6: WIND-INDUCED VIBRATIONS OF PIPING

A 1-in. ϕ Schedule 40 pipe is to run up a process tower, and it is necessary to determine what span intervals are needed to avoid vibration resonance caused by vortex shedding induced by wind external to the pipe. Piping designers have the line supported at 30-foot even intervals.

The first problem with the layout are the even intervals for the supports. Piping spans subjected to vibration should be in uneven intervals to prevent sine wave oscillations that would be symmetric and periodic, and thus self-destructive.

Experimental data from Blevins[6] support the following formulations:

$\zeta = 0.15$ damping factor for small pipe (4-in. $\phi <$)

$\zeta = 0.025$ for large pipe ($>$6-in. ϕ)

$$\frac{A}{D} = \frac{0.07\,C}{(\delta_r + 1.9)S^2}\left[0.30 + \frac{0.72}{(\delta_r + 1.9)S}\right]^{0.5} \quad (2\text{-}39)$$

$L = 30$ ft, 1-in. ϕ Sch. 40 pipe

For a uniformly loaded member with simple supports,

$$\delta = \frac{5WL^3}{384EI}$$

where $I = 0.0874$ in.^{4}; $W = 1.68$ lb/ft (30)ft

$= 50.40$ lb

$$\delta = 12.569 \text{ in.}$$

From Table 2-5,

$$f = \frac{3.55}{(12.569)^{0.5}} = 1.001 \text{ cycles/sec}$$

$$\delta_r = \frac{4m\pi\zeta}{\rho D} \quad (2\text{-}38)$$

For air at 60°F,

$$\rho = 0.076 \text{ lb}_m/\text{ft}^3$$

$$m = \frac{0.140\,\frac{\text{lb}}{\text{in.}}}{32.2\,\frac{\text{ft}}{\text{sec}^2}} = 0.004\,\frac{\text{lb}_m}{\text{in.}}$$

Air velocity under investigation = 25 mi/hr = 36.68 ft/sec

$\delta_r = 13.037$; $N_{Re} = 2.54 \times 10^4$

From Figure 2-33,

$$S = 0.18$$

Thus, $\frac{A}{D} = 0.126$

$$\text{Damping} = \frac{16m\pi\zeta s^2}{\rho D^2} = 12.681$$

Values from Figure 2-34 indicate we are close to resonance, as we are within an L/D ratio of 5 and L/D $\approx$ 30. Thus, we should experience resonance at 25 mph for the 1-in. ϕ Sch 40 bare pipe. The line should have more supports added at uneven intervals closer than 30 ft and the previous analysis repeated for a range of wind velocities. Such a problem can be approached with a computer program based on experimental data.

As is obvious, vortex shedding vibrations is still a subjective phenomenon based on empirical data, but this example should assist one in protecting piping surrounded by vortices.

NOTATION

A = area, in.^{2}
C = compliance, in./lb or deg/ft-lb
D = diameter, in.
E_D = modulus of elasticity in down condition, psi
E_o = modulus of elasticity in operating condition, psi
F = force, lbs
G = modulus of rigidity, psi
I = moment of inertia, ft^4
J = polar moment of inertia, ft^4
K = stiffness, either translational (lb/in.) or rotational (ft-lb/deg)
K_b = stress concentration factor for bending
K_n = stress concentration factor for pure tension or compression
KTX = translational stiffness along X-axis, lb/in.
KRX = rotational stiffness about X-axis, ft-lb/deg
KTY = translational stiffness along Y-axis, lb/in.
KRY = rotational stiffness about Y-axis, ft-lb/deg
KTZ = translational stiffness along Z-axis, lb/in.
KRZ = rotational stiffness about Z-axis, ft-lb/deg
L = length, in.
M = moment, ft-lb
P = force (lb) or moment (ft-lb) in stiffness matrix
P_i = internal pressure, psig
P_o = external pressure, psig
P_R = internal pressure evaluated at radius R, psig
R = reaction, lb
R_D = reaction in down (non-operating) condition, lb
R(x, y, z) = vector resultant operator
r_i = inside radius, in.
r_o = outside radius, in.
T = torsion, ft-lb
t = thickness, in.
U = displacement, in.
W = weight of fluid, lb_m
w = weld size, in.
Z = section modulus, in.^{3}
Z_m = section modulus of mean section radius, in.^{3}
σ_B = bending stress, psi
σ_c = circumferential stress, psi
σ_L = longitudinal stress, psi
σ_p = pressure stress, psi
σ_R = radial stress, psi
σ_T = torsional stress, psi
σ_y = yield stress, psi
τ = shear stress, psi
F,M + X, + Y, + Z = forces or moments acting only in +X, +Y, or +Z direction, respectively
F,M − X, − Y, − Z = forces or moments acting only in −X, −Y, or −Z direction, respectively

REFERENCES

1. Faires, V. M., *Design of Machine Elements,* The Macmillan Company, New York, 1965.
2. Przemieniecki, J. S., *Theory of Matrix Structural Analysis,* McGraw-Hill Book Co., New York, 1968.
3. Wichman, K. R., Hopper, A. G., Mershon, J. L., *Welding Research Council Bulletin 107, Local Stresses in Spherical and Cylindrical Shells Due to External Loadings,* Welding Research Council, New York, 1979.
4. Expansion Joint Manufacturers Association, Inc., *Standards of the Expansion Joint Manufacturers Association,* Inc., New York.
5. Hesse, W. J., Mumford, Jr., N. V. S., *Jet Propulsion for Aerospace Applications,* Second Edition, Pitman Publishing Corporation, New York, 1964.
6. Blevins, R. D., *Flow Induced Vibration,* van Nostrand Reinhold Company, New York, 1977.
7. Lienhard, J. H., "Synopsis of Lift, Drag and Vortex Frequency Data For Rigid Circular Cylinders," Washington State University, College of Engineering Research Division Bulletin 300, 1966.

3

Heat Transfer in Piping and Equipment

Providing thermal energy to process systems and maintaining desired temperatures are key responsibilities of mechanical design. Although they border on chemical engineering, the concern here is with the mechanical aspects of process systems, and not with the processes themselves. (Chapters 2 and 4 illustrate how mechanical design borders civil engineering in a similar manner.)

Process systems require thermal energy for various reasons, and the most common are to accelerate chemical reactions; to heat products and services so the products remain liquid and do not clog piping or equipment, such as with asphalt and roofing materials, viscous fuel oils, and syrups; and to cool products and services, for example to protect epoxy from polymerizing.

In piping systems there are three ways to transfer heat to the process service—tubular tracers mounted externally to the pipe, jacketing the process pipe with a larger pipe forming an annulus in which the heat transfer fluid flows, and electrically tracing the pipe. We will discuss the first two types of transfer systems.

JACKETED PIPE VERSUS TRACED PIPE

The difference between traced pipe and jacketed pipe is obviously the heat transfer area available on each. The two types are depicted in Figures 3-1a and b. Jacketed systems offer more heat transfer area, but are expensive and can be difficult to maintain. One common problem is cracks that develop from the thermal stresses that are incurred. Such cracks, which are difficult to locate and repair, can cause the heat transfer and process fluids to mix, which can have catastrophic results. However, the disadvantages of jacketed pipe must be weighed with the economics of adding tracers. A proven guideline is to use jacketed pipe for process fluids with viscosities of 5,000 centipoises or more. Such high-viscosity fluids are quite common with coating mixes used in manufacturing roofing tiles. Tracing such viscous mixtures with several tracers has proven to be so inferior to jacketed pipe that the disadvantages of jacketed systems are offset. With a viscosity of 4,000 centipoises, one should consider jacketed pipe.

Most jacketed pipe is limited in commercially available sizes. Normally 8-in. by 10-in. is the largest size

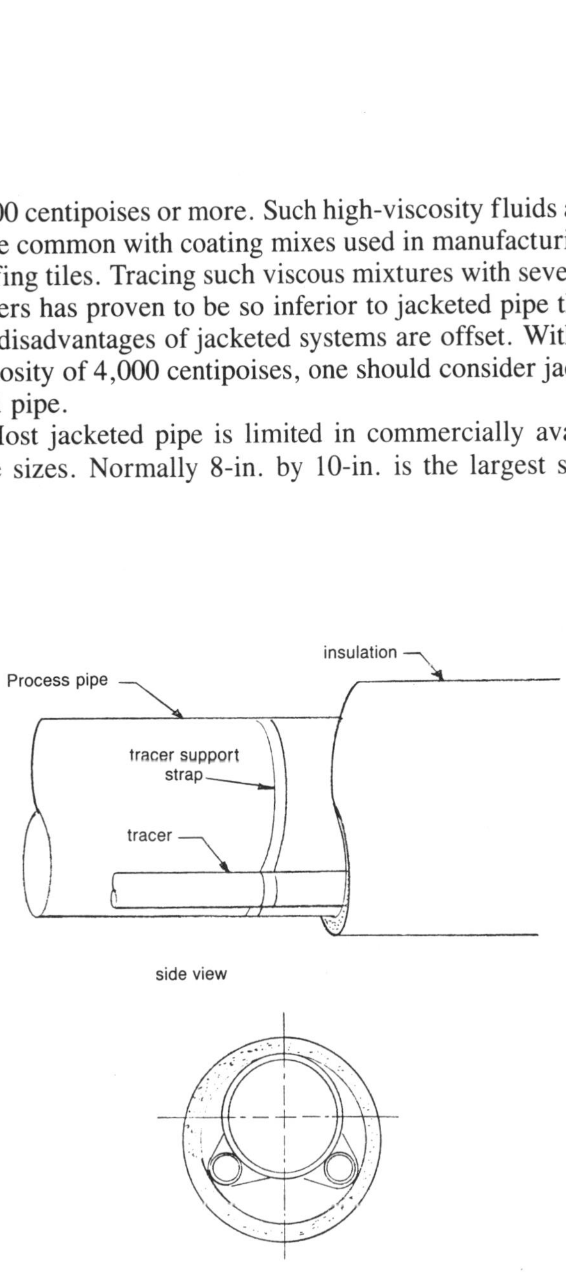

Figure 3-1A. Traced pipe.

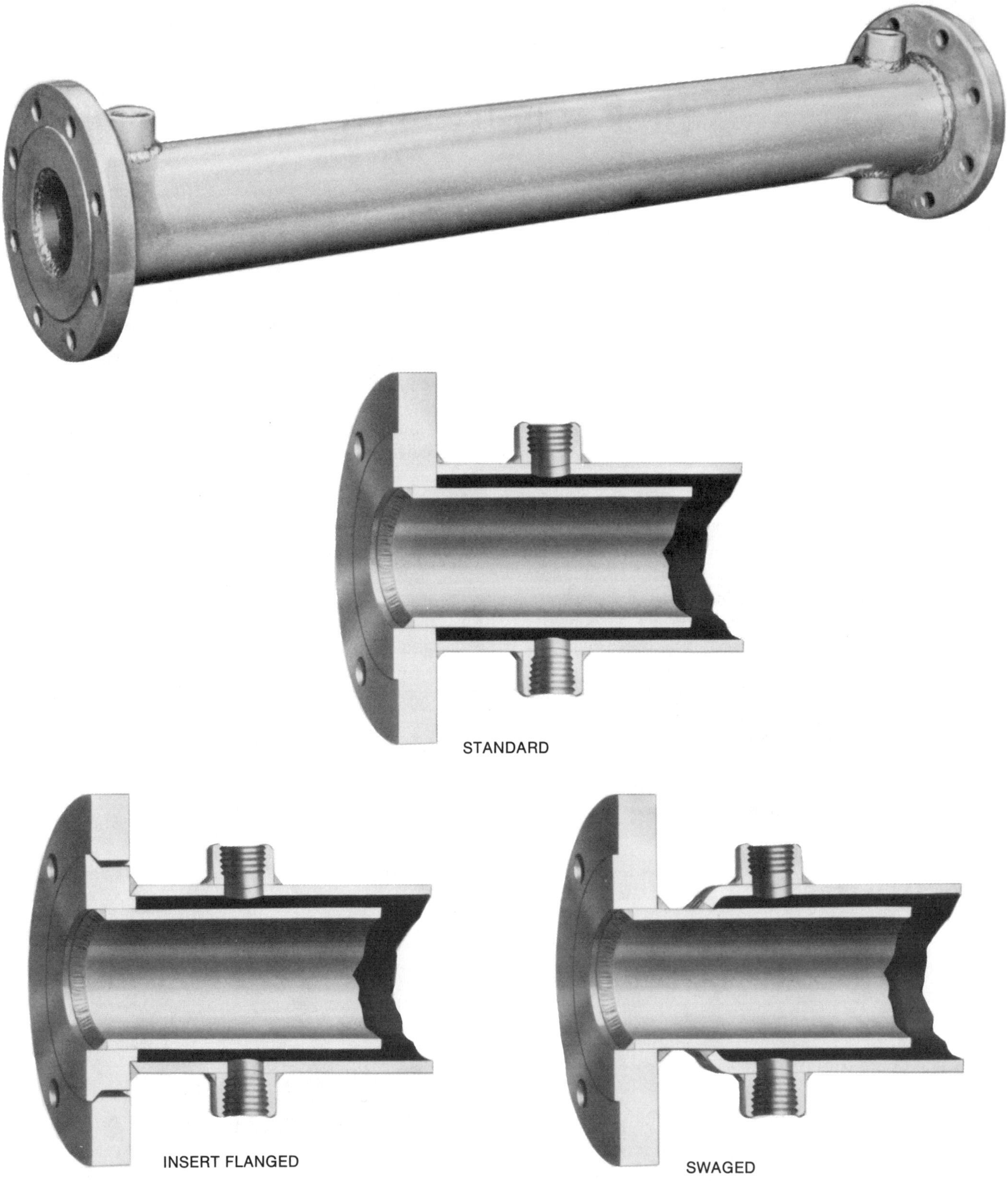

Figure 3-1B. Jacketed pipe. (Courtesy of Parks-Cramer Co.)

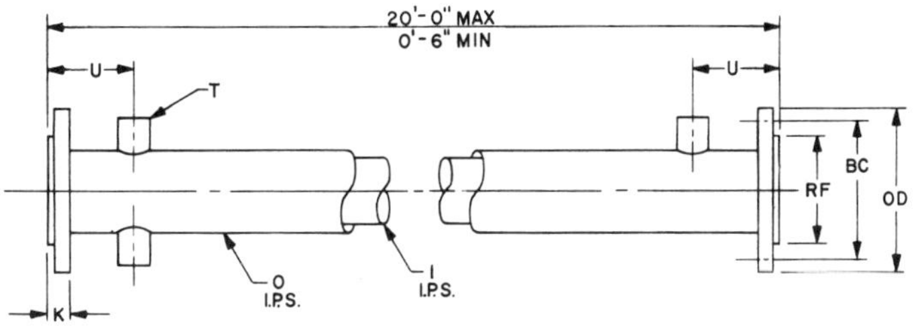

DIMENSIONS

	COMMON TO ALL				150 LB.*						300 LB.*					
						Holes						Holes				
SIZE	I IPS	O IPS	U	T NPT	OD	No.	Dia.	BC	RF	K	OD	No.	Dia.	BC	RF	K
½x1¼	½	1¼	2.56	¾	4.25	4	0.62	3.12	2.00	0.75	4.88	4	0.75	3.50	2.00	0.88
¾x1½	¾	1½	2.56	¾	5.00	4	0.62	3.88	2.88	0.75	6.12	4	0.88	4.50	2.88	0.88
1x2	1	2	2.56	¾	6.00	4	0.75	4.75	3.62	0.75	6.50	8	0.75	5.00	3.62	0.88
1¼x2	1¼	2	3.44	¾	6.00	4	0.75	4.75	3.62	0.62	6.50	8	0.75	5.00	3.62	0.88
1½x2½	1½	2½	3.31	¾	7.00	4	0.75	5.50	4.12	0.69	7.50	8	0.88	5.88	4.12	1.00
2x3	2	3	4.69	1	7.50	4	0.75	6.00	5.00	0.75	8.25	8	0.88	6.62	5.00	1.12
3x4	3	4	4.44	1	9.00	8	0.75	7.50	6.19	0.94	10.00	8	0.88	7.88	6.19	1.25
4x6	4	6	4.31	1	11.00	8	0.88	9.50	8.50	1.00	12.50	12	0.88	10.62	8.50	1.44
6x8	6	8	4.31	1	13.50	8	0.88	11.75	10.62	1.12	15.00	12	1.00	13.00	10.62	1.62
8x10	8	10	4.88	1½	16.00	12	1.00	14.25	12.75	1.19	17.50	16	1.12	15.25	12.75	1.88

*Flanges of higher pressure class and other facings available.

Figure 3-1B. Continued.

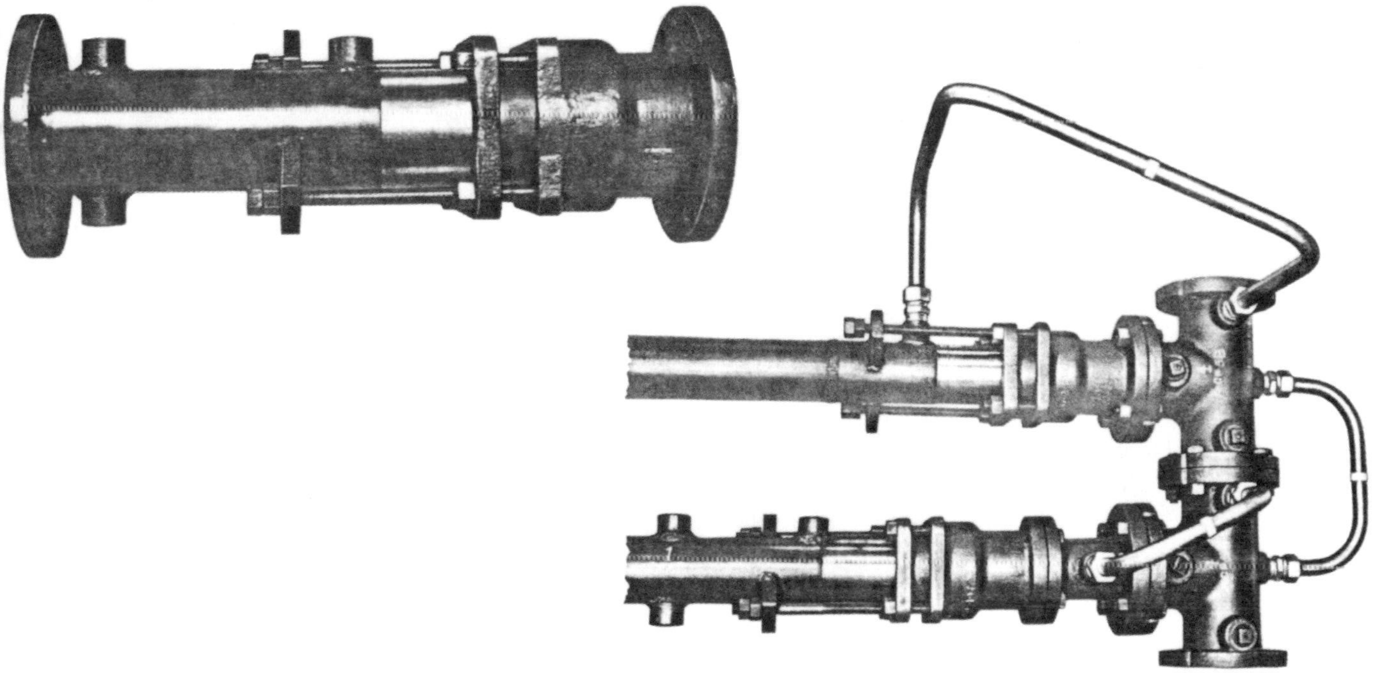

Figure 3-1C. Expansion joints for jacketed pipe. (Courtesy of Parks-Cramer Co.)

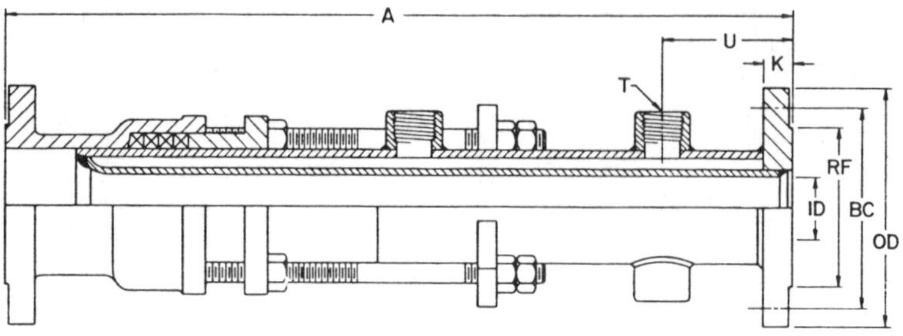

DIMENSIONS
150 LB., DUCTILE IRON, STEEL

	FLANGE DIMENSIONS*											
		OD		Holes					A			T
SIZE	ID IPS	ins.	mm.	No.	Dia.	BC	RF	K	ins.	mm.	U	NPT
1¼x2	1¼	6.00	152	4	0.75	4.75	3.62	0.75	24.00	610	3.44	¾
1½x2½	1½	7.00	178	4	0.75	5.50	4.12	0.75	25.00	635	3.31	¾
2x3	2	7.50	190	4	0.75	6.00	5.00	0.88	25.38	645	4.69	1
3x4	3	9.00	229	8	0.75	7.50	6.19	1.00	26.00	660	4.44	1
4x6	4	11.00	279	8	0.88	9.50	8.50	1.12	26.50	673	4.31	1
6x8	6	13.50	343	8	0.88	11.75	10.62	1.19	27.38	695	4.31	1
8x10	8	16.00	406	12	1.00	14.25	12.75	1.38	34.25	870	4.88	1¼

All dimensions in inches (ins.) unless otherwise noted.
*Flanges of higher pressure class and other facings available.

Figure 3-1C. Continued.

carried in stock, but larger sizes can be specially fabricated. When a jacketed system is selected, a careful stress analysis should be made to ensure that the system is not overstressed. (Chapter 2 covers such stress analyses.)

TRACING PIPING SYSTEMS

When process fluids have low to moderate values of viscosity ($\leqslant$ 4,500 cp), it is best to trace them with tubes containing hot or cold fluids. The tracing can be done with or without heat transfer cement around the tracer tubes (Figure 3-2). We will consider two methods for analyzing both systems.

Usually, steam or hot oil is used to trace systems. Hot oil is used when the fluid to be traced is hotter than saturated steam at typical operating pressures, which would be about 350°F and above. Hot oil systems are somewhat simpler than steam-traced systems, because steam traps and condensate return lines are unnecessary. However, hot oil can be expensive and if there is ample auxiliary steam available for tracing, steam is favorable for moderate- to low-temperature systems. When there is much piping to be traced, steam at the available temperature and pressure may condense into hot water before tracing the entire system. For these situations, only hot oil can be used. Thus, hot oil is used in tracing applications where steam is either not practical or not available.

There are many types of hot oils marketed by various chemical companies as heat transfer fluids.

It is most desirable and should be mandatory to use heat transfer cement in tracing tubes on process piping, because it provides more heat transfer area. Heat transfer cements are available in all major countries and in some of the larger Third World countries. However, there are times of expediency in which traced systems must be installed without the cement.

Traced Piping Without Heat Transfer Cement

The modes of heat transfer in a system without heat transfer cement are natural convection through the air space inside the insulation, and to a much lesser extent, direct radiation between the tracer and pipe or equipment. Since the tracer tube and pipe surface have very little surface contact, conduction is minimal. Any effect of film resistance to heat transfer between the air space outside the insulation and the inside insulation surface is negligible.

The procedure for tracer design without heat transfer cement is outlined in the following steps (see Figure 3-3 for parameters):

1. Assume a value of air space temperature equal to or greater than the minimum temperature of the process temperature inside the pipe.

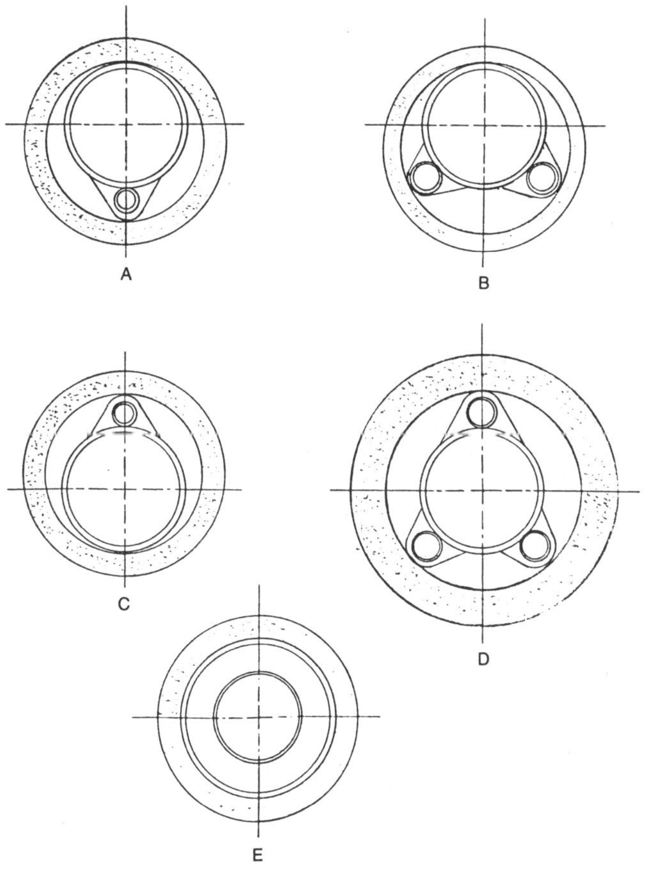

Figure 3-2. Various traced pipe configurations: (A) single traced pipe, with tracer under pipe, with heat transfer cement (HTC); (B) process pipe with two tracers with HTC; (C) one tracer on top of process pipe with HTC; (D) process pipe with three tracers with HTC; (E) jacketed pipe.

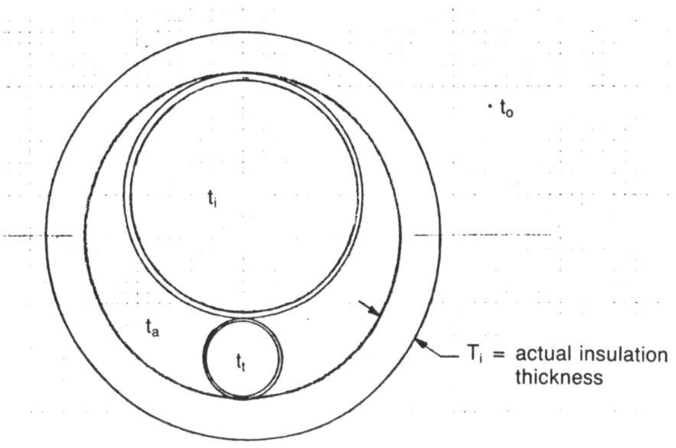

Figure 3-3. Traced pipe with one tracer under bottom without HTC.

2. Estimate the natural convection coefficient, h_c, from the tracer to the air space from Figure 3-4.
3. Calculate the equivalent cylindrical insulation thickness, T_e, as

$$T_e = \left(\frac{D_i + 2T_i}{2}\right) \ln \left(\frac{D_i + 2T_i}{D_i}\right) \tag{3-1}$$

4. Determine the outside film coefficient of the insulation to atmosphere, h_o, from Figure 3-5 and calculate U_o from the following:

$$\frac{1}{U_o} = \frac{T_e}{k_i} + \frac{1}{h_o} \tag{3-2}$$

where D_i = inside diameter of pipe insulation, ft
h_o = outside film coefficient from insulation to atmosphere, Btu/hr-ft^2-°F
k_i = thermal conductivity of insulation, Btu/hr-ft-°F
T_e = equivalent thickness of cylindrical insulation, ft
T_i = actual insulation thickness, ft
U_o = overall heat transfer coefficient from the air space to the atmosphere, Btu/hr-ft^2-°F

5. Formulate a heat balance around the air space, solving for the temperature of the air space, t_a:

$$Q_1 = (U_o)(A_o)(t_a - t_o) \tag{3-3}$$
$$Q_2 = (h_t)(A_t)(n)(t_t - t_a) \tag{3-4}$$
$$Q_2 \geqslant (\gamma)(Q_1) \tag{3-5}$$

where A_o = outside insulation surface area, ft^2/ft
A_t = outside surface area of tracer tube, ft^2/ft
h_t = convection film coefficient from tracer or heat transfer cement (HTC) to air space, Btu/hr-ft^2-°F

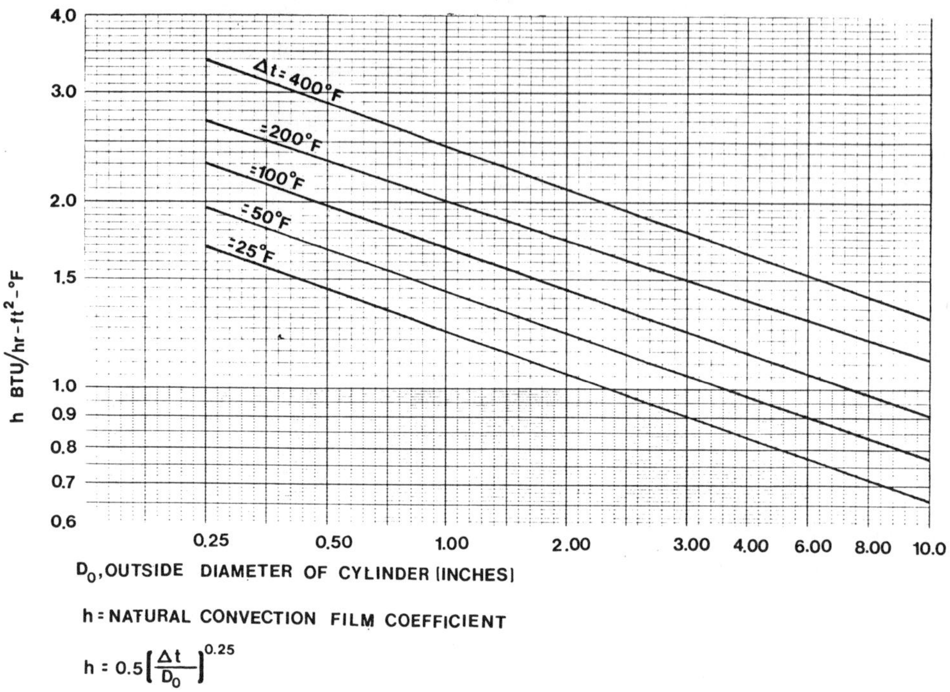

Figure 3-4. Natural convection on horizontal cylinders.

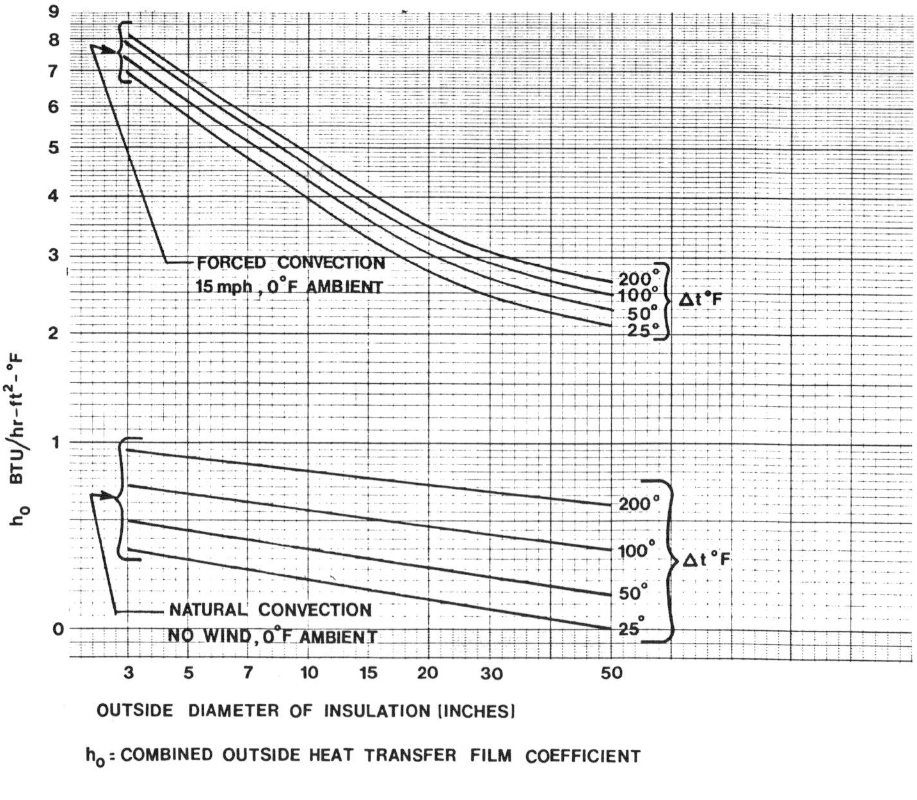

Figure 3-5. Heat transfer outside horizontal pipes.

Q_1 = heat transfer per lineal foot from air space to atmosphere, Btu/hr-ft
Q_2 = heat transfer per lineal foot from tracer to air space, Btu/hr-ft
t_o = temperature of outside air, °F
t_t = temperature of tracer fluid, °F
γ = safety factor; 1.3 for piping systems without HTC, 2.0 for piping systems with HTC, 1.5 for vessels without HTC, 2.5 for vessels with HTC

If $t_a \geqslant t_i$, then the system is adequate. The maximum spacing of tracer tubes for cylindrical vessels is calculated in the same manner except that a flat plate approximation ($T_e = t_i$) is used to compute the heat losses, or Q values.

Traced Piping With Heat Transfer Cement

One mode of heat transfer in a system with heat transfer cement is conduction from the tracer tube through the pipe or vessel wall to the point of the wall most distant from the tracer. The thermal distribution of such a system is shown in Figure 3-6. The other mode of heat transfer is the natural convection from the tracer and the pipe or vessel wall to the air space. Thus, the air space temperature is lower than the minimum process pipe or vessel wall temperature. The contribution of radiation from the tracer and pipe or vessel to the inside wall of the insulation is negligible, as is the film resistance to heat transfer on the inside insulation wall.

The procedure for tracer design with heat transfer cement is as follows:

1. Determine the scheme of tracers to be applied using Figure 3-7.
2. Calculate the metal wall area (equals wall thickness) A_m; the outside surface area of insulation, A_o, the outside surface area of pipe, A_p, and the outside surface area of tracer tube or heat transfer cement (HTC).
3. Assume a value of the minimum pipe wall temperature, t_p, equal to or greater than the minimum process fluid temperature.
4. Assume a value of air space temperature, t_a.
5. Estimate the natural convection coefficient, h_c, from the HTC to air space.
6. Calculate T_e using Equation 3-1.
7. Determine the outside film coefficient of the insulation to the atmosphere, h_o, from Figure 3-5 and calculate U_o from Equation 3-2.
8. Calculate the average pipe wall temperature $\bar{t}_p$ and estimate the natural convection coefficient from the pipe or vessel to air space, h_p, from Figure 3-4.

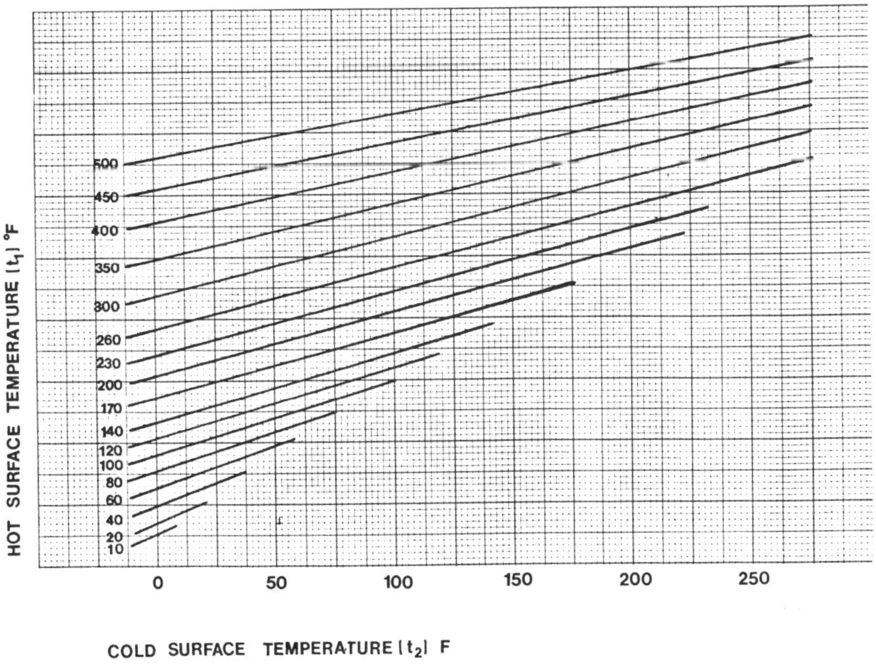

$$h_r = \left(\frac{0.173}{t_1 - t_2}\right)\left[\left(\frac{t_1+460}{100}\right)^4 - \left(\frac{t_2+460}{100}\right)^4\right]$$

Figure 3-6. Heat transfer by radiation.

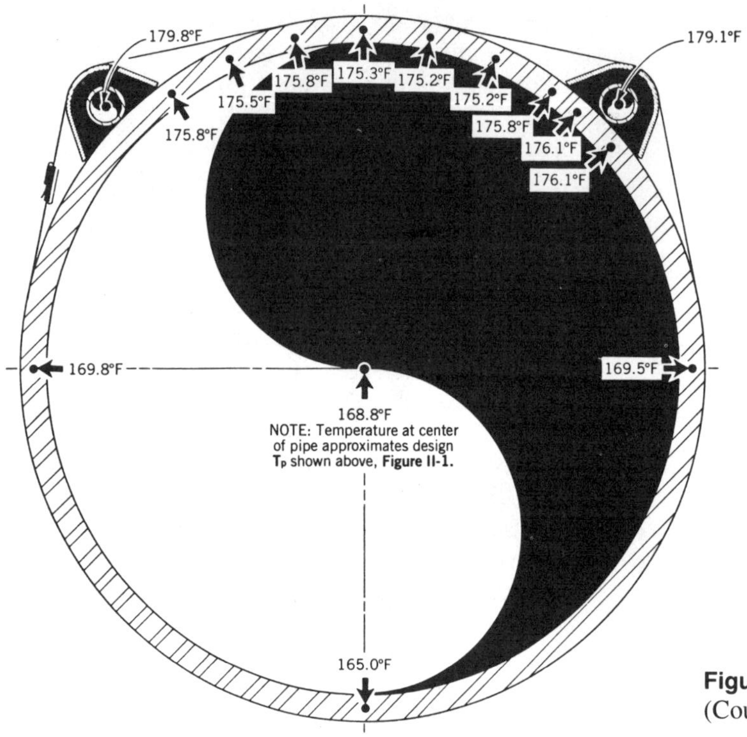

Figure 3-7. Temperature distribution of a two tracer system. (Courtesy of Thermon Manufacturing Co.)

9. Formulate a heat balance around the pipe or vessel wall and air space and perform an iteration analysis solving for t_a and t_p with the following steps:

$$Q_1 = (U_o)(A_o)(t_a - t_o) \tag{3-6}$$
$$Q_2 = (h_t)(A_t)(t_t - t_a) \tag{3-7}$$
$$Q_3 = (h_p)(A_p)(t_p - t_a) \tag{3-8}$$

$$Q_4 = (2)(n_t)\left(\frac{k_m}{t_m}\right) A_m(t_{P1} - t_{P2}) \tag{3-9}$$

$$Q_4 \geqslant (\gamma)(Q_3) \tag{3-10}$$
$$Q_2 + Q_3 \geqslant Q_1 \tag{3-11}$$

where A_m = cross-sectional pipe wall area (equals pipe thickness), ft²/ft
A_p = outside surface area of pipe, ft²/ft
h_p = convection film coefficient from pipe to air space, Btu/hr-ft²-°F
k_m = thermal conductivity of vessel shell material, Btu/hr-ft²-°F
n_t = number of tracers, dimensionless
Q_3 = heat transfer per lineal foot from pipe to air space, Btu/hr-ft
Q_4 = heat transfer per lineal foot from tracer to pipe, Btu/hr-ft
t_a = temperature of air space, °F
t_o = ambient temperature, °F
t_m = length of heat flow through metal, ft
t_{P1} = pipe temperature at point nearest tracer, °F
t_{P2} = pipe temperature at point farthest from tracer, °F

Likewise for traced systems with HTC, for traced vessels, the maximum tracer tube spacing for traced vessels is calculated by the same procedure, except that the flat plate approximation ($T_e = t_i$) is used to compute the heat losses, or Q values.

Condensate Return

Steam differs from hot oil in that condensate is formed by loss of heat energy. During energy shortages, the use of condensate return lines is normally justified. Considering the use of 1/2-in. tracers, normally a 1-in. condensate subheader will handle condensate from 2–8 tracers, a 1 1/2-in. header from 9–20 tracers, and a 2-in. subheader from 21–50 tracers. With a condensate collection and return system the steam supply pressure should be at least 100 psig. Even though these rules of thumb are well tested in field practice, the reader is encouraged to calculate the condensate load for his particular needs. Consider the following analysis:

Total heat loss from steam tracer $= Q_T$
For systems with HTC (by adding Equations 3-7 and 3-9),

$$Q_T = Q_4 + Q_2 \quad (3\text{-}12)$$

For systems without HTC,

$$Q_T = Q_2 \quad (3\text{-}4)$$

The steam in the tracer is assumed to enter the system as saturated steam at an initial temperature or pressure. Considering the amount of heat loss over a given temperature range, the condensate load from n tracers on a given process pipe is

$$\dot{m} = \frac{Q_T}{nh_{fg}}, \text{ lb/hr} \quad (3\text{-}13)$$

where h_{fg} = enthalpy of vaporization (also called latent heat of vaporization), Btu/lb

The steam in the tracer is assumed to transfer energy as heat for a given mass of steam under constant pressure. A typical condensate return system is shown in Figure 3-8.

When collecting condensate, care must be taken to prevent water hammer caused by the mixing of condensate at different temperatures and pressures. To prevent water hammer in condensate systems, spargers and steam separation kegs should be considered.

To size the condensate return lines, as well as the tracers themselves, use the methods presented in Chapter 1 for line sizing. In systems where a large quantity of condensate is formed by steam flashing, a condensate return pump may be required. Normally, condensate return pumps are the horizontal centrifugal type. Pumps and their applications are presented in Chapter 6.

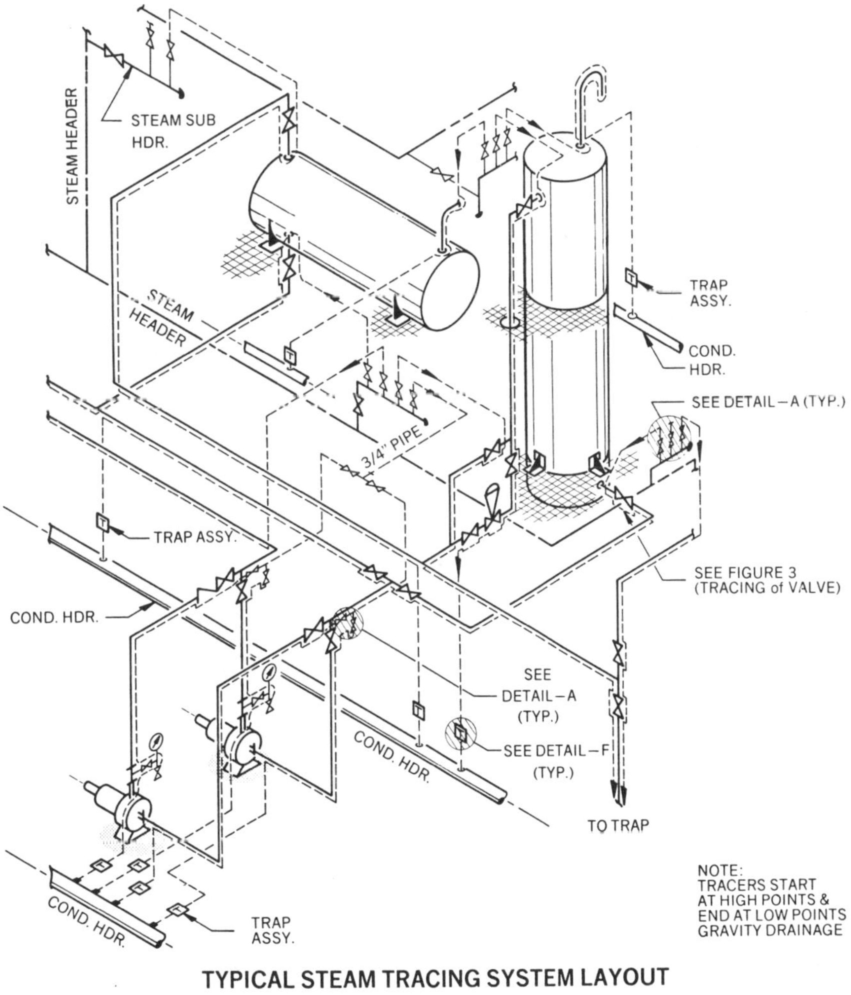

Figure 3-8. Condensate return header in tracer system. (Courtesy of Thermon Manufacturing Co.)

Jacketed Pipe

Figure 3-9 illustrates details of jacketed pipe. For jacketed systems, it is customary to assume a temperature drop over a given length of pipe for hot oil. In applications of hot oil heating a viscous fluid such as asphalt, 100°F drop per 100 ft, or 1°F per foot, is quite common. If one is not familiar with a given service, then a heat balance must be made, like those done for tracers. However, using a temperature drop over a given length of pipe simplifies the analysis and has been proven in practice, because all examples cited are from actual, successful operating systems. The following steps illustrate one such method of designing jacketed pipe:

1. Compute the overall heat transfer coefficient, U, by the following relation:

$$U = \left[\frac{r_4}{h_{1\text{-}2}} + \frac{r_4 \ln (r_3/r_2)}{k_{2\text{-}3}} + \frac{1}{h_{3\text{-}4}}\right]^{-1} \tag{3-14}$$

where $N_{Re} = \dfrac{VD_\rho}{\mu}$ (3-15)

$$h_{1\text{-}2} = 1.86(N_{Re})^{1/3}(N_{Pr})^{1/3}\left(\frac{D}{L}\right)^{1/3}\left(\frac{k}{D}\right)\left(\frac{N}{N_w}\right)^{0.14} \tag{3-16}$$

$$N_{Nu} = 0.020\, N_{Re}^{0.8}\, N_{Pr}^{1/3}\left(\frac{D_2}{D_1}\right)^{0.53} \tag{3-17}$$

for $D_i/D_o > 0.2$

For an annulus, the hydraulic radius, R_H, is

$$R_H = \frac{D_{Li} - d_{so}}{4} \tag{3-18}$$

in which

$$D = 4R_H \tag{3-19}$$

$$N_{Pr} = \frac{NC_p}{k} \tag{3-20}$$

and thus

$$h_{3\text{-}4} = \frac{N_{Nu}k}{D} \tag{3-21}$$

2. After solving for the overall heat transfer coefficient, determine the amount of heat transfer from the relation

$$q = UA(LMTD) \tag{3-22}$$

where A = outside area of inner tube, ft², and the LMTD is based on the assumed rate of heat loss per unit length of pipe.

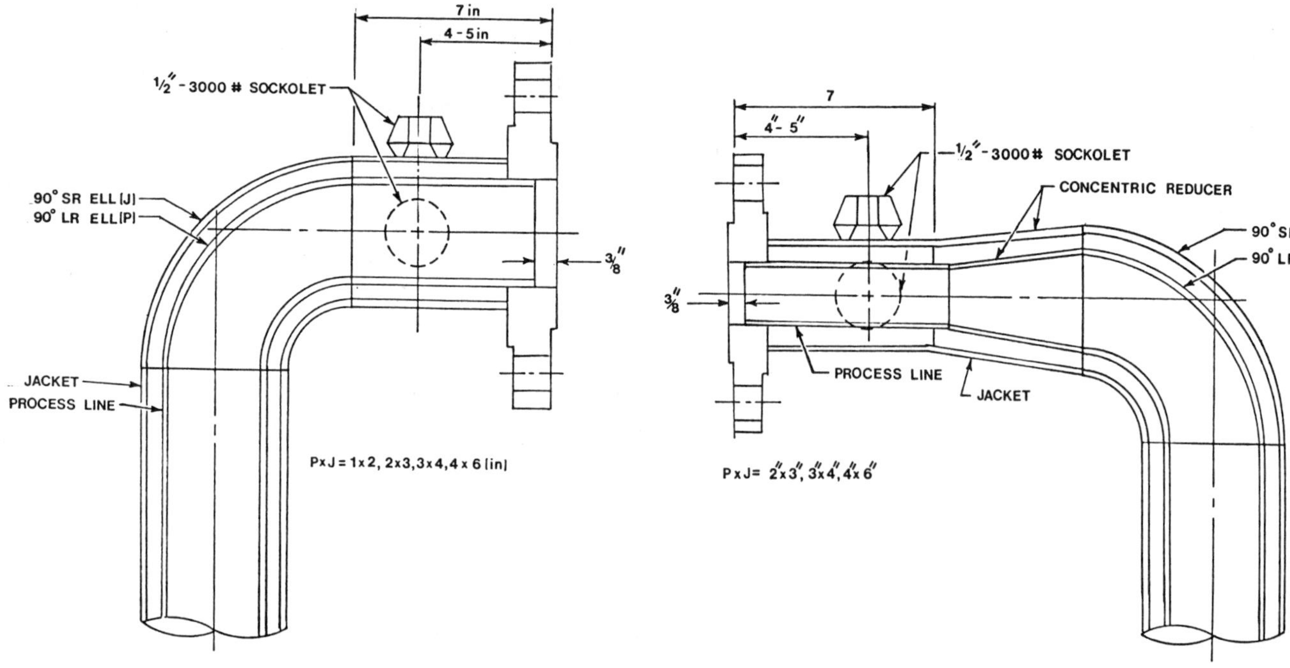

Figure 3-9. Standard fabrication details for jacketed piping.

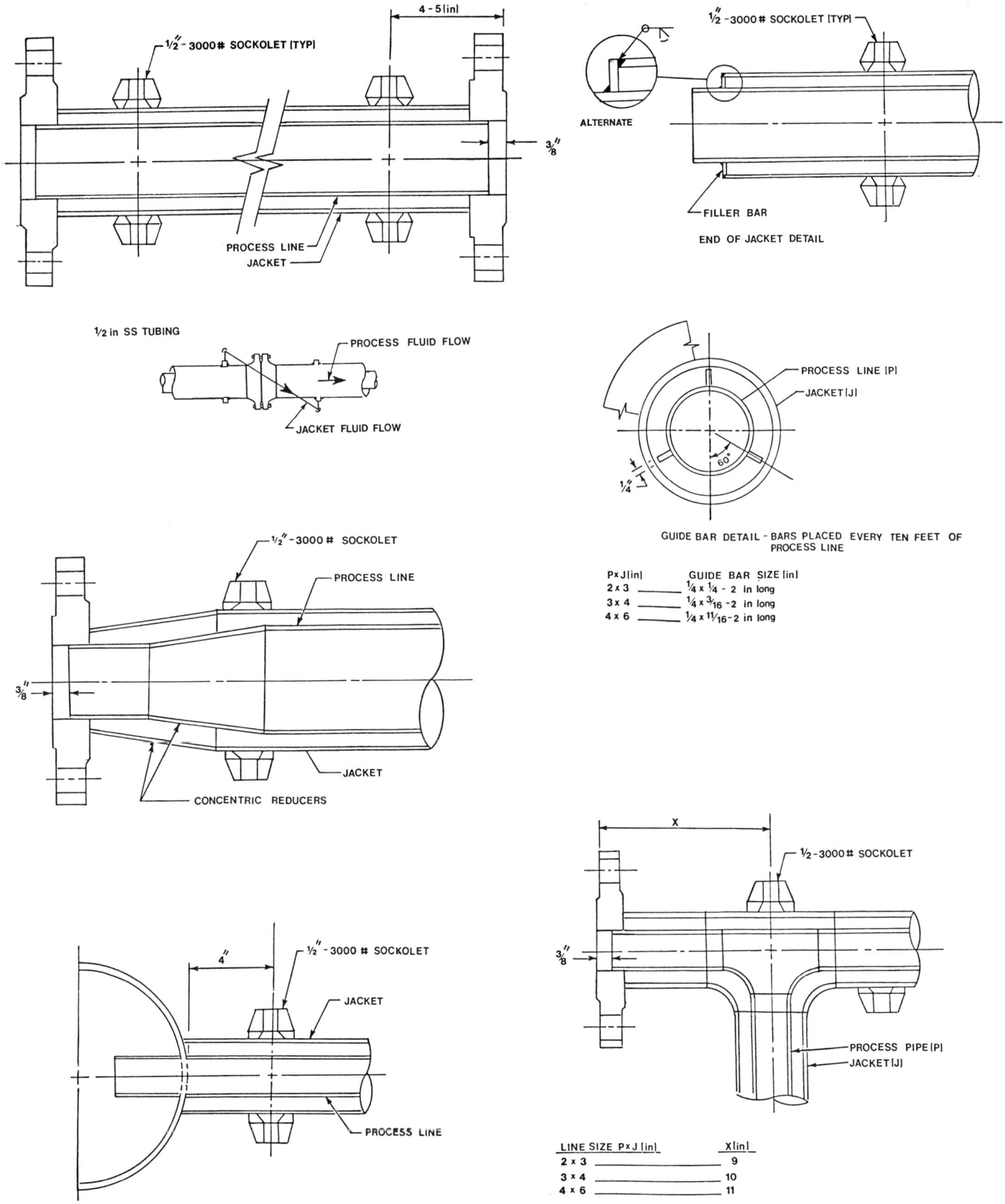

Figure 3-9. Continued.

The LMTD is solved using the following formula:

$$\text{LMTD} = \frac{(\text{GTTD}) - (\text{LTTD})}{\ln\left(\frac{\text{GTTD}}{\text{LTTD}}\right)} \tag{3-23}$$

where LMTD = logarithmic mean temperature difference
GTTD = greater terminal temperature difference
LTTD = lesser terminal temperature difference

To facilitate manual calculations refer to Figure 3-10 [1]. The concept of the logarithmic mean temperature difference is widely explained in most basic engineering textbooks, so its explanation will not be presented here. The reader is referred to Kern [2] or Ludwig [3] for a formal description of the significance of the LMTD.

3. Once the amount of heat transferred is determined from Equation 3-22, assuming a given temperature

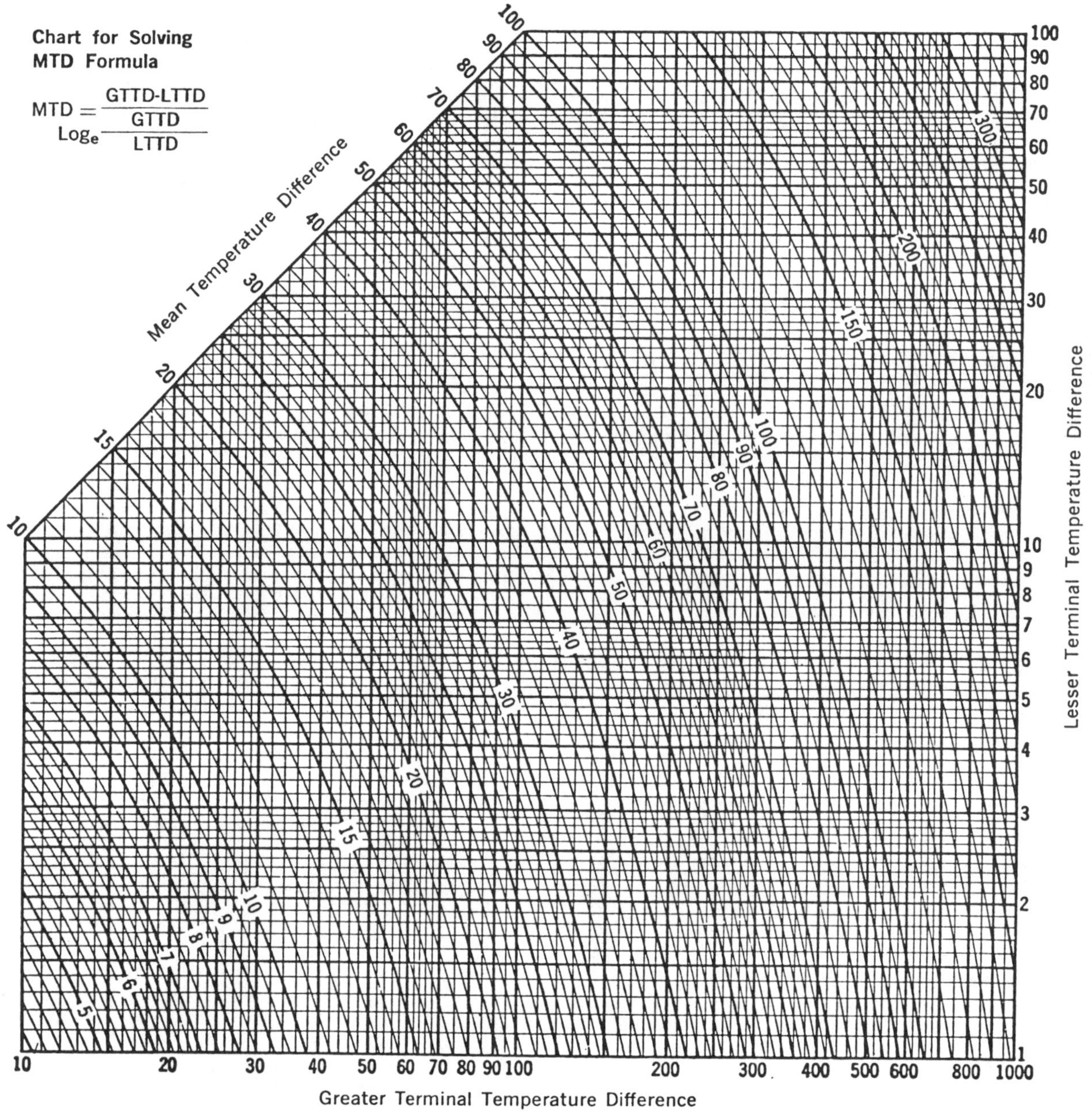

Figure 3-10. LMTD chart. (© 1978 by Tubular Exchanger Manufacturers Association. Reprinted by permission.)

drop, the amount of flow rate of the heating fluid is determined by

$$q = \dot{m}C_p\Delta t \quad (3\text{-}24)$$

where $\dot{m}$ = hot fluid flow rate, lb/hr
C_p = specific heat of hot fluid, Btu/lb-°F
Δt = hot fluid temperature drop

From this formulation we determine the flow rate required.

4. Using Figure 3-11, the amount of pressure drop in the annulus is determined and added to the pressure drop in the whole system (which includes the piping connecting the annuli). The pressure drop for the piping other than the annuli is determined by using the methods presented in Chapter 1. Chapter 6 shows how to select and size the pumps to handle fluids that usually require jacketed services, such as hot oil.

Once the flow rate is determined, the hydraulic analysis made, and the pressure drop judged adequate for the size of pumps selected, the jacketed system details can be designed.

Typical jacketed piping components are depicted in Figure 3-9. In extensively jacketed systems, valves can be procured that have jacketed spaces built in. These types of valves are recommended for services where jacketed pipe is required ($\mu \geqslant 5{,}000$ cp). Some of these valves are shown in Figure 3-9.

Vessel and Equipment Traced Systems

Systems that require piping to be either traced or jacketed likewise require similar components for vessels. The complexity of traced components depends on the viscosity of the process fluids being handled. For high-viscosity, non-Newtonian fluids special items must be added to vessels, such as agitators that are composed of blades and usually powered by electric motors. There are many reasons to use agitators, and one of the most common is to keep suspended particles in a non-Newtonian fluid evenly distributed to prevent particle settlement on the tank bottoms.

There are two basic types of heating and cooling devices used for vessels—internal and external jackets that fit on the inside and outside of the vessel, respectively. These jacket types are shown in Figure 3-12. The hot fluid (normally steam or hot oil) enters one side of the coil and flows through the baffle (inside the tank) or jacket (outside the tank) and exits through another side heating the vessel's contents. This can be seen in Figure 3-12.

Before we analyze in detail these various components, we must first look at the overall heat requirements of the vessel to determine how much heating surface is required. The controlling criterion in determining the amount of heating panel surface area of a vessel is the transient state, i.e., how much surface area is required to heat a given mass of fluid of specified properties to a specified temperature within a specified time. Figure 3-13 illustrates a control mass inside a pressure vessel. Consider two transient boundary conditions in the vessel—the fluid resting at steady state and the fluid moving through the tank at a given mass flow rate. Thus, the following two criteria must be established before the heat transfer area required for a process vessel can be determined:

1. A vessel shown in Figure 3-13(a) contains a static fluid of X gallons at an initial temperature, Y°F. How many degrees of temperature per hour will the fluid mass rise for a given surface area of clamped-on jacketed coils?
2. Using clamped-on jacketed coils shown in Figure 3-13(b), how many degrees of temperature per hour will be transferred to a given mass of fluid of defined properties flowing through the vessel at a constant mass flow rate with an initial temperature of Y°F.

These two criteria are established by considering the following relationships:

$$Q - mC_p\Delta t \quad (3\text{-}25)$$

and

$$Q = UA(LMTD) \quad (3\text{-}26)$$

Equating Equation 3-25 to 3-26, we obtain

$$\Delta t = \frac{UA(LMTD)}{mC_p} \quad (3\text{-}27)$$

The U value, or overall heat transfer coefficient, is calculated on the basis of whether the panel of heat tracing tubes are clamped on outside the vessel or located inside the vessel. These overall heat transfer U values are determined through extensive laboratory tests and accumulated field experience. The U value used in calculations should be that recommended by the heat transfer panel manufacturer, as various panel designs are available and the calculation of the U value analytically can

½" x 1" Sch. 40 Jacketed Pipe

Curves based on 100 feet of jacketed pipe (five 20'-0" lengths) and include five 1" O.D. x .065" wall tubing jump-overs plus entrance and exit losses. Water @ 60°F. (16°C.)

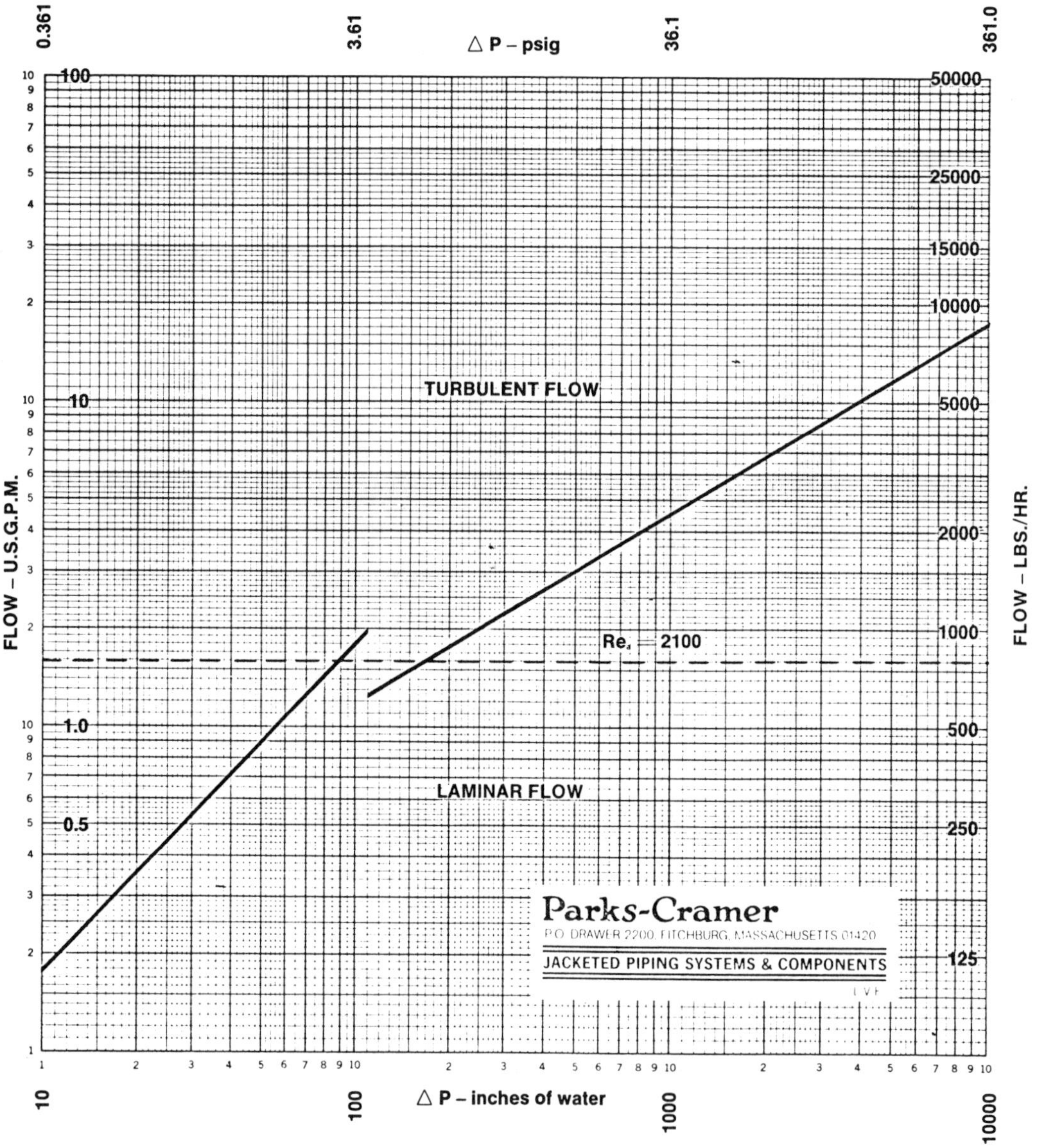

Figure 3-11A. Engineering data pressure drop through jacket. (Courtesy of Parks-Cramer Co.)

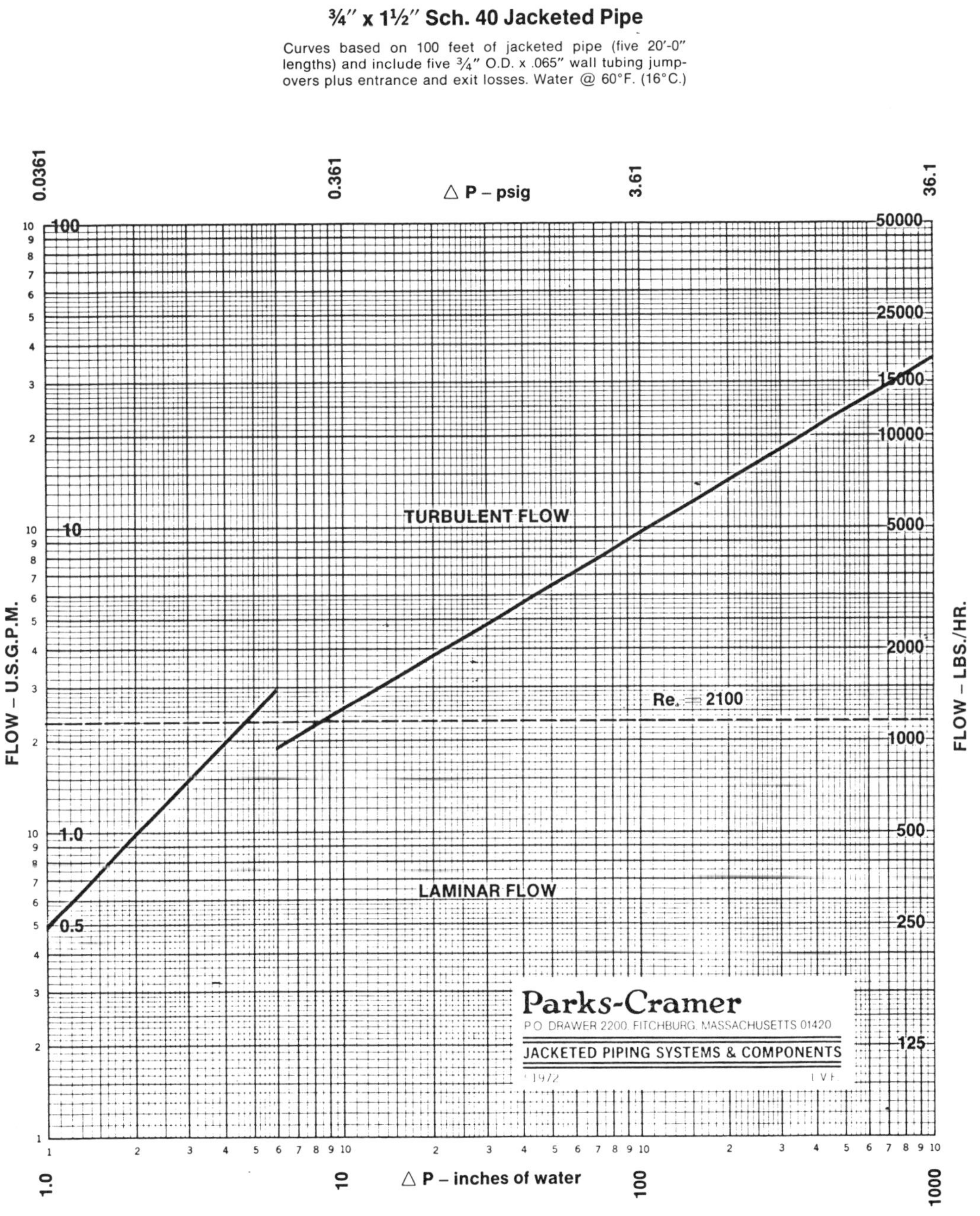

Figure 3-11B. Engineering data pressure drop through jacket. (Courtesy of Parks-Cramer Co.)

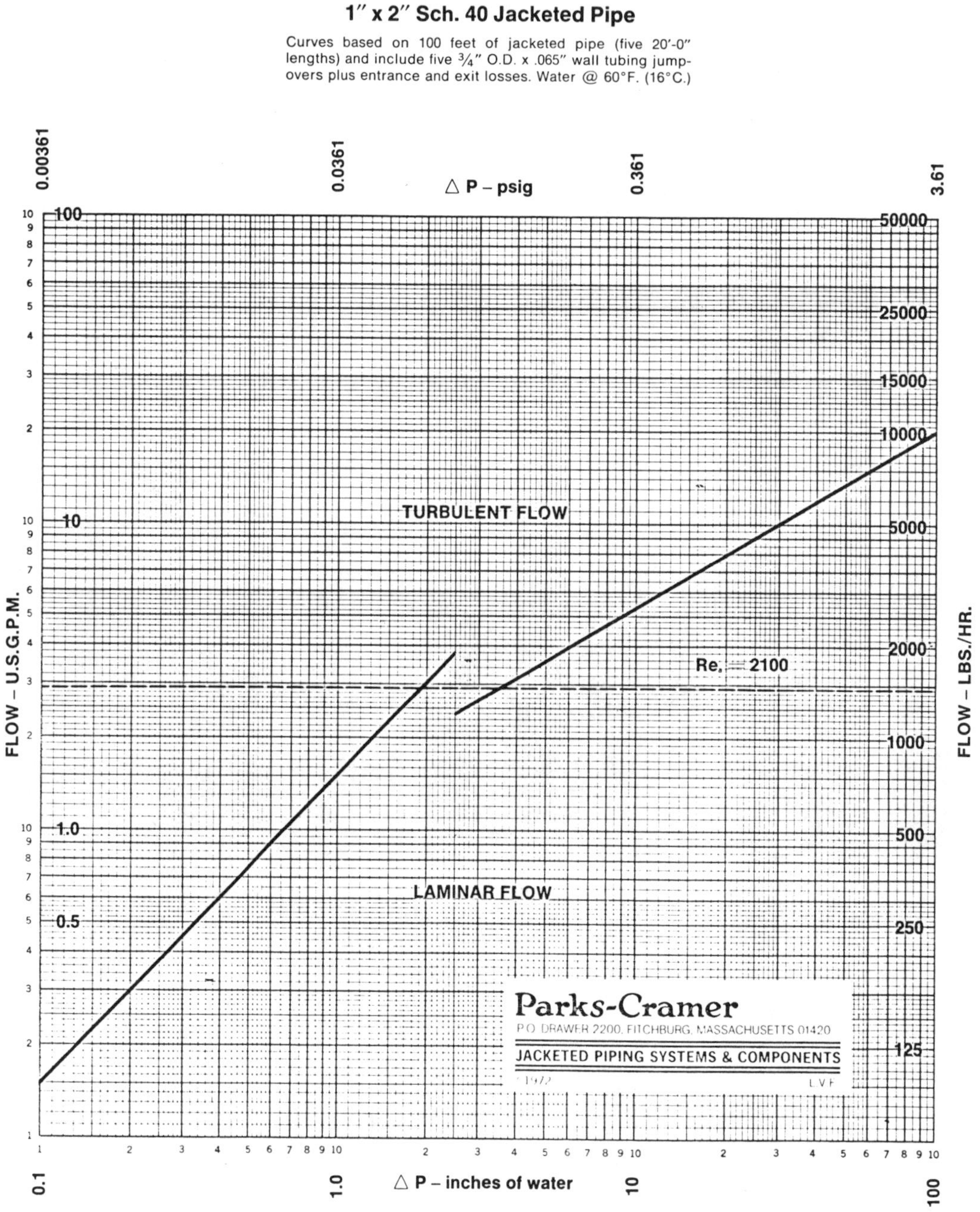

Figure 3-11C. Engineering data pressure drop through jacket. (Courtesy of Parks-Cramer Co.)

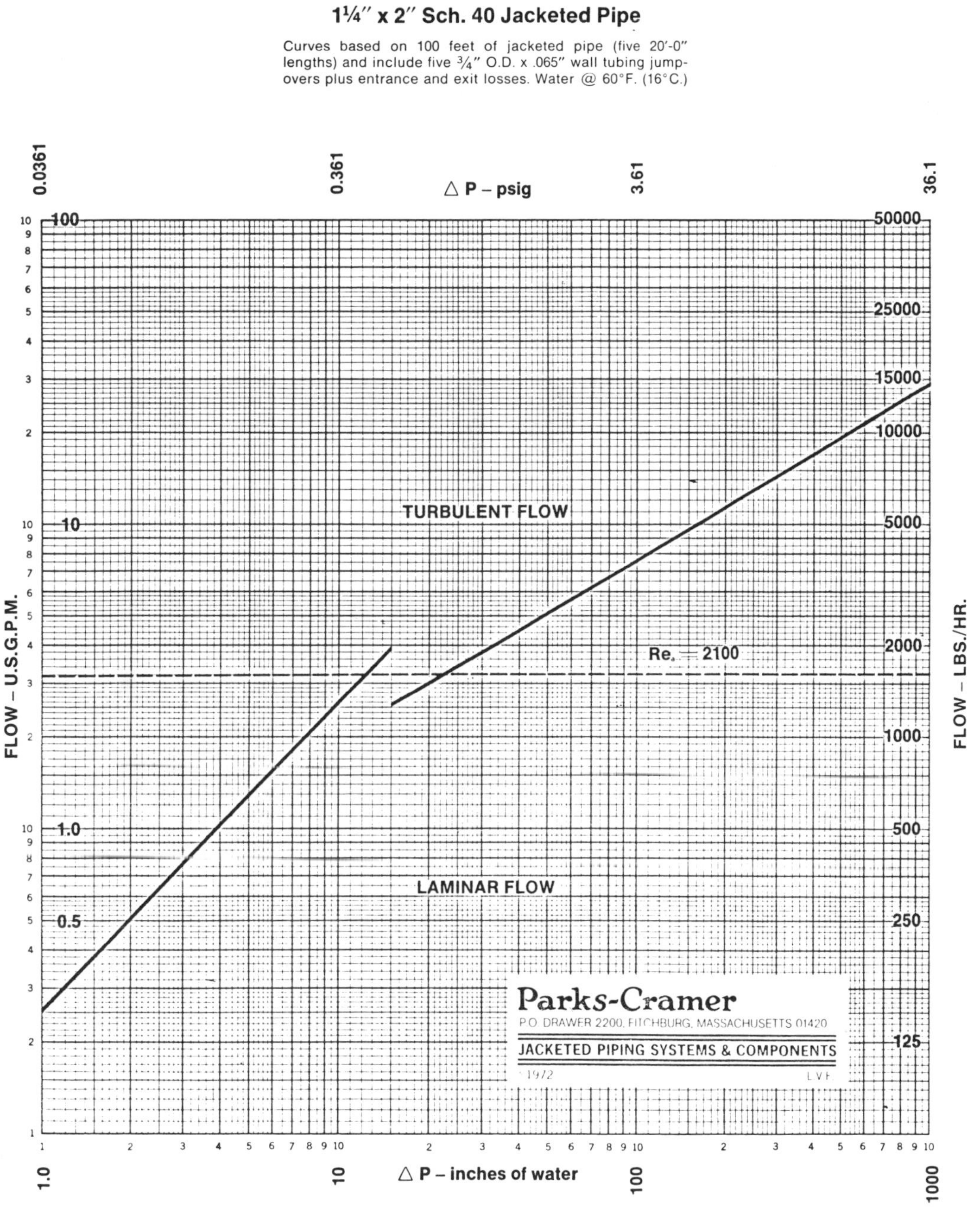

Figure 3-11D. Engineering data pressure drop through jacket. (Courtesy of Parks-Cramer Co.)

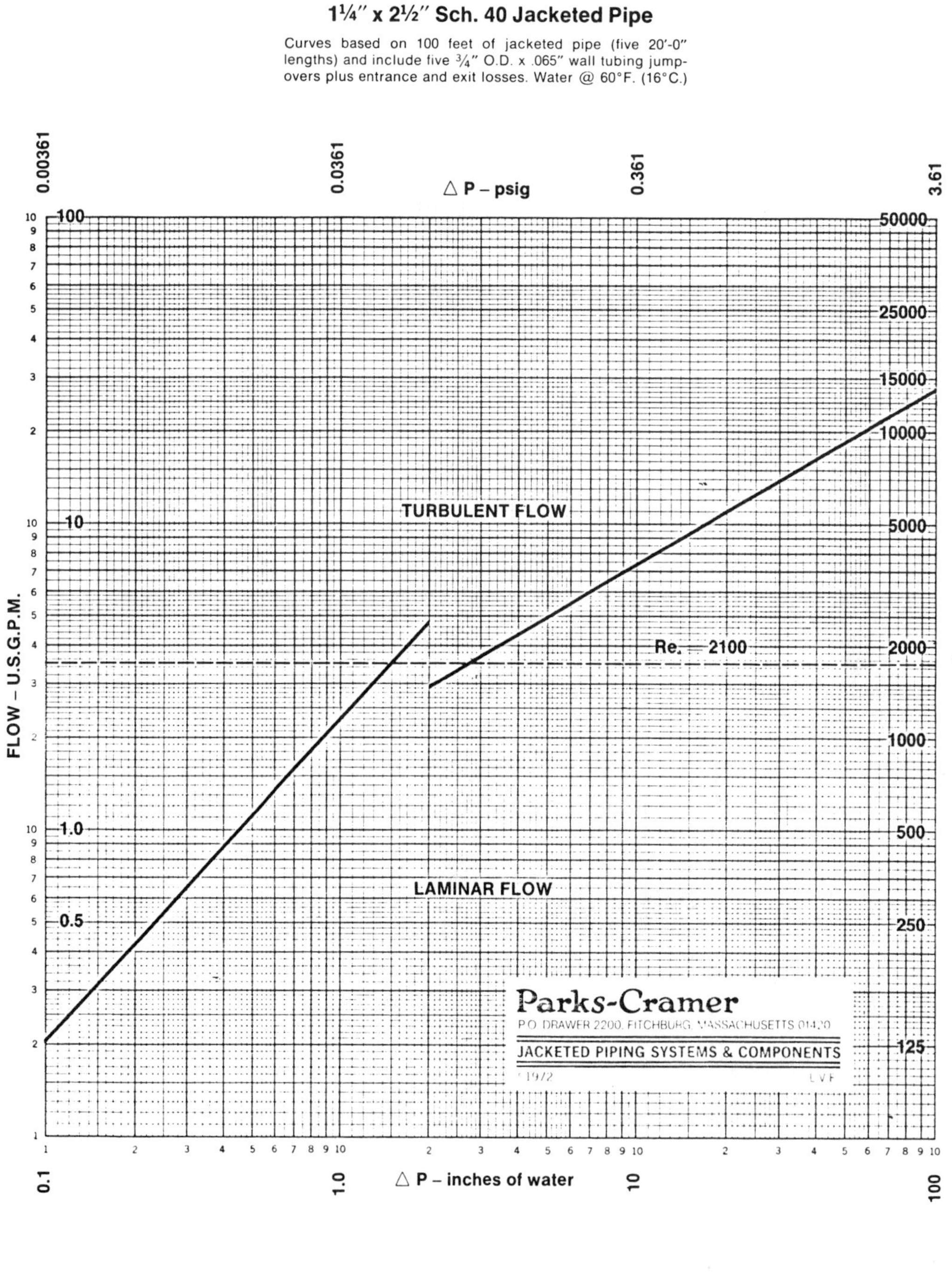

Figure 3-11E. Engineering data pressure drop through jacket. (Courtesy of Parks-Cramer Co.)

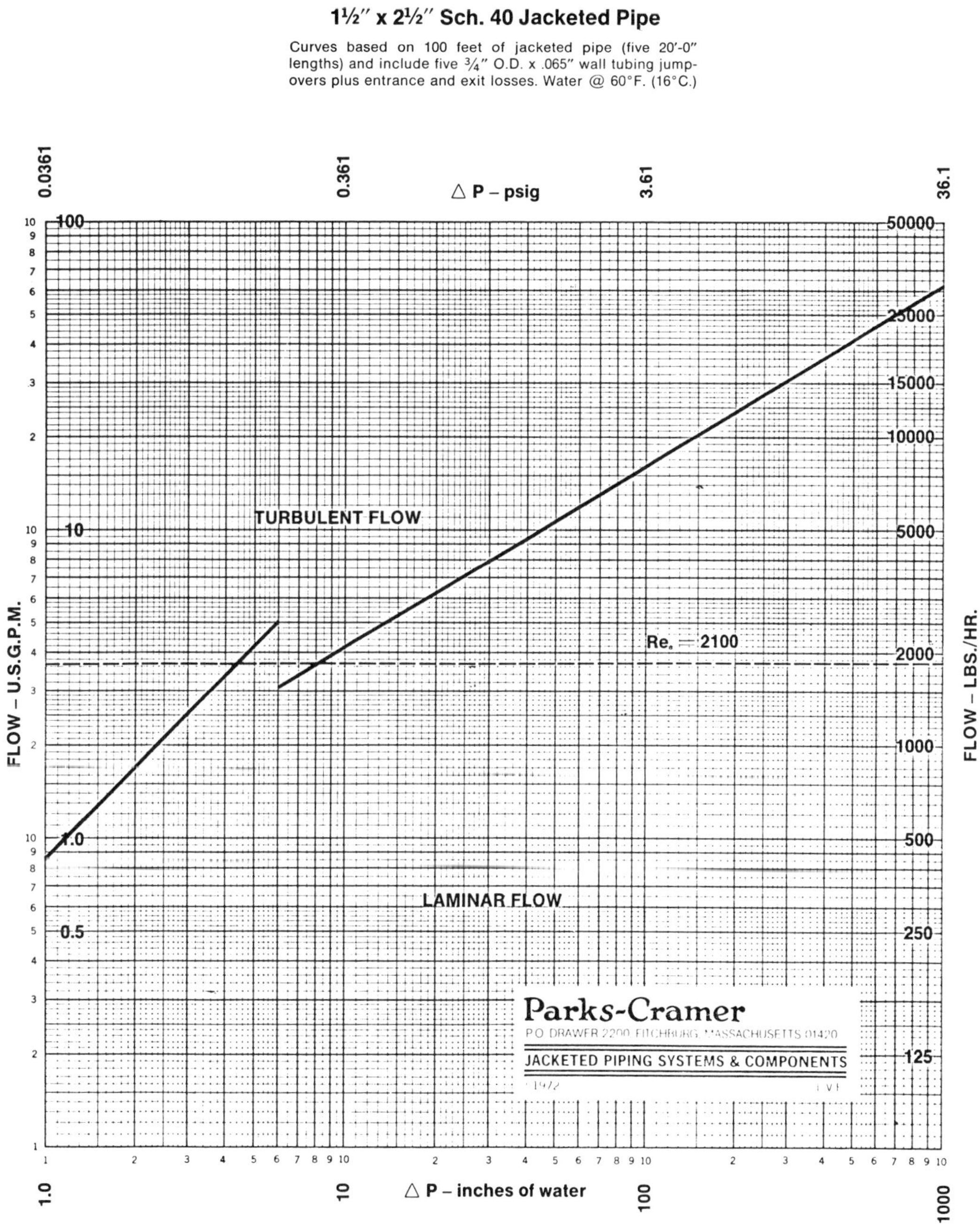

Figure 3-11F. Engineering data pressure drop through jacket. (Courtesy of Parks-Cramer Co.)

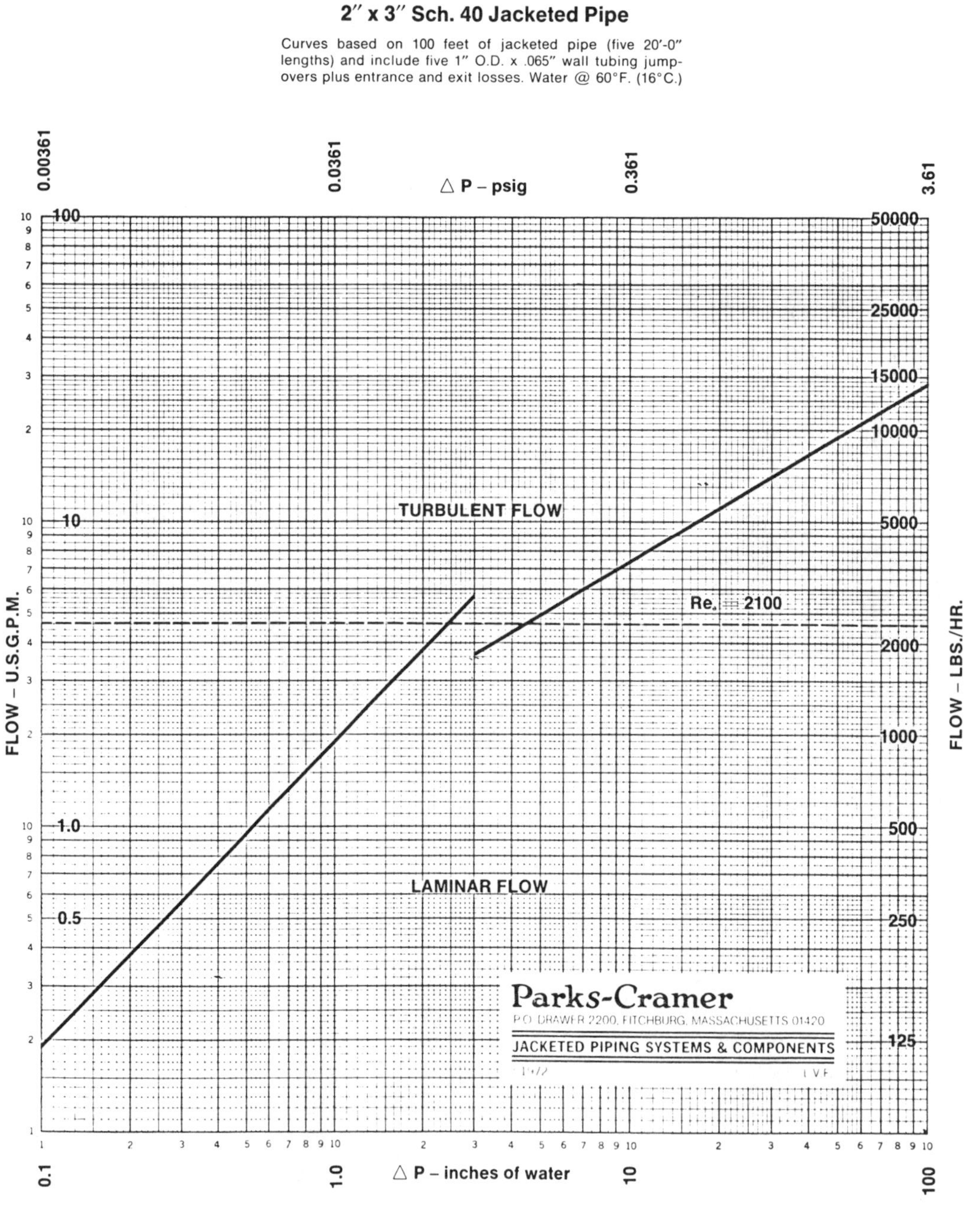

Figure 3-11G. Engineering data pressure drop through jacket. (Courtesy of Parks-Cramer Co.)

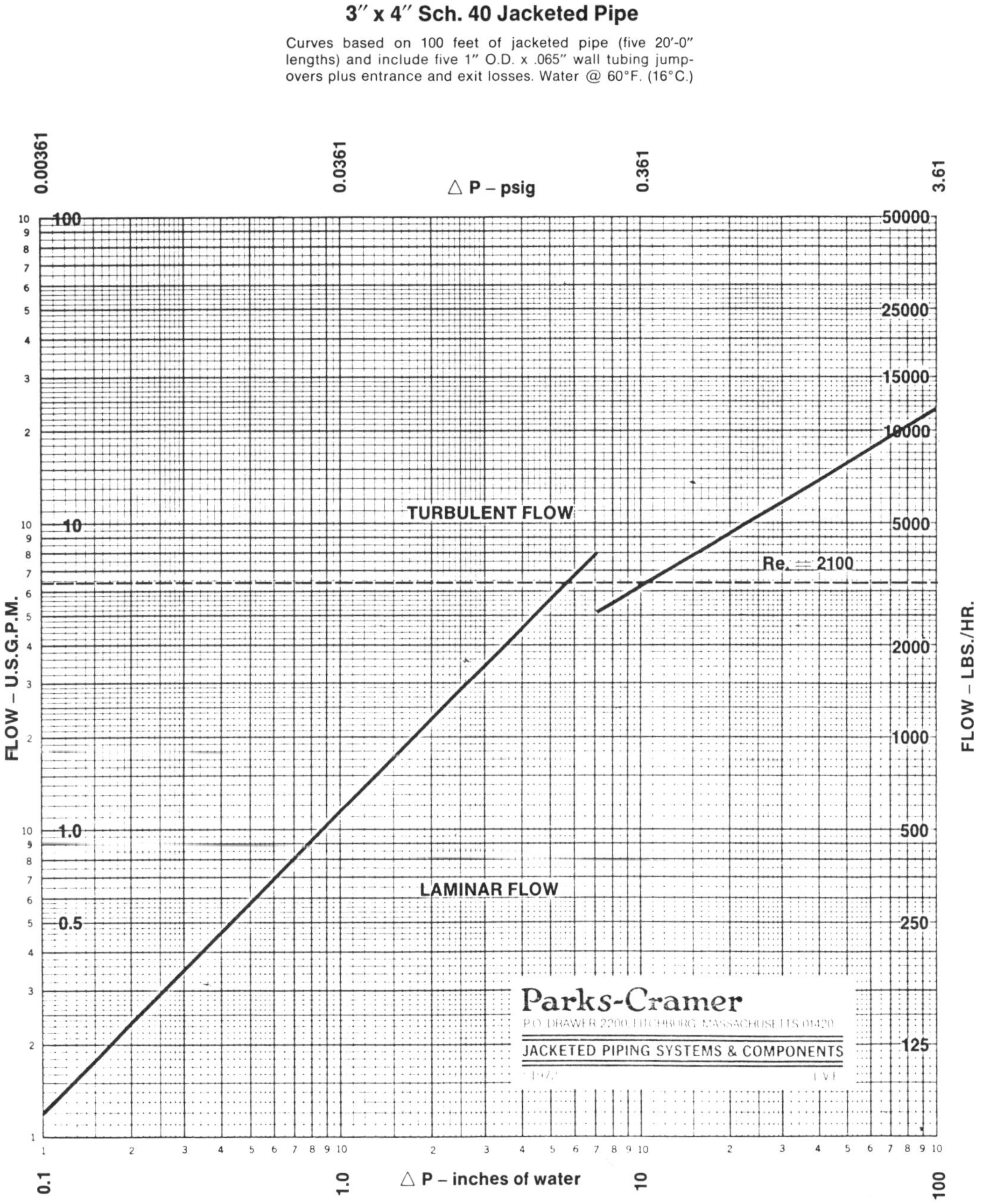

Figure 3-11H. Engineering data pressure drop through jacket. (Courtesy of Parks-Cramer Co.)

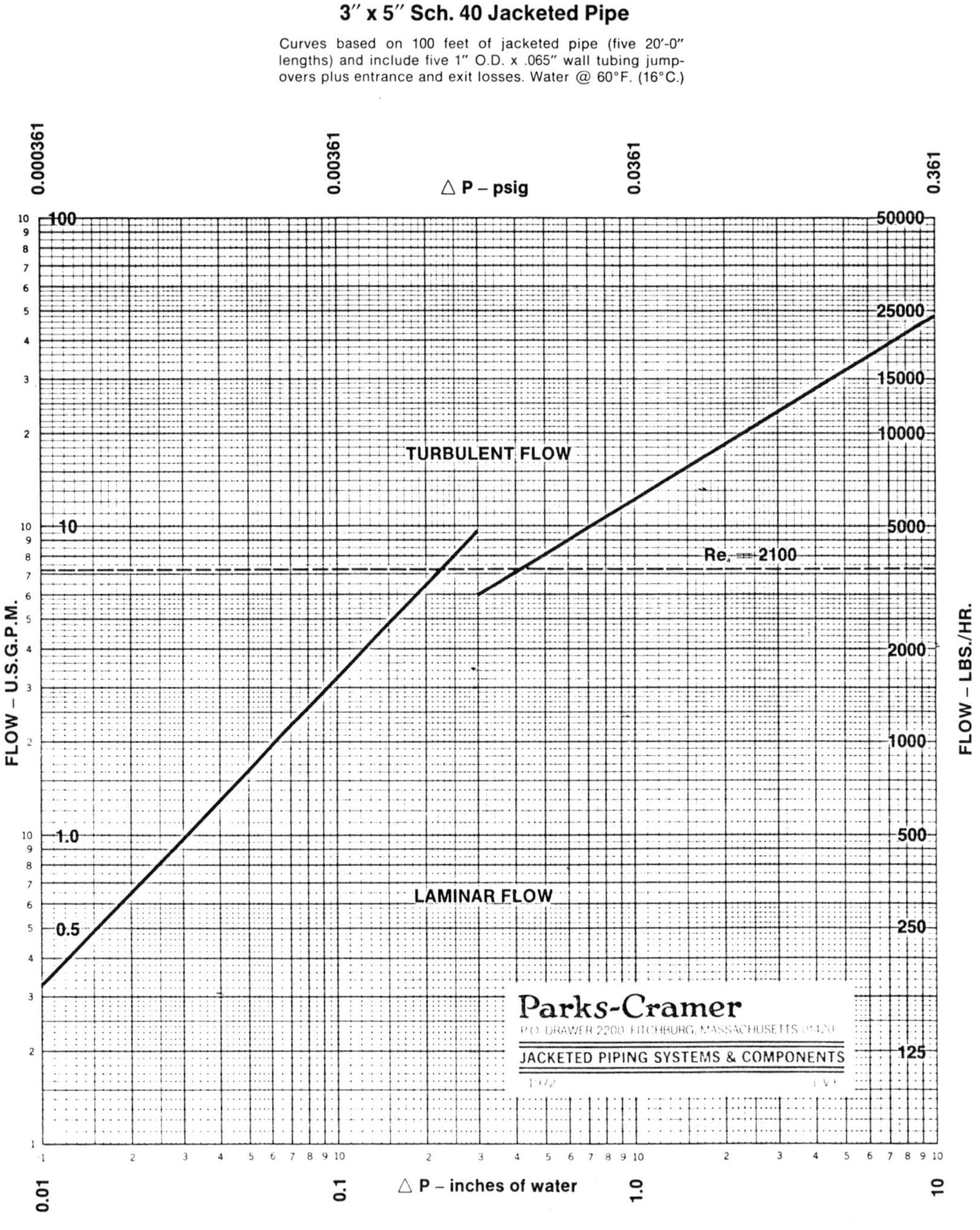

Figure 3-11l. Engineering data pressure drop through jacket. (Courtesy of Parks-Cramer Co.)

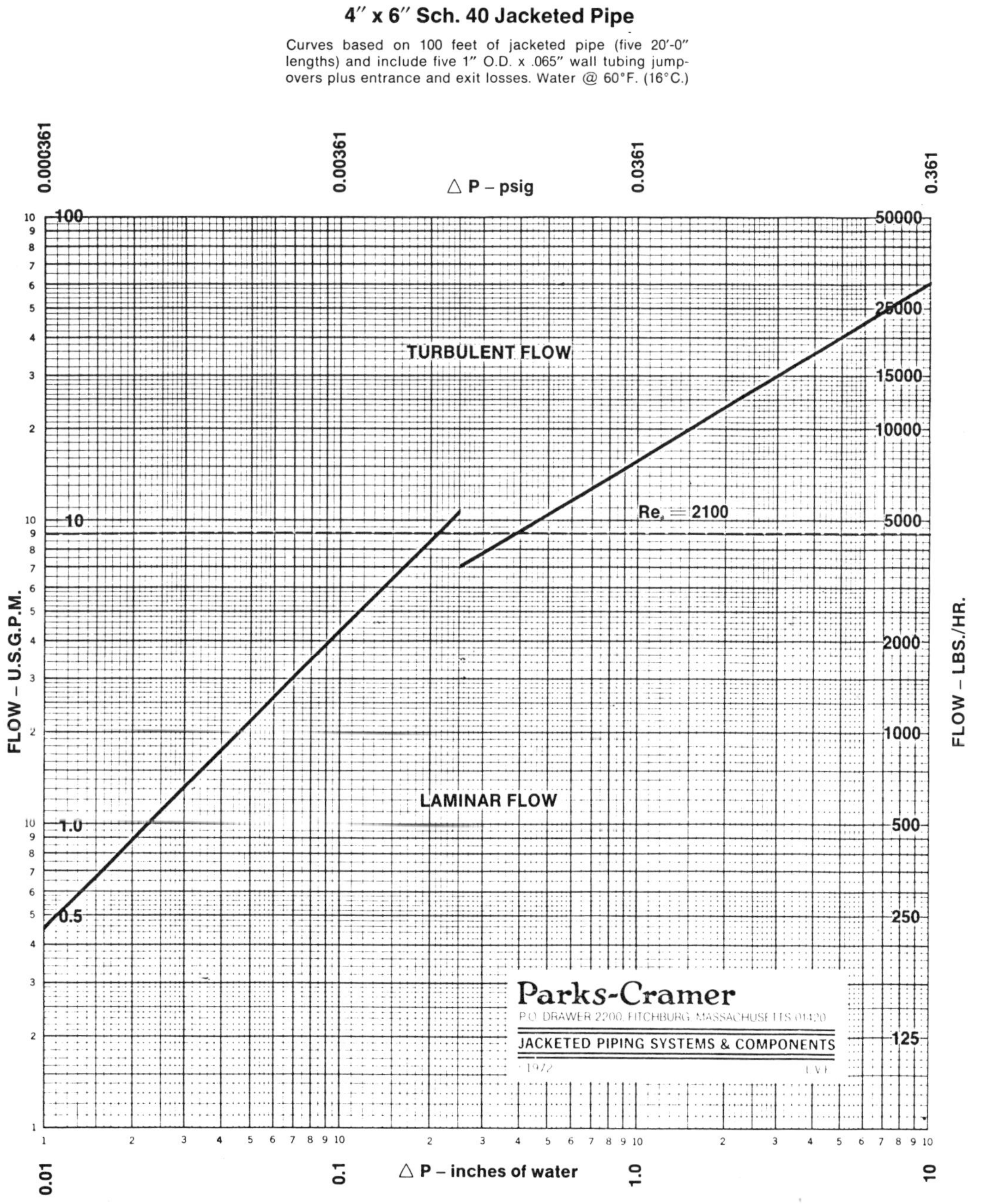

Figure 3-11J. Engineering data pressure drop through jacket. (Courtesy of Parks-Cramer Co.)

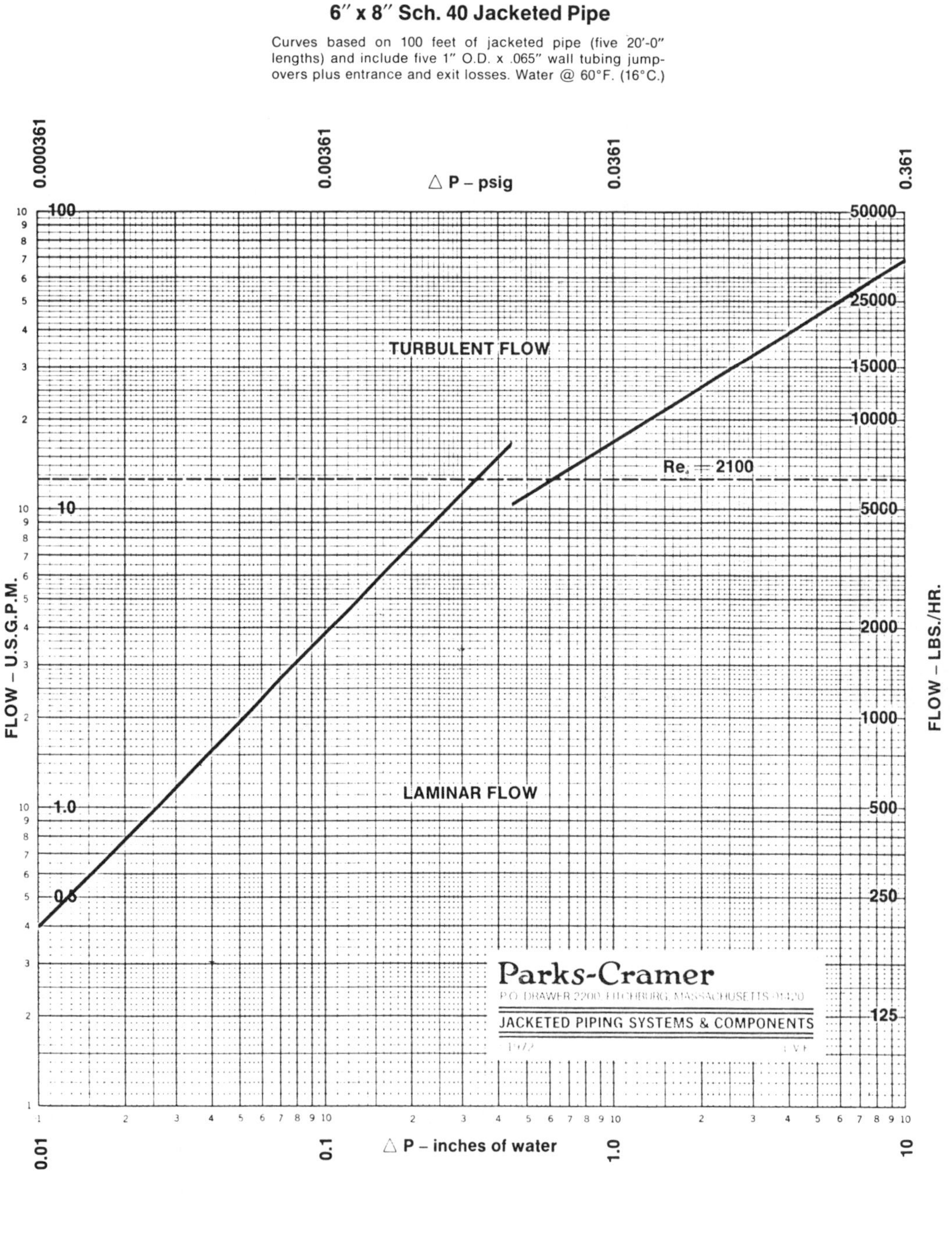

Figure 3-11K. Engineering data pressure drop through jacket. (Courtesy of Parks-Cramer Co.)

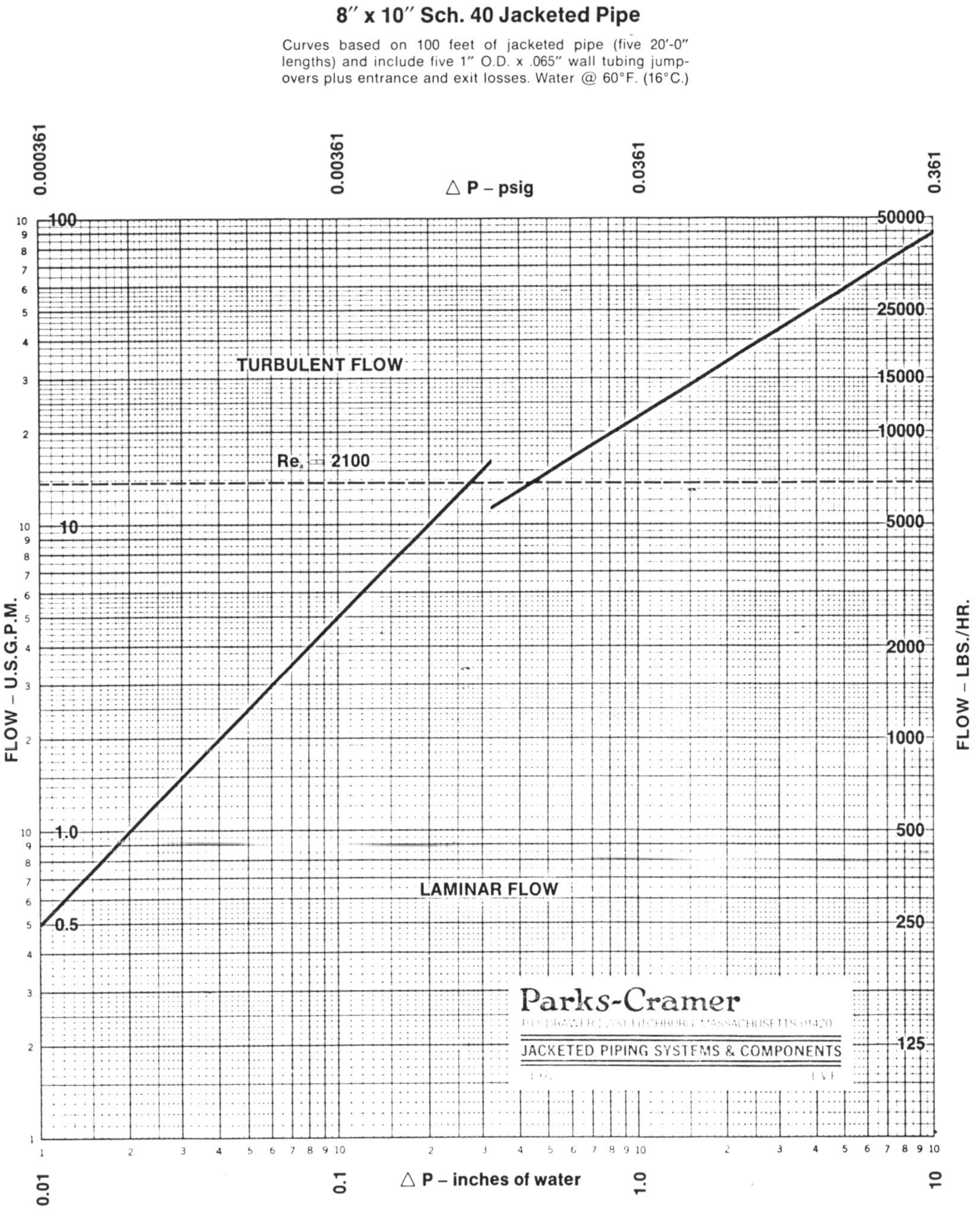

Figure 3-11L. Engineering data pressure drop through jacket. (Courtesy of Parks-Cramer Co.)

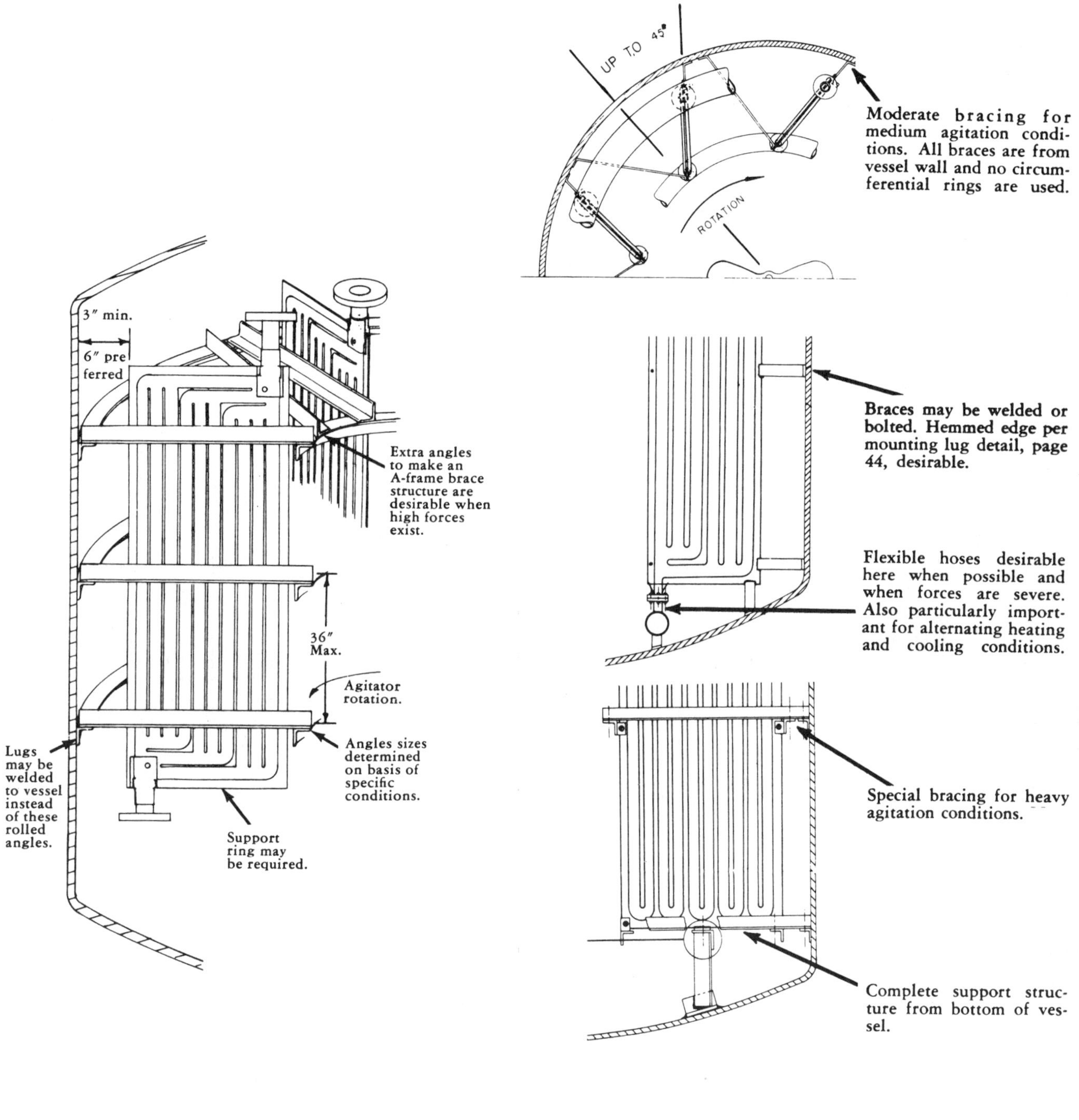

Figure 3-12A. Heat transfer internal plate or panel baffles inside a vessel. (Courtesy of Tranter, Inc.)

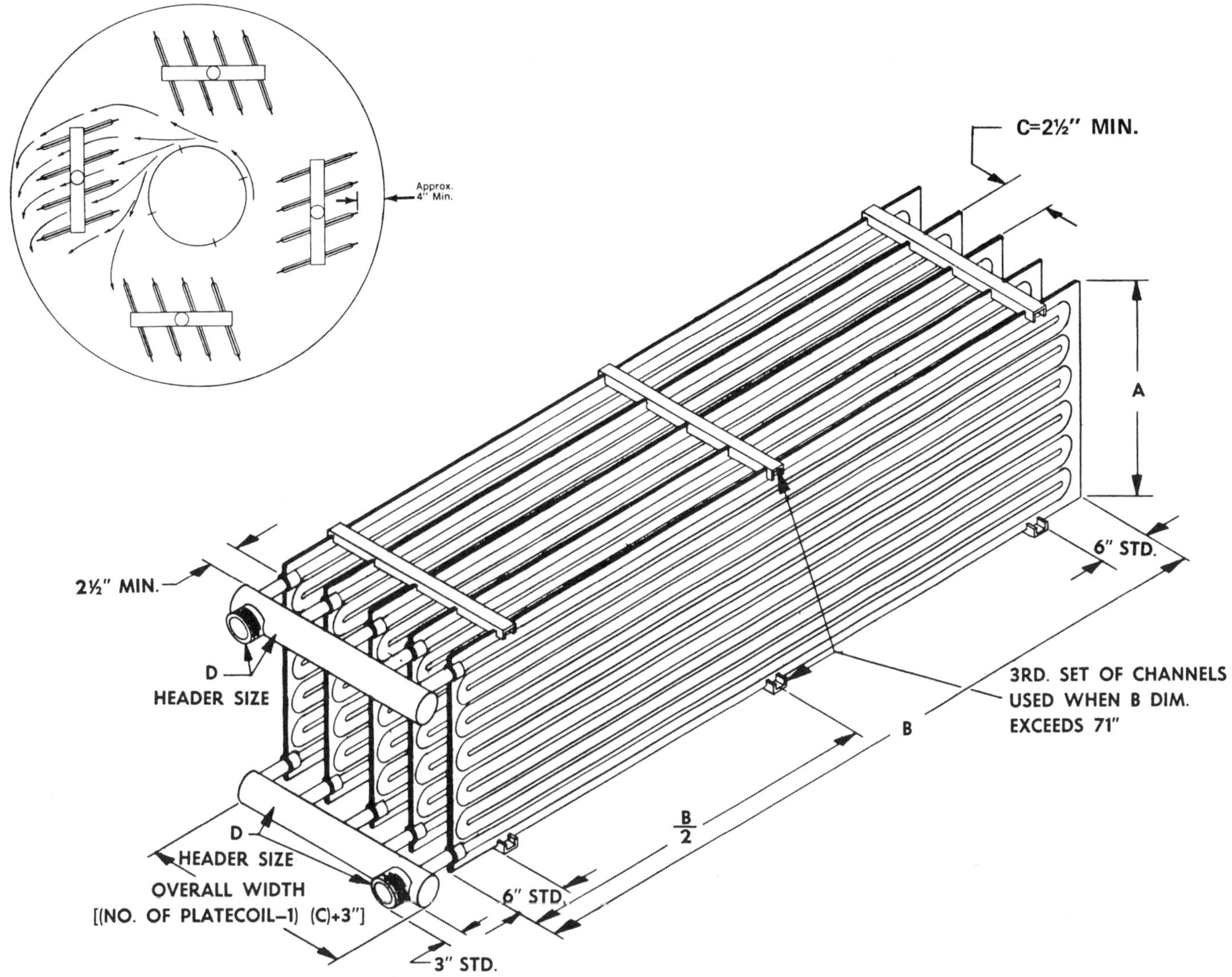

Figure 3-12B. Schematic depicting how heat transfer panel plates heat up or cool down process fluid in tank. (Courtesy of Tranter, Inc.)

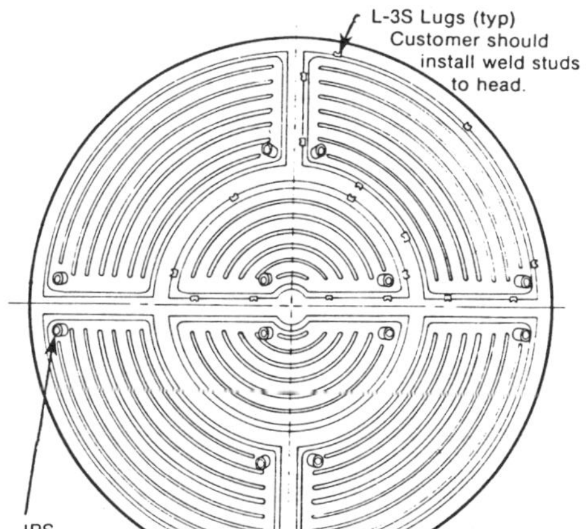

Figure 3-12C. Heat transfer panel plates designed to fit on curved surfaces. (Courtesy of Tranter, Inc.)

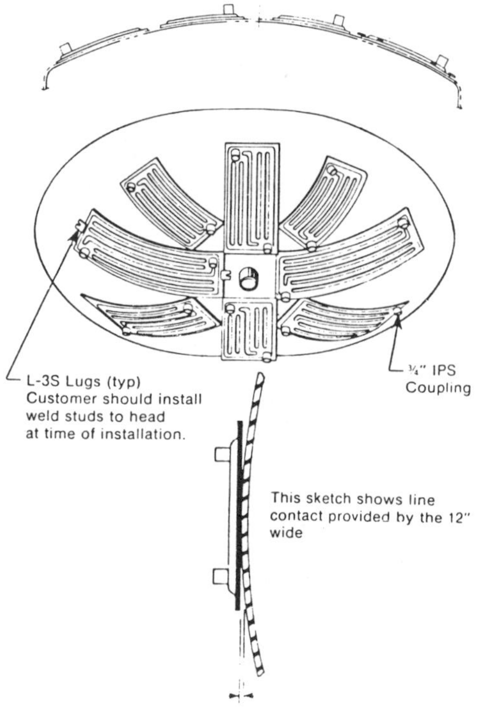

Figure 3-12C (continued). Heat transfer panel plates designed to fit on curved surfaces. (Courtesy of Tranter, Inc.)

lead to erroneous results if the exact details of the panel design are not known. Most panel manufacturers hesitate to reveal such detailed features, so the U values on the low side are recommended for situations where the panel manufacturer does not have a recommended U value. With high-viscosity fluids such as tar and asphalt at temperatures of 300–700°F, a good U value for internal baffle panels is 9.60 and for external clamped-on panels a value of 4.00 is reasonable.

After a U value has been selected, Equation 3-26 is solved, revealing the net temperature change per hour.

The second criterion involves the mass flow rate of the same fluid through the vessel. To establish this criterion, Equations 3-25 and 3-26 are solved together to determine the temperature rise. The analysis of both criteria is graphically illustrated later in Example 3-4.

Once both transient conditions 1 and 2 are satisfied by the selected heat transfer area, the detailed design of the baffle panels (both external clamp-on and internal) can be designed. This is best shown by example and done so in Example 3-5.

Further applications of Equations 3-25 and 3-26 are given in Example 3-6. In this example a material handling problem is analyzed in which both steady state and transient heat transfer conditions are considered. After reviewing Examples 3-5 and 3-6, the reader is encouraged to *always* consider transient conditions of heat transfer in similar situations. Transient criteria, as revealed, usually govern to a large degree.

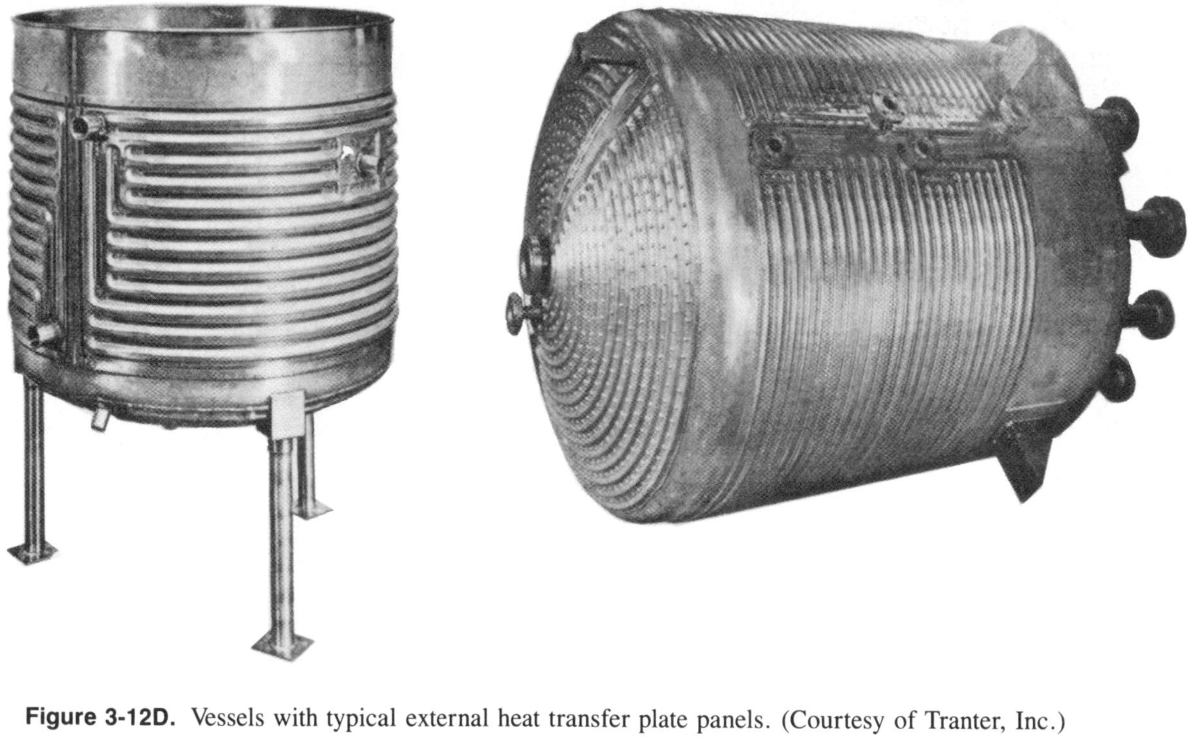

Figure 3-12D. Vessels with typical external heat transfer plate panels. (Courtesy of Tranter, Inc.)

Figure 3-12D. Continued.

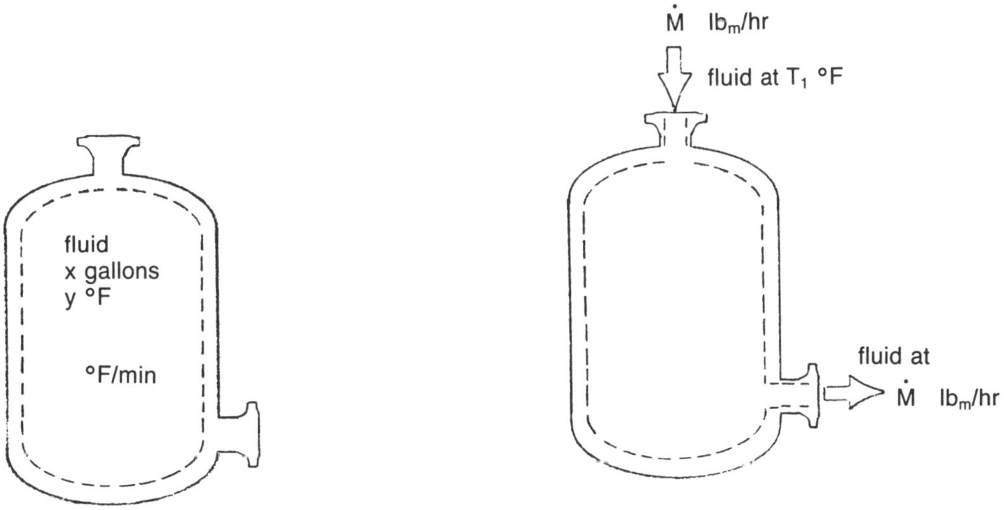

A B

Figure 3-13. Two schemes in which the heat transferred must be considered: (A) control mass scheme; (B) control volume scheme.

HEAT TRANSFER IN RESIDUAL SYSTEMS

Heat Transfer Through Cylindrical Shells

Heat transfer through pipe supports, vessel skirts, and empty branch piping connections to hot or cold headers can cause critical stress problems as well as damage to equipment. Excessive thermal deflections can result in unacceptable loads on rotating equipment and vessel nozzles. In cryogenic service, vessel skirts can fail by brittle fracture if the transition temperature point between alloy steels and carbon steel is not considered.

This section discusses the analysis procedures for analyzing heat transfer in such residual components as vessel skirts and pipe supports. The methods used have been tested with empirical data and have been used for several years in design practice. For derivations to the following method on heat transfer through cylinders, the reader is referred to the author's paper [4].

Vessel skirts are normally insulated on the inside and outside surfaces as depicted in Figure 3-14. In cryogenic applications, there are many reasons why a heat transfer analysis of the skirt is desirable. The primary reason is the one previously cited—to protect carbon steel components from fracture failure. Another reason involves economics—a tall skirt made of alloy steel is much more expensive than a similar skirt made mostly of carbon steel. Also, we will see how the skirt can actually deflect as a result of this heat exchange.

Consider the skirt in Figure 3-14. The vessel is at either an elevated temperature or a cold temperature denoted at the shell-skirt juncture as t_s. Thermal conduction and convection are the controlling modes of heat transfer. The convection can either be considered as natural or free convection, or in the case of wind, forced convection. It has been found that using the free convection coefficient is the most desirable in many cases, since vessels are normally surrounded by other equipment and structures, making free convection more applicable.

Assume that the temperature inside the skirt is the same as ambient temperature and wind chill factors are not present. Air seepage under the skirt and open apertures on the shell allow for equilibrium to be established with the outside temperature.

The first step is to determine the free convection film coefficient for the outside surface of the pressure vessel skirt insulation. In normal conditions, the air temperature inside the vessel skirt, t_i, is assumed five degrees lower than the outside ambient, t_5. The free convection film coefficient is found by iteration using the following equations:

$$U_4 = \left[\frac{r_1 \ln (r_2/r_1)}{k_{2\text{-}1}} + \frac{r_4 \ln (r_3/r_2)}{k_{3\text{-}2}} + \frac{r_4 \ln (r_4/r_3)}{k_{3\text{-}4}} + \frac{1}{h_{4\text{-}5}}\right]^{-1} \quad (3\text{-}28)$$

$$t_4 = (U_4/h_{4\text{-}5})(t_i - t_5) + t_5 \quad (3\text{-}29)$$

$$N_{Gr} = [d^3\gamma^2 gB \mid \Delta t \mid (3{,}600)^2]/\mu^2 \quad (3\text{-}30)$$

$$\Delta t = t_4 - t_5$$

$$N_{Nu} = C(N_{Gr}N_{Pr})^m \quad (3\text{-}31)$$

$$h'_{4\text{-}5} = (k_{air}N_{Nu})/d \quad (3\text{-}32)$$

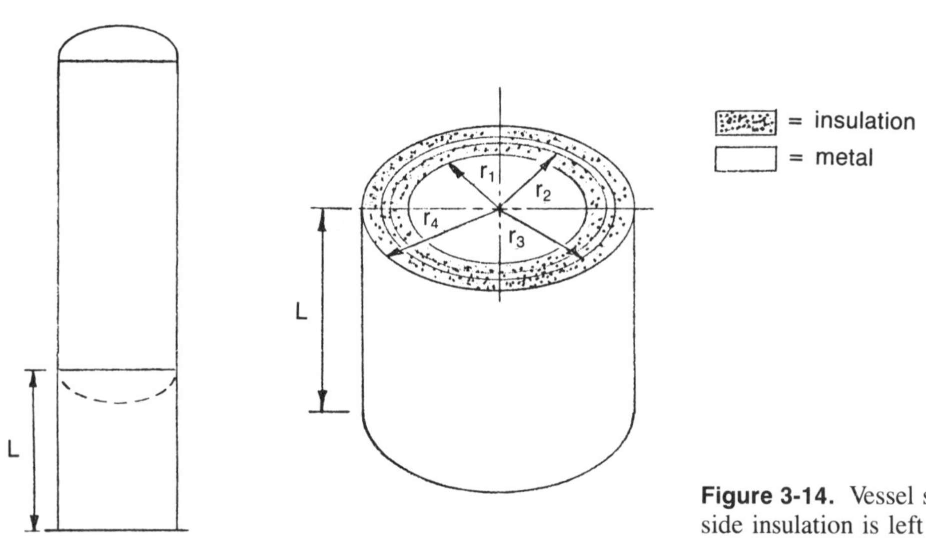

Figure 3-14. Vessel skirt insulation detail. Sometimes the inside insulation is left off.

For free convection of cross flow around cylinders, the following constants hold [5]:

$10^4 \leqslant N_{Gr}N_{Pr} \leqslant 10^9$, C = 0.525, m = 1/4

$10^9 \leqslant N_{Gr}N_{Pr} \leqslant 10^{12}$, C = 0.129, m = 1/3

These relationships are valid for applications for the refining, petrochemical, and gas processing industries.

Now, for a cylinder with insulation on both sides, we use the final value of $h_{4\text{-}5}$ after performing iterations from Equations 3-28 through 3-32 in the following equations:

$$Q = \left(\frac{2\pi k_{2\text{-}1}}{k_m A_m}\right)\left[\frac{1}{|\ln (r_1/r_2)|} + \frac{1}{\ln (r_4/r_3)}\right] \tag{3-33}$$

$$Z = \left|\left(\frac{2\pi}{k_m A_m}\right)\left[r_4 h_{4\text{-}5}(t_4 - t_5) - k_{2\text{-}1}\left[\frac{t_4}{\ln (r_4/r_3)} - \frac{t_i}{\ln (r_1/r_2)}\right]\right]\right| \tag{3-34}$$

$$\bar{Z} = Z/Q \tag{3-35}$$

Substituting these parameters into the following equation, we obtain the temperature distribution down the skirt length:

$$t_x = \frac{2(t_s - \bar{Z})\, e^{xQ^{0.5}}}{1 + e^{2xQ^{0.5}}} + \bar{Z} \tag{3-36}$$

The difference between the process temperature inside the vessel and the outside ambient temperature is the main driving force of heat transfer. It is analogous to electrical EMF driving force or the potential energy of height differential from which a fluid is dropped and turned into kinetic energy.

The degree in significance of convection is inversely proportional to the insulation thickness. The air around the outside insulation surface is in a state of local turbulence and for this reason the variance of the Grashof number down the outside insulation wall is insignificant. Experimental measurements confirm this fact. The reader will see in Examples 3-7 and 3-8 how to apply this method to vessel skirts.

Piping that is supported by piping sections is treated in a similar manner to vessel skirts. Such piping supports are shown in Figure 3-15 in which the pipe supports and branch lines are subject to thermal gradients from a hot or cold process header. Figure 3-15a shows a stub piece used as a piping header support. The temperature gradient through the stub piece must be analyzed to determine if the Teflon slide beneath the base plate will be protected from the elevated temperature inside the process header. If the process header is in cryogenic service, the stub piece must be analyzed to assure the design engineer that the carbon steel structural members are adequately protected from temperatures below the transition temperature.

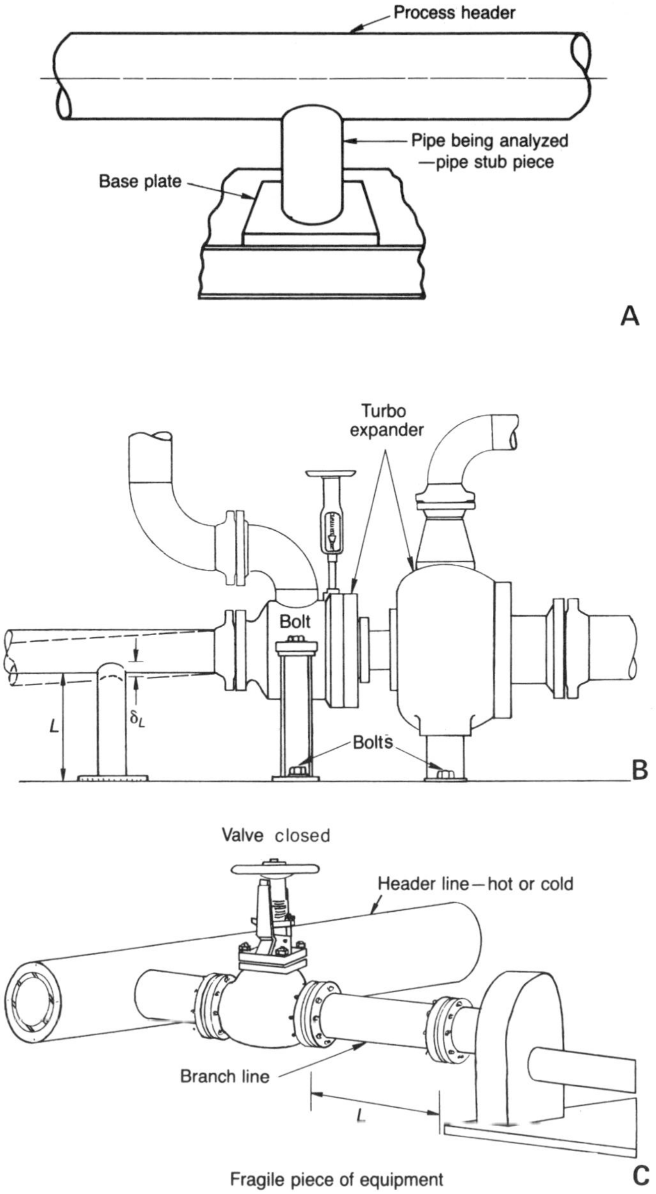

Figure 3-15. (A) Stub piece used as header support; (B) process line is connected to a turboexpander. The line is supported by a short section of pipe welded to a base plate; (C) branch line from a header (hot or cold) connected through a shut-off valve to a fragile piece of process equipment.

Shown in Figure 3-15b is a common situation in which a process line connected to a turboexpander is supported by a section of pipe welded to a base plate. If the pipe stub deflects enough (shown by δ_L), the thermal deflection could induce a sufficient bending moment on the turbine to cause serious mechanical damage.

Figure 3-15c shows a branch line running from a hot or cold pipe header to a fragile piece of equipment. Even though the valve on the branch line is closed, the residual temperature distribution through the branch line may be enough to cause the pipe to deflect and damage the equipment.

Referring to Figure 3-16, the procedure for determining the temperature distribution through the empty branch pipe or pipe support is similar to the case of a vessel skirt. First, solve for the free convection film coefficient on the exterior surface of the pipe insulation. To do this, use the equation for the overall heat transfer coefficient:

$$U_3 = \left(\frac{r_3 \ln \left(\frac{r_2}{r_1}\right)}{k_m} + \frac{r_3 \ln \left(\frac{r_3}{r_2}\right)}{k_i} + \frac{1}{h_o} \right)^{-1} \quad (3\text{-}37)$$

$$t_3 = \left(\frac{U_3}{h_o}\right)(t_i - t_o) + t_o \quad (3\text{-}38)$$

$$\Delta t = t_3 - t_o$$

$$N_{Gr} = [d^3\gamma^2 g\beta(|\Delta t|)(3{,}600)^2]/\mu^2 \quad (3\text{-}30)$$

$$N_{Nu} = C(N_{Gr}N_{Pr})^m \quad (3\text{-}31)$$

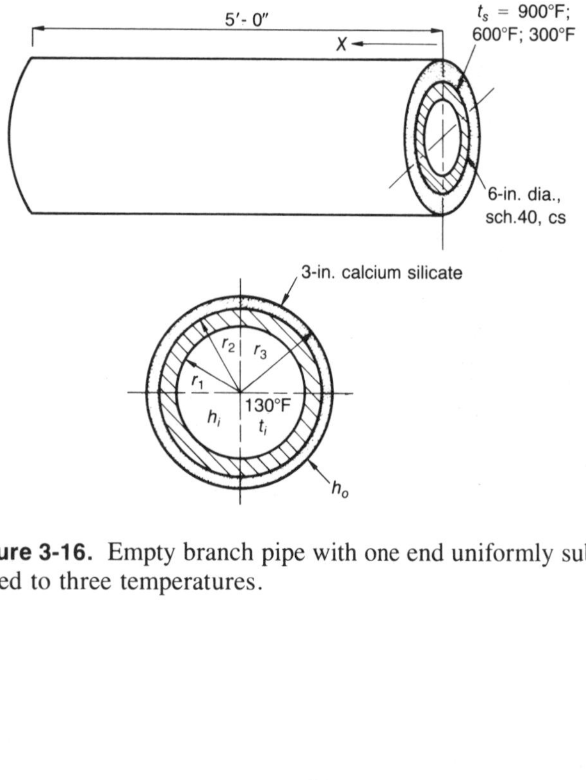

Figure 3-16. Empty branch pipe with one end uniformly subjected to three temperatures.

where c and m are determined as previously for skirts

$$h_o' = (k_{air}N_{Nu})/d \quad (3\text{-}32)$$

$$t_3' = (U_3/h_o')(t_i - t_o) + t_o \quad (3\text{-}29)$$

$$\Delta t_3' = t_3 - t_3' \leqslant 2°F \quad (3\text{-}39)$$

Once $\Delta t_3'$ criterion is met, we can proceed with the final iterative value for the film coefficient, h_o. With this final value, we solve for the parameters Q, Z, and $\overline{Z}$ as follows:

$$Q = \frac{2\pi k_i}{k_m A_m \ln\left(\frac{r_3}{r_2}\right)} \quad (3\text{-}40)$$

$$Z = \left(\frac{2\pi}{k_m A_m}\right)\left[r_3 h_o(t_3 - t_o) - \frac{k_i t_3}{\ln\left(\frac{r_3}{r_2}\right)}\right] \quad (3\text{-}41)$$

$$\overline{Z} = \frac{Z}{Q}$$

Once Q and $\overline{Z}$ are known, we solve for the temperature distribution with

$$t_x = \frac{2(t_s - \overline{Z})e^{xQ^{0.5}}}{1 + e^{2xQ^{0.5}}} + \overline{Z} \quad (3\text{-}36)$$

You will notice that the form of the final solution, Equation 3-36, is the same for the skirt problem with insulation on the inside and outside shell surfaces as the pipe problem with insulation on only the outside surface. The difference in the solutions is because of the boundary conditions, i.e., a cylinder with insulation on both inside and outside surfaces versus a cylinder with just insulation on the outside surface alone. The solutions to the basic differential equations are affected by these differences in boundary conditions. For further information on this subject, the reader is referred to the author's paper [4].

For cases of tapered vessel skirts the cylinder section can be approximated by using an average diameter. This approximation is very close to actual results because skirts should not taper more than 15° (see Chapter 4).

As a consequence of heat transfer along vessel skirts and pipe connections, thermal deflections will occur. The deflection equations are the same regardless of what case is considered, whether it is a shell with insulation on the inside and outside surfaces or a shell with only external insulation. The values of Q and $\overline{Z}$ are determined from the appropriate equations of each respective case.

The thermal deflection equations are dependent on the type of material considered since the coefficient of thermal conductivity is the governing property of the particular material being considered. Taking a differential element of a shell, we solve for the amount of thermal deflection by

$$dL = \alpha(t)t(x)\,dx \qquad (3\text{-}42)$$

Since the temperature varies over the shell length, we integrate Equation 3-42 to obtain the total deflection, δ, as

$$\delta = \int dL = \int_0^L \alpha(t)t(x)\,dx \qquad (3\text{-}43)$$

The function, $\alpha(t)$, is the coefficient of thermal expansion for the particular material being considered. Values of the thermal expansion were curve fitted over a large range of temperature and a relation in terms of temperature was obtained for various materials. The function for t(x) is obtained from Equation 3-36 and is substituted with $\alpha(t)$ in Equation 3-43. Then, the product of $\alpha(t)t(x)$ is integrated over a length L and we obtain the thermal deflection function for each particular material. For carbon steel, the expanded thermal deflection equation is as follows:

$$\delta_{cs} = \frac{2[5.89 + (2.496 \times 10^{-3})2\bar{Z}](t_s - \bar{Z}) \arctan(e^{LQ^{0.5}})}{10^6 Q^{0.5}}$$
$$+ \frac{(2.496 \times 10^{-3})(t_s - \bar{Z})^2(e^{2LQ^{0.5}} - 1)}{(10^6)Q^{0.5}(1 + e^{2LQ^{0.5}})}$$
$$+ \frac{(2.496 \times 10^{-3})\bar{Z}^2 L}{10^6}$$
$$- \frac{(6.536 \times 10^{-7})}{(10^6)} \Bigg\{ 8(t_s - \bar{Z})^3$$
$$\times \left[\frac{\operatorname{sech}(LQ^{0.5}) \tanh(LQ^{0.5}) + \arctan[\sinh(LQ^{0.5})]}{2Q^{0.5}}\right]$$
$$+ \frac{8\bar{Z}(t_s - \bar{Z})^2}{2Q^{0.5}} \left(\frac{e^{2LQ^{0.5}}}{1 + e^{2LQ^{0.5}}}\right)$$
$$+ 4\bar{Z}(t_s - \bar{Z}) \left[\frac{2}{Q^{0.5}} \arctan(e^{LQ^{0.5}})\right] + \bar{Z}^3 L \Bigg\} \qquad (3\text{-}44)$$

For practical applications in the refining, petrochemical, and gas processing industries, sufficient accuracy may be obtained by omitting the last term beginning with (6.536×10^{-7}) in the calculations.

Similarly, for stainless steel, the thermal deflection equation is as follows:

$$\delta_{ss} = \frac{2(t_s - \bar{Z})[8.96 + (4.11 \times 10^{-3})\bar{Z}] \arctan(e^{LQ^{0.5}})}{(10^6)Q^{0.5}}$$
$$+ \frac{(2.055 \times 10^{-3})\bar{Z}^2 L}{10^6}$$
$$+ \frac{(2.055 \times 10^{-3})(t_s - \bar{Z})^2(e^{2Q^{0.5}L} - 1)}{10^6 Q^{0.5}(1 + e^{2LQ^{0.5}})}$$
$$+ \frac{(1.06 \times 10^{-6})}{10^6} \Bigg\{ 8(t_s - \bar{Z})^3$$
$$\times \left[\frac{\operatorname{sech}(LQ^{0.5}) \tanh(LQ^{0.5}) + \arctan[\sinh(LQ^{0.5})]}{2Q^{0.5}}\right]$$
$$+ \frac{8\bar{Z}(t_s - \bar{Z})^2}{2Q^{0.5}} \left(\frac{e^{2LQ^{0.5}}}{1 + e^{2LQ^{0.5}}}\right)$$
$$+ 4\bar{Z}(t_s - \bar{Z}) \left[\frac{2}{Q^{0.5}} \arctan(e^{LQ^{0.5}})\right] + \bar{Z}^3 L \Bigg\} \qquad (3\text{-}45)$$

Like Equation 3-44, Equation 3-45 can be adequately handled using only the first three terms. The use of these equations will be demonstrated in the examples.

Residual Heat Transfer Through Pipe Shoes

Heat transfer through plate surfaces is much simpler than more complex surfaces, because they can be handled with one-dimensional equations that are simple to use. Based on Figure 3-17, we consider the heat balance down through the shoe as follows:

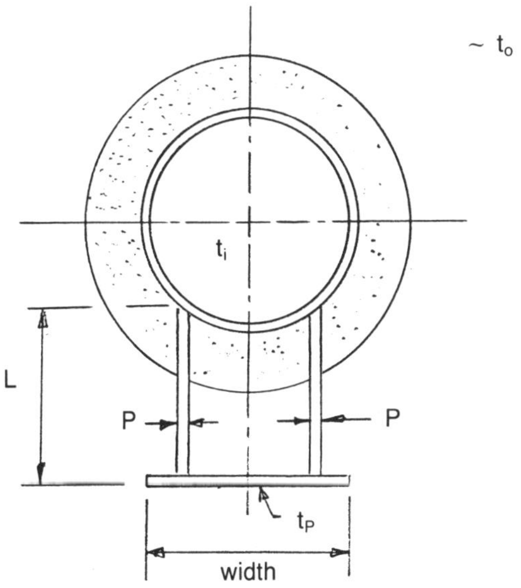

Figure 3-17. Pipe supported on a shoe.

$$\begin{pmatrix}\text{Heat conducted through} \\ \text{shoe to base plate}\end{pmatrix} = \begin{pmatrix}\text{Heat loss by convection from} \\ \text{shoe to outside air}\end{pmatrix}$$

Writing in equation form, we have for one-dimensional, steady state flow:

$$k_m A_m \left(\frac{\Delta t}{L}\right) = h_o A_p(\Delta t) \tag{3-46}$$

For the conduction process, $\Delta t = t_i - t_p$
For the convection process, $\Delta t = t_p - t_o$
Substituting into Equation 3-46, we have

$$k_m A_m \left(\frac{t_i - t_p}{L}\right) = h_o A_p(t_p - t_o)$$

Solving for t_p, we have

$$t_p = \frac{k_m A_m t_i + h_o A_p L t_o}{(k_m A_m + h_o A_p L)}, \ ^\circ F \tag{3-47}$$

where A_m = (P × length of shoe) × 2, $in.^2$
A_p = Base width × length of shoe, $in.^2$
h_o = free convection coefficient for shoe to air, Btu/hr-ft^2-°F
k = thermal conductivity of shoe material, Btu/hr-ft-°F
L = shoe height, in.

Like the analysis for cylinders, the free convection coefficient, h_o, can be substituted with a forced convection coefficient. However, most pipe shoes are surrounded by enough obstructions to prevent a direct wind from blowing on the shoe for any length of time. Figures 3-18a and 3-18b show thermal gradients for various simple pipe supports.

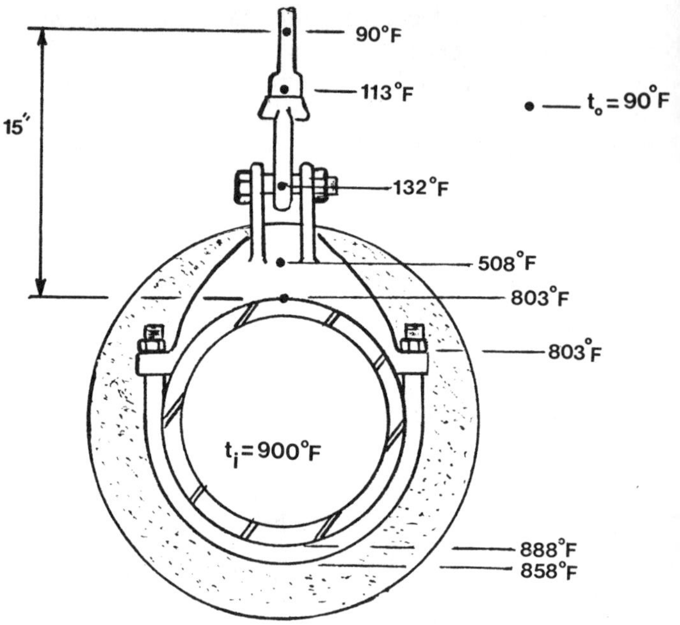

Figure 3-18A. Thermal gradient through pipe clamp, clevis, and supporting rod.

EXAMPLE 3-1: STEAM TRACING DESIGN

Determine the steam tracing requirements for an 8-in. Schedule 40 gas-vapor line with a minimum process temperature of 140°F. The piping insulation is 2½ in. calcium silicate, 9-inch nominal IPS. The system is to be designed for an ambient temperature of 0°F and a 15 mph wind. The tracing medium will be 150 psig steam, and ½-in. copper tubing without HTC will be used for tracing.

We first try using two tracers running alongside bottom of process pipe. Calculating the areas we have the following:

$A_o = 3.63\ ft^2$

$A_t = 0.131\ ft^2$

From Figure 3-4, h_t = 2.5
Calculating T_e,

$$\ln\left(\frac{D_i + 2t}{D_i}\right) = 0.419$$

$$T_e = \left(\frac{1.22}{2}\right) 0.419 = 0.256\ ft$$

$k_i = (2.0)(0.04) = 0.08$

From Figure 3-5, $h_o = 4.5$ (assuming $\Delta t = 50°F$)

$U_o = 0.292$

Formulating a heat balance for the system we have the following:

Q_1 (air space to atm) = (0.292)(3.63)(140) = 148 Btu/hr

Q_2 (tracer to air space) = (2.5)(0.131)(2)(226) = 147 Btu/hr

The assumed number of tracers is inadequate for γ = 1.3.

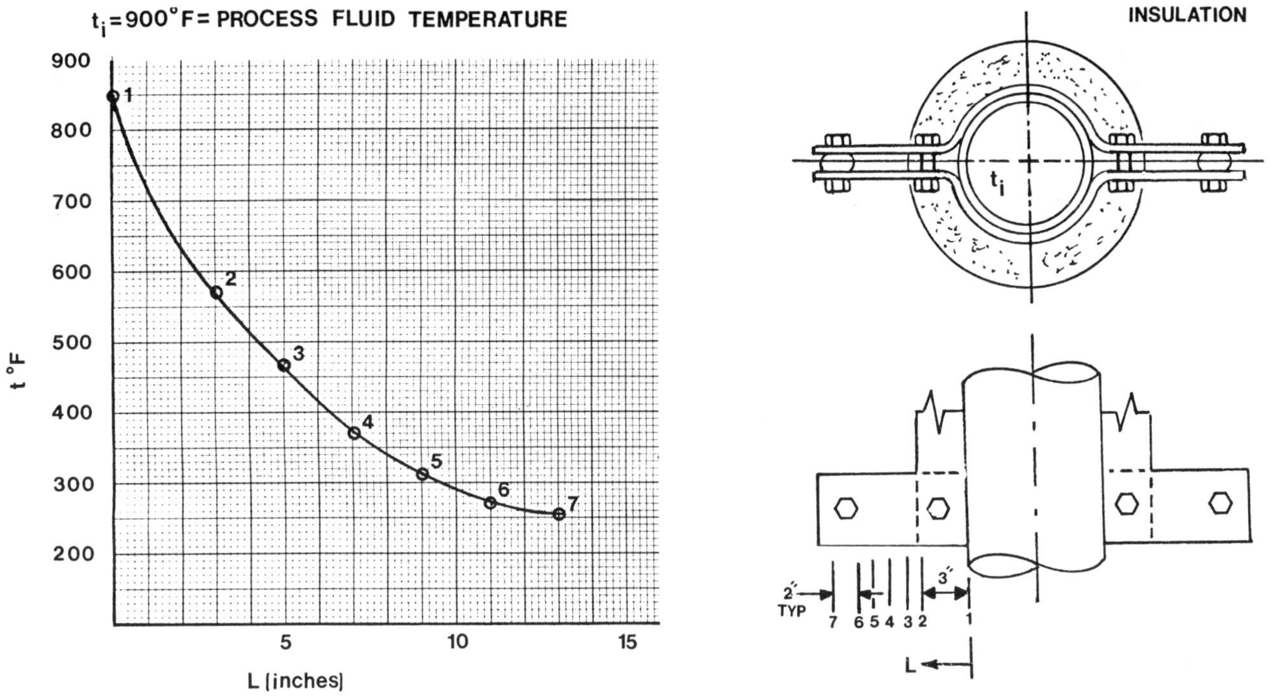

Figure 3-18B. Thermal gradient through pipe clamp support.

Trying three tracers, we have

$Q_1 = 148$ Btu/hr
$Q_2 = 221$ Btu/hr

Since $Q_2 > (\gamma)Q_1$, the system is adequate using three tracers.

EXAMPLE 3-2: HOT OIL TRACING DESIGN

A 3-in. schedule header contains asphalt which is to be maintained at least to 445°F. The 3-in. header is to be traced with hot oil (C_p = 0.50 Btu/lb-°F, ρ = 58.7 lb/ft³ at 475°F). Determine the size and number of hot oil tracers required to maintain the asphalt at a minimum temperature of 450°F. For asphalt, C_p = 0.368 Btu/lb-°F at 500°F.

For most applications, ½-in. copper tubing is the standard size for tracing operations. We select a ½-in. 18 BWG gauge steel tube, A_t = 0.131 ft²/ft, k_m = 27.5 Btu-ft/hr-ft²-°F. First we will try one tracer,

$D_i = 3.50 + 0.50 = 4.00$ in. $= 0.333$ ft

$D_o = 8.00$ in. $= 0.667$ ft;
$D_o + 2t_i = 8.0 + 2(2.0) = 12.0$ in. $= 1.0$ ft

$$T_e = \frac{0.667 + 2(0.167)}{2} \ln\left[\frac{0.667 + 2(0.167)}{0.667}\right] = 0.203 \text{ ft}$$

$A_o = 2.095$ ft²/ft; $A_p = 0.916$ ft²/ft; $A_t = 0.131$ ft²/ft

$$A_m = \frac{0.216}{12} = 0.018 \text{ ft}; \ t' = \frac{t_m}{2n_t}$$

$$= \frac{0.690 \text{ ft}}{2(1.0)} = 0.345 \text{ ft}$$

$$h_{fc} = 0.33\left[\frac{(0.016)(5940)^{0.56}}{(1.0)(0.043)^{0.56}}\right] = 3.992$$

$$h_a = h_{fc} + \epsilon h_r = 3.992 + (0.90)(1.185) = 5.059$$

$$\frac{1}{U_o} = \frac{0.203}{0.1} + \frac{1}{5.059} = U_o = 0.449$$

Now performing a heat balance we have

t_a = 350°F and t_{p2} = 490°F. Using Equations 3-6 through 3-11 with 70°F ambient,

$$q_1 = (0.449)(2.095)(350 - 70) = 263.383 \text{ Btu/hr-ft}$$

$$h_t = 0.5\left(\frac{550 - 350}{0.5}\right)^{0.25} = 2.236$$

$$h_p = 0.5\left(\frac{550 - 350}{3.50}\right)^{0.25} = 1.375$$

$$q_2 = (2.236)(0.131)(550 - 350) = 58.583 \text{ Btu/hr-ft}$$

$$q_3 = (1.375)(0.916)(520 - 350) = 214.115 \text{ Btu/hr-ft}$$

$$q_4 = (2)\left(\frac{27.5}{0.345}\right)(0.018)(550 - 490) = 172.174 \text{ Btu/hr-ft}$$

$q_4 < 2q_3 \Rightarrow$ No balance

Consider $t_a = 300°F$ and $t_{p_2} = 450°F$,

$$h_t = 0.5\left(\frac{550 - 300}{0.5}\right)^{0.25} = 2.364$$

$$h_p = 0.5\left(\frac{550 - 300}{3.50}\right)^{0.25} = 1.454$$

$$q_1 = (0.449)(2.095)(300 - 70) = 216.351 \text{ Btu/hr-ft}$$

$$q_2 = (2.364)(0.131)(550 - 300) = 77.421 \text{ Btu/hr-ft}$$

$$q_3 = (1.454)(0.916)(550 - 300) = 266.373 \text{ Btu/hr-ft}$$

$$q_4 = (2)\left(\frac{27.5}{0.345}\right)(0.018)(550 - 450) = 286.957 \text{ Btu/hr-ft}$$

$q_4 < 2q_3 \Rightarrow$ No balance

Consider $t_a = 350°F$ and $t_{p_2} = 450°F$

$$q_1 = (0.449)(2.095)(350 - 70) = 263.383 \text{ Btu/hr-ft}$$

$$q_2 = (2.236)(0.131)(550 - 350) = 58.583 \text{ Btu/hr-ft}$$

$$q_3 = (1.375)(0.916)(500 - 350) = 188.925 \text{ Btu/hr-ft}$$

$$q_4 = (2)\left(\frac{27.5}{0.345}\right)(0.018)(550 - 450) = 286.957 \text{ Btu/hr-ft}$$

$q_4 < 2q_3 \Rightarrow$ No balance

Consider $t_a = 345°F$ and $t_{p_2} = 445°F$

$$h_t = 0.5\left(\frac{550 - 345}{0.5}\right)^{0.25} = 2.25$$

$$h_p = 0.5\left(\frac{550 - 345}{3.50}\right)^{0.25} = 1.383$$

$$q_1 = (0.449)(2.095)(345 - 70) = 258.680 \text{ Btu/hr-ft}$$

$$q_2 = (2.25)(0.131)(550 - 345) = 60.424 \text{ Btu/hr-ft}$$

$$q_3 = (1.383)(0.916)(497.50 - 345.00) = 193.191 \text{ Btu/hr-ft}$$

$$q_4 = (2)\left(\frac{27.5}{0.345}\right)(0.018)(550 - 445) = 301.304 \text{ Btu/hr-ft}$$

$q_4 < 2q_3 \Rightarrow$ No balance

Since we have reached the minimum desirable temperatures for t_a and t_{p_2}, it is clear that the system will not balance using one 1/2-in. tracer. Therefore, we will use two 1/2-inch tracers. Referring to Figure 3-2b we consider the following:

$$D_i = 4.645 \text{ in.} = 0.387 \text{ ft}$$

$$D_o = 8.645 \text{ in.} = 0.720 \text{ ft}$$

$$t_i = 2 \text{ in.} = 0.167 \text{ ft}$$

$$T_e = \frac{D_i + 2t_i}{2} \ln\left(\frac{D_i + 2t_i}{D_i}\right) = 0.224$$

$$A_o = 2\pi(0.360) = 2.262 \text{ ft}^2\text{/ft}$$

$$h_{fc} = 0.33\left[\frac{(0.016)(5{,}940)^{0.56}}{(0.721)^{0.44}(0.043)^{0.56}}\right] = 1.996$$

$$h_{os} = h_{fc} + \epsilon h_r = 1.996 + (0.90)(1.185) = 3.063$$

$$\frac{1}{U_o} = \frac{0.224}{0.1} + \frac{1}{3.063} \Rightarrow U_o = 0.390$$

$$t_m = 2\pi(2.323) - 2(0.886)(2.323) = 10.479 \text{ in.} = 0.873 \text{ ft}$$

$$t' = \frac{t_m}{2n_t} = \frac{0.873}{2(2)} = 0.218 \text{ ft}$$

$A_t = 0.131$ ft²/ft; $A_m = 0.018$ ft; $A_p = 0.916$ ft²/ft

Consider $t_a = 350°F$ and $t_{p_2} = 490°F$ $h_t = 2.236$; $h_p = 1.375$

$$q_1 = (0.390)(2.262)(350 - 70) = 247.010 \text{ Btu/hr-ft}$$

$$q_2 = (2.236)(0.131)(550 - 350) = 58.583 \text{ Btu/hr-ft}$$

$$q_3 = (1.375)(0.916)(520 - 350) = 214.115 \text{ Btu/hr-ft}$$

$$q_4 = (2)(2)\left(\frac{27.5}{0.218}\right)(0.018)(550 - 490) = 544.954 \text{ Btu/hr-}$$

Consider $t_a = 350°F$ and $t_{p2} = 500°F$

$$q_1 = (0.390)(2.262)(350 - 70) = 247.010 \text{ Btu/hr-ft}$$

$$q_2 = (2.236)(0.131)(550 - 350) = 58.583 \text{ Btu/hr-ft}$$

$$q_3 = (1.375)(0.916)(525 - 350) = 220.413 \text{ Btu/hr-ft}$$

$$q_4 = (2)(2)\left(\frac{27.5}{0.218}\right)(0.018)(550 - 500) = 454.128 \text{ Btu/hr-ft}$$

$q_4 > 2q_3$ and $q_2 + q_3 > q_1$

Therefore, the system is balanced.
For 100 ft of pipe,

$$q_4 = (100)\text{ft}(454.128) \text{ Btu/hr-ft} = 45{,}412.8 \text{ Btu/hr}$$

or,

$$q = 22{,}706.422 \text{ Btu/hr for each tracer}$$

$$\dot{m} = \frac{22{,}706.422 \text{ Btu/hr}}{(0.5) \text{ Btu/lb-°F } (50)\text{°F}} = 908.257 \text{ lb/hr}$$

$$Q = \frac{(908.257)\dfrac{\text{lb}}{\text{hr}}\left(7.48\dfrac{\text{gal}}{\text{ft}}\right)\left(\dfrac{1 \text{ hr}}{60 \text{ min}}\right)}{(58.7)\dfrac{\text{lb}}{\text{ft}^3}} = 1.929 \text{ gpm}$$

Thus, we see that two 1/2-in. tracers containing hot oil flowing at 1.929 gpm is adequate to maintain the asphalt at a minimum temperature of 450°F. The next step in the design is to do a hydraulic analysis using the principles of Chapter 1 and size pumps to handle the hot oil. (Chapter 6 discusses how to select the pumps required to distribute the oil in the system.)

EXAMPLE 3-3: JACKETED PIPE DESIGN

A jacketed pipe shown in Figure 3-1b is to be analyzed. The process fluid to be heated is a film coating mixture used in the manufacture of roofing tiles.

$$U = \frac{1.0}{\dfrac{r_4}{h_{1\text{-}2}} + \dfrac{r_4 \ln(r_3/r_2)}{k_{2\text{-}3}} + \dfrac{1}{h_{3\text{-}4}}}$$

For film coating in inner tube,

$$\rho = 95.909 \text{ lb/ft}^3;\ C_p = 0.34 \frac{\text{Btu}}{\text{lb-°F}}$$

$$\mu = 938.08\ C_p \text{ at } 500°F;\ k = 0.1 \frac{\text{Btu}}{\text{hr-ft-°F}}$$

$$V = 150\frac{\text{gal}}{\text{min}}\left(\frac{\text{ft}^3}{7.48 \text{ gal}}\right)\left(\frac{1 \text{ min}}{60 \text{ sec}}\right)\left(\frac{1}{12.73 \text{ in.}^2}\right)\left(\frac{144 \text{ in.}^2}{\text{ft}^2}\right)$$

$$V = 3.781 \text{ ft/sec}$$

$$N_{Re} = \frac{VD\rho}{\mu}$$

$$N_{Re} = \frac{(3.781)\dfrac{\text{ft}}{\text{sec}}(4.026)\text{in.}\left(\dfrac{1 \text{ ft}}{12 \text{ in.}}\right)(95.909)\dfrac{\text{lb}}{\text{ft}^3}\left(\dfrac{60 \text{ sec}}{\text{hr}}\right)}{(938.08)C_p\left(\dfrac{2.4\dfrac{\text{lb}}{\text{ft-hr}}}{C_p}\right)}$$

$N_{Re} = 3.242 << 2{,}100 \Rightarrow$ Laminar flow

For laminar flow, we have [2]

$$h_{1\text{-}2} = 1.86(N_{Re})^{1/3}(N_{Pr})^{1/3}\left(\frac{D}{L}\right)^{1/3}\left(\frac{k}{D}\right)\left(\frac{\mu}{\mu_w}\right)^{0.14} \quad (3\text{-}16)$$

In most instances, the ratio $\mu/\mu_W \approx 1.0$. Thus, we have

$$N_{Pr} = \frac{\mu C_p}{k}$$

$$= \frac{(938.08)C_p\left(\dfrac{2.4\dfrac{\text{lb}}{\text{ft-hr}}}{C_p}\right)(0.34)\dfrac{\text{Btu}}{\text{lb-°F}}}{0.1\dfrac{\text{Btu}}{\text{hr-ft-°F}}}$$

$$= 7{,}654.733$$

$$D = 4.026 \text{ in.}\left(\frac{1 \text{ ft}}{12 \text{ in.}}\right) = 0.336;\ L = 100 \text{ ft}$$

$$h_{1\text{-}2} = 1.86(3.242)^{1/3}(7{,}654.733)^{1/3}\left(\frac{0.336}{100}\right)^{1/3} \times \left(\frac{0.1}{0.336}\right)(1.0)\frac{\text{Btu}}{\text{hr-ft}^2\text{-°F}}$$

$$h_{1\text{-}2} = 2.415\frac{\text{Btu}}{\text{hr-ft}^2\text{-°F}}$$

For hot oil in annulus, $D_i/D_o = 0.664 > 0.2$, and Perry [6] recommends

$$N_{Nu} = 0.020\ N_{Re}^{0.8}\ N_{Pr}^{1/3}\left(\frac{D_2}{D_1}\right)^{0.53}$$

For the annulus,

$$R_H = \text{hydraulic radius} = \frac{3.033 - 2.250}{2} = 0.392 \text{ in.}$$

$$D = 4R_H = 1.566 \text{ in.} = 0.131 \text{ ft}$$

For hot oil flowing at 0.5 gpm

$$N_{Re} = (40.107)\frac{(0.5)(58.7)}{(0.15)(1.566)} = 5{,}011.241$$

$$N_{Pr} = \frac{\mu C_p}{k}$$

$$= \frac{(0.15)C_p\left(\frac{2.4 \text{ lb/ft-hr}}{C_p}\right)(0.5)\frac{\text{Btu}}{\text{lb-°F}}}{0.071\frac{\text{Btu}}{\text{hr-ft}^2\text{°F/ft}}}$$

$$= 2.535$$

$$N_{Nu} = 0.020(5{,}011.241)^{0.8}(2.535)^{1/3}\left(\frac{6.065}{4.500}\right)^{0.53} = 29.121$$

$$h_{3\text{-}4} = \frac{N_{Nu}k}{D} = \frac{(29.121)(0.071)\frac{\text{Btu}}{\text{hr-ft}^2\text{-°F/ft}}}{0.131 \text{ ft}}$$

$$h_{3\text{-}4} = 15.783\frac{\text{Btu}}{\text{hr-ft}^2\text{-°F}}$$

$$U = \left[\frac{(3.033)}{2.415} + \frac{(3.033)\ln(2.250/2.013)}{29} + \frac{1}{15.783}\right]^{-1}$$

$$U = 0.751\frac{\text{Btu}}{\text{ft}^2\text{-hr-°F}}$$

A = heat transfer area = outside surface area of inner tube

$A = 1.178 \text{ ft}^2\text{/ft} = 117.800 \text{ ft}^2$ for 100 ft of pipe

In hot oil applications it is common to assume that the hot oil decreases in temperature 100°F per 10 feet in jacketed and traced systems. For the film coating mixture,

$t_{CH} = 500°\text{F}$ and $t_{CC} = 450°\text{F}$

For hot oil,

$t_{OH} = 550°\text{F}$ and $t_{OC} = 450°\text{F}$

LMTD = 72.135°F

Now,

$$q = UA(LMTD)$$

$$q = 0.751\frac{\text{Btu}}{\text{ft-hr-°F}}(117.80)\text{ft}^2(72.135)\text{°F}$$

$$q = 6{,}381.625\frac{\text{Btu}}{\text{hr}} \text{ is the heat transfer required}$$

$$q = \dot{m}C_p\Delta t$$

For hot oil, $\Delta t = t_{OH} - t_{OC} = 100°\text{F}$ and,

$$C_p = 0.5\frac{\text{Btu}}{\text{lb-°F}}$$

$$\dot{m} = \frac{6{,}381.625\frac{\text{Btu}}{\text{hr}}}{0.5\frac{\text{Btu}}{\text{lb-°F}}(100)\text{°F}} = 127.632\frac{\text{lb}}{\text{hr}}$$

Now,

$$Q = \frac{127.632\frac{\text{lb}}{\text{hr}}\left(\frac{7.48 \text{ gal}}{1 \text{ ft}^3}\right)\left(\frac{1 \text{ hr}}{60 \text{ min}}\right)}{58.7\frac{\text{lb}}{\text{ft}^3}}$$

$$= 0.271 \text{ gpm required}$$

For 1/2 gpm,

$$q = (235.428)\frac{\text{lb}}{\text{hr}}(0.5)\frac{\text{Btu}}{\text{lb-°F}}(100)\text{°F}$$

or

$$q = 11{,}771.400 \text{ Btu/hr}$$

Thus, 0.5 gpm is a sufficient flow rate to transfer the required heat to the film coating mixture.

EXAMPLE 3-4: THERMAL EVALUATION OF A PROCESS TANK

A coating surge tank contains 6,000 gal of fill coating mix (see Figure 3-19). Two problems must be solved: (a) how many degrees per hour can be obtained from a clamped-on jacketed system, when the fill coating mixture is static; (b) how many degrees per hour can be obtained from a clamped-on jacketed system, when the fill

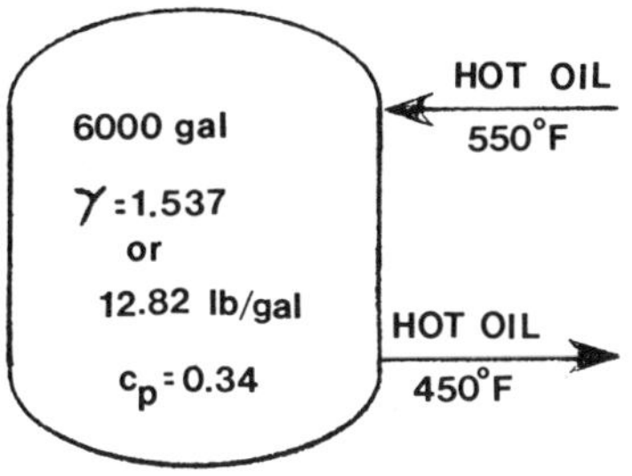

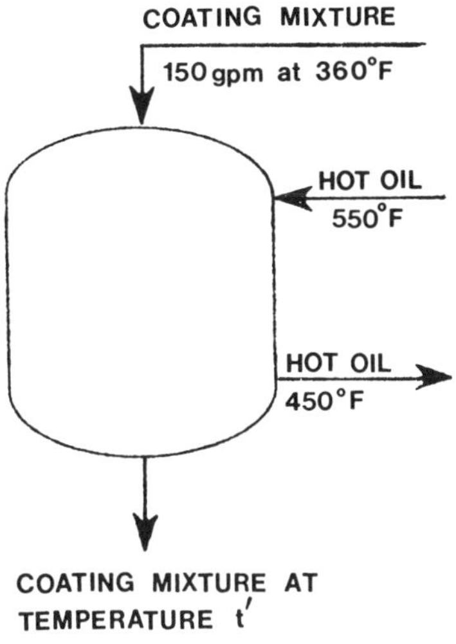

Figure 3-19. Coating surge tank.

coating mixturc is flowing through thc tank at 150 gpm at 360°F.

From Figure 3-10, we have

$$\text{LMTD} = \frac{(550 - 360) - (450 - 400)}{l_n \left[\dfrac{550 - 360}{450 - 400}\right]} = 104.869°\text{F}$$

with $Q = mC_p\Delta t$

and $Q = UA(\text{LMTD})$

$$\Delta t = \frac{UA(\text{LMTD})}{mC_p}$$

$$m = (6{,}000)\text{gal}(12.82)\,\frac{\text{lb}}{\text{gal}} = 76{,}920.00 \text{ lb}$$

Using heat transfer panels shown in Figure 3-12 we compute the total available heat transfer area as follows:

$$\begin{aligned}
\text{Flanged and dished head} &= (92)(0.8)\\
&= 73.60 \text{ ft}^2\\
\text{Shell} &= (379.347)(0.8)\\
&= 303.478 \text{ ft}^2\\
\text{4-internal heat transfer panels} &= 4(107)(12)/144\\
&= 35.667 \text{ ft}^2
\end{aligned}$$

The overall heat transfer coefficient, U-value, supplied by the panel manufacturer for applications to the fill coating mix is as follows:

Process Conditions (as determined by process engineers or client for desired capability of tank):
Initial temperature of coating mixture = 360°F
Final temperature of coating mixture = 400°F
For internal panels, U = 9.52 Btu/hr-ft^2-°F
For external clamp-on jacketing, U = 4.00 Btu/hr-ft^2-°F

Substituting into the previous equation for Δt we have

$$\Delta t = \frac{(9.52)(35.667)(\text{LMTD}) + (4.0)(377.078)(\text{LMTD})}{(76{,}920.00)(0.34)}$$

$$\Delta t = 7.410°\text{F/hr}$$

Referring to Figure 3-19 we can now determine how many degrees per hour the fill coating mix will rise using external clamp-on jacketing on surfaces of the flanged and dished head, the vessel shell, and four internal panels just considered:

$$Q = (60)\,\frac{\text{min}}{\text{hr}}\,(150)\,\frac{\text{gal}}{\text{min}}\,(12.82)\,\frac{\text{lb}}{\text{gal}} \times 0.34\,\frac{\text{Btu}}{\text{lb-°F}}\,(t - 360)°\text{F}$$

$$Q = 39{,}229.20(t - 360) \text{ Btu/hr}$$

Now,

$$UA(\text{LMTD}) = 39{,}229.20(t - 360)$$

$$(9.52)(35.667)(\text{LMTD}) + (4.0)(377.078)(\text{LMTD}) = 39{,}229.20(t' - 360)$$

$$1{,}847.862(\text{LMTD}) = 39{,}229.20(t' - 360)$$

$$\text{LMTD} = 21.230(t' - 360)$$

Now,

$$\text{LMTD} = \frac{(550 - 360) - (450 - t')}{l_n \left[\dfrac{550 - 360}{450 - t}\right]} = 21.23(t' - 360)$$

Solving for t,

$$1.0 = \frac{(21.23t' - 7{,}642.80)}{(110.394t' - 39{,}841.956)} \ln(450 - t')$$

$$\text{Letting } y = \frac{(21.23t' - 7{,}642.80)}{(110.394t' - 39{,}841.956)}$$

we have $1.0 = \ln(450 - t')^y$

or $e^{1.0} = e^{\ln(450-t')^y}$

in which $2.718 = (450 - t')^y$

After several iterations, $t' = 366.12°F$

Thus, the temperature rise is

$$\Delta t = 366.12°F - 360°F = 6.12°F$$

The amount of heat required for the system is

$$Q = UA(LMTD)$$

$$Q = (9.52)(35.667)(LMTD) + (4.0)(377.078)(LMTD)$$

or

$$Q = 1{,}847.862(LMTD)$$

Now,

$$LMTD = \frac{(550 - 360) - (450 - 366.12)}{\ln\left[\frac{550 - 360}{450 - 366.12}\right]} = 129.789°F$$

in which $Q = 1{,}847.862(129.789) = 239{,}832.162$ Btu/hr is the heat transferred to the coating mix.

EXAMPLE 3-5: THERMAL DESIGN OF PROCESS TANK

The coating surge tank of Example 3-4 is now analyzed for detailed heat transfer requirements. The flow rates through the various types of heat transfer jacket plates are desired.

Internal Baffle Plates Film Coefficient

Some of the plates used are shown in Figure 3-12. Looking at Figures 3-20, 3-21, and 3-22 we determine the hydraulic radius as follows:

$$WP = \text{Wetted Perimeter} = \left(\frac{y}{\sin\theta}\right) 4 + 2w$$

$$WP = \left(\frac{0.4125}{\sin 45°}\right)(4) + 2(0.4125) = 3.158 \text{ in.}$$

A = cross-sectional flow area = 0.681 in.^{2} = 4a(W)

where W = a = 0.4125 in.
W = effective heat transfer length (see page 145)

$$A = 2Y^2 + 2wY$$

$$A = 2(0.4125)^2 + 2(0.4125)^2 = 0.681 \text{ in.}^2$$

$$R_H = \frac{A}{WP} = \frac{0.681 \text{ in.}^2}{3.158 \text{ in.}} = 0.216 \text{ in.}$$

$$D = 4R_H = 0.862 \text{ in.}$$

The equivalent circular cross-sectional area = $\pi(0.431)^2$ = 0.584 in.^{2}

The hot oil properties are as follows:

$$t_m = \frac{550°F + 450°F}{2} = 500°F$$ (since we anticipate a 100°F temperature drop in the plates)

Velocity of hot oil through baffles = 7.913 ft/sec

$$\rho = 58.7 \text{ lb/ft}^3$$

$$k = 0.071 \text{ Btu/hr-ft}^2\text{-°F/ft}$$

$$C_p = 0.50 \text{ Btu/lb-°F}$$

$$\mu = 0.15\, C_p \left(\frac{2.4 \text{ lb/ft-hr}}{1\, C_p}\right) = 0.360 \text{ lb/ft-hr}$$

The maximum pressure drop permitted through the internal baffles, which are connected in parallel, is normally 10 psi, thus

$$N_{Re} = \frac{V D \rho}{\mu} = \frac{(7.913)\frac{ft}{sec}(0.862) \text{ in.}\left(\frac{1 \text{ ft}}{12 \text{ in.}}\right)(58.7)\frac{lb}{ft^3}\left(\frac{3600 \text{ sec}}{1 \text{ hr}}\right)}{0.360 \frac{lb}{ft\text{-}hr}}$$

$$N_{Re} = 333{,}661$$

$$N_{Pr} = \frac{\mu C_p}{k} = \frac{(0.360)\frac{lb}{ft\text{-}hr}(0.50)\frac{Btu}{lb\text{-}°F}}{(0.071)\frac{Btu}{hr\text{-}ft^2\text{-}°F/ft}} = 2.535$$

From Kern [1],

$$h_{1\text{-}2} = (0.027)\left(\frac{K}{D}\right)(N_{Re})^{0.8}(N_{Pr})^{1/3}\left(\frac{\mu}{\mu_w}\right)^{0.14}$$

$$h_{1\text{-}2} = (0.027)\frac{(0.071)\dfrac{\text{Btu}}{\text{hr-ft}^2\text{-}^\circ\text{F/ft}}}{(0.862)\text{in.}\left(\dfrac{1\text{ ft}}{12\text{ in.}}\right)} \times (333{,}661)^{0.8}(2.535)^{1/3}(1.0)$$

$h_{1\text{-}2} = 953.82 \dfrac{\text{Btu}}{\text{hr-ft}^2\text{-}^\circ\text{F}}$ is the film coefficient inside the jacket baffles located inside the tank

Film Coefficient External to Baffles—Forced Convection

The coating mixture inside the tank is in a state of flow across the baffle plates made possible by agitator blades powered by an electric motor. From Perry [6],

$$h_{3\text{-}4} = 0.09\left(\frac{k}{D_o}\right)\left(\frac{L_e^2 N_r \rho}{\mu}\right)^{0.65}\left(\frac{C_p \mu}{k}\right)^{1/3} \times \left(\frac{L_p}{Dj}\right)^{1/3}\left(\frac{2}{n_b}\right)^{0.2}\left(\frac{\mu}{\mu_f}\right)^{0.4} \quad (3\text{-}48)$$

n_b = number of internal baffle plates = 4

For coating mix, $10^3 < N_{Re} < 2 \times 10^6$
The properties of the mix are as follows:

$\rho = 95.909\ \text{lb/ft}^3$

$\mu = 6{,}000\ C_p = 14{,}400\ \text{lb/ft-hr}$

$k = 0.1\ \text{Btu/hr-ft-}^\circ\text{F}$

Other properties related to the internal baffle plate approximated as a string of cylinders with diameters equal to four times the hydraulic radius of the trapezoid plate sections shown in Figures 3-20 and 3-21.

N_h = angular velocity of agitator, revolutions/hr

$$N_r = 56\ \frac{\text{rev}}{\text{min}}\left(\frac{60\text{ min}}{1\text{ hr}}\right) = 3360\ \text{rev/hr}$$

L_p = diameter of agitator plate = 65 in. = 5.42 ft

Dj = 10.0 ft

D_o = 0.862 in.

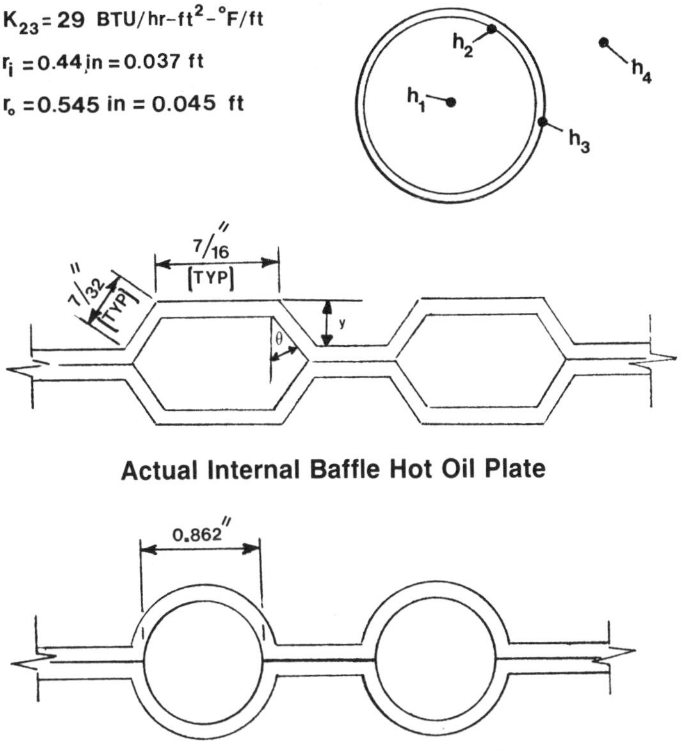

Figure 3-20. Cross section of panel plate tube approximated as a cylinder whose surface area is equivalent to the heat transfer of contact area.

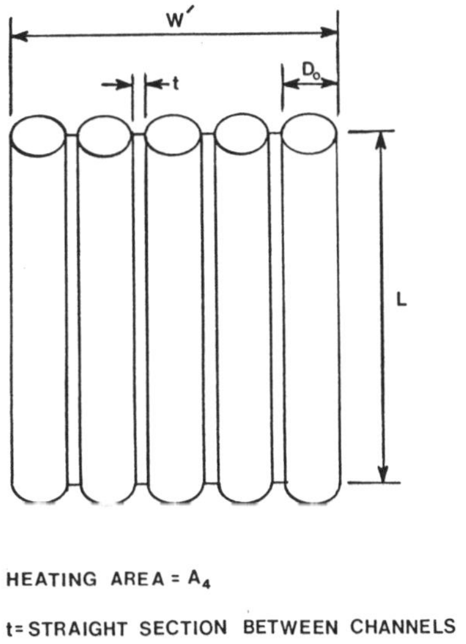

Figure 3-21. String of tubes.

$k = 0.071$ Btu/hr-ft^2-°F/ft

$$h_{3\text{-}4} = 0.09\left(\frac{0.1}{0.862}\right)\left[\frac{(5.42)^2(3{,}360.0)(95.909)}{14{,}400}\right]^{0.65}$$

$$\times\left[\frac{(0.34)(14{,}400)}{(0.1)}\right]^{1/3}\left(\frac{5.42}{10.0}\right)^{1/3}\left(\frac{2}{4}\right)^{0.2}(1.0)$$

$h_{3\text{-}4} = 18.334$ Btu/hr-ft^2-°F for all four baffle plates

Now, to solve for the overall heat transfer coefficient, U, we must develop the appropriate equation.

Let A_4 = the overall area of plate = LW

L = length of plate, W = width of plate

Referring to Figures 3-20 and 3-21 we have

$q = UA_4(t_1 - t_4)$ for baffle plate

$A_4 = LW$ = wetted area $= 2\pi\sum rL$

where W = wetted perimeter (WP)

$\sum r = \frac{n}{2}D_o + r'$, n = number of flow circuits

To account for the residual heat transferred through the plate connecting the hexagon tubes we consider r′ to be the equivalent radius of a cylinder that is the total surface area of the baffle plate separating the hexagon tubes. Now,

$\sum t$ = total distance between channels (Figure 3-21)

$$2\pi r' = \sum t$$

$$r' = \frac{\sum t}{2\pi}$$

From above,

$$q = UA_4(t_1 - t_4) = 2\pi U\sum r(t_1 - t_4)$$

$$= \frac{2\pi(t_1 - t_4)\,n}{\dfrac{r_o}{r_ih_{1\text{-}2}} + \dfrac{r_o\ln(r_o/r_i)}{k_{2\text{-}3}} + \dfrac{1}{h_{3\text{-}4}}}$$

in which

$$U = \frac{n}{\sum r\left[\dfrac{r_o}{r_ih_{1\text{-}2}} + \dfrac{r_o\ln(r_o/r_i)}{k_{2\text{-}3}} + \dfrac{1}{h_{3\text{-}4}}\right]}$$

or

$$U = \frac{n}{\left[\left(\dfrac{n}{2}\right)D_o + \dfrac{\sum t}{2\pi}\right]\left[\dfrac{r_o}{r_ih_{1\text{-}2}} + \dfrac{r_o\ln(r_o/r_i)}{k_{2\text{-}3}} + \dfrac{1}{h_{3\text{-}4}}\right]}$$

for these baffle plates, n = 7, for which

$$U = \frac{7.0}{\left[\left(\dfrac{7}{2}\right)(0.862) + \dfrac{5(1.5)}{2\pi}\right]}$$

$$\frac{7.0}{\left[\dfrac{0.431}{(0.326)(949.883)} + \dfrac{(0.045)\ln(0.431/0.326)}{29} + \dfrac{1}{18.334}\right]}$$

$$U = 29.492\ \frac{\text{Btu}}{\text{hr-ft}^2\text{-°F}}$$

The baffle plate area for all four baffles is determined from the baffle plate manufacturer's dimensions, as follows:

Length of channel per baffle = 51.723 ft

$$A = \text{surface area} = 2\pi(0.431)\text{in.}\left(\frac{1\text{ ft}}{12\text{ in.}}\right)(51.723\text{ ft})$$

$A = 11.672$ ft^2

For all four baffle plates,

$A = 4(11.672)$ ft^2 = 46.690 ft^2

Heat Duty of Internal Baffle Plates

For hot oil the anticipated temperature through each plate is 100°F, as stated earlier, thus

$$q = UA\Delta T$$

$$q = (29.492)\frac{\text{Btu}}{\text{hr-ft}^2\text{-°F}}(46.690)\text{ft}^2\,(100)\text{°F}$$

$$q = 137{,}698{,}148\ \frac{\text{Btu}}{\text{hr}}$$

Outside Heat Transfer Jacket Plates

In the case of external jacket plates, the heat transfer parameters are based on the dimension, W, shown in

Figure 3-22, because it is this surface that is in contact with the vessel wall. Consequently we can analyze the configuration in Figure 3-22 as a tube with circumference of W. Hence, we have the following:

$$2\pi r_1 = W = 1.375 \text{ in.}$$

where r_1 = tube inside radius, in.

from above,

$$r_1 = \frac{1.375}{2\pi} = 0.219 \text{ in. or } D_1 = 0.438 \text{ in.}$$

The tube equivalent flow rate for the length W mapped into a circle or radius r_1 is by the continuity condition of fluid mechanics,

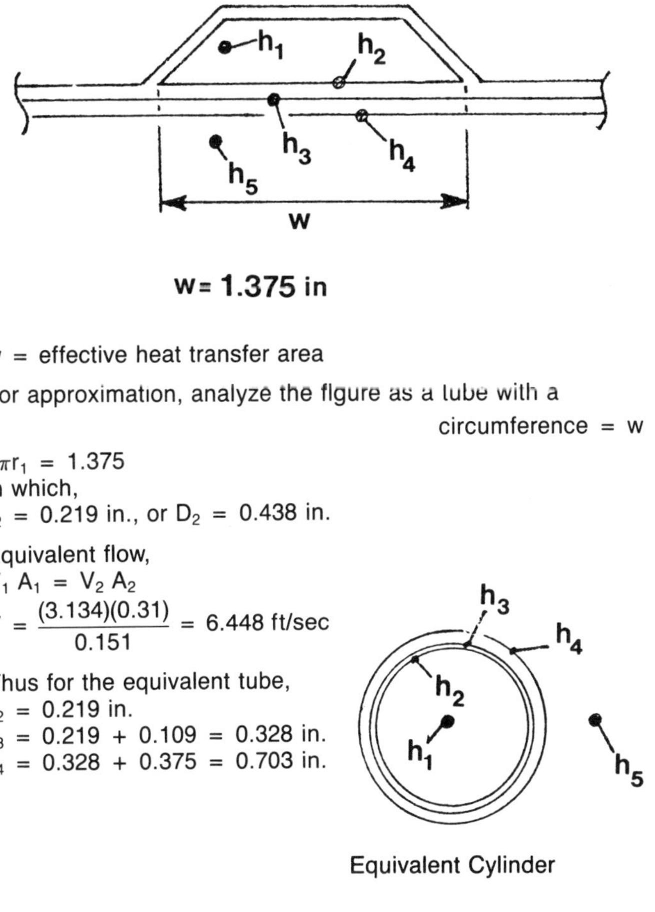

Figure 3-22. Panel total flow cross section. Contact length w is mapped into an equivalent circular tube whose circumference equals w.

$$V_1A_1 = V_2A_2$$

where

A_1 = tube cross flow area

$$A_1 = \pi(0.219)^2 = 0.151 \text{ in.}^2$$

Velocity of hot oil through outside plates = V_2

V_2 = 3.134 ft/sec (determined from process data)

internal cross flow area of baffle plate = 0.62 in.^{2} (from plate manufacturer's data)

$$V_1 = \frac{(3.134)\dfrac{\text{ft}}{\text{sec}}(0.62)\text{ in.}^2}{0.151 \text{ in.}^2} = 6.448\frac{\text{ft}}{\text{sec}}$$

$$N_{Re} = \frac{V\,D\,\rho}{\mu}$$

$$= \frac{(6.448)\dfrac{\text{ft}}{\text{sec}}(0.438)\text{ in.}\left(\dfrac{1\text{ ft}}{12\text{ in.}}\right)(58.7)\dfrac{\text{lb}}{\text{in.}^3}\left(\dfrac{3600\text{ sec}}{1\text{ hr}}\right)}{(0.360)\dfrac{\text{lb}}{\text{ft-hr}}}$$

$$N_{Re} = 138{,}150.85$$

Solving for the over-all heat transfer coefficient, U, we have

$$N_{Pr} = \frac{\mu\,C_p}{k} = 2.535$$

$$h_{1\text{-}2} = (0.027)\left(\frac{0.071}{0.438}\right)(138{,}150.85)^{0.8}(2.535)^{1/3}(1.0)$$

$$h_{1\text{-}2} = 77.260\frac{\text{Btu}}{\text{hr-ft}^2\text{-}^\circ\text{F}}$$

Perry [6] gives the correlation for heat transfer for jacketed walls to the agitated liquid as follows:

$$h_{4\text{-}5} = a\left(\frac{k}{D_j}\right)\left(\frac{L_p^2 N_r \rho}{\mu}\right)^b\left(\frac{C_p\mu}{k}\right)^{1/3}\left(\frac{\mu_b}{\mu_w}\right)^m \tag{3-49}$$

where, $h_{4\text{-}5}$ = film coefficient at vessel wall (see Figure 3-22)
D_j = inside diameter of the vessel, ft = 10.0 ft
k = 0.1 Btu/hr-ft-°F
L_p = diameter of agitator = 5.42 ft
N_r = angular velocity, or rotation of agitator

= 3,360 rev/hr
ρ = 95,909 lb_m/ft^3

μ = 14,400 lb/ft-hr
C_p = specific heat of coating mix
μ_b = viscosity at bulk temperature, lb/ft-hr
μ_w = viscosity at wall temperature, lb/ft-hr
a, b, M = see Table 3-1

Laboratory tests were made on the coating mix and the results showed that $\mu_b/\mu_w = 0.65$. Since the coating mix is a non-Newtonian fluid, it is strongly recommended that the physical properties be determined by a qualified laboratory, the ratio μ_b/μ_w should *never* be assumed to be 1.0 for a non-Newtonian fluid without laboratory tests of fluid samples.

For a disk, flat-blade turbine agitator we find values for a, b, and M from Table 3-1 as follows:

$a = 0.54,\ b = 2/3,\ M = 0.14$

since $40 \leqslant 138{,}151 \leqslant 3 \times 10^5$

substituting above values into Equation 3-49 we have the following:

$$h_{4\text{-}5} = (0.54)\left(\frac{0.1}{10.0}\right)\left[\frac{(5.42)^2(3{,}360.0)(95.909)}{(14{,}400)}\right]^{2/3} \times \left[\frac{(0.34)(14{,}400)}{(0.1)}\right]^{1/3}(0.65)^{0.14}$$

$$h_{4\text{-}5} = 14.060\ \frac{\text{Btu}}{\text{hr-ft}^2\text{-}^\circ\text{F}}$$

Thus,

$$U = \left[\frac{r_4}{r_2 h_{1\text{-}2}} + \frac{r_4 \ln(r_3/r_2)}{k_{2\text{-}3}} + \frac{r_4 \ln(r_4/r_3)}{k_{3\text{-}4}} + \frac{1}{h_{4\text{-}5}}\right]^{-1}$$

Table 3-1
Values of Constants for Equation 3-49

Agitator	a	b	M	Reynolds Number Range
Disk, flat-blade turbine	0.54	2/3	0.14	$40 \leqslant N_{Re} \leqslant 3 \times 10^5$
Pitched-blade turbine	0.53	2/3	0.24	$80 \leqslant N_{Re} \leqslant 200$
Propeller	0.54	2/3	0.14	$N_{Re} \approx 2000$
Paddle	0.36	2/3	0.21	$300 \leqslant N_{Re} \leqslant 3 \times 10^5$
Helical ribbon	0.633	1/2	0.18	$8 \leqslant N_{Re} \leqslant 10^5$

$$U = \left[\frac{(0.703)}{(0.219)(77.26)} + \frac{(0.73)\ln(0.328/0.219)}{29} + \frac{(0.703)\ln(0.703/0.328)}{29} + \frac{1}{14.06}\right]^{-1}$$

$$U = 8.141\ \frac{\text{Btu}}{\text{hr-ft}^2\text{-}^\circ\text{F}}$$

From manufacturer's drawings, the shell jacket plate heat transfer area, A, is

$A = 37{,}043.82\ \text{in.}^2$

Now,

A′ = area of channels in all nine jacket shell plates clamped-on to outside of shell

$A' = 257.249\ \text{ft}^2$

$\Delta t = 100^\circ$F for hot oil-coating mix service

$$q = (8.141)\ \frac{\text{Btu}}{\text{hr-ft}^2\text{-}^\circ\text{F}}\ (257.249)\text{ft}^2\ (100)^\circ\text{F}$$

or

$$q = 209{,}414.44\ \frac{\text{Btu}}{\text{hr}}$$

Heat Duty of Jacket Plates Clamped to Bottom Vessel Head

The bottom head panel sections are depicted in Figure 1-17. In Chapter 1, Example 1-2 we analyzed the hydraulics for the hot oil flow through the panels mounted on the tank. From this analysis we determined the following velocities required to obtain 10 psi pressure drop through the panels:

inner panel = V = 7.315 ft/sec
outer panel = V = 5.237 ft/sec

Heat Duty for Bottom Head Inner Panels

Similar to the shell panel plates above, we must compute the equivalent tube diameter and equivalent velocity. As determined above the equivalent radius is

$r_1 = 0.219$ in.

Since the bottom baffles have the same flow area as the shell plates, the cross flow area of the equivalent tube is

$$A' = \pi(0.219)\text{ in.}^2 = 0.151\text{ in.}^2$$

The cross flow area inside the plate channel is found from the manufacturer's catalog to be 0.31 in.^{2}. Since the equivalent tube circumference is equal to the contact dimension, w, as above we must compute the equivalent velocity. Thus

$$\text{Equivalent velocity} = V_E = \frac{(7.315)\,\dfrac{\text{ft}}{\text{sec}}\,(0.31)\text{in.}^2}{0.151\text{ in.}^2}$$

$$V_E = 10.050\text{ ft/sec}$$

$$N_{Re} = \frac{(15.050)(0.438)\left(\dfrac{1}{12}\right)(58.7)(3{,}600)}{0.360} = 322{,}453.78$$

$$N_{pr} = 2.535$$

$$h_{1\text{-}2} = (0.027)\left(\frac{0.071}{0.438}\right)(322{,}453.78)^{0.8}(2.535)^{1/3}(1.0)$$

$$h_{1\text{-}2} = 152.211\,\frac{\text{Btu}}{\text{hr-ft}^2\text{-}^\circ\text{F}}$$

The vessel-side film coefficient is the same as for the shell panels,

$$h_{4\text{-}5} = 14.914\,\frac{\text{Btu}}{\text{hr-ft}^2\text{-}^\circ\text{F}}$$

Thus,

$$U = \left[\frac{(0.703)}{(0.219)(152.211)} + \frac{(0.73)\ln(0.703/0.328)}{29} + \frac{(0.703)\ln(0.703/0.328)}{29} + \frac{1}{14.914}\right]$$

$$U = 8.590\,\frac{\text{Btu}}{\text{hr-ft}^2\text{-}^\circ\text{F}}$$

Let A = heat transfer area of bottom head plates from the manufacturer's data the flow path length is 388.231 in. for one half of the head, hence,

$$A = (388.231)\text{ in. }(1.375)\text{ in. }(2) = 1{,}067.635\text{ in.}^2$$

or

$A = 7.414\text{ ft}^2$ = area of the two inner head plates

q = heat duty

$$q = UA\,\Delta t = (8.590)\,\frac{\text{Btu}}{\text{hr-ft}^2\text{-}^\circ\text{F}}\,(7.414)\text{ft}^2\,(100)^\circ\text{F}$$

in which

$$q = 6{,}368.71\,\frac{\text{Btu}}{\text{hr}}\quad\text{for both two inner plates}$$

Heat Duty for Bottom Outer Panels

$$\text{Equivalent velocity} = \frac{(5.237)(0.31)}{0.151} = 10.751\text{ ft/sec}$$

$$N_{Re} = \frac{(10.751)(0.438)\left(\dfrac{1}{12}\right)(58.7)(3{,}600)}{0.360} = 230{,}355.341$$

$$N_{Pr} = 2.535$$

$$h_{1\text{-}2} = (0.027)\left(\frac{0.071}{0.438}\right)(230{,}355.341)^{0.8}(2.535)^{1/3}(1.0)$$

$$h_{1\text{-}2} = 116.303\,\frac{\text{Btu}}{\text{hr-ft}^2\text{-}^\circ\text{F}}$$

Similarly as for inner panels,

$$h_{4\text{-}5} = 334.996\,\frac{\text{Btu}}{\text{hr-ft}^2\text{-}^\circ\text{F}}$$

Thus,

$$U = \left[\frac{0.703}{(0.219)(116.303)} + \frac{(0.703)\ln\left(\dfrac{0.328}{0.219}\right)}{29} + \frac{(0.703)\ln\left(\dfrac{0.703}{0.328}\right)}{29} + \frac{1}{14.914}\right]^{-1}$$

$$U = 6.394\,\frac{\text{Btu}}{\text{hr-ft}^2\text{-}^\circ\text{F}}$$

A = heat transfer area

$$A = 4(1{,}014.389)\text{ in. }(1.375)\text{ in.} = 5{,}579.140\text{ in.}^2$$

or

$A = 38.744\text{ ft}^2$ for all four outer panels

$$q = UA\,\Delta t = (6.394)(38.744)(100) = 24{,}772.333\,\frac{\text{Btu}}{\text{hr}}$$

Total Heat Duty of Tank

At maximum flow rates the total heat duty is as follows:

$$q = \underbrace{137{,}698.148}_{\text{internal panels}} \frac{\text{Btu}}{\text{hr}} + \underbrace{209{,}414.44}_{\text{shell side panels}} \frac{\text{Btu}}{\text{hr}}$$

$$+ \underbrace{6{,}368.71}_{\text{two inner panels on bottom head}} \frac{\text{Btu}}{\text{hr}} + \underbrace{24{,}772.333}_{\text{four outer panels on bottom head}} \frac{\text{Btu}}{\text{hr}}$$

$$q = 378{,}253.631 \frac{\text{Btu}}{\text{hr}}$$

From Example 3-4 the total heat duty required is

$$q_{min} = 239{,}832.162 \frac{\text{Btu}}{\text{hr}}$$

Now,

$$q = \dot{m}\, C_p\, \Delta t$$

$$\dot{m} = \frac{378{,}253.631 \frac{\text{Btu}}{\text{hr}}}{(0.50) \frac{\text{Btu}}{\text{lb-°F}} (100)\text{°F}} = 7{,}565.073 \frac{\text{lb}}{\text{hr}}$$

$$\dot{m}_{min} = \frac{239{,}832.162 \frac{\text{Btu}}{\text{hr}}}{(0.50) \frac{\text{Btu}}{\text{lb-°F}} (100)\text{°F}} = 4{,}796.64 \frac{\text{lb}}{\text{hr}}$$

$$Q_{max} = \frac{7{,}565.073 \frac{\text{lb}}{\text{hr}}}{58.7 \frac{\text{lb}}{\text{ft}^3}} = 128{,}877 \frac{\text{ft}^3}{\text{hr}} \left(\frac{1 \text{ hr}}{60 \text{ min}}\right)\left(\frac{7.48 \text{ gal}}{\text{ft}^3}\right)$$

$$= 16.067 \text{ gpm}$$

$$Q_{min} = \frac{4{,}796.640 \frac{\text{lb}}{\text{hr}}}{58.7 \frac{\text{lb}}{\text{ft}^3}} = 81{,}714 \frac{\text{ft}^3}{\text{hr}} \left(\frac{1 \text{ hr}}{60 \text{ min}}\right)\left(\frac{7.48 \text{ gal}}{\text{ft}^3}\right)$$

$$= 10.187 \text{ gpm}$$

Thus, the minimum hot oil flow rate in pipe header supplying the total hot oil to surge tank is 10 gpm, the actual flow rate is 16 gpm.

EXAMPLE 3-6: TRANSIENT AND STATIC HEAT TRANSFER DESIGN

Roofing shingles are made by adding asphalt, filler material, granules, talc and adhesives to a plastic-glass sheet, which is the basic component of the roof shingle. The process is shown schematically in Figure 3-23. Granules are added last, after adhesives and talc. The sheet must be cooled so that workers can handle it with gloves. Cooling is accomplished by water sprays, circulating water through the rollers, and using radiant heat transfer to the surroundings. The sheet, once cooled to the desired temperature, is cut into specified dimensions by mechanical cutters and then packaged into boxes for shipment.

There are two aspects to this problem—static heat transfer and transient heat transfer. First, we solve the static conditions and then the transient case to determine how fast the sheet can be cooled with the cooling system designed in the static case.

Static Heat Transfer Analysis

The static criteria to be determined are as follows:

1. Specific heat of the composite sheet (Table 3-2).
2. Mean temperature of the sheet leaving the granule section (Table 3-3).

Table 3-2
Specific Heat of Composite Sheet Leaving Granule

Component	wt., lb	% by wt.	Component C_p	% of C_p
Glass mat	6.30	1.9785	0.2	0.3957
Asphalt	87.32	27.4220	0.4	10.9688
Filler	120.58	37.8682	0.217	8.2174
Granules	96.08	30.1740	0.20	6.0348
Talc	6.11	1.9200	0.2	4.0128
Adhesives	2.03	0.6375	0.50	0.3188
	318.42	100.00		29.9483

$$C_p = \frac{29.9483}{100} = 0.299 \text{ Btu/lb-°F}$$

Table 3-3
Mean Temperature of Sheet Leaving Granule Section

$Q = mC_p \Delta t$ for each component

Component	Temp. of component prior to mixing	C_p	% by wt.
Glass	80°F	0.2	1.9785
Asphalt	400°F	0.4	27.422
Filler	400°F	0.22	37.8682
Granules	80°F	0.2	30.17398
Talc	80°F	0.2	1.92
Adhesives	400°F	0.5	0.6375

$$(1.9785)(0.2)(80 - t_m) + (27.422)(0.4)(400 - t_m) + (37.8682)(0.22)(400 - t_m)$$

$$= (30.17398)(0.2)(t_m - 80) + (1.92)(0.2)(t_m - 80) + (0.6375)(0.5)(t_m - 80)\ 31.656 - 0.396\ t_m + 4387.520 - 10.969\ t_m + 3332.402 - 8.331\ t_m$$

$$= 6.035\ t_m - 482.784 + 0.384\ t_m - 30.720 + 0.319\ t_m - 25.50$$

where $t_m = 313.633°F$

3. Heat to be removed from sheet and amount of water required.

 a. $t_m = 313.63°F$ (from Table 3-3)
 $\Delta t = 313.63 - 212 = 101.63°F$
 Specific heat of sheet = 0.299 Btu/lb-°F
 Weight of sheet = 0.9375 lb/ft²

 Thus, the amount of heat to be removed per square foot is

 $mC_p\Delta t = (0.9375)(0.299)(101.63) = 28.49$ or 29 Btu/ft²

 b. Sensible heat loss through rollers $= 3.0 \times 10^6$ Btu/hr

 c. Heat loss through forced convection and radiation of heat passing through air medium is determined as follows:

 $\Delta t = 313.63 - 90 = 223.63°F$ = temperature difference between sheet and ambient air

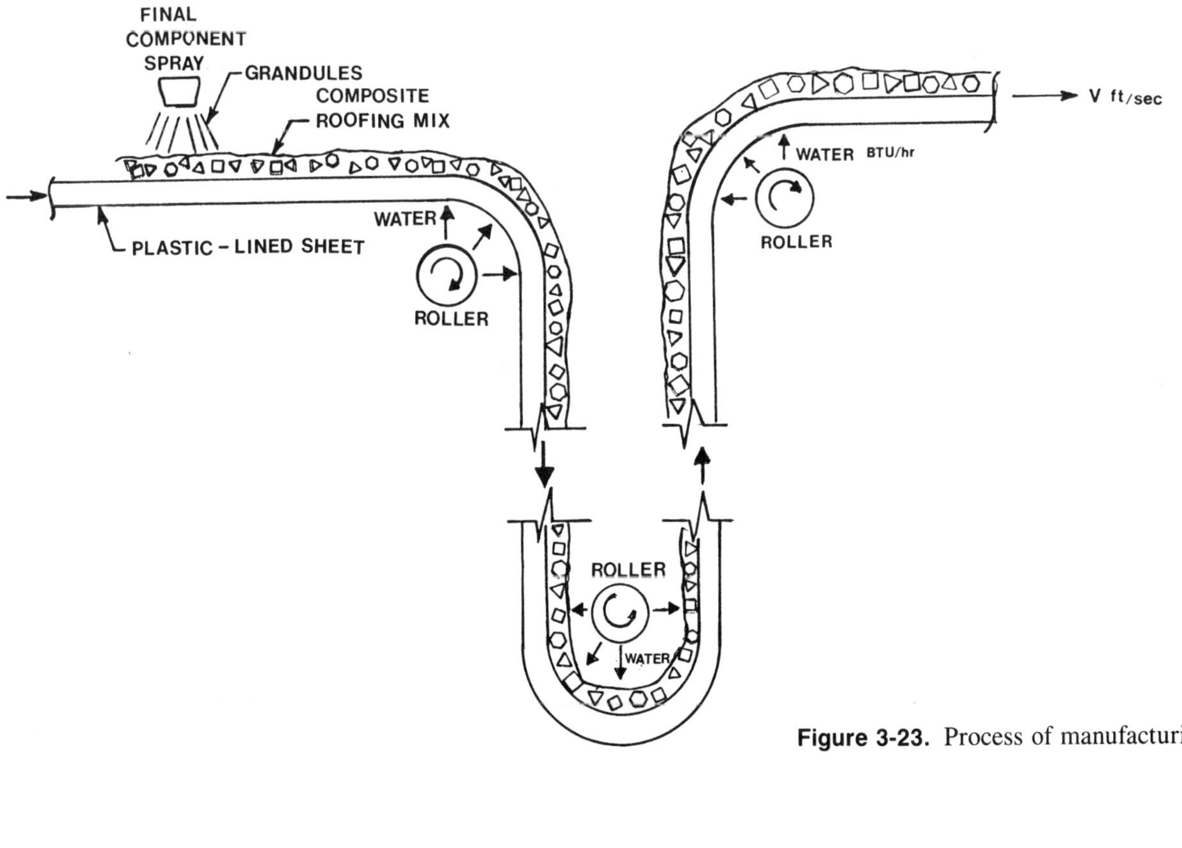

Figure 3-23. Process of manufacturing roof shingles.

For convection,

$$Q = hA\Delta t$$

For flowing air, $h_{min} = 2$ Btu/hr/ft²/°F
$h_{max} = 50$ Btu/hr/ft²/°F

Use $h_c = 23$ Btu/hr/ft²/°F

$$A = \frac{(68.5)}{12}(1) = 5.708 \text{ ft}^2$$

For radiation,

$$h_r = F_e F_A \left[\frac{\sigma(t_1^4 - t_2^4)}{(t_1 - t_2)}\right]$$

$F_A = 1.0$; $F_e = \epsilon \approx 0.90$; $\sigma = 0.173 \times 10^{-8}$

$$h_r = (0.90)(1.0) \times \left[\frac{(0.173 \times 10^{-8})[(313.63 + 460)^4 - (90 + 460)^4]}{(313.63 - 90)}\right]$$

$$h_r = 1.857 \text{ Btu/hr/ft}^2/°\text{F}$$

Total Heat Removal

$$h_T = h_c + h_r = 23.000 + 1.857 = 24.857 \text{ Btu/hr/ft}^2/°\text{F}$$

$$Q = h_T A(\Delta t) = (24.857)(5.708)(223.63) = 31{,}729.464 \text{ Btu/hr}$$

Water Required for Cooling

Let Q_V = heat removed by vaporization

The heat removed for a sheet 6 feet wide moving at 100 ft/min is

$$Q_V = (100)(6) \text{ ft}^2/\text{min } (29) \text{ Btu/ft}^2 = 17{,}400 \text{ Btu/min}$$

For a sheet velocity of 600 ft/min,

$$Q_V = 104{,}400.00 \text{ Btu/min}$$

$$Q_V = 6{,}264{,}000.00 \text{ Btu/hr}$$

Total heat removal = 6,264,000 + 31,729.464 + 3,000,000
Q = 9,295,729.464 Btu/hr

For 600 shingles/hr (or 144,000 Btu/hr) the heat removal would be = (144,000)(0.9375)(0.299)(313.63 − 125)

= 7.614 mm Btu/hr

Thus, $\text{Total heat removal} > \text{Heat removal required}$

and the cooling system is adequate.

For vaporization, q = 29 Btu/ft²

$$Q_V = 104{,}400.00 \text{ Btu/min}$$

For water, h_{fg} = 1,000 Btu/bb

At 600 fpm, we have,

$$\frac{104{,}400}{1{,}000} = 104.40 \text{ lb/min} = \text{amount of water required}$$

Thus,

$$\text{gpm} = \frac{104.40 \frac{\text{lb}}{\text{min}}}{8.34 \frac{\text{lb}}{\text{gal}}} = 12.518 \text{ gpm}$$

Thus, the water pump to be used is to be sized for 13 gpm at a terminal exit pressure of 200 psi.

Transient Heat Transfer Analysis

This method is based on the Fourier analysis of unsteady-state heat conduction. The following assumptions are made:

1. The composite sheet is approximated by a material of average conductivity.
2. The sheet is infinitely long and is an isotropic rigid solid.

Figure 3-23 shows a view of the roofing slab. Assuming that the material is a composite sheet approximated by an integral sheet of average properties, the temperature distribution is at all times symmetric about the midplane of the slab, thus x = 0 at the center of the sheet. From Fourier's law of conduction,

$$Q = -k\frac{\partial t}{\partial x}$$

The heat transfer across x = 0 is zero and at the midplane of x = 0 the sheet behaves as a perfect insulator—an adiabatic surface. Consequently, the solution to this problem applies to a slab that is perfectly insulated at one face, initially at a known temperature, t_o, and then exposed on one face to a fluid at a constant temperature, t_f.

Temperature of the sheet = 314°F = t_o

Temperature of the spray water = 90°F = t_f

Here we are spraying water on the sheet and we wish to determine the time required for the sheet to reach 125°F.

$$\theta_o = 314 - 90 = 224°F$$

$$\theta_{x/L=0} = 125 - 90 = 35°F$$

$$\left.\frac{\theta}{\theta_o}\right|_{x=L} = \frac{35}{224} = 0.156$$

Thickness of sheet = 3/16 in. in which L = 3/32 in. = 0.094 in.

For water,
Surface coefficient (worst condition) = 300 Btu/hr-ft²-°F

$$\frac{k}{h_L} = \frac{0.30}{(300)(0.094)} = 0.011$$

N_{Fo} = Fourier number

From Figure 3-24, $N_{Fo} = 0.90 = \frac{\alpha\tau}{L^2}$

$$\alpha = \frac{k}{\rho\, C_p} = \frac{0.30}{(0.9375)(0.299)} = 1.070 \text{ ft}^2\text{/hr}$$

$$\tau = \frac{0.90(0.094)^2}{1.070} = 0.007 \text{ hr} = 0.446 \text{ min}$$

or τ = 26.756 ≈ 27 sec for 90° water

Approximate length of sheet exposed to nozzle spray = 150 ft
Velocity of sheet = V_s

For V_s = 600 ft/min,

$$\tau = \frac{150 \text{ ft}}{600 \text{ ft/min}} = 0.25 \text{ min} = 15 \text{ sec}$$

For V_s = 400 ft/min,

$$\tau = \frac{150 \text{ ft}}{400 \text{ ft/min}} = 0.375 \text{ min} = 22.5 \text{ sec}$$

The length of the cooling section and the velocity of the sheet are both fixed. The only parameter not fixed is the temperature of the water.

Thus,

$$\tau = \frac{N_{Fo}L^2}{\alpha} = 0.00417$$

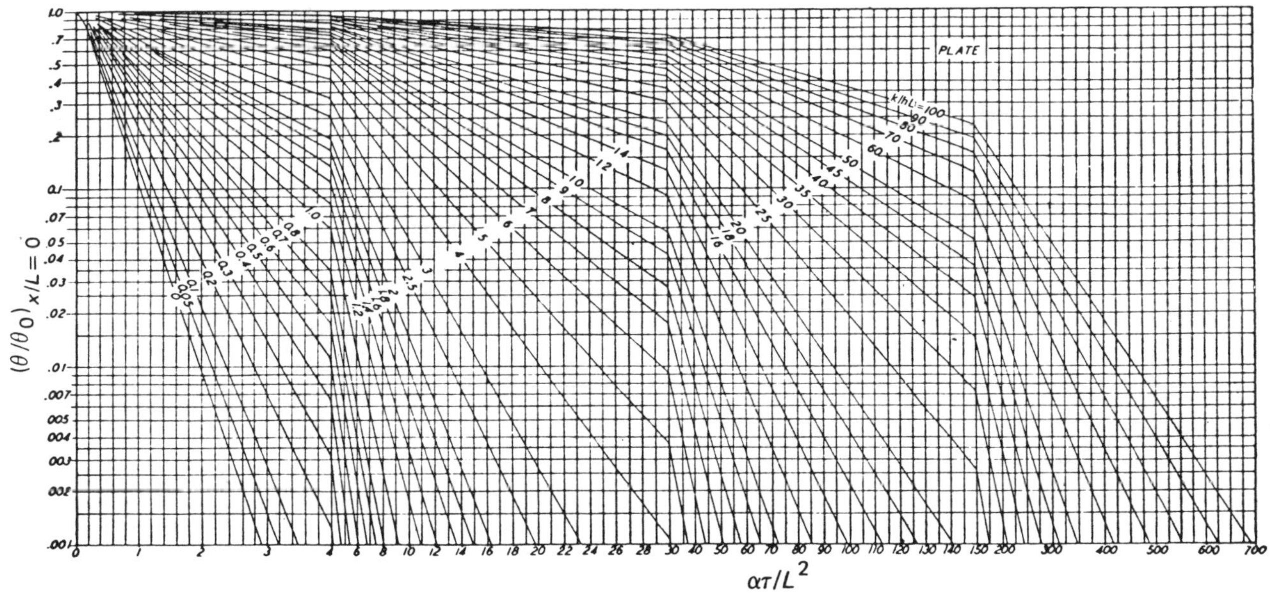

Figure 3-24. Heisler's main chart for the infinite slab [7].

$$(N_{Fo})_{req'd} = \frac{(0.00417)(1.070)}{(0.094)^2} = 0.505$$

For a Fourier number of 0.505,

$$\left(\frac{\theta}{\theta_o}\right)_{x/L=0} = 0.420$$

Let t_w = required water temperature

$$\frac{125 - t_w}{314 - t_w} = 0.420$$

in which $t_w = -11.86°F$ and is well below the freezing temperature of water.

Thus, for a cooling section of 150 feet long, the sheet moving at 600 ft/min cannot be cooled to 125°F since the theoretical value of t_w is below freezing.

At V_s = 400 ft/min,
τ = 0.375 min = 0.0063 hr

$$(N_{Fo})_{req'd} = \frac{(0.0063)(1.070)}{(0.094)^2} = 0.7568$$

$$\left(\frac{\theta}{\theta_o}\right)_{x/L=0} = 0.180$$

$$\frac{125 - t_w}{314 - t_w} = 0.18$$

and

t_w = 83.51°F for a sheet velocity of 400 ft/min

Thus, the sheet can be reasonably cooled while moving at 400 ft/min. If a velocity of 600 ft/min is desired, additional water sprays must be added. However, one must balance the sheet velocity against the cutting machines and workers' capability to handle the additional material. It is found in most roof shingle plants that 400 ft/min is an optimum velocity. As demonstrated, the transient heat transfer analysis is mandatory in evaluating a system.

EXAMPLE 3-7: HEAT TRANSFER THROUGH VESSEL SKIRTS

Calculate the temperature distribution down the length of a vessel skirt. The vessel contains a cold process fluid that varies in temperature because of cyclic process conditions. Three operating temperatures are to be analyzed, −200°F, −100°F, and −50°F. The skirt is made of Type 304 stainless steel and is insulated on the inside and outside as shown in Figure 3-28. The insulation is sized for the most extreme process temperature that the vessel will be exposed to, −200°F. Data used in the example are given in Figure 3-25.

First, determine the natural convection film coefficient for the skirt. The temperature inside the skirt, t_i, is assumed to be five degrees lower than the ambient temperature, t_5.

$$U_4 = \left[\frac{r_1 \ln (r_2/r_1)}{k_{2\text{-}1}} + \frac{r_4 \ln (r_3/r_2)}{k_{3\text{-}2}} + \frac{r_4 \ln (r_4/r_3)}{k_{3\text{-}4}} + \frac{1}{h_{4\text{-}5}}\right]^{-1}$$

$$= \left[7.115 + \frac{1}{h_{4\text{-}5}}\right]^{-1}$$

Assume $h_{4\text{-}5}$ = 0.275

U_4 = 0.093

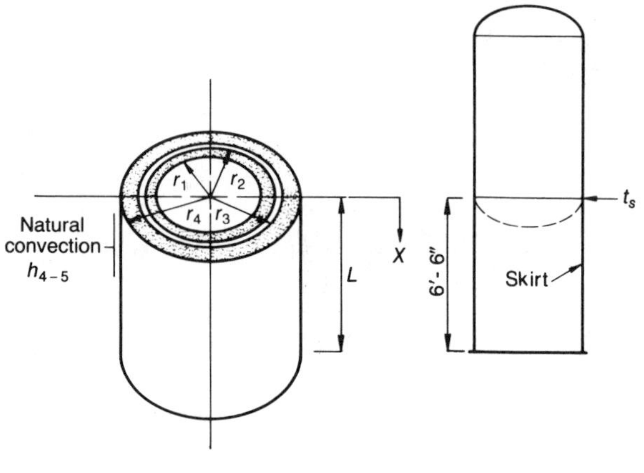

t_i = 55°F
t_5 = 60°F
r_1 = 30 7/8-in. = 2.573 ft
r_2 = 36 7/8-in. = 3.073 ft
r_3 = 37 5/8-in. = 3.135 ft
r_4 = 43 5/8-in. = 3.635 ft
$A_m = (r_3^2 - r_2^2) = 1.209\ ft^2$
β = 1/(460 + 60) = 0.001923
γ = 0.07633 lb/ft³
μ = 0.04339 lb/ft h
k_{air} = 0.01466 Btu/h ft °F
k_{3-2} = 8.0 Btu/h ft °F
$k_{2-1} = k_{3-4}$ = 0.14 Btu/h ft °F
N_{Pr} = 0.712

Figure 3-25. Cryogenic pressure vessel with internal and external insulation on the skirt.

$$t_4 = \left(\frac{U_4}{h_{4\text{-}5}}\right)(t_i - t_5) + t_5$$

$$= \left(\frac{0.093}{0.275}\right)(-5) + 60$$

$$= 58.31°F$$

$$t_4 - t_5 = 58.31 - 60 = -1.69°F$$

$$N_{Gr} = [d^3\gamma^2 g\beta(|\Delta t|)(3{,}600)^2]/\mu^2$$

$$= \frac{[(7.27)^3(0.07633)^2(32.2)(0.001923)(1.69)(3{,}600)^2]}{(0.04339)^2}$$

$$N_{Gr} = 1{,}613{,}720{,}723$$

where $d = 2r_4 = 2(3.635) = 7.27$ ft

$$N_{Gr}N_{Pr} = (1{,}613{,}720{,}723)(0.712) = 1{,}148{,}969{,}155 > 10^9$$

$C = 0.129$ and $m = 1/3$

$$N_{Nu} = C(N_{Gr}N_{Pr})^m$$

$$= 0.129(1{,}148{,}969{,}155)^{1/3} = 135.11$$

$$h_{4\text{-}5}' = \frac{(k_{air}\,N_{Nu})}{d} = \frac{(0.01466)(135.11)}{7.27} = 0.2725$$

$$t_4' = \left(\frac{0.093}{0.2725}\right)(-5) + 60 = 58.29°F$$

$$\Delta t_4' = 58.31 - 58.29 = 0.02$$

$$h_{4\text{-}5} = 0.275$$

For a cylinder with insulation on both sides,

$$Q = \left[\frac{(2\pi k_{2\text{-}1})}{k_m A_m}\right]\left[\frac{1}{\left|\ln\left(\frac{r_1}{r_2}\right)\right|} + \frac{1}{\ln\left(\frac{r_4}{r_3}\right)}\right]$$

$$= \left[\frac{2\pi(0.14)}{(8)(1.209)}\right]\left[\frac{1}{\left|\ln\left(\frac{2.573}{3.073}\right)\right|} + \frac{1}{\ln\left(\frac{3.635}{3.135}\right)}\right]$$

$$Q = 1.1267\ \text{ft}^{-2}$$

$$Z = \left|\frac{2\pi}{k_m A_m}\left[r_4\, h_{4\text{-}5}\,(t_4 - t_5) - k_{2\text{-}1}\left[\frac{t_4}{\ln\left(\frac{r_4}{r_3}\right)} - \frac{t_i}{\ln\left(\frac{r_1}{r_2}\right)}\right]\right]\right|$$

$$= \left|\frac{2\pi}{(8)(1.209)}\left[(3.635)(0.275)(-1.69) - 0.14 \times \left[\frac{58.31}{\ln\left(\frac{3.635}{3.135}\right)} - \frac{55.00}{\ln\left(\frac{2.573}{3.073}\right)}\right]\right]\right|$$

$$Z = 65.102°F/ft^2$$

$$\overline{Z} = \frac{Z}{Q} = \frac{65.102}{1.1267} = 57.781°F$$

For $t_s = -200°F$,

$$t_x = \frac{2(t_s - \overline{Z})e^{xQ^{0.5}}}{1 + e^{2xQ^{0.5}}} + \overline{Z}$$

$$t_x = \frac{(-515.56)(2.89)^x}{1 + (2.89)^{2x}} + 57.781$$

Similarly for:

$t_s = -100°F$,

$$t_x = \frac{(-515.56)(2.89)^x}{1 + (2.89)^{2x}} + 57.781$$

and for $t_s = 50°F$,

$$t_x = \frac{(-215.56)(2.89)^x}{1 + (2.89)^{2x}} + 57.781$$

Figure 3-26 shows these distribution curves.

The axial deflection of the skirt will now be calculated using the first three terms of the stainless steel deflection equation (Equation 3-45). The hyperbolic terms in the equation are not necessary when the steel temperature is greater than −300°F or less than 1000°F,

$$\delta_{ss} = \frac{2(t_s - \overline{Z})[8.96 + (4.11 \times 10^{-3})\overline{Z}]\arctan(e^{LQ^{0.5}})}{(10^6)Q^{0.5}}$$

$$+ \frac{(2.055 \times 10^{-3})\overline{Z}^2 L}{10^6}$$

$$+ \frac{(2.055 \times 10^{-3})(t_s - \overline{Z})^2(e^{2QL^{0.5}} - 1)}{10^6\, Q^{0.5}(1 + e^{2LQ^{0.5}})}$$

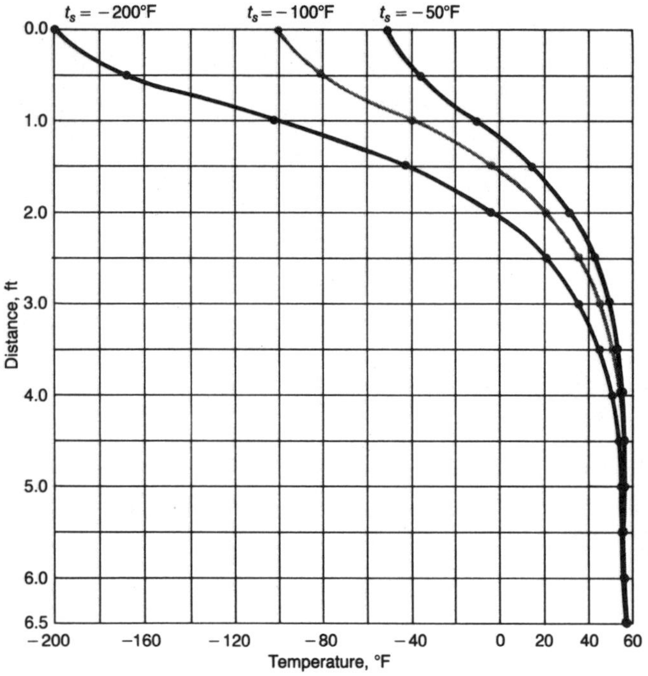

Figure 3-26. Temperature distribution for the three shell-skirt junction temperatures.

$$- \frac{(1.06 \times 10^{-6})}{10^6}\left[8(t_s - \bar{Z})^3\right.$$

$$\times \left[\frac{\text{sech}(LQ^{0.5})\tanh(LQ^{0.5}) + \arctan[\sinh(LQ^{0.5})]}{2Q^{0.5}}\right]$$

$$+ \frac{8\bar{Z}(t_s - \bar{Z})^2}{2Q^{0.5}}\left(\frac{e^{2LQ^{0.5}}}{1 + e^{2LQ^{0.5}}}\right)$$

$$\left. + 4\bar{Z}(t_s - \bar{Z})\left[\frac{2}{Q^{0.5}}\arctan(e^{LQ^{0.5}})\right] + \bar{Z}^3L\right]$$

For $t_s = -200°F$, $Q = 1.1267$, $\bar{Z} = 57.781$ and $L = 6.5$

$$\delta_{ss} = 0.00701 + 0.00004 + 0.00013$$

$$\delta_{ss} = 0.00718 \text{ ft} = 0.08616 \text{ in.}$$

From Figure 3-26 we see that for the worst case of $t_s = -200°F$ that $-20°F$ is obtained at $x = 1.75$ ft. At about 2.0 ft and below, the skirt could be of carbon steel construction and considerable material savings could be obtained.

EXAMPLE 3-8: RESIDUAL HEAT TRANSFER

A section of carbon steel process pipe is shown in Figure 3-15c. Three conditions will be analyzed for process fluids at 900°F, 600°F and 300°F, the basic analysis is the same as used in Example 3-7 beginning with the iteration procedure to find the natural convection film coefficient. Note that it is assumed that the temperature inside the empty pipe header, t_i, is 130°F and that the ambient temperature, t_o, is 60°F.

$$U_3 = [(r_3 \ln(r_2/r_1)/k_m) + r_3 \ln(r_3/r_2)/k_i + {}^1/h_o]$$
$$= [(0.526 \ln(0.267/0.2527)/25) + 0.526 \ln(0.526/0.276)/0.027 + {}^1/h_o]^{-1}$$

$$U_3 = [12.565 + {}^1/h_o]^{-1}$$

Let $h_o = 1.0$ Btu/hr-ft^2-°F

$$U_3 = 0.0737$$

$$t_3 = (U_3/h_o)(t_i - t_o) + t_o$$
$$= (0.0737/1.0)(130 - 60) + 60 = 65.16°F$$

$$\Delta t = t_3 - t_o = 5.16°F$$

$$N_{Gr} = [d^3\alpha^2 g\beta(|\Delta t|)(3{,}600)^2]/\mu^2$$
$$= [(1.052)^3(0.07633)^2(0.001923)(32.2) \times (3{,}600)^2(5.16)]/(0.04339)^2$$
$$= 14{,}920{,}198.65$$

$$N_{Gr}N_{Pr} = (14{,}920{,}198.65)(0.712) = 10{,}623{,}181.44$$

$$N_{Nu} = C(N_{Gr}N_{Pr})^m$$

where,

$C = 0.525$ and $m = {}^1/_4$
$N_{Nu} = 0.525\,(10{,}623{,}181.44)^{1/4} = 29.97$
$h_o' = (k_{air}/d)N_{Nu} = (0.01466/1.052)\,29.57$
$= 0.4177$
$t_3' = (U_3/h_o')(t_i - t_o) + t_o = (0.0737/0.4177)(130 - 60) + 60$
$t_3' = 72.35°F$
$\Delta t_3' = t_3 - t_3' = 65.16 - 72.35 = -7.19°F$
too large, try another trial valve for h_o.

Let $h_o = 0.49$ Btu/hr-ft-°F
$U_3 = 1/(12.565 + 1/0.49) = 0.0687$ Btu/hr-ft^2-°F
$t_3 = (0.06847/0.49)\,70 + 60 = 69.781°F$
$\Delta t = 9.781°F$
$N_{Gr} = 28{,}279{,}559.99$
$N_{Gr}N_{Pr} = 20{,}135{,}046.71$
$N_{Nu} = 35.17$
$h_o' = 0.4901$ Btu/hr-ft-°F
$t_3' = (0.06847/0.4901)\,70 + 60 = 69.779°F$
$\Delta t_3' = 69.781 - 69.779 = 0.002°F < 0.1$

$h_o = 0.49$ Btu/hr-ft-°F
$Q = 2\pi k_i/[k_M A_M \ln (r_3/r_2)]$
$Q = 2\pi(0.027)/[25(0.0387) \ln (0.526/0.276)] = 0.272 \text{ ft}^{-2}$
$Z = | [2\pi/(k_m A_m)][r_3 h_o(t_3 - t_o) - k_i t_3/\ln (r_3/r_2)] |$
$Z = | [2\pi/[25 (0.0387)][0.526(0.49)(69.78 - 60) - 0.027 (69.78)/\ln (0.526/0.276)] |$
$Z = | -2.607 | = 2.607°\text{F/ft}^2$
$\overline{Z} = Z/Q = 9.587°\text{F}$
$t_x = [2 (t_s - \overline{Z})\, e^{xQ^{0.5}}/(1 + e^{2xQ^{0.5}})] + \overline{Z}$
$= [2 (t_s - 9.587)e^{x(0.272)^{0.5}}/(1 + e^{2x(0.272)^{0.5}})] + 9.587$

For $t_s = 900°$F,

$$t_x = \frac{1{,}780.83\,(1.313)^x}{[1 + (1.313)^{2x}]} + 9.587$$

For $t_s = 600°$F,

$$t_x = \frac{1{,}180.83\,(1.313)^x}{[1 + (1.313)^{2x}]} + 9.587$$

For $t_s = 300°$F

$$t_x = \frac{580.83\,(1.313)^x}{[1 + (1.313)^{2x}]} + 9.587$$

Curves depicting t_x are shown in Figure 3-27. Unlike Example 3-7, the slopes of the curves change much less, almost approaching straight lines.

Axial temperature gradients along a section of piping produce thermal deflections. The pipe support will now be analyzed for thermal deflections.

The surface temperature, t_s, of the branch pipe at the point of the contact with the header is 600°F. The average temperature inside the pipe may be calculated from the 600°F curve in Figure 3-27 which shows a temperature at a distance x of five feet to be 294°F.

$$t_i = (600 + 294)/2 = 447°\text{F}$$

Through the process of iteration, $h_o = 0.68$ Btu/hr-ft-°F at the average internal temperature of 447°F. This was obtained using the natural convection iteration technique described in Example 3-7.

Using the same techniques, $Q = 0.2719 \text{ ft}^{-2}$, $Z = 66.7916°\text{F/ft}^2$, and $\overline{Z} = 245.6476°$F.

To calculate the axial deflection, substitute these values into the expanded thermal deflection equation for carbon steel, Equation 3-39. Note: Values for the arctan used in the equation must be calculated in radians. Calculate the arctan m degrees and convert to radians in which the relationship is 2π radians = 360 degrees. Using equation 3-39,

$$\delta_{cs} = \frac{(14.233)(354.352)(1.497)}{(0.521 \times 10^6)} + \frac{(2.496 \times 10^{-3})(125{,}565.623)(182.902)}{(0.521 \times 10^6)(184.902)}$$
$$+ 0.0008 - (6.536 \times 10^{-13})\,[558{,}494{,}713.0 + 2706.95 + 2000884.26 + 74{,}115{,}250.45]$$

$$\delta_{cs} = 0.0155 \text{ ft} = 0.1860 \text{ in. axial deflection}$$

This example shows that residual heat through a closed branch line can be significant enough to cause thermal movements, which can result in high stresses. These thermal deflections are particularly important when space is limited and the piping system has little flexibility.

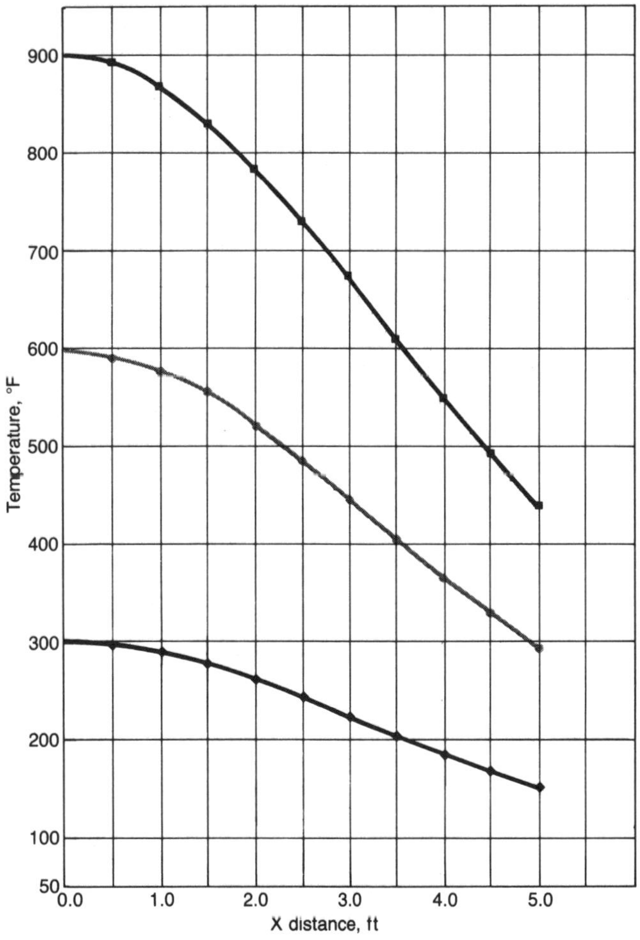

Figure 3-27. Temperature of a branch pipe connected to a header through a closed valve plotted from the pipe to valve connection every six inches for a distance of five feet in Figure 3-15C.

EXAMPLE 3-9: HEAT TRANSFER THROUGH PIPE SHOE

A 12-in process header shown in Figure 3-28 is supported by a shoe 14-in. long. The process fluid is at 750°F and it is desired to determine the temperature of the bottom of the shoe base plate where Teflon is mounted to accommodate pipe movement. The Teflon cannot withstand a temperature in excess of 400°F. Referring to Figure 3-28 and using Equation 3-47 we have

$$t_p = \frac{k_m A_m t_i + h_o A_p L t_o}{(k_m A_m + h_o A_p L)} \text{ °F}$$

h_o = 3.0 Btu/hr-ft²-°F for carbon steel in still air

k_m = 26.0 Btu/hr-ft-°F

L = 10.0 in.

$A_m = (0.375)(14) = 5.25 \text{ in.}^2$

$A_p = (8.0)(14) = 112 \text{ in.}^2$

$t_o = 90°F$

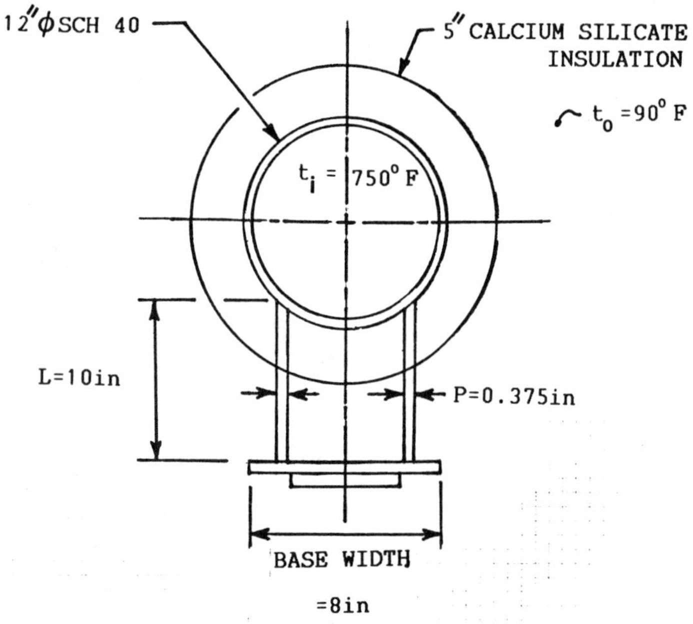

Figure 3-28. Heat transfer through pipe shoe.

$$t_p = \frac{(26.0)\dfrac{\text{Btu}}{\text{hr-ft-°F}}\left(\dfrac{5.25 \text{ in.}^2}{144 \text{ in.}^2}\right)\text{ft}^2 (750)\text{ °F} + (3.0)\dfrac{\text{Btu}}{\text{hr-ft}^2\text{-°F}}\left(\dfrac{112 \text{ in.}^2}{144 \text{ in.}^2}\right)\text{ft}^2\left(\dfrac{10}{12}\right)\text{ft } (90)\text{°F}}{\left[(26.0)\dfrac{\text{Btu}}{\text{hr-ft-°F}}\left(\dfrac{5.25 \text{ in.}^2}{144 \text{ in.}^2}\right)\text{ft}^2 + (3.0)\dfrac{\text{Btu}}{\text{hr-ft}^2\text{-°F}}\left(\dfrac{112 \text{ in.}^2}{144 \text{ in.}^2}\right)\text{ft}^2\left(\dfrac{10.0}{12}\right)\text{ft}\right]}$$

$t_p = 306.303°F$

Thus, the Teflon on the base is adequately protected. The amount of heat loss through the shoe base plate is

$$q = h_o A_p (t_p - t_o)$$

$$= (3.0)\frac{\text{Btu}}{\text{hr-ft}^2\text{-°F}}\left(\frac{112 \text{ in.}^2}{144 \text{ in.}^2}\right)\text{ft}^2 (306.303 - 90)\text{°F}$$

$$= 504.706 \text{ Btu/hr}$$

NOTATION

A_m = area of metal in pressure vessel shell or pipe, ft²
A_o = outside surface area of insulation, ft²/ft
A_p = outside surface area of pipe, ft²/ft
A_t = outside surface area of tracer tube or HTC, ft²/ft
C_p = specific heat, Btu/lb-°F
d = outside diameter of a pressure vessel, ft
D = diameter, ft, in.
D_i = inside diameter of pipe insulation, ft, in.
D_o = outside diameter, in.
D_{ti} = inside of outer ring of annulus, in.
d_{so} = outside diameter of inner ring of annulus, in.
D_t = inside diameter of tracer tube, in.
g = acceleration of gravity, 32.2 ft/sec²
h_o = natural convection coefficient at OD of the piping insulation, Btu/hr-ft²-°F
h_o' = corrected value for h_o, Btu/hr-ft²-°F
h_p = convection coefficient, pipe to air space, Btu/hr-ft²-°F
h_t = convection coefficient, tracer or HTC to air space, Btu/hr-ft²-°F
$h_{4\text{-}5}$ = convection coefficient between the outside vessel insulation and ambient air, Btu/hr-ft²-°F

h_{4-5}' = corrected convection coefficient, Btu/hr-ft²-°F
k_i = insulation conductivity, Btu/hr-ft-°F
k_i = thermal conductivity of air inside empty pipe, Btu/hr-ft-°F
k_m = thermal conductivity of vessel skirt or pipe, Btu/hr-ft-°F
L = length of branch pipe, ft
N_{Gr} = Grashof number, dimensionless
N_{Nu} = Nusselt number, dimensionless
N_{Pr} = Prandl number, dimensionless
N_{Re} = Reynolds number, dimensionless
Q = heat transfer factor, ft^{-2}
Q_1 = heat transfer from air space to atmosphere, Btu/hr
Q_2 = heat transfer from tracer to air space, Btu/hr
Q_3 = heat transfer from pipe to air space, Btu/hr
Q_4 = heat transfer from tracer to pipe, Btu/hr
t_a = air space temperature, °F
t_i = process fluid temperature, °F
t_i = air temperature inside the vessel skirt, pipe support or branch pipe, °F
t_s = surface temperature of the branch pipe at contact point with the header, or operating temperature in a pressure vessel, °F
t_x = temperature at distance x along the vessel skirt, pipe support or branch pipe, °F
t_o, t_5 = ambient temperature, °F
t_3, t_3' = temperature and corrected temperature at OD of the pressure vessel insulation, °F
t_4, t_4' = temperature and corrected temperature at OD of the pressure vessel insulation, °F
Δt = $t_3 - t_o$ in piping example, $t_4 - t_5$ in vessel skirt example, °F
$\Delta t'$ = $t_4 - t_4'$ in vessel skirt example, °F
$\Delta t_3'$ = $t_3 - t_3'$ in piping example, °F
U_3 = overall heat transfer coefficient at OD of pipe insulation, Btu/hr-ft²-°F
U_4 = overall heat transfer coefficient at r_4, Btu/hr-ft²-°F
X = distance of plotted temperature points along the vessel skirt or piping, ft
$\underline{Z}$ = heat transfer factor, °F/ft²
$\overline{Z}$ = heat transfer factor, Z/Q, °F

Greek Symbols

β = volumeric coefficient of thermal expansion, ′/K
δ_{cs}, δ_{ss} = axial deflection of carbon or stainless steel skirt or pipe, in.
γ = safety factor for traced pipe
μ = absolute viscosity, lb/ft-hr
ρ = density, lb/ft³

REFERENCES

1. Tubular Exchanger Manufacturers Association, *Standards of the Tubular Manufacturers Association (TEMA),* sixth edition, New York, N.Y., 1978.
2. Kern, Donald, *Process Heat Transfer,* McGraw-Hill Book Company, 1950.
3. Ludwig, Ernest E., *Applied Process Design for Chemical and Petrochemical Plants,* volume 3, second edition, Gulf Publishing Company, Houston, Texas, 1983.
4. Escoe, A. Keith, "Heat Transfer in Vessels and Piping," *Hydrocarbon Processing,"* January, 1983, vol. 62, no. 1, Gulf Publishing Company, Houston, Texas.
5. Chapman, Allen E., *Heat Transfer,* third edition, Macmillan Publishing Company, New York, 1974.
6. Perry, Robert H. and Don Green, *Perry's Chemical Engineers' Handbook,* sixth edition, McGraw-Hill Book Company, New York, 1984.
7. Heisler, M. P., "Temperature Charts for Induction and Constant Temperature Heating," Transactions of the A.S.M.E., vol. 69 (1947), pp. 227–236.

4

The Engineering Mechanics of Pressure Vessels

The specifying, design, and construction of pressure containing vessels varies all over the globe. Each adopted code that has been used for any significant length of time has proven to be workable because its use has resulted in safe, economic designs. The main differences in codes are the theories of yield that are used for determining maximum allowable stresses, material specifications, and basic procedures.

With increasing international competition and cyclic economic conditions, there is a growing need to emphasize economics and familiarity of foreign codes, and avoid unnecessary overdesign that relies on only one set of codes and standards. This chapter emphasizes the optimization of economics and safety. If you choose to be conservative in your design, you can be; however, if you are bidding in a highly competitive market, you can use these methods to produce a safe, economical design.

International competition and economic conditions have caused engineers to restructure their thinking that a good design uses only enough material that produces a safe and economical product. Thus, this chapter's philosophy is to optimize engineering design within code rules, whatever the code. Overly conservative design that results in excessive material use becomes unproductive and expensive when one is competing in the world market today.

A thorough treatment of vessel engineering and its concomitant aspects of static and dynamic phenomena would fill several volumes. To present this broad subject with clarity, various physical phenomena are briefly discussed and references are made to sources that give detailed theoretical explanations. It is not this book's purpose to give a treatise of static and dynamic problems, but rather descriptions of proven practices. The theory of these problems is always available, but proven solutions are not—hence, the reason for this book.

The first problem you face in designing a vessel containing pressure is how to physically make the components and assemble them. In the petroleum refining industry (CPI—Chemical Process Industry) and allied industries, the most practical and economical method is welding. We will refer to welding later in more detail, but first we will look at the vessel from a pure engineering viewpoint assuming perfect welds with given efficiencies. Some have proposed bonding pressure vessels together with glue, as is done with aircraft components. The main disadvantages to bonding are

1. Clean surfaces are required for assembly.
2. Glues that exhibit high tensile and compressive strengths are very expensive.
3. Chemical bonding, especially in thick-walled vessels, takes much longer than any welding process.

Another form of assembly that has been even more seriously considered than bonding is threading components and screwing them in place. Even though this may appear to be simple, the process becomes enormously expensive with large diameters. Thus, welding is the most practical and economical means of assembling pressure vessels for the foreseeable future.

DESIGNING FOR INTERNAL PRESSURE

The two factors that must be considered in the design for internal pressure are *component thickness* and *quality of welds*. Before either of these two factors can be addressed, you must know what the vessel is to contain. This chapter only considers gases and liquids. Vessels, silos and bins containing solids are discussed in Chapter 5.

In the design for liquids under pressure, the most severe condition of coincident pressure and temperature expected in operation must be considered in computing shell thickness. This is fairly universal in codes throughout the world. The intent of the statement is that the most frequently occurring liquid level should be considered. For example, if a vessel is filled to a certain level "A" 75% of the time and a higher level "B" 25% of the time, level "A" should be used for design purposes. The normal liquid level to be used for vessel design and its quantitative value should be determined by the process engineer. For upset conditions each code allows an increase in allowable stresses under temporary conditions, and you should consult whichever code is to be used for exact amounts allowed. It is recommended that a value of 30 psig or 10% be added to the operating pressure for design pressure. This practice varies with each company throughout the industry.

Once the internal pressure is determined it must be decided how the vessel is to be welded. The factors affecting this decision are as follows:

1. Size of vessel—whether rolled plate or seamless pipe is used.
2. The toxic nature of the fluid to be contained.
3. The economics of fabrication as to whether a full joint efficiency is necessary.

One can appreciate the degree of types of welds required for a vessel. A slug catcher, which acts as a scrubber handling a non-toxic substance, does not require the same caution as a vessel containing cyanide gas.

The quality of a weld joint is determined by a radiographic inspection. Full radiography includes a complete X-ray inspection (100% for butt weld and 90% for single-welded butt joint) and spot radiography implies 85% for butt joints. See Table 4-1 for maximum allowable efficiencies for arc and gas welded joints. The reader is strongly urged to consult whatever code happens to govern. Listed in Table 4-2 are the joint efficiencies for the various welded combinations for pressure vessels under ASME Section VIII, Division I[1].

Any discussion on designing for internal pressure must include maximum allowable working pressure, which is the maximum gauge pressure permissible at the top of the completed vessel in its operating position for a designated temperature. This pressure, MAWP, is normally specified on two conditions—new and cold (ambient) (NAC), and design. "New and cold" implies the MAWP for a new vessel (non-corroded) at atmospheric condition, and "design" implies the vessel corroded at design temperature and pressure. The value of the MAWP at the two conditions gives the exact range of temperature and pressure that the vessel can withstand if the owner decides to use it in another application. The reader is cautioned to consult his respective code on the practice of using a vessel for another application. The following example illustrates how the MAWP is applied:

An ASME Section VIII Division I vessel is made of SA 240—304 SS, design pressure = 500 psig, design temperature = 150°F. The vessel has a shell thickness of 1.00 in. and a joint efficiency of 1.0.

$$\text{MAWP (NAC)} = \frac{(18{,}800) \times (1.00) \times (1.00)}{(21.00) + (0.6) \times (1.00)}$$

$$= 870.4 \text{ psig}$$

$$\text{MAWP (Design)} = \frac{(18{,}300) \times (1.00) \times (1.00)}{(21.00) + (0.6) \times (1.00)}$$

$$= 847.2 \text{ psig}$$

The 18,300 psi is obtained by linear interpolation of the allowable stress values in Table UHA-23 of the ASME Code.

The vessel owner knows the maximum allowable pressure for the shell at the new and cold condition as well as the design condition. It is a common practice to limit the MAWP by the head or shell and not by the flanges or openings, only the MAWP is determined by the flanges or openings when the vessel is to be reapplied in another application or a design oversight is made.

Finally, in computing the minimum thickness of the shell or head, mechanical allowances must be considered. In the manufacture of heads, the metal is thinned on forming the section (a forgery process). This forming allowance must be considered when the *nominal* thickness is specified. When a *minimum* thickness is specified to the head manufacturer, the forming allowance is not considered because it is the manufacturer's responsibility to ensure the minimum thickness.

DESIGNING FOR EXTERNAL PRESSURE

The design for external pressure of vessels is fairly standard in the ASME and codes of other nations. The procedures for determining minimum shell thickness, spacing, and section properties of stiffening rings are straightforward and simple. Because there is much published material on external pressure design, the subject is not discussed here. The reader is urged to consult the pressure vessel code to be used.

Table 4-1
Maximum Allowable Joint Efficiencies for Arc and Gas Welded Joints [1]

No.	Type of Joint Description	Limitations	Degree of Examination (a) Fully Radio-graphed	(b) Spot Examined	(c) Not Spot Examined
(1)	Butt joints as attained by double-welding or by other means which will obtain the same quality of deposited weld metal on the inside and outside weld surfaces to agree with the requirements of UW-35. Welds using metal backing strips which remain in place are excluded.	None	1.00	0.85	0.70
(2)	Single-welded butt joint with backing strip other than those included under (1).	(a) None except as in (b) below (b) Butt weld with one plate offset—for circumferential joints only.	0.90	0.80	0.65
(3)	Single-welded butt joint without use of backing strip.	Circumferential joints only, not over 5/8 in. thick and not over 24 in. outside diameter	. . .	. . .	0.60
(4)	Double full fillet lap joint	Longitudinal joints not over 3/8 in. thick. Circumferential joints not over 5/8 in. thick	. . .	. . .	0.55
(5)	Single full fillet lap joints with plug welds conforming to UW-17	(a) Circumferential joints for attachment of heads not over 24 in. outside diameter to shells not over 1/2 in. thick (b) Circumferential joints for the attachment to shells of jackets not over 5/8 in. in nominal thickness where the distance from the center of the plug weld to the edge of the plate is not less than 1 1/2 times the diameter of the hole for the plug.	. . .	. . .	0.50
(6)	Single full fillet lap joints without plug welds	(a) For the attachment of heads convex to pressure to shells not over 5/8 in. required thickness, only with use of fillet weld on inside of shell; or (b) for attachment of heads having pressure on either side, to shells not over 24 in. inside diameter and not over 1/4 in. required thickness with fillet weld on outside of head flange only.	. . .	. . .	0.45

Table 4-2
Joint Efficiencies for Arc and Gas Welded Joints Per ASME

Joint Types H, C, and L
T1 = Type 1 Joint (ASME UW-12)
T2 = Type 2 Joint (ASME UW-12)

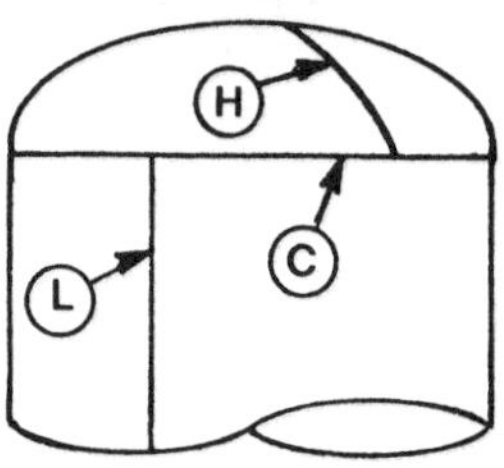

Asterisk (*) denotes which joint type governs.

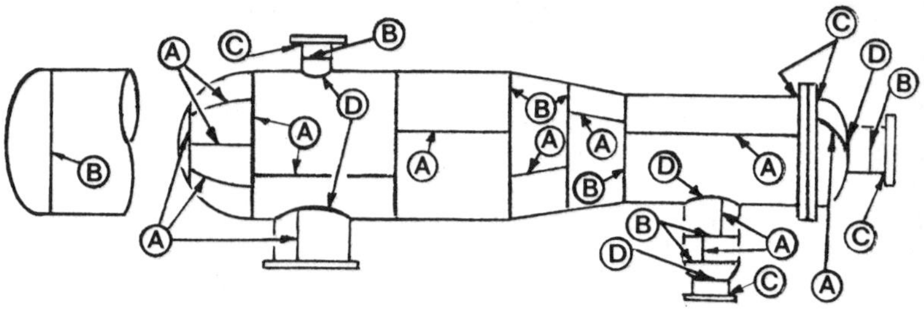

Illustration of weld joint locations Typical of Categories A, B, C, and D—see ASME Section VIII Division I.

Welded Head (Non-Hemispherical)—Welded Shell											
Radiograph Type			Head Thk. Calcu.		Shell Thickness Calculations						
			E			E. Cir. Stress			E. Long Stress		
L*	C*	H*	T1	T2	% σ_{all}	T1	T2	% σ_{all}	T1	T2	% σ_{all}
Full	Full	Full	1.00	0.90							
Full	Full	Spot	0.85	0.80					1.00	0.90	
Full	Full	None									
Full	Part	Full	1.00	0.90		1.00	0.90				
Full	Part	Spot									
Full	Part	None									
Full	Spot	Full									
Full	Spot	Spot	0.85	0.80							
Full	Spot	None									
Full	None	Full									
Full	None	Spot									
Full	None	None									
Spot	Full	Full	1.00	0.90							
Spot	Full	Spot	0.85	0.80							
Spot	Full	None									
Spot	Part	Full	1.00	0.90							
Spot	Part	Spot			100	0.85	0.80	100	0.85	0.80	100
Spot	Part	None									
Spot	Spot	Full									
Spot	Spot	Spot	0.85	0.80							
Spot	Spot	None									
Spot	None	Full									
Spot	None	Spot									
Spot	None	None									
None	Full	Full	1.00	0.90							
None	Full	Spot	0.85	0.80							
None	Full	None									
None	Part	Full	1.00	0.90							

* L Weld governs in circumferential stress calculations.

Table 4-2 continued

Welded Head (Non-Hemispherical)—Welded Shell

Radiograph Type			Head Thk. Calcu. E		Shell Thickness Calculations						
						E. Cir. Stress			E. Long Stress		
L*	C*	H*	T1	T2	% σ_{all}	T1	T2	% σ_{all}	T1	T2	% σ_{all}
None	Part	Spot									
None	Part	None									
None	Spot	Full									
None	Spot	Spot	0.85	0.80	100	0.85	0.80	100	0.85	0.80	100
None	Spot	None									
None	None	Full									
None	None	Spot									
None	None	None	0.70	0.65		0.70	0.65		0.70	0.65	

* L Weld governs in circumferential stress calculations.

Welded Head (Hemispherical)—Welded Shell

Radiograph Type Joint			Head Thickness Calculations E					Shell Thickness Calculations					
								E. Cir. Stress			E. Long Stress		
L*	C*	H*	H = T1 C = T1	H = T1 C = T2	H = T2 C = T1	H = T2 C = T2	% σ_{all}	T1	T2	% σ_{all}	T1	T2	% σ_{all}
Full	Full	Full	1.0	0.90	0.90	0.90							
Full	Full	Spot						1.00	0.90		1.00	0.90	
Full	Full	None											
Full	Spot	Full											
Full	Spot	Spot	0.85	0.80	0.80	0.80							
Full	Spot	None											
Full	None	Full											
Full	None	Spot											
Full	None	Spot											
Spot	Full	Full	1.00	0.90	0.90	0.90							
Spot	Full	Spot											
Spot	Full	None											
Spot	Spot	Full											
Spot	Spot	Spot	0.85	0.80	0.80	0.80	100	0.85	0.80	100	0.85	0.80	100
Spot	Spot	None											
Spot	None	Full											
Spot	None	Spot											
Spot	None	None											
None	Full	Full	1.00	0.90	0.90								
None	Full	Spot											
None	Full	None											
None	Spot	Full											
None	Spot	Spot	0.85	0.80	0.80								
None	Spot	None											
None	None	Full											
None	None	Spot											
None	None	None	0.70	0.65	0.65	0.65		0.70	0.65		0.70	0.65	

Table 4-2 continued

Welded Hemispherical Head—Seamless Shell

Radiograph Type		Head Thickness Calculations: E					Shell Thickness Calculations				
H	C	H = T1, C = T1	H = T1, C = T2	H = T2, C = T1	H = T2, C = T2	% σ_{all}	E. Cir. Stress T1	% σ_{all}	E. Long Stress T1	E. Long Stress T2	% σ_{all}
Full	Full	1.00	0.90	0.90	0.90	↑	↑	↑	1.00	0.90	↑
Full	Spot	↑	↑	↑	↑						
Full	None								0.85	0.80	
Spot	Full								1.00	0.90	
Spot	Spot	0.85	0.80	0.80	0.80	100	1.0	100			100
Spot	None								0.85	0.80	
None	Full								1.00	0.90	
None	Spot	↓	↓	↓	↓			↓	0.85	0.80	
None	None	0.70	0.65	0.65	0.65	↓	↓	80	0.70	0.65	↓

Seamless Hemispherical Head—Welded Shell

Radiograph Type		Head Thick. Calcu.: E			Shell Thickness Calculations: E. Cir. Stress			E. Long Stress		
L*	C*	T1	T2	% σ_{all}	T1	T2	% σ_{all}	T1	T2	% σ_{all}
Full	Full	1.00	0.90	↑			↑	1.00	0.90	↑
Full	Spot	0.85	0.80		1.00	0.90		0.85	0.80	
Full	None				↑	↑				
Spot	Full	1.00	0.90					1.00	0.90	
Spot	Spot	0.85	0.80	100	0.85	0.80	100	0.85	0.80	100
Spot	None									
None	Full	1.00	0.90					1.00	0.90	
None	Spot	0.85	0.80		↓	↓		0.85	0.80	
None	None	0.70	0.65	↓	0.70	0.65	↓	0.70	0.65	↓

* C Weld governs on head and longitudinal stress calculations.
* L Weld governs on shell circumferencial stress calculations.

Seamless Hemispherical Head—Seamless Shell

	Head Thickness Calculations: E			Shell Thickness Calculations				
C	T1	T2	% σ_{all}	E. Cir. Stress	% σ_{all}	E. Long Stress T1	E. Long Stress T2	% σ_{all}
Full	1.00	0.90	↑	↑	100	1.00	0.90	↑
Spot	0.85	0.80	100	1.0		0.85	0.80	100
None	0.70	0.65	↓	↓	80	0.70	0.65	↓

* C Weld governs for head and longitudinal stress calculations.

Table 4-2 continued

Seamless (Non-Hemispherical) Head-Seamless Shell

Head Thick. Calcu.			Shell Thickness Calculations				
			E. Cir.		E. Long Stress		
C*	E	% σ_{all}	Stress	% σ_{all}	T1	T2	% σ_{all}
Full	↑		↑	100	1.00	0.90	↑
Part	1.0		1.0		0.85	0.80	100
Spot		85		85			
None	↓	80	↓	80	0.70	0.65	↓

Seamless (Non-Hemispherical) Head-Welded Shell

Radiograph Type		Head Thick. Calculations		Shell Thickness Calculations					
				E. Cir. Stress			E. Long Stress		
L	C*	E	% σ_{all}	T1	T2	% σ_{all}			% σ_{all}
Full	Full	↑	100	1.00	0.90	↑	1.00	0.90	↑
Full	Part						↑	↑	
Full	Spot		85	↑	↑				
Full	None								
Spot	Full		100						
Spot	Part	1.00		0.85	0.80	100	0.85	0.80	100
Spot	Spot		85						
Spot	None								
None	Full		100						
None	Part								
None	Spot		85	↓	↓		↓	↓	
None	None	↓	80	0.70	0.65	↓	0.70	0.65	↓

Welded (Non-Hemispherical)—Seamless Shell

Radiograph Type		Head Thick. Calculations			Shell Thickness Calculations				
		E			E. Cir.		E. Long Stress		
H*	C*	T1	T2	% σ_{all}	Stress	% σ_{all}	T1	T2	% σ_{all}
Full	Full	1.00	0.90	↑	↑	100	1.00	0.90	↑
Full	Part								
Full	Spot	↑	↑			85	0.85	0.80	
Full	None								
Spot	Full						1.00	0.90	
Spot	Part					100			100
Spot	Spot	0.85	0.80	100	1.0		0.85	0.80	
Spot	None					85			
None	Full					100	1.00	0.90	
None	Part								
None	Spot	↓	↓			85	0.85	0.80	
None	None	0.70	0.65	↓	↓	80	0.70	0.65	↓

* H Weld governs in head calculations.
* C Weld governs in longitudinal stress calculations.

DESIGN OF HORIZONTAL PRESSURE VESSELS

The analysis of horizontal pressure vessels converges on the design for internal pressure and vessel supports. This chapter only considers metal, cylindrical vessels, and focuses on the supports of horizontal pressure vessels.

L. P. Zick [2] of the Chicago Bridge and Iron Company developed the method of analyzing supports for horizontal cylindrical shells in 1951. We will not derive the method, but rather summarize it in a general discussion along with guidelines and useful practices that make the design of such items more straightforward.

Horizontal vessels should be designed to withstand internal and external pressures, and support reactions produced by the vessel weight and additional loads from ladders, platforms, piping, etc. Zick [2] showed that supporting horizontal vessels by more than two saddles is not only inefficient, but incurs additional undesirable problems. Figures 4-1 and 4-2 illustrate a horizontal vessel supported by two saddles.

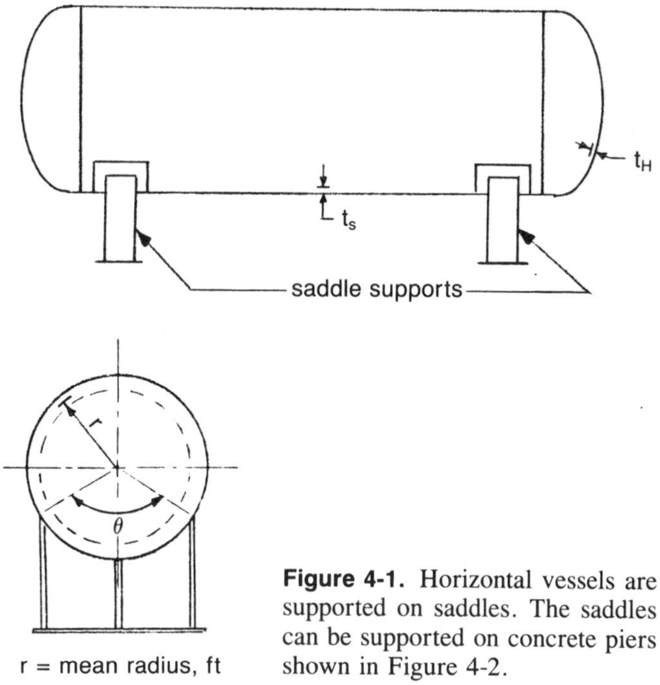

Figure 4-1. Horizontal vessels are supported on saddles. The saddles can be supported on concrete piers shown in Figure 4-2.

Figure 4-2. Horizontal vessel with saddles on concrete piers.

LONGITUDINAL BENDING STRESSES

A horizontal vessel supported on two beams is the same as a beam overhanging two supports. The maximum longitudinal bending stresses occur at the supports themselves and at the center of the vessel, as shown in Figure 4-3. Zick [2] and Brownell and Young [3] give a detailed derivation of the equations for longitudinal bending stresses at the saddle and at mid-span. This analysis is summarized in the following:

At Saddle

σ_l = longitudinal bending stress at saddle
σ_{all} = allowable stress in tension, psi
σ_{all} = B = allowable stress in compression, psi

For tension,

$$\sigma_l = E\sigma_{all} + \sigma_p$$

where E = welding joint efficiency
σ_p = pressure stress, psi

The allowable stress for compression is based on the accepted formula for buckling of short cylindrical columns, which is

$$\sigma_l \leq \left(\frac{E}{29}\right)\left(\frac{t}{r}\right)\left[2 - \left(\frac{2}{3}\right)(100)\left(\frac{t}{r}\right)\right]$$

or the allowable stress in compression is

$$\sigma_l \leq B/2$$

where r = radius of cylindrical shell, in.
t = thickness of cylindrical shell, in.
E = modulus of elasticity of shell, psi
B = B factor in the ASME vessel code, psi

Referring to Figure 4-3, σ_1 occurs at either $[\beta/6 + \theta/2]$ degrees or zero degrees at the shell acting in the longitudinal direction. This only applies to *unstiffened* shells. The vessel must meet the allowable with or without pressure.

$$\sigma_1 = \frac{3\,Q\,L}{\pi r^2(t_s - CA)}\left[\frac{4A}{L}\left(1 - \frac{1 - \frac{A}{L} + \frac{R^2 - H^2}{2AL}}{1 + \frac{4H}{3L}}\right)\right]$$

$$\left(\frac{\pi\left(\frac{\sin\Delta}{\Delta} - \cos\Delta\right)}{\Delta + \sin\Delta\cos\Delta - 2\,\frac{\sin^2\Delta}{\Delta}}\right) \qquad (4\text{-}1)$$

where A, H, L, Q, r, and t_s are defined in Figure 4-3.

CA = corrosion allowance, in.

θ = angle of contact of saddle with shell, degrees (Figure 4-1)

$$\Delta = \frac{\pi}{180}\left(\frac{5\theta}{12} + 30\right) \qquad (4\text{-}2)$$

2Δ = arc, in radians, of unstiffened shell in plane saddle effective against bending

At Mid-span

σ = longitudinal bending stress at midspan

The longitudinal bending stress at midspan has the same limitations as σ_1

$$\sigma_2 = \frac{3QL}{\pi r^2(t - CA)}\left[\frac{1 + 2\,\frac{R^2 - H^2}{L^2}}{1 + \frac{4H}{3L}} - \frac{4A}{L}\right] \qquad (4\text{-}3)$$

Tangential Shear Stress

1. For shell stiffened by ring in the plane of the saddle,

$$\sigma_{3Max} = \frac{(0.318)Q}{r\,(t - CA)}\left(\frac{L - 2A - H}{L + H}\right) \qquad (4\text{-}4)$$

$\sigma_3 \leq 0.08\sigma_{all}$

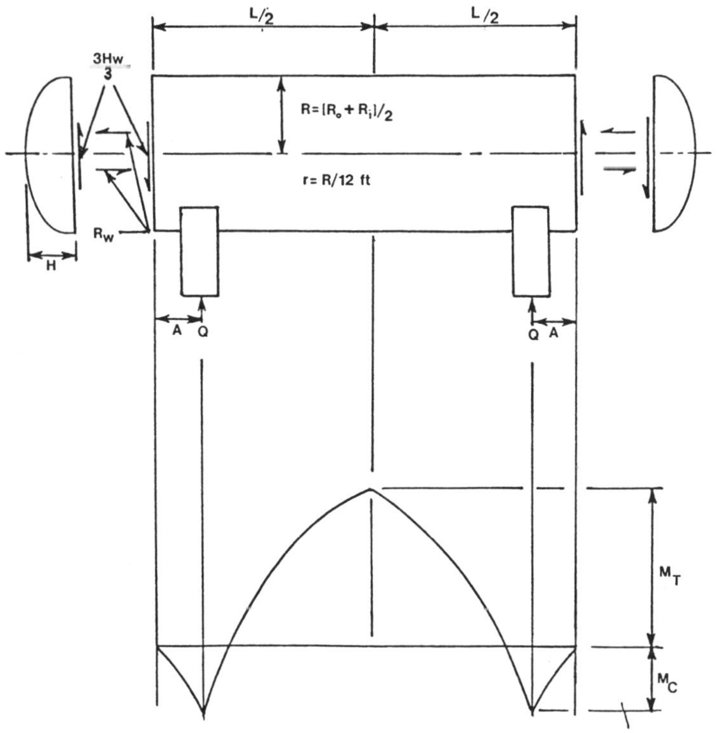

Figure 4-3. Bending moment diagram for a horizontal vessel developed by Zick [2].

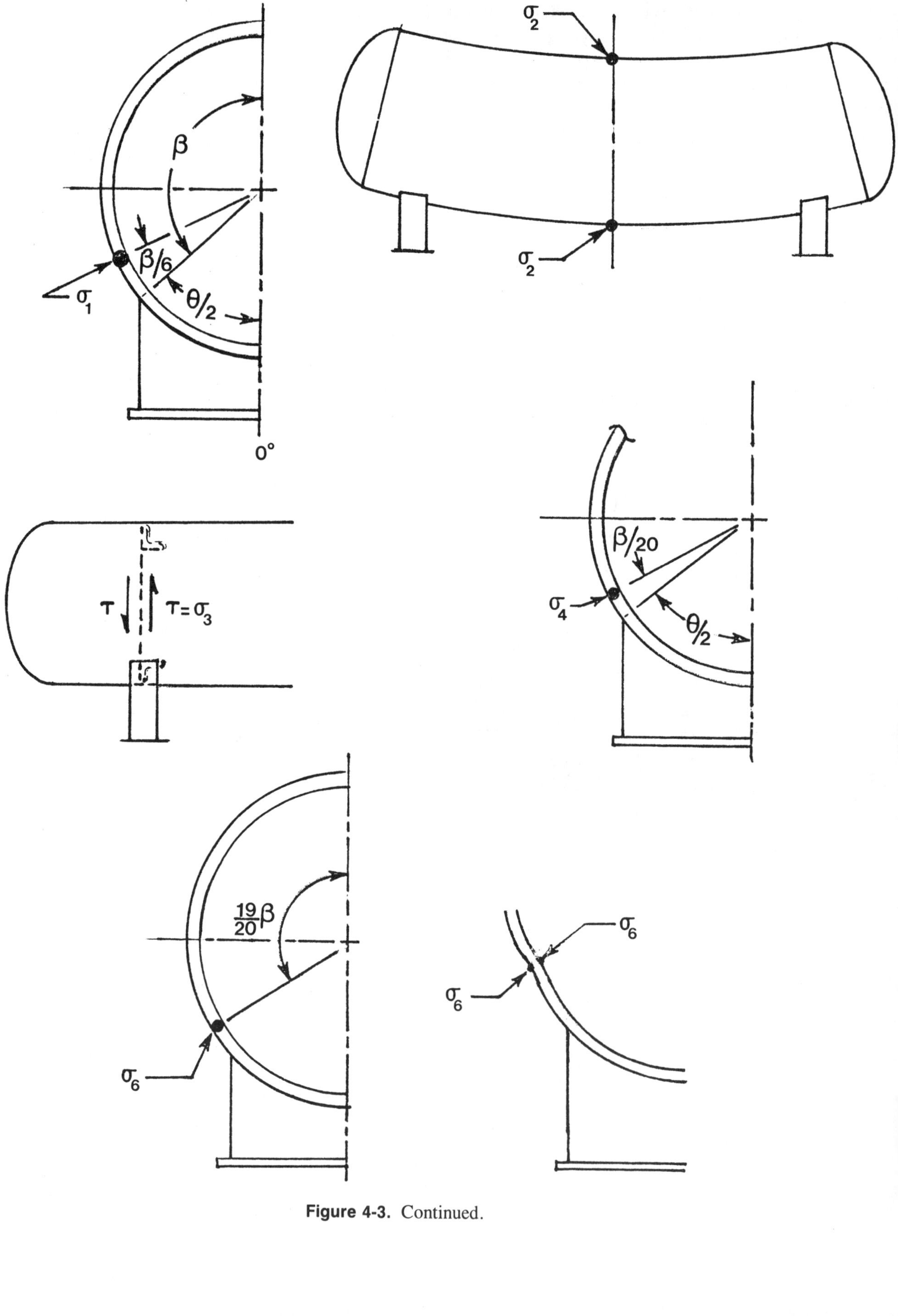

Figure 4-3. Continued.

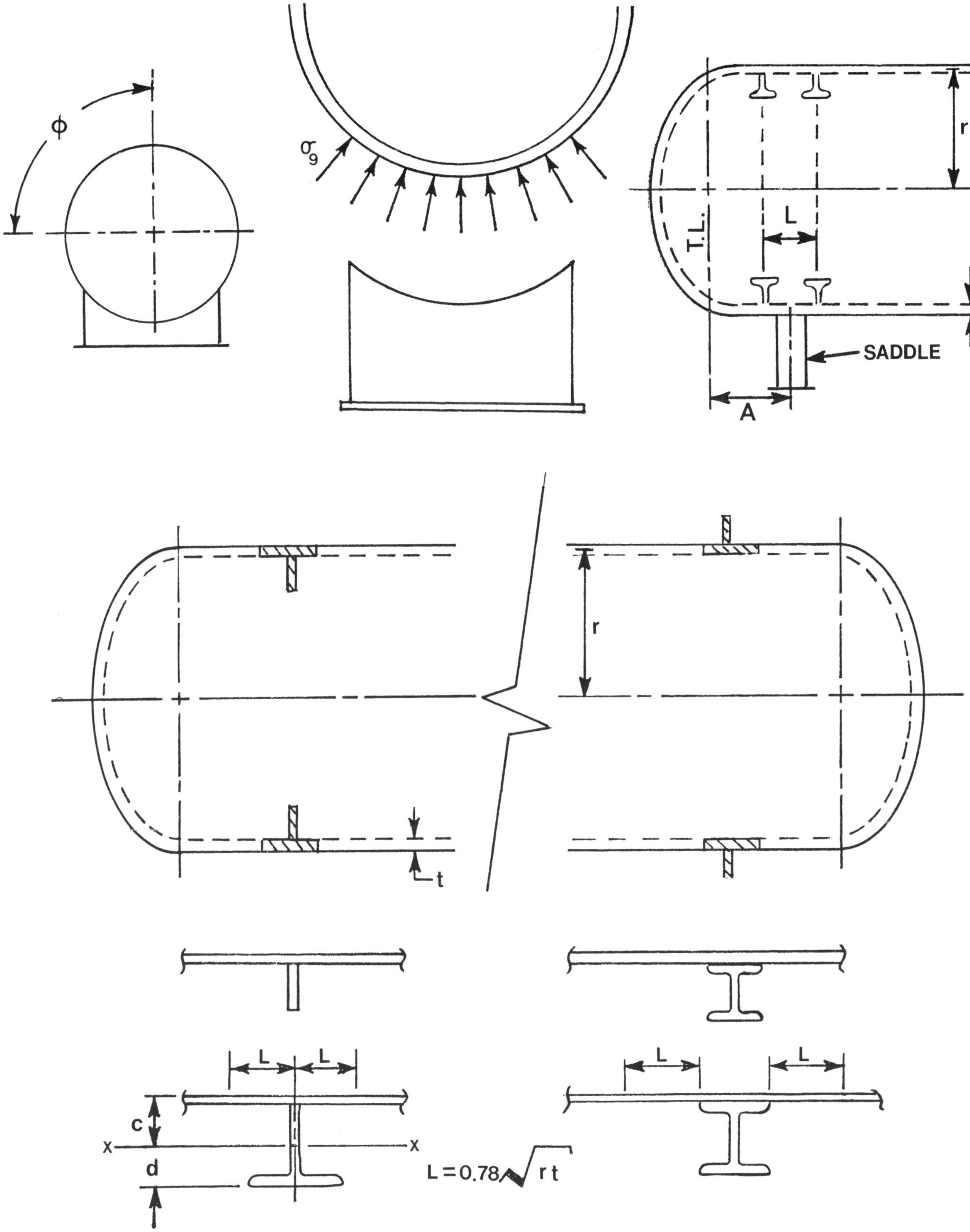

Figure 4-3. Continued.

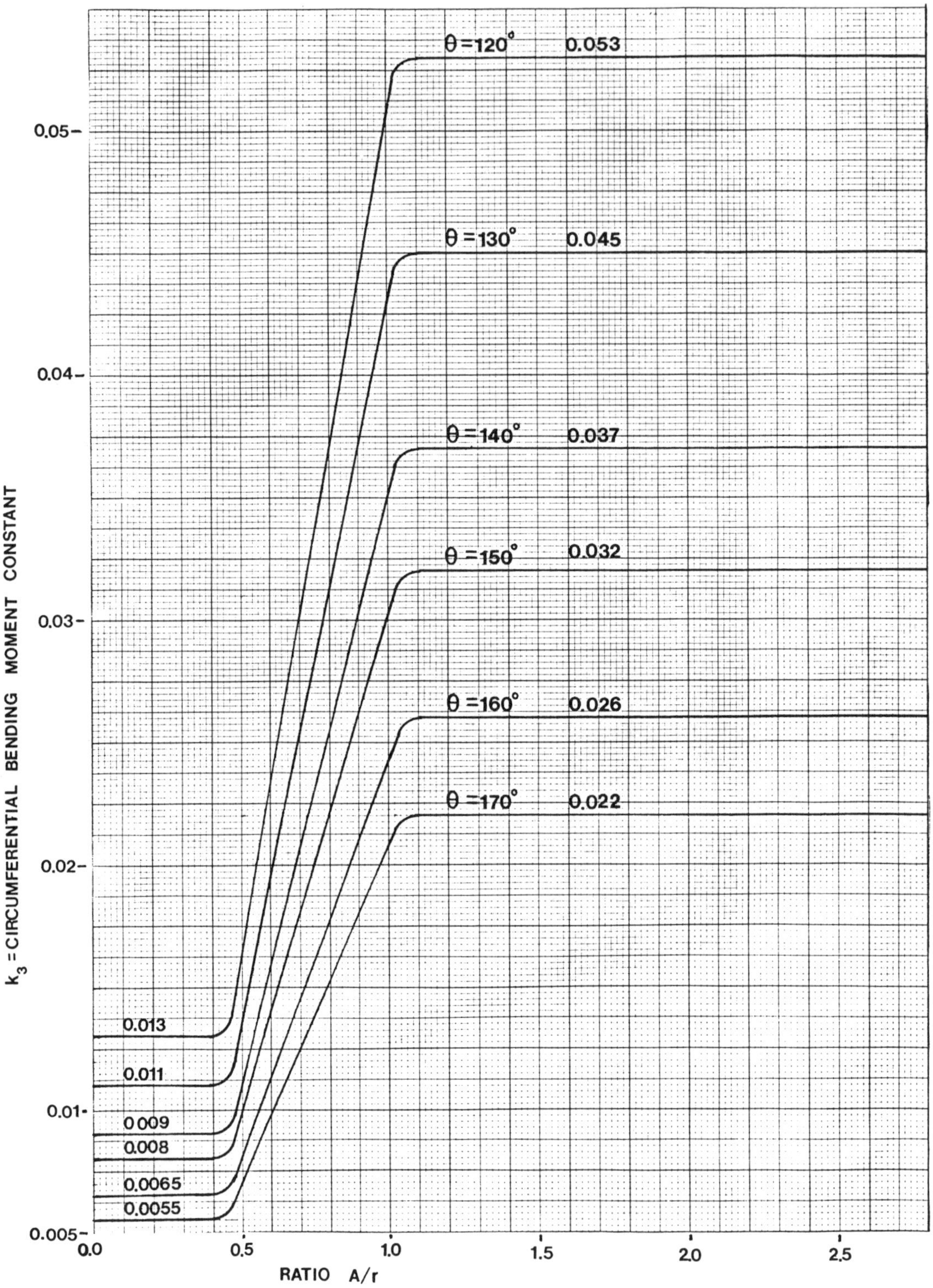

Figure 4-3. Continued.

2. Unstiffened shell with saddles away from head (A >R)

$$\sigma_{4Max} = \frac{Q}{r\,(t - CA)}\left(\frac{L - H - 2A}{L + H}\right) \times \left(\frac{\sin\alpha}{\pi - \alpha + \sin\alpha\cos\alpha}\right) \quad (4\text{-}5)$$

where

$$\alpha = \pi - \frac{\pi}{180}\left(\frac{\theta}{2} + \frac{\beta}{20}\right), \ \theta \ \& \ \beta \text{ in degrees} \quad (4\text{-}6)$$

σ_4 = tangential shear stress, located at an angle of $\beta/20$

3. Shell stiffened by head,

σ_5 = shear stress in head

This stress occurs only when the shell is stiffened by the head and when the head is located less than one shell radius from the saddle. The maximum shear is located at an angle of $(19/20)\beta$ as shown.

$$\sigma_5 = \frac{Q}{r\,(t_h - CA)}\left[\frac{\sin\alpha}{\pi}\left(\frac{\alpha - \sin\alpha\cos\alpha}{\pi - \alpha + \sin\alpha\cos\alpha}\right)\right] \quad (4\text{-}7)$$

4. Shear stress in shell,

$$\sigma_6 = \frac{Q}{r\,(t_5 - CA)}\left[\frac{\sin\alpha}{\pi}\left(\frac{\alpha - \sin\alpha\cos\alpha}{\pi - \alpha + \sin\alpha\cos\alpha}\right)\right] \quad (4\text{-}8)$$

σ_6 is the same as σ_5 and also is located at $(19/20)\beta$. With the shell stiffened by the head, then

$\sigma_6 \leq 0.8\sigma_{all}$

Circumferential Stress at Horn of Saddle

For shell stiffened by head the maximum circumferential stress at horn of saddle is,

$$\text{If } L > 8R,\ \sigma_7 = \frac{-Q}{4(t - CA)\,(b + 1.56(r(t - CA))^{0.5})} - \frac{3QK_6}{2(t - CA)^2} \quad (4\text{-}9)$$

$$\text{If } L < 8R,\ \sigma = \frac{-Q}{4(t - CA)\,(b + 1.56(r(t - CA))^{0.5})} - \frac{12K_6QR}{L(t - CA)^2} \quad (4\text{-}10)$$

If $A/r > 1$, then $K_6 = 0.4222e^{-0.0177\theta}$ (4-11)
where θ = deg
If $A/r < 0.5$, then $K_6 = K_3/4$
Otherwise, use Figure 4-3.

$$K_3 = \frac{1}{\pi}\left[\frac{\beta\sin\beta}{2} - \frac{\sin\beta}{\beta}\cos\beta + \frac{\sin\beta}{\beta} + \left(\frac{\cos\beta - \dfrac{\sin\beta}{\beta}}{4}\right) \times \left[\frac{4 - 6\left(\dfrac{\sin\beta}{\beta}\right)^2 + 2\cos^2\beta}{\dfrac{\sin\beta\cos\beta}{\beta} + 1 - 2\left(\dfrac{\sin\beta}{\beta}\right)^2}\right]\right] \quad (4\text{-}12)$$

σ_7 = circumferential compressive stress

This stress is located at the horn of the saddle If $\sigma_7 < \sigma_{all}$, it is not necessary to take credit for the wear plate.

$\sigma_7 \leq 1.50\,\sigma_{all}$

Additional Stress in Head When Used as a Stiffener

$$\sigma_8 = \frac{3Q}{8r(t_h - CA)}\left[\frac{\sin^2\alpha}{\pi - \alpha + \sin\alpha\cos\alpha}\right] \quad (4\text{-}13)$$

$\sigma_8 \leq \sigma_{all}$

Wear Plates—Ring Compression in Shell Over Saddle

$$\sigma_9 = \frac{Q}{(t - CA)(b + 1.56(r(t - CA))^{0.5})}\left(\frac{1 + \cos\alpha}{\pi - \alpha + \sin\alpha\cos\alpha}\right) \quad \sigma_9 \leqslant 0.5\sigma_y \quad (4\text{-}14)$$

σ_9 = ring compressive stress in shell over the saddle

This stress is located at $\phi = \pi$

This stress is compressive and acts in a radial direction between the saddle and shell. The limitation of this stress is

$0.5\sigma_{yield}$

where σ_{yield} = the yield strength of the saddle material (metal or concrete)

Location of Saddle Supports

For thin wall vessels with large diameters, it is desirable to locate the saddles close to the head, where A = ID/4, using the stiffness of the head. Although arbitrary on what a thin shell is, and Zick [2] does not define the term, a shell is generally regarded as "thin" when $D/t \geq 100$, where D = shell diameter and t = shell thickness. For shells where $D/t \leq 100$ and the distance from the head tangent to head tangent is rather large (approximately $L/r > 10$), the saddles are best spaced when the longitudinal bending stress at the saddle, σ_1, equals the longitudinal bending stress at midspan, σ_2. Under no circumstances should the distance from the saddle center line to the head tangent, A, exceed 0.25L.

A listing of allowable stress criteria is given in Table 4-3. Each of the previously mentioned stress values should be evaluated with this table and the appropriate code.

Table 4-3
Allowable Stress Values

Longitudinal Bending Stress				
$\sigma_1 + \sigma_P$	$= \sigma_1$	$\dfrac{PR}{2t_s}$	$< \sigma_{all}$ E, where E = joint efficiency	
$\sigma_2 + \sigma_P$	$= \sigma_2$	$\dfrac{PR}{2t_s}$	$< \sigma_{all}$ E	
Tangential Shear Stress				
σ_3,	σ_4	$<$	$0.8\ \sigma_{all}$	
σ_5,	σ_6	$<$	$0.8\ \sigma_{all}$	
Circumferential Stress at Horn Saddle				
σ_7	$<$	$1.5\ \sigma_{all}$		
Circumferential Stress at Bottom of Shell				
σ_9	$<$	$0.5\ (\sigma_{y(min)})$	→ Compressive Yield	

Wear Plate Design

One of the first things to consider when designing a horizontal vessel is the need for wear plates. Too often these plates are "automatically" included with no thought given to their necessity in each application. Wear plates involve material and labor expense and are a waste if not needed.

Wear plates are not required if two criteria are met: The circumferential stress at the horn of the saddle must be less than 1.5 times the allowable stress, and the ring compression stress in the shell over the saddle must be less than one half the minimum yield strength. These criteria can be written as follows:

$\sigma_7 < 1.5\ \sigma_{all}$
$\sigma_9 < 0.5\ \sigma_{y\ min}$

Table 4-4 shows minimum allowable shell thicknesses required for horizontal vessels without wear plates. The values are based on using a fluid 1.75 times the weight of water, and the metal has a minimum yield of 30,000 psi and an allowable stress of 17,500 psi. For vessels in seismic regions wear plates should always be used to minimize stress concentrations at saddle plate-shell juncture.

Zick Stiffening Rings

When the Zick stresses in a vessel become excessive and the location of the saddles no longer is a factor because the stresses are below the allowable stress, then two options are available—increase the vessel wall thickness or add stiffening rings. Almost always it is more desirable to add stiffening rings because it is cheaper to add a few rings than go to a larger size shell thickness, particularly with expensive alloys. Also, if the vessel is subjected to external pressure, the Zick rings can act as external pressure stiffening rings as well as Zick rings.

Referring to Figure 4-3, if two Zick stiffening rings are located on each side of the saddle, then

$L_{min} = 1.56\sqrt{rt}$, ft

$L_{max} = r$, ft

The stress in the ring is

$$\sigma_{10} = \frac{-K_7Q}{nA_r} + \frac{K_6Qr}{nZ}\lambda$$

Where
Z = I_{x-x}/c for ring in the plane of the saddle, $in.^3$
Z = I_{x-x}/d at saddle horn at tip or flange of stiffener ring, $in.^3$
I_{x-x} = moment of inertia of stiffening ring about axis x−x, $in.^4$ (includes wear plate thickness if one is used)
A_r = cross-sectional area of stiffening ring, $in.^2$
n = number of stiffener rings per saddle
r = mean vessel radius, in.
K_6 = previously defined

Table 4-4
Minimum Shell Thickness Required for Horizontal Vessels Without Wear Plates

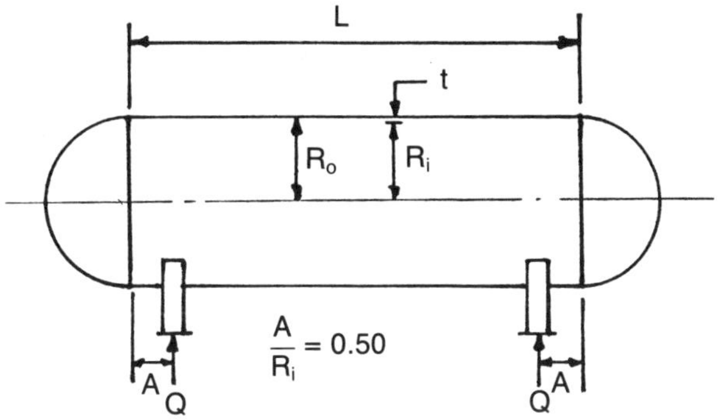

L(ft) \ ID (in.)	24	30	36	42	48	54	60	66	72
8	1/8 in.								
10									
12							5/16 in.		
14			3/16 in.		1/4 in.				3/8 in.
16									
18	—								
20	—	—							
30	—	—	—	—		5/16 in.			
40	—	—	—	—	—	—	—		7/16 in.
50	—	—	—	—	—	—	—	—	—
60	—	—	—	—	—	—	—	—	—
65	—	—	—	—	—	—	—	—	—

L(ft) \ ID (in.)	78	84	90	96	102	108	114	120	132	144	156
8				1/2 in.			—	—	—	—	—
10	3/8 in.						—	—	—	—	—
12								11/16 in.		—	—
14											
16						5/8 in.					
18	7/16 in.			9/16 in.					13/16 in.	7/8 in.	1 in.
20											
30		1/2 in.					11/16 in.	3/4 in.			
40										15/16 in.	
50											—
60	—	—	—						7/8 in.	1 in.	—
65	—	—	—	—				13/16 in.			—

Notes
1. The above table is based on the following:
   a. Vessel is fully loaded with a fluid of specific gravity of 1.5.
   b. The ratio of the shell outside radius, R_o, to shell thickness, t, is $R_o/t \geqslant 72$.
   c. Vessel weight is computed with note (a.) and hemispherical heads.
   d. Vessel material has the following properties:
      $\sigma_{min\ yield} = 30{,}000$ psi and $\sigma_{allow} = 17{,}500$ psi
2. In seismic zones 3 and 4 wear plates should be used.

In compression, σ_{10} is negative,

$$\sigma_{10} + B \leqslant 0.5\sigma_{cy}$$

In tension, σ_{10} is positive,

$$\sigma_{10} + \sigma_p \leqslant \sigma_{all} \text{ [tension]}$$

where B = ASME compressive stress (see ASME Section 8 Division 1)
σ_{cy} = compressive yield stress (see ASME Section 8 Division 2)
σ_p = internal pressure stress (includes wear plate thickness if one is used)

In defining the parameter K_7, it must be noted that the Zick stiffening rings can fit on either the inside or outside of the vessel shell. Many clients object to the rings being external to the vessel surface because of aesthetics. However, after insulation is applied, the rings are no longer visible. We will consider the rings in both ways. The constant K_7 is defined as follows:

For a ring in the plane of the saddle—

$$\lambda = +1.0$$

$$K_7 = \begin{cases} 0.340, & \theta = 120° \\ 0.303, & \theta = 150° \\ 0.250, & \theta = 180° \end{cases}$$

For rings adjacent to saddle—
For internal rings,

$$\lambda = -1.0$$

$$K_7 = \begin{cases} 0.271, & \theta = 120° \\ 0.219, & \theta = 150° \\ 0.140, & \theta = 180° \end{cases}$$

For external rings,

$$\lambda = -1.0$$

$$K_7 = \begin{cases} 0.271, & \theta = 120° \\ 0.219, & \theta = 150° \\ 0.140, & \theta = 180° \end{cases}$$

STEEL SADDLE PLATE DESIGN

Once the shell conditions have been met, the saddle plates must be analyzed. The main phenomenon encountered with saddle plates is local buckling with the plates undergoing bending, compression, shear, or any combination of these loadings. Normally, saddle plates are strengthened with stiffener or web plates, but often too many are used, which increases labor and material costs. In the past, saddle plates have been purposely over-designed to guard against uncertainty. This is no longer required, since literature on flat plate theory has increased with mounting experimental data. One such organization that has engaged in extensive research is U.S. Steel [4].

Figure 4-4 shows a typical saddle configuration for a horizontal vessel. Section A-A shows that only an effective portion of the member will resist compression, shear, and bending loads because when the member is loaded, the outside fibers of the web plates and the center of the saddle plate—shown by the shaded areas in Figure 4-4—go into the plastic range. The rest of the plate area is still in the elastic range because of residual stresses that were created by non-uniform heating during rolling or welding. Presently, this "effective" area can be determined only by experiment. Equations 4-9, 4-10, and 4-14 are used in saddle design as follows:

$$b_e = K_i t_s \tag{4-15}$$

where b_e = effective width, in.
K_i = plate buckling coefficient for either compression, shear, bending, or a combination of these loads (see Figures 4-5, 4-6, 4-7, 4-8, 4-9, and 4-10).
i = c, b, s, or a combination of these characters,
where K_c = plate buckling coefficient for compression, dimensionless
K_b = plate buckling coefficient for bending, dimensionless
K_s = plate buckling coefficient for shear, dimensionless

We now have

$$\sigma_7 = \frac{-Q}{4(t - CA)[b_e + 1.56(r(t - CA))^{0.5}]} - \frac{3QK_7}{2(t - CA)^2}, \text{ If } L > 8R \tag{4-9}$$

$$\sigma_7 = \frac{-Q}{4(t - CA)[b_e + 1.56(r(t - CA))^{0.5}]} - \frac{12K_7QR}{L(t - CA)^2}, \text{ If } L < 8R \tag{4-10}$$

$$\sigma_9 = \frac{Q}{4(t - CA)[b_e + 1.56(r(t - CA))^{0.5}]} \times \left(\frac{1 + \cos\alpha}{\pi - \alpha + \sin\alpha\cos\alpha}\right) \tag{4-14}$$

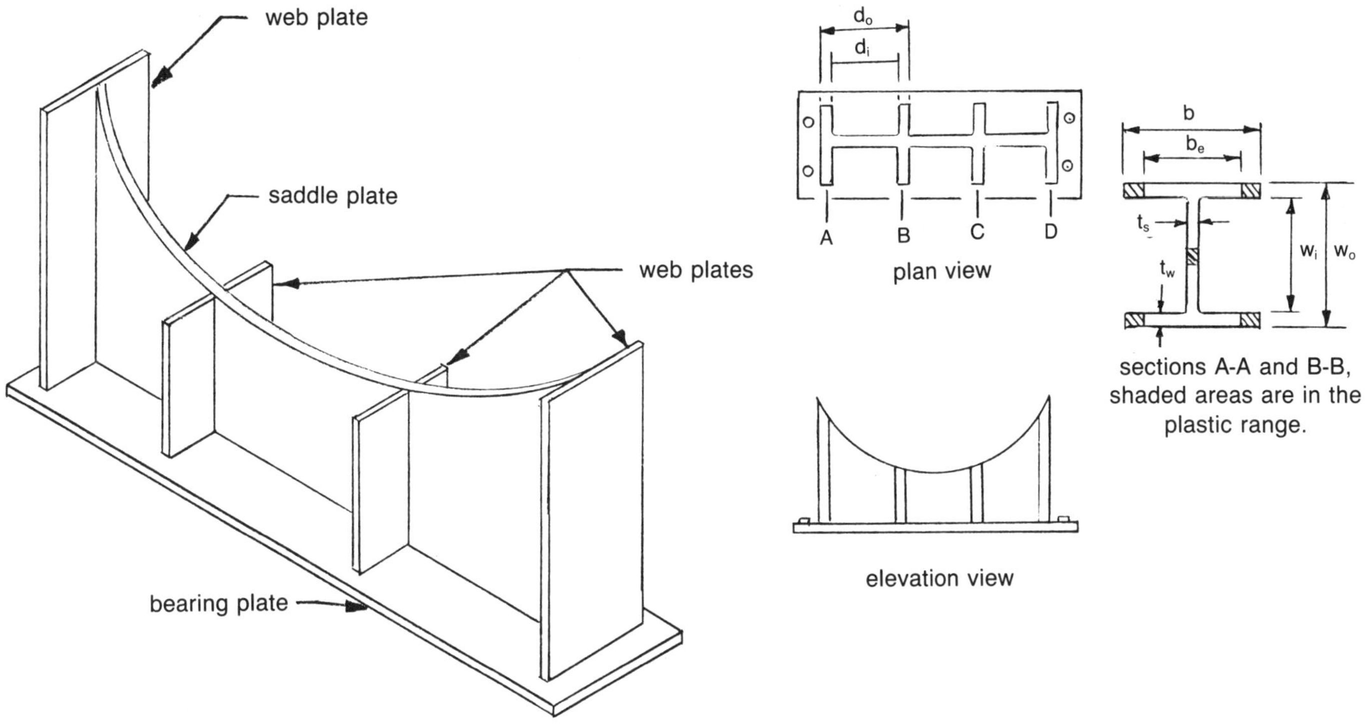

Figure 4-4. Horizontal vessel saddle support detail.

Figures 4-5 to 4-10 are courtesy of United States Steel Corporation. USSC makes no warranties, express or implied, and no warranty as to the merchantability, fitness for any particular purpose, or accuracy of the information contained in any material reproduced herein from its *Steel Design Manual.* In the event of any liability arising out of the publication of such material herein, consequential damages are excluded.

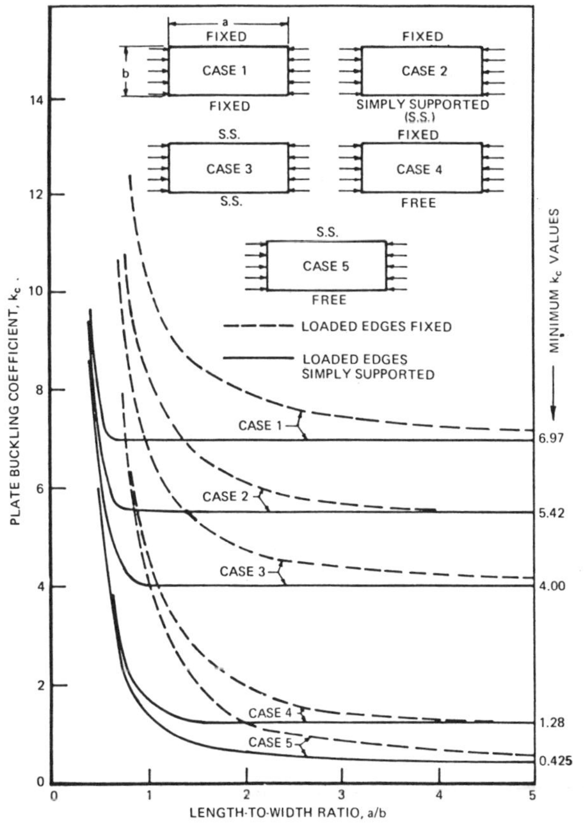

Figure 4-5. Buckling coefficients for flat plates under uniform compression. (Courtesy of U.S. Steel [4].)

LOADING	RATIO OF BENDING STRESS TO UNIFORM COMPRESSION STRESS, f_b/f_c	MINIMUM BUCKLING COEFFICIENT,* k_1: UNLOADED EDGES SIMPLY SUPPORTED	MINIMUM BUCKLING COEFFICIENT,* k_1: UNLOADED EDGES FIXED
$f_2 = -f_1$	∞ (PURE BENDING)	23.9	39.6
$f_2 = -2/3 f_1$	5.00	15.7	
$f_2 = -1/3 f_1$	2.00	11.0	
$f_2 = 0$	1.00	7.8	13.6
$f_2 = 1/3 f_1$	0.50	5.8	
$f_2 = f_1$	0.00 (PURE COMPRESSION)	4.0	6.97

*VALUES GIVEN ARE BASED ON PLATES HAVING LOADED EDGES SIMPLY SUPPORTED AND ARE CONSERVATIVE FOR PLATES HAVING LOADED EDGES FIXED.

Figure 4-6. Buckling coefficients for flat plates under compression and bending. (Courtesy of U.S. Steel [4].)

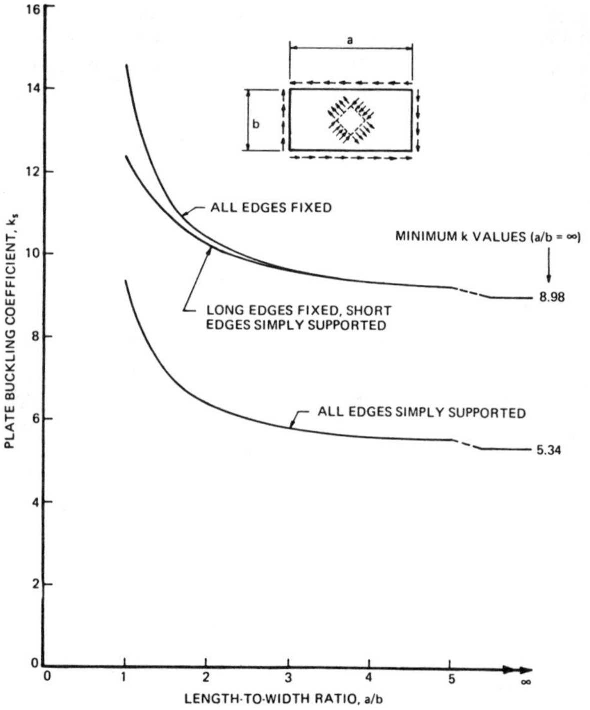

Figure 4-7. Buckling coefficients for flat plates in shear. (Courtesy of U.S. Steel [4].)

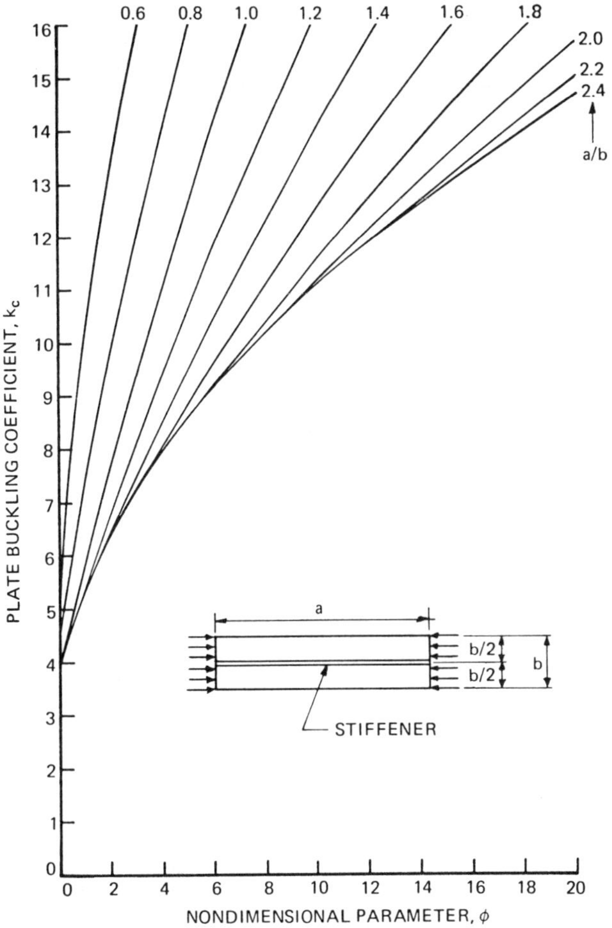

Figure 4-8. Buckling coefficients for stiffened plates under uniform compression (one longitudinal stiffener at mid-point). (Courtesy of U.S. Steel [4].)

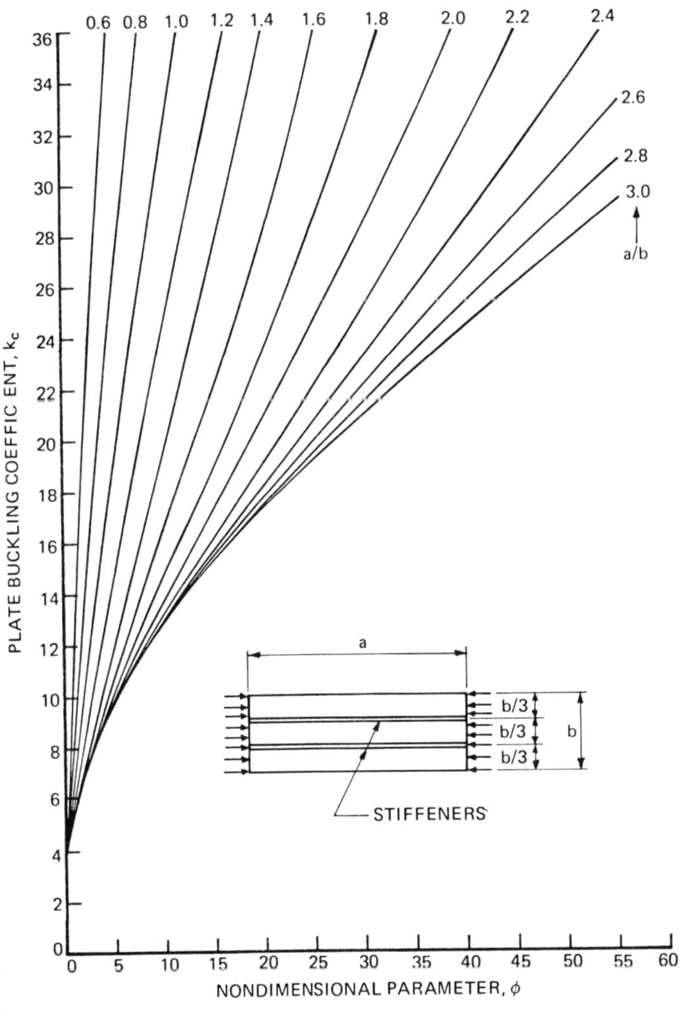

Figure 4-9. Buckling coefficients for stiffened plates under uniform compression (two longitudinal stiffeners at third points). (Courtesy of U.S. Steel [4].)

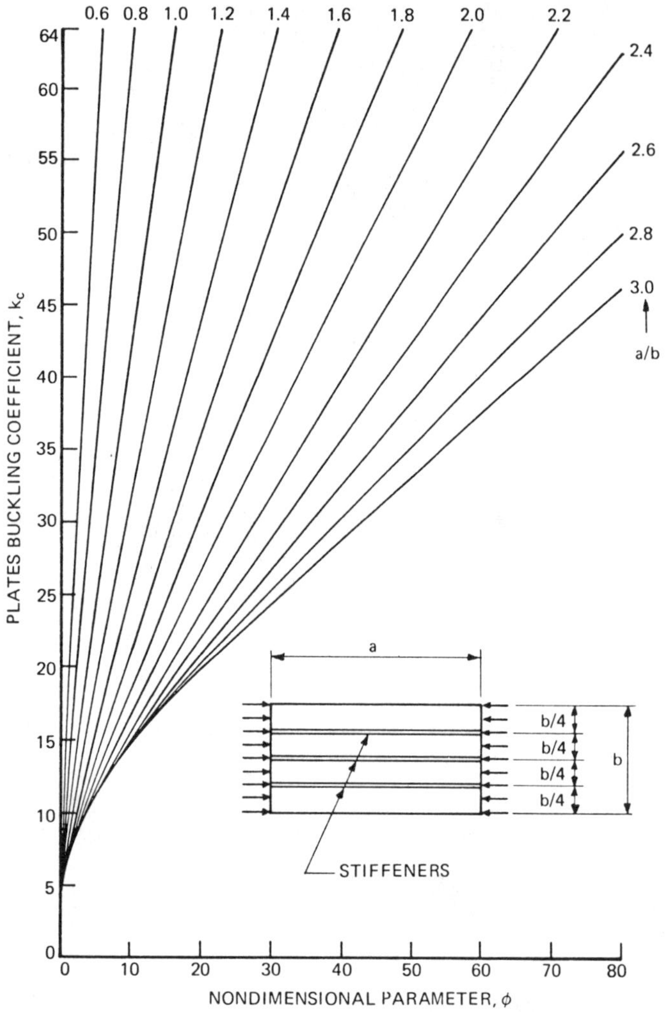

Figure 4-10. Buckling coefficients for stiffened plates under uniform compression (three longitudinal stiffeners at quarter points). (Courtesy of U.S. Steel [4].)

If no web plates are used then $b_e = t_s$. It is very common for engineers and designers to use the web plate width, b, instead of b_e. *This is wrong.* The only time $b_e = b$ is when $t_s = b$, as is true for a solid concrete saddle. With steel this never happens, as values of b can be as great as 24 in. and obtaining plate that thick is impossible (at least on this planet).

Values of b_e depend upon K_i and t_s. Since the value of t_s is known, the real independent variable in Equation 4-15 is K_i. Once again referring to Figure 4-4, we analyze the saddle configuration for end (boundary) conditions. Section B-C is considered fixed-fixed in Figure 4-5, since it is stiffened by sections A-B and C-D. Sections A-B and C-D are considered fixed-free since the outer web plate is not stiffened by another section. The fixed-free condition is the most critical because it is more susceptible to buckling, and thus has a lower value of the plate buckling coefficient than the fixed-fixed case. It is interesting to note that the plate buckling coefficient for uniform compression for the fixed-free case, when multiplied by t_s, yields approximately the effective width, b_e, that is used for residual stress. In other words, if a member is known or suspected to have residual stress and is subjected to compression, bending, shear, or a combination, the plate buckling coefficient is equal to the effective width that is determined by the residual stress criterion, which is as follows:

$$b_e = \frac{d_i t_s}{[d_i t_s + 2t_w(b - 1)]} \tag{4-16}$$

The general equation in which the saddle plate stress distribution is defined is as follows:

$$\sigma_{cr} = \frac{K_i \pi^2 E}{12(1 - \nu^2)\left(\frac{d_i}{t_s}\right)^2} \tag{4-17a}$$

where d_i = saddle plate length normal to vertical axis of stiffener (web) plate, shown in Figure 4-4.
E = modulus of elasticity, psi
ν = Poisson's ratio
b_e = effective saddle width, in.
t_s = saddle plate thickness, in.

Substituting the elastic buckling stress in Equation 4-17a into

$$\sigma_{cr}' = \sigma_y - \frac{\sigma_y^2}{4\sigma_{cr}}$$

$$= \sigma_y - \frac{3(1-\nu^2)\,\sigma_y^2\left(\frac{b}{t_s}\right)^2}{K_c\pi^2 E}, \text{ If } \sigma_{cr} \geqslant \sigma_y/2 \qquad (4\text{-}17b)$$

gives the relationship of the plate buckling stress in the inelastic range. This equation is based on the conservative assumption that a plate will always buckle before the yield stress is reached. However, U.S. Steel [4] states that plates will deform plastically without buckling because of strain hardening. This process is similar to the "elastic shakedown" described in Chapter 2.

In most applications, as already cited, saddle plates are reinforced with stiffener plates. A simplified analysis can be made to design saddles by using

$$F_B = n(A_s + 2b_e t_s)\sigma_s \qquad (4\text{-}18)$$

where F_B = buckling load for compressive loading, LB_f
A_s = section area of stiffener, $in.^2$
n = number of stiffeners
t_s = thickness of saddle plate, in.
d_e = effective width of saddle plate that is perpendicular to the web plate, in.

with

$$d_e = d_i(0.25 + 0.91\lambda^2)$$

$$\lambda = \left(\frac{t_s}{d_i}\right)\left(\frac{E}{\sigma_{cr}}\right)^{0.5}$$

σ_s = maximum unit load the stiffener can carry as a column, psi

Horizontal Reaction on Saddle

As shown in Figure 4-11, the load Q has a horizontal component exerted on the horns of the saddle. The saddle must be designed to prevent the horns of the saddle from separating. To accomplish this make sure that the minimum cross-sectional area at the lowest point on the saddle can resist the horizontal force component. This force is as follows:

$$F = Q\left[\frac{1 + \cos\beta - 0.5\sin^2\beta}{\pi - \beta + \sin\beta\cos\beta}\right] \qquad (4\text{-}19)$$

The effective cross section to resist the horizontal force is A_E, shown in Figure 4-11 and calculated as follows:

$$A_E = \left(\frac{R}{3}\right)t_s$$

where R = outside vessel radius

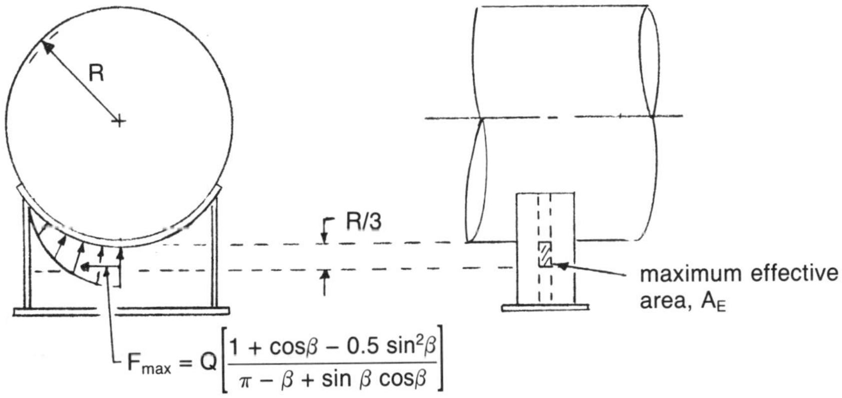

Figure 4-11. The load distribution on a saddle.

SADDLE BEARING PLATE THICKNESS

Designing bearing plates for saddles requires knowing what type of foundation the vessel will rest on. For concrete the following analysis applies.

Consider a bearing plate with the dimensions shown in Figure 4-12. From ACI Standard 318-77 Par. 10.16.1, the allowable bearing strength on concrete is

$$\sigma = \phi(0.85\ \sigma_c'\ A_1)\left(\frac{A_2}{A_1}\right)^{0.5} \quad (4\text{-}20)$$

For bearing on concrete (ACI 9.3.2.e)

$$\phi = 0.70$$
$$\phi_c' = 3000 \text{ psi}$$

in which

$$\sigma = (0.70)(0.85)(3{,}000)A_1 \times \left(\frac{(A_1 + 9) + 3(L_2 + L_1)}{A_1}\right)^{0.5} \quad (4\text{-}21)$$

Using a factor of safety of 1.6, Equation 4-21 becomes

$$\sigma' = \frac{\sigma}{1.6} = 1{,}115.63\ A_1 \left(\frac{(A_1 + 9) + 3(L_2 + L_1)}{A_1}\right)^{0.5}$$

Using a minimum yield strength of 30,000 psi we have the allowable stress for bending, per AISC recommendation,

$$\sigma_{all} = 0.66\ \sigma_y = 0.66(30{,}000) = 19{,}800 \text{ psi}$$

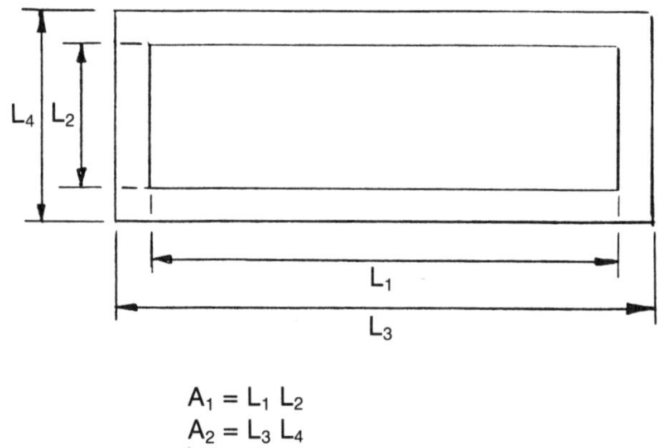

Figure 4-12. Bearing plate dimensions.

Table 4-5
Bearing Plate Thickness Values for Various Saddle Loads

L_1	L_2	Q_{max} (lbs)	t (in.)	BP (psi)	σ' (psi)
17	4	2,858	0.165	42.029	108,852.563
26	4	5,043	0.178	48.490	162,100.694
30	4	8,103	0.210	67.525	185,744.857
33	4.25	11,131	0.241	79.365	213,447.893
36	4.25	16,007	0.277	104.621	232,042.324
39	5.75	20,418	0.350	91.050	320,269.131
42	5.75	25,387	0.376	105.122	344,024.233
45	5.75	33,523	0.417	129.557	367,777.375
48	5.75	40,154	0.442	145.486	391,528.914
51	5.75	51,042	0.484	174.056	415,279.125
54	5.75	59,549	0.508	191.784	439,028.224
57	5.75	68,777	0.531	209.846	462,776.382
60	5.75	84,203	0.573	244.067	486,523.736
63	5.75	101,759	0.614	280.908	510,270.399
66	5.75	114,664	0.637	302.145	534,016.463
69	6.75	128,417	0.715	275.721	637,918.463
72	6.75	143,003	0.738	294.245	665,075.973
76	6.75	174,748	0.794	340.639	701,285.275
84	6.75	210,035	0.828	370.432	773,701.873
95	6.75	250,290	0.850	390.316	873,271.364

Thus, the maximum stress in the bearing plate is

$$\sigma_p = \frac{M}{Z} = \frac{\frac{1}{2}\left(\frac{Q}{2L_1}\right)\left(\frac{b}{2}\right)}{\left(\frac{t^2}{6}\right)} \quad (4\text{-}22)$$

or

$$t = \left(\frac{Q\ L_2}{24{,}600\ L_1}\right)^{0.5}, \text{ in.} \quad (4\text{-}23)$$

where $b = L_2$
Q = load on saddle, lbs

$$BP = \text{bearing pressure} = \frac{Q}{L_1L_2} < \sigma'$$

Table 4-5 depicts values of bearing plate thickness for various saddle loads.

DESIGN OF SELF-SUPPORTED VERTICAL VESSELS

Today's tall, cylindrical process towers are self-supporting, i.e., they are supported by a cylindrical or conical shell (skirt) with a large base ring attached to a concrete foundation or steel structure with anchor bolts embedded in the concrete or steel. Normally, a vertical vessel must be at least thirty feet tall to be classified as a "tower." This height is used because thirty feet is the old first wind-zone demarcation in code use. However, smaller vertical vessels are governed by the same design criteria, but are not usually referred to as towers.

The various phenomena that affect towers in normal operation make their design complex and worthy of experienced engineering personnel. Therefore, towers should never be taken lightly by any design office, because a failure could result in massive loss of material and possibly lives.

MINIMUM SHELL THICKNESS REQUIRED FOR COMBINED LOADS

High-speed electronic computers now provide detailed, exact solutions to complex mathematical problems, and so have replaced the "strong arm" approximations of yesterday. An example is solving the equations of the moments of inertia and section modulus. Before the advance of computers, the following expressions were used to quicken computations on a slide rule or a small electronic calculator:

$$I = \pi R^3 t;\ \text{exact} = I = \frac{\pi}{64}(D_o^4 - D_i^4) \quad (4\text{-}24)$$

$$Z = \pi R^2 t;\ \text{exact} = Z = \frac{\pi}{32}\left(\frac{D_o^4 - D_i^4}{D_o}\right) \quad (4\text{-}25)$$

$$A = \pi r t;\ \text{exact} = A = \frac{\pi}{4}(D_o - D_i) \quad (4\text{-}26)$$

Using R as the mean radius minimizes the error and using R as the outside radius results in considerable error. Solving for the thickness or stress with the exact formulations involves iterative analysis, which is a key attribute of today's computers.

The minimum shell thickness required for internal or external pressure alone is often not sufficient for additional stresses induced by bending moments and weight loads. Bending stress is a result of static wind, dynamic wind gusts, vibration or seismic response spectra. In design the engineer takes the largest bending moments induced by one of the following: wind, vibration or seismic.

Referring to Figure 4-13, we analyze the stress element depicted. The maximum stress resulting from internal pressure occurs along the x-axis, i.e., the hoop stress is twice the longitudinal stress. Wind, vibration, and/or seismic forces cause the shell to bend about the z-axis, so the internal stress in the circumferential direction is combined with the bending and tensile (or compressive) stresses. Writing this expression we obtain,

$$\sigma = \pm\left(\frac{PD}{4t}\right) \pm \left(\frac{M}{Z}\right) \pm \left(\frac{W}{A}\right) \quad (4\text{-}27)$$

where Z = section modulus of the shell cross section, in.^3
A = area of the shell cross section, in.^2

Substituting Equations 4-24, 4-25, and 4-26 into Equation 4-27 produces

$$\sigma = \pm\left(\frac{PD}{4t}\right) \pm \left(\frac{32\,D_oM}{\pi(D_o^4 - D_i^4)}\right) \pm \left(\frac{4W}{\pi(D_o^2 - D_i^2)}\right) \quad (4\text{-}28)$$

$$\sigma = \pm\left(\frac{PD}{4t}\right) \pm \left(\frac{16\,D_oM}{\pi t(D_o + D_i)(D_o^2 + D_i^2)}\right) \pm \left(\frac{2W}{\pi t(D_o + D_i)}\right) \quad (4\text{-}29)$$

or

$$t = \pm\left(\frac{PD}{4\sigma}\right) \pm \left(\frac{16\,D_oM}{\pi\sigma(D_o + D_i)(D_o^2 + D_i^2)}\right) \pm \left(\frac{2W}{\pi\sigma(D_o + D_i)}\right) \quad (4\text{-}30)$$

Referring again to Figure 4-13, we summarize the following:

1. For the tension or windward side,

$$\sigma = \left(\frac{PD}{4t}\right) + \left(\frac{16\,D_oM}{\pi t(D_o + D_i)(D_o^2 + D_i^2)}\right) - \left(\frac{2W}{\pi t(D_o + D_i)}\right) \quad (4\text{-}31)$$

or

$$t = \left(\frac{PD}{4t}\right) + \left(\frac{16\,D_oM}{\pi\sigma(D_o + D_i)(D_o^2 + D_i^2)}\right) - \left(\frac{2W}{\pi\sigma(D_o + D_i)}\right) \quad (4\text{-}32)$$

2. For the compression or leeward side,

$$\sigma = \left(\frac{PD}{4t}\right) - \left(\frac{16\,D_oM}{\pi t(D_o + D_i)(D_o^2 + D_i^2)}\right) - \left(\frac{2W}{\pi t(D_o + D_i)}\right) \quad (4\text{-}33)$$

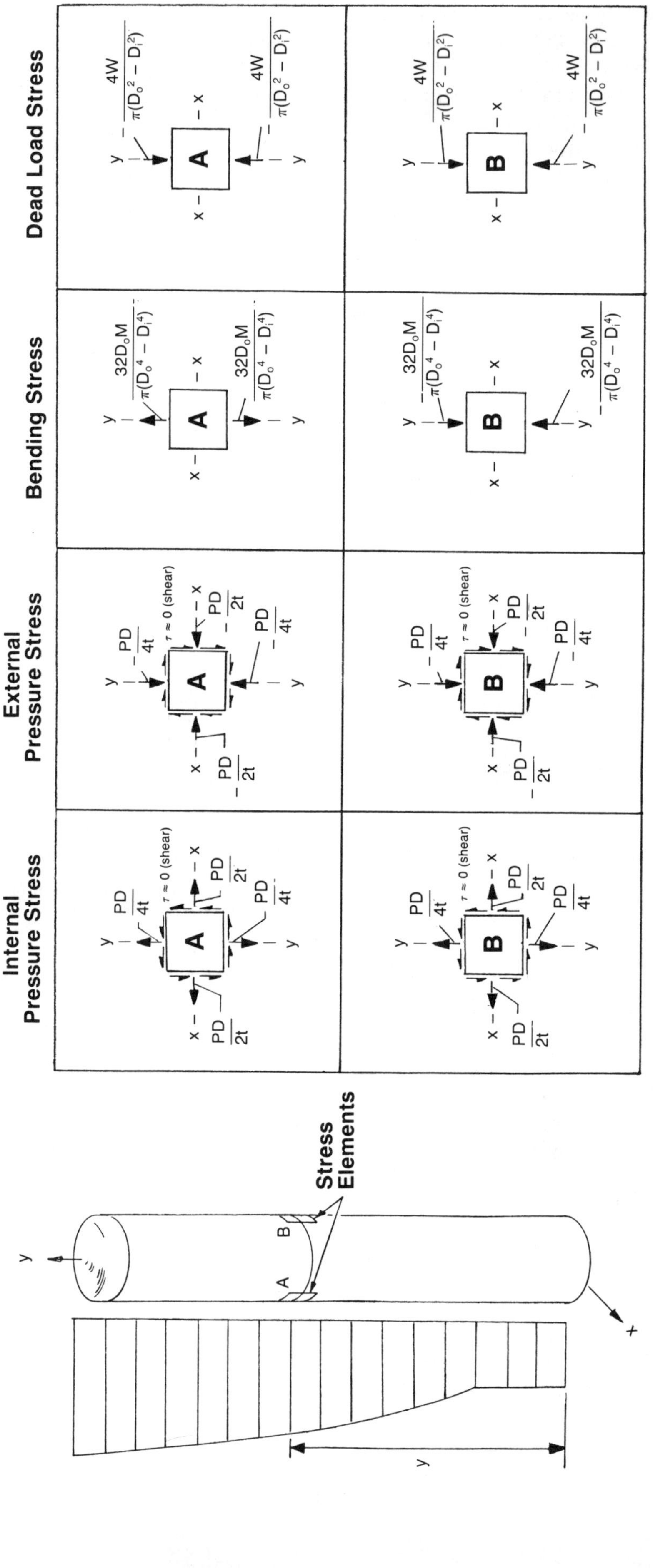

Figure 4-13. Stress elements in tower shell subjected to combined loadings and sign conventions.

or

$$t = \left(\frac{PD}{4\sigma}\right) - \left(\frac{16\ D_oM}{\pi\sigma(D_o + D_i)(D_o^2 + D_i^2)}\right) - \left(\frac{2W}{\pi\sigma(D_o + D_i)}\right) \tag{4-34}$$

3. For vacuum vessels the maximum stress occurs on the compressive side, such that Equations 4-29 and 4-30 become

$$\sigma = -\left(\frac{PD}{4t}\right) - \left(\frac{16\ D_oM}{\pi t(D_o + D_i)(D_o^2 + D_i^2)}\right) - \left(\frac{2W}{\pi t(D_o + D_i)}\right) \tag{4-35}$$

and

$$t = -\left(\frac{PD}{4\sigma}\right) - \left(\frac{16\ D_oM}{\pi\sigma(D_o + D_i)(D_o^2 + D_i^2)}\right) - \left(\frac{2W}{\pi\sigma(D_o + D_i)}\right) \tag{4-36}$$

SUPPORT SKIRT DESIGN

The design of vessel skirts is one area in which designers disagree philosophically. Lump-sum contractors seek to use higher allowables and thus less material, whereas the cost-plus contractors seek to standardize designs and use lower pressure vessel code allowables. As with wear plates on horizontal vessels, most lump-sum contractors would elect to omit them whenever possible to save material and labor costs. This philosophy is becoming increasingly popular with recent economic upheavals and increasing international competition.

Types of skirt supports are shown in Figure 4-14. Figure 4-14b shows the most common and desirable skirt, since the shear is eliminated by the type of attachment. This type is used primarily on short vertical vessels.

The skirt is designed to resist loadings caused by bending and the tower weight. Writing the expression that describes this we obtain

$$\sigma = \frac{M}{Z} + \frac{W}{A}$$

Substituting Equations 4-25 and 4-26, as before, we obtain

$$\sigma = \frac{16MD_o}{\pi t(D_o + D_i)(D_o^2 + D_i^2)} + \frac{2W}{\pi t(D_o + D_i)} \tag{4-37}$$

$$t = \frac{16MD_o}{\pi\sigma(D_o + D_i)(D_o^2 + D_i^2)} + \frac{2W}{\pi\sigma(D_o + D_i)} \tag{4-38}$$

Once again, Equations 4-37 and 4-38 must be solved by iteration. Normally, these equations do not govern the skirt thickness, as the reaction of external bolting and

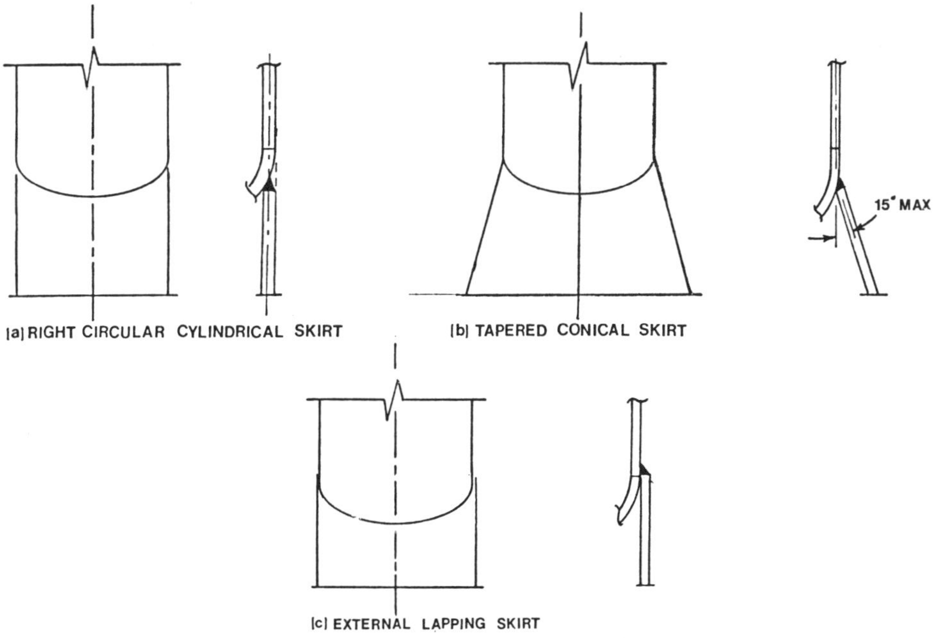

Figure 4-14. Skirt designs.

compression rings is not considered. The stresses in the skirt shell that result in compressive loading on the compression ring and bolting chair can be quite high in applications where external chairs shown in Figure 4-15 are used. See Brownell and Young [3], for a derivation of the reaction expression. The skirt thickness required to resist the reaction of external chairs or compression ring for a chair of the type in Figure 4-15 is determined as follows:

$$t = 1.76\left[\frac{F_i B}{m(G_H)\sigma_{all}}\right]^{2/3} r^{1/3} \qquad (4\text{-}39)$$

where
- t = skirt thickness, in.
- r = radius of skirt, in.
- m = bolt spacing, in. or 2B in Figure 4-15
- F_i = uplift bolt load, lb
- B = radial distance from outside of skirt to bolt circle, in., Figure 4-15
- G_H = gusset height, in.

Equation 4-39 is normally the controlling criteria for a skirt with external chairs. However, for a skirt with or without external chairs, Equation 4-38 must be satisfied.

ANCHOR BOLTS

Anchor bolts are one of the most important aspects of tower design, and, unfortunately, are often not taken seriously enough. Consequently, many problems related to towers during construction or operation can be linked to anchor bolts.

Wind and seismic loads are dynamic and result in cyclic loading of the anchor bolts. For this reason, I will only present the method for analyzing preloaded anchor bolts. Initial preload is significant since pre-torquing the bolts reduces the variable stress range the bolts experience during cyclic loading. The tower weight and bolt load allow firm contact between the compression ring and concrete or steel such that the support base rotates about the neutral axis of the contact area, as shown in Figure 4-16. Referring to this figure we see that under a moment M at the base plate-concrete juncture the maximum and minimum stress is

$$\sigma_c = \left(\frac{NF_i}{A_c}\right) + \left(\frac{W}{A_c}\right) \pm \left(\frac{MD_o}{2I_c}\right) \qquad (4\text{-}40)$$

where
$$A_c = \frac{\pi}{4}(D_o^2 - D_i^2)$$
$$I_c = \frac{\pi}{64}(D_o^4 - D_i^4)$$

- F_i = see below
- W = vessel weight, lb
- W_o = operating weight, lb
- W_E = empty weight, lb
- N = number of anchor bolts

The minimum initial bolt load required to maintain compression between the base plate and compression ring exist when $\sigma_c = 0$. Thus, using Equation 4-40 and substituting $\sigma_c = 0$ we have

$$F_i = \frac{8M\,D_o}{N(D_o^2 + D_i^2)} - \frac{W_E}{N} \qquad (4\text{-}41)$$

The required bolt area is

$$A_b = \frac{\left(\frac{4M}{BC}\right) - W_E}{N\sigma_a} \qquad (4\text{-}42)$$

where
- BC = bolt circle diameter, in.
- σ_a = allowable anchor bolt stress, psi
- M = in.-lb

Equation 4-42 is one of the major differences in designing a tower under a lump-sum contract versus cost-plus. Most cost-plus designers use vessel code allowable stress values that are based on a factor of safety of 4:1. This large a factor of safety is intended for components containing *pressure*. Thus, using vessel code allowable stresses for bolts leads to large anchor bolts, which is undesirable because more concrete is required and larger bolts are much harder to torque, requiring bigger wrenches and being more susceptible to galling. To keep anchor bolt sizes down follow AISC guidelines for bolting, since anchor bolts are purely structural in nature.

Table 4-6 provides the allowable stresses for bolting per AISC [5]. Type A325 bolts and ASTM A193-B7 high-strength bolts are used in most applications. A307 bolts are used where bolt loadings are not large and the bolt size need not be massive. When bolt sizes get large (2¾ to 3 in.) or it is desirable to reduce the bolt size, then A193-B7 or A325 bolts are used. One can see from Table 4-6 that A325 has more than twice the allowable stress value as A307 bolts. The extra cost of the high-strength material will still be less than the additional concrete and labor costs associated with a larger bolt. Certainly, if one pays more for high-strength steel, he should be permitted to use the larger allowable, as given by AISC. Normally, 40,000 psi is used with A325 and 193-B7 bolts.

The spacing of anchor bolts is another critical parameter. Spacing the anchor bolts too close to one another

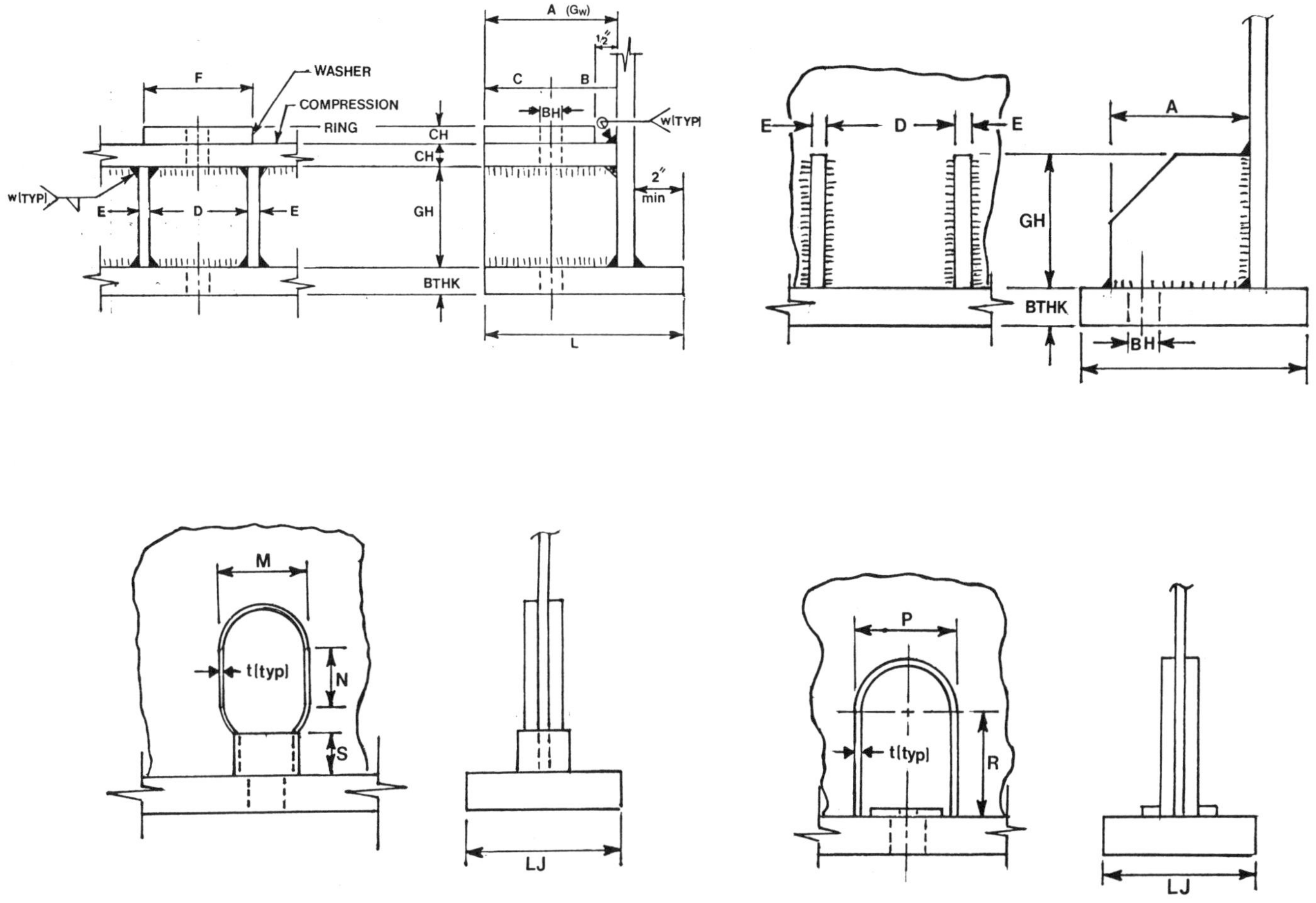

NOTES:
all dimensions in inches BTHK to be evaluated by equations 4-57 or 4-60 all welds to be size "t"

BOLT SIZE	A	B	C	BH	D	E	F	GH	CH	L	LJ	M	N	P	R	t	S
1	3 1/2	1 3/4	1 3/4	1 1/2	2	3/8	3 1/4	6	5/8	5 1/2	5	3 1/4	3	3 1/2	7	3/16	3
1 1/8	3 3/4	1 7/8	1 7/8	1 5/8	2 1/8	3/8	3 3/8	6	5/8	5 3/4	5 1/4	3 1/2	3	3 3/4	7	3/16	3
1 1/4	4	2	2	1 3/4	2 1/2	3/8	3 3/4	6	3/4	6	5 1/2	3 3/4	3 3/8	4	7 1/4	1/4	3
1 3/8	4 1/4	2 1/8	2 1/8	1 7/8	2 5/8	1/2	4 1/8	7	3/4	6 1/4	5 3/4	4	3 3/4	4 3/8	7 1/2	1/4	3
1 1/2	4 1/2	2 1/4	2 1/4	2	2 3/4	1/2	4 1/4	8	3/4	6 1/2	6	4 1/4	4 1/8	4 1/2	8	1/4	4
1 3/4	5	2 1/2	2 1/2	2 1/4	3 1/4	5/8	5	9	1	7	6 1/2	4 3/4	4 3/4	4 3/4	9	5/16	5
2	5 1/2	2 3/4	2 3/4	2 1/2	3 1/2	3/4	5 1/2	10	1	7 1/2	7	5 1/4	5 1/2	6	12	3/8	6
2 1/4	5 3/4	3	2 3/4	2 3/4	3 3/4	3/4	5 3/4	12	1	7 3/4	7 1/2	5 3/4	5 7/8	6 1/2	13	3/8	7
2 1/2	6 1/4	3 1/4	3	3	4 1/8	7/8	6 7/8	13	1 1/8	8 1/4	7 3/4	6 3/4	6	7	14	7/16	8
2 3/4	6 1/2	3 1/2	3	3 1/4	4 1/2	1	7 1/2	14	1 1/4	8 1/2	8	6 3/4	6 3/8	7 1/2	16	1/2	9
3	7	3 3/4	3 1/4	3 1/2	4 3/4	1 1/8	8	16	1 1/4	9	8 1/4	7	7	7 3/4	18	1/2	10

Figure 4-15. Typical designs and dimensions of chair and base plates.

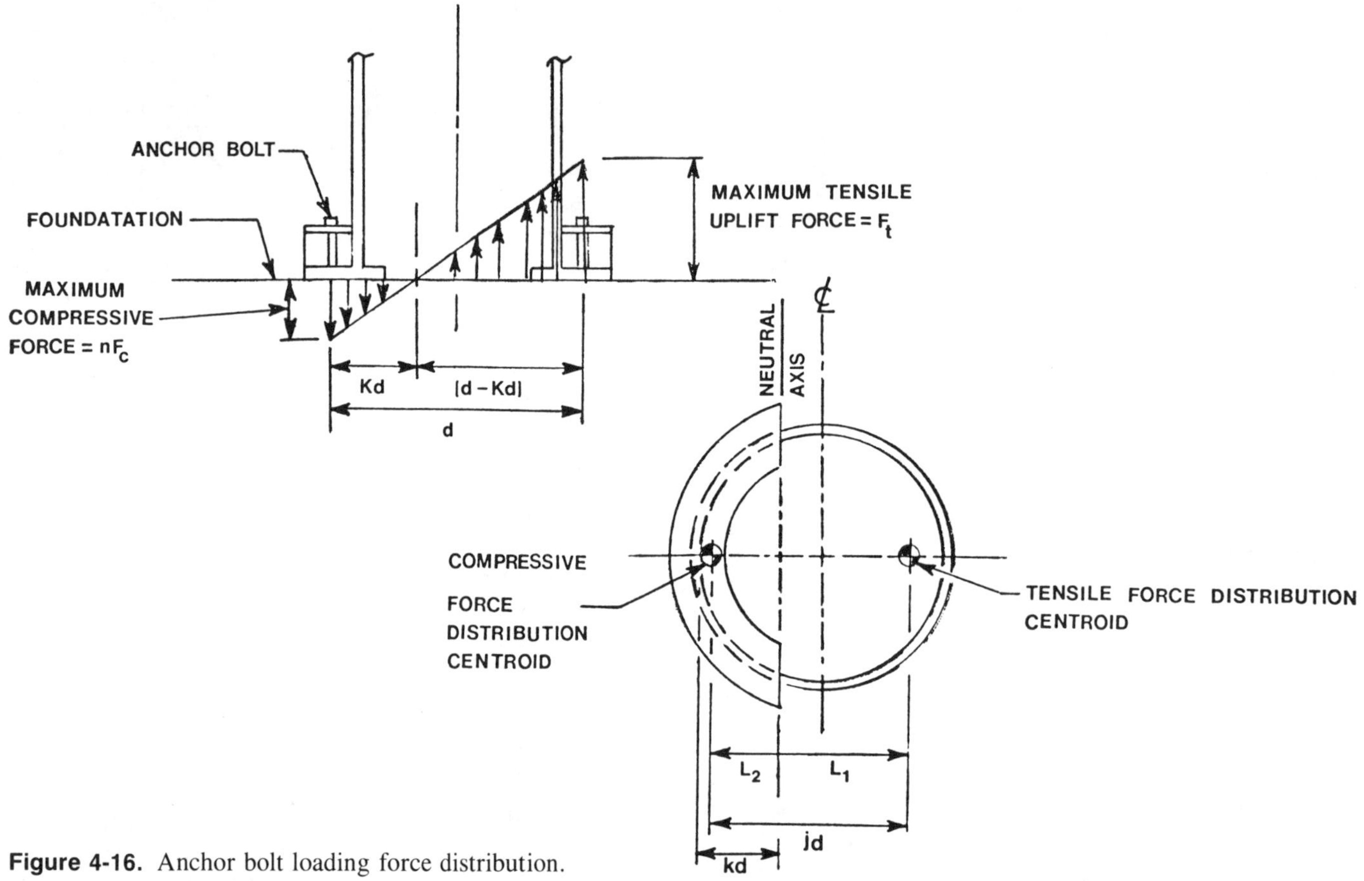

Figure 4-16. Anchor bolt loading force distribution.

prevents the strength of the bolting in the concrete from becoming fully developed. It is advisable to set the bolts at least 18 inches apart. To accommodate this minimum spacing a wider base ring with gusset plates can be used or the skirt can be tapered with a conical skirt. As shown in Figure 4-14, with a tapered skirt the apex angle should not exceed 15°.

BASE PLATE THICKNESS DESIGN

Base plate design involves the loadings passed on from the tower to the foundation. The base plate is a circular ring plate used to distribute these loads around the circumference of the bolt circle. Anchor bolts normally vary in diameter from one to three inches—bolts smaller than one inch are more likely to strip or shear off; bolts larger than three inches require large wrenches and create excessive problems for construction personnel. For these reasons it is desirable to attempt to adhere to the one to three inch size range.

In the case of a concrete foundation, the relative strength of the concrete to steel becomes a significant factor. The modulus of elasticity of steel is approximately 30 × 10⁶ psi and that of concrete approximately 2.0 to 4.0 × 10⁶ psi. Defining the ratio of the two as n, we write

$$n = \frac{E_s}{E_c} \tag{4-43}$$

since $E_c = \frac{\sigma_c}{\epsilon_c}$

and $E_s = \frac{\sigma_s}{\epsilon_s}$

and $\epsilon_s = \epsilon_c$ because of the base plate-concrete bond we have

$$\sigma_{s(\text{induced})} = n\sigma_{c(\text{induced})} \tag{4-44}$$

Listed in Table 4-7 are values of the moduli ratio n and the various concrete mixes from Brownell and Young [3]. Figure 4-16 shows a detail of the compressive force of the concrete, F_c, multiplied by the value of n shown opposite the maximum tensile stress, F_T of the base plate steel.

Table 4-6
Bolts, Threaded Parts, and Rivets
Tension [5]
Allowable loads in kips

Bolts and Rivets
Tension on gross (nominal) area

ASTM Designation	F_t Ksi	5/8	3/4	7/8	1	1 1/8	1 1/4	1 3/8	1 1/2
		Area (Based on Nominal Diameter), in.^{2}							
		0.3068	0.4418	0.6013	0.7854	0.9940	1.227	1.485	1.767
A307 bolts	20.0	6.1	8.8	12.0	15.7	19.9	24.5	29.7	35.3
A325 bolts	44.0	13.5	19.4	26.5	34.6	43.7	54.0	65.3	77.7
A490 bolts	54.0	16.6	23.9	32.5	42.4	53.7	66.3	80.2	95.4
A502-1 rivets	23.0	7.1	10.2	13.8	18.1	22.9	28.2	34.2	40.6
A502-2,3 rivets	29.0	8.9	12.8	17.4	22.8	28.8	35.6	43.1	51.2

(Nominal Diameter, d, in.)

The above table lists ASTM specified materials that are generally intended for use as structural fasteners.
For dynamic and fatigue loading, only A325 or A490 high-strength bolts should be specified. See AISC Specification. Appendix B, Sect. B3.
For allowable combined shear and tension loads, see AISC Specification Sect. 1.6.3.

Threaded Fasteners [5]
Tension on gross (nominal) area

ASTM Designation	F_y Ksi	F_u Ksi	F_t Ksi	5/8	3/4	7/8	1	1 1/8	1 1/4	1 3/8	1 1/2
				Area (Based on Nominal Diameter), in.^{2}							
				0.3068	0.4418	0.6013	0.7854	0.9940	1.227	1.485	1.767
A36	36	58	19.1	5.9	8.4	11.5	15.0	19.0	23.4	28.4	33.7
A572, Gr. 50	50	65	21.5	6.6	9.5	12.9	16.9	21.4	26.4	31.9	38.0
A588	50	70	23.1	7.1	10.2	13.9	18.1	23.0	28.3	34.3	40.8
A449											
$d \leq 1$	92	120	39.6	12.1	17.5	23.8	31.1	—	—	—	—
$1 < d \leq 1\frac{1}{2}$	81	105	34.7	—	—	—	—	34.5	42.6	51.5	61.3

(Nominal Diameter, d, in.)

The above table lists ASTM specified materials available in round bar stock that are generally intended for use in threaded applications such as tie rods, cross bracing and similar uses.
The tensile capacity of the threaded portion of an upset rod shall be larger than the body area times $0.6F_y$.
F_u = specified minimum tensile strength of the fastener material.
$F_t = 0.33F_u$ = allowable tensile stress in threaded fastener.

Table 4-7
Design of Supports for Vertical Vessels [3]

Values of Constants C_t, C_c, Z, and J as a Function of k

k	C_c	C_t	Z	J
0.050	0.600	3.008	0.490	0.760
0.100	0.852	2.887	0.480	0.766
0.150	1.049	2.772	0.469	0.771
0.200	1.218	2.661	0.459	0.776
0.250	1.370	2.551	0.448	0.779
0.300	1.510	2.442	0.438	0.781
0.350	1.640	2.333	0.427	0.783
0.400	1.765	2.224	0.416	0.784
0.450	1.884	2.113	0.404	0.785
0.500	2.000	2.000	0.393	0.786
0.550	2.113	1.884	0.381	0.785
0.600	2.224	1.765	0.369	0.784

Average Values of Properties of Three Concrete Mixes

Water Content U.S. Gallons per 94-lb Sack of Cement	σ_c' 28-day Ultimate Compressive Strength, psi	n $\frac{30 \times 10^6}{E_c}$	σ_c Allowable Compressive Strength, psi
7 1/2	2000	15	800
6 3/4	2500	12	1000
6	3000	10	1200
5	3750	8	1400

Equation 4-44 is shown as a linear proportion by the straight line shown in Figure 4-16. Even though the tensile strength of the bolt is, by Equation 4-44, equal to the ratio n times the concrete allowable compressive strength, it is not necessarily evenly distributed about the neutral axis as shown in Figure 4-16. This "offset factor," known as the "k Factor," is determined from

$$\frac{\sigma_s}{(d - kd)} = \frac{n\sigma_c}{kd}$$

$$\text{or } k = \frac{1}{1 + \dfrac{\sigma_s}{n\sigma_c}} \tag{4-45}$$

Equation 4-45 is solved by iteration using the following steps: Take values for C_c, C_t, Z, and j in Table 4-7 for a given value of Z. Normally, k = 0.333, C_c = 1.588, C_t = 2.376, Z = 0.431, and j = 0.782 to start the process. Then the following equations are solved:

$$F_t = \frac{M - (W_E)(Z)\left(\dfrac{BC}{12}\right)}{(j)\left(\dfrac{BC}{12}\right)} \tag{4-46}$$

$$t_1 = \frac{(A_b)N}{\pi(BC)} \tag{4-47}$$

$$f_s = \frac{F_t}{(t_1)\left(\dfrac{BC}{2}\right)C_t} \tag{4-48}$$

$$f_c = f_t + W_E$$

BPW = base plate width, in.

$$BPW = \frac{(D_o) - (D_i)}{2} \tag{4-49}$$

$$t_2 = BPW - t_i \tag{4-50}$$

$$E_R = \frac{E_s}{(C_s)(1{,}000)} \tag{4-51}$$

E_s = modulus of elasticity of base plate metal, psi
C_s = compressive strength of concrete, psi, denoted in Table 4-7 as σ_c

$$SFC = \frac{f_c}{(t_2 - (E_R)(t_1))\left(\dfrac{BC}{2}\right)(C_c)} \tag{4-52}$$

where SFC = compressive stress on concrete at the bolt circle and

$$k = \frac{1}{1 + \dfrac{f_s}{(E_R)(SFC)}} \tag{4-53}$$

After computing an initial value of k, this process should be repeated five times in order to converge on a value for k.

Once a value for k is determined, we now solve for the maximum induced stress at the outer periphery in the concrete,

$$\sigma_{c(max)} = (SFC)\left[\frac{2(k)(BC) + BPW}{2(k)(BC)}\right] \tag{4-54}$$

$$\text{using } L = \frac{D_o - (\text{skirt OD})}{2}\text{, in.} \tag{4-55}$$

we solve for the base plate thickness, BTHK,

$$BTHK = L\left[\frac{3\sigma_{c(max)}}{\sigma_{all}}\right]^{0.5} \tag{4-56}$$

where σ_{all} = allowable working stress for base plate metal, psi

By using Equation 4-56 one assumes no gusset plates on the base plate-skirt connection. To reduce the required base plate thickness in Equation 4-56 the additional strength of gusset plates can be used, because with the gusset plate stiffening the base plate at the skirt juncture, the base plate between the gusset plates can be considered as a rectangular, uniformly loaded plate with two edges simply supported (at gusset plates), a third edged fixed (skirt side), and the fourth edge free. The deflections and bending moments are tabulated by Timoshenko [6] and are shown in Table 4-8. The process of using gusset plates to stiffen the base plate is begun by making the number of gusset plates equal to the number of anchor bolts. Doing this we write

NG = number of gusset plates
let NG = N bolts

$$SG = \frac{\pi(BC)}{NG + N\text{ bolts}} \tag{4-57}$$

$$RAT = \frac{L}{SG}$$

$$M_x = |E_x|\,\sigma_{c(max)}(SG)^2 \tag{4-58}$$

$$M_y = |E_y|\,\sigma_{c(max)}(L)^2 \tag{4-59}$$

Table 4-8
Maximum Bending Moments in a Bearing Plate with Gussets [6]

l/b	$M_x \begin{pmatrix} x = b/2 \\ y = l \end{pmatrix}$	$M_y \begin{pmatrix} x = b/2 \\ y = l \end{pmatrix}$
0	0	$-0.500f_cl^2$
1/3	$0.0078f_cb^2$	$-0.428f_cl^2$
1/2	$0.0293f_cb^2$	$-0.319f_cl^2$
2/3	$0.0558f_cb^2$	$-0.227f_cl^2$
1	$0.0972f_cb^2$	$-0.119f_cl^2$
3/2	$0.123f_cb^2$	$-0.124f_cl^2$
2	$0.131f_cb^2$	$-0.125f_cl^2$
3	$0.133f_cb^2$	$-0.125f_cl^2$
∞	$0.133f_cb^2$	$-0.125f_cl^2$

b = gusset spacing (x direction) inches.
l = bearing-plate outside radius minus skirt outside radius (y direction) inches.

Where $\sigma_{c(max)}$ is determined by Equation 4-55, using the greater of M_x and M_y we have

$$BTHK = \left[\frac{6M}{\sigma_{all}}\right]^{0.5} \tag{4-60}$$

where M = M_x or M_y, in.-lb/in.

This iteration can be repeated as many times as desired to reduce the base plate thickness. In normal practice, it is unusual to use more gusset plates than anchor bolts.

The bearing pressure on the base plate must be checked to prevent exceeding the allowable compressive stress of the concrete. Computing the uplift force on each anchor bolt we have,

$$F_i = \frac{96MD_o}{N(D_o^2 + D_i^2)} - \frac{W_E}{N}, \text{ lb} \tag{4-61}$$

$$\sigma_c = \frac{N_{fi}}{A_c} + \frac{W_E}{A_c} + \frac{12MD_o}{2I_c}, \text{ psi} \tag{4-62}$$

$\sigma_c \leqslant 1{,}200$ psi for weight and wind
where, M = ft-lbs

where $A_c = \dfrac{\pi[(D_o)^2 - (D_i)^2]}{4}$

$I_c = \dfrac{\pi[(D_o)^4 - (D_i)^4]}{64}$

COMPRESSION RING AND GUSSET PLATE DESIGN

Typical designs and dimensions of chair and base plate designs are shown in Figure 4-15. The compression plate thickness is determined by

$$t_{CR} = \left[\frac{(F_i)C}{4(\sigma_{all})(A - BS)}\right] \tag{4-63}$$

where A and C are dimensions in Figure 4-15
BS = nominal bolt size × 1.25
F_i = bolt uplift force, determined by Equation 4-61
σ_{all} = allowable stress of compression ring metal, psi

The gusset plate thickness is determined by the following:

$$18{,}000\, G_W\, t_G^3 - (F_i)t_G^2 - \frac{G_H^2(F_i)}{1{,}500} = 0 \tag{4-64}$$

where t_G = gusset plate thickness, in.
G_W = gusset plate width, in. (A in Figure 4-15)
G_H = gusset plate height, in. (see Figure 4-15)

The minimum skirt-to-base plate weld size is determined by

$$\xi = \left[\frac{4M}{\pi D_{sk}^2}\right] + \left[\frac{W_E}{\pi D_{sk}}\right]$$

$$F_W = (1.33)(\sigma_{all})(0.55)$$

and

$$W = \frac{\xi}{2F_W} = \text{weld size} \tag{4-65}$$

where σ_{all} = smaller of the allowable stress values for the base plate and skirt metals
M = moment at base plate induced by wind or seismic forces, in.-lb
D_{sk} = outside diameter of skirt, in.

ANCHOR BOLT TORQUE

There have been many recipes proposed for the computation of bolt torque over the years. The mystery of bolt loads is unveiled by such authorities as Bickford [7] and Faires [8]. Their extensive research into bolt loading produced the following recommended formulation:

$$T = CDF_i, \text{ in.-lb} \tag{4-66}$$

where C = 0.20 bare steel
C = 0.15 for lubricated bolt
D = nominal bolt diameter, in.
F_i = anchor bolt uplift force, lbs

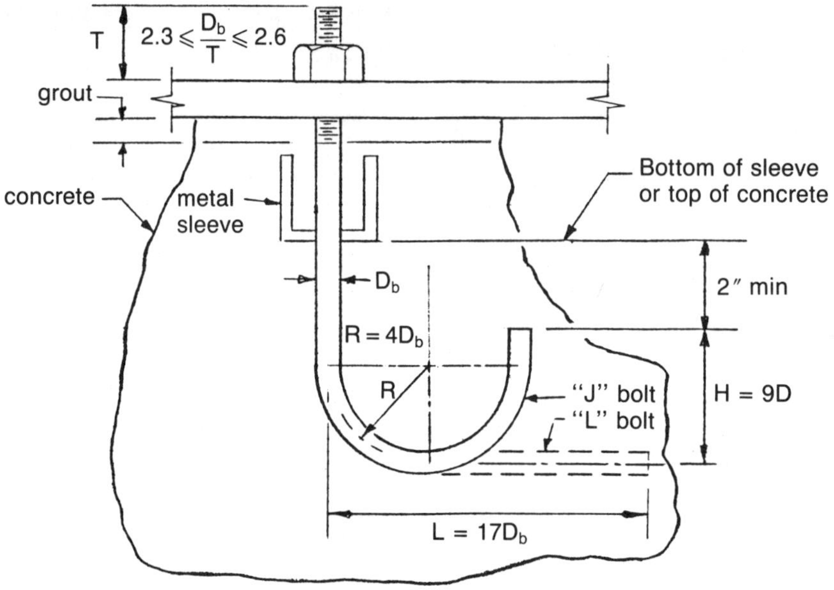

Figure 4-17. "J" and "L" type anchor bolts are used for small vessels.

In most tower applications, Fel-Pro C5A is a very common bolt lubricant. The field of bolt design and bolt lubricants is almost as involved as tower design and the interested reader is referred to the excellent work of Bickford [7].

Figure 4-17 shows the two most common types of anchor bolts, "J" and "L." For large towers where large loads are anticipated, the bolt in Figure 4-18 is used.

WIND ANALYSIS OF TOWERS

Analyzing wind loading on towers requires combining loads induced by wind, internal or external pressure, and weight. Such an analysis must be made to ensure that the tower shell thickness is sufficient to withstand the combined loads.

Wind and seismic analyses are completed separately, with their respective bending moments being used to determine the tower shell thickness values at each section.

Before examining the design criteria, let's consider the following terms:

σ_w = stress due to wind or earthquake
σ_p = stress due to internal or external pressure
σ_g = stress due to weight

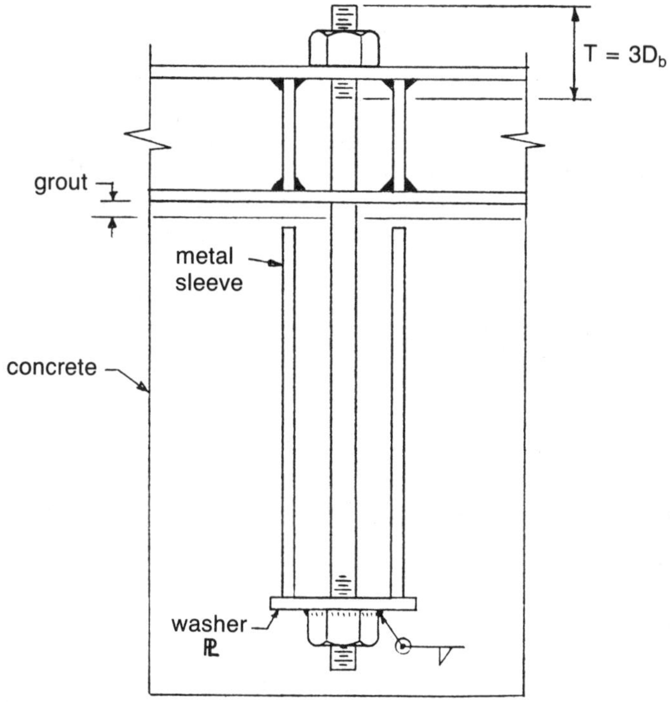

Figure 4-18. Straight type bolt—used for large vessels, especially towers.

Referring to Figure 4-13, we see that the stress element in the shell is affected by the combined loads more in the longitudinal direction than the circumferential direction. However, for the longitudinal axis the internal or external pressure stress is governed by the relation

$$\sigma_p = \frac{PD}{4t} \tag{4-67}$$

where D = mean diameter of vessel, in.
P = internal or external pressure, psi
σ_p = longitudinal stress, psi
t = shell thickness less corrosion allowance, in.

There are two conditions where Equation 4-67 is used to combine stress values:

1. *Combination of wind (or earthquake) load, internal pressure, and weight of vessel.*
For windward side,

$$\sigma_{WS} = \sigma_W + \sigma_p - \sigma_{wt} \tag{4-68}$$

For leeward side

$$\sigma_{LS} = \sigma_p - \sigma_w - \sigma_{wt} \tag{4-69}$$

Comparing Equation 4-68 with Equation 4-69 we have

$$|\sigma_W + \sigma_p - \sigma_{wt}| > |\sigma_p - \sigma_w - \sigma_{wt}| \tag{4-70}$$

if and only if $|\sigma_p| > |\sigma_{wt}|$ (4-71)

which is true for most applications, when the internal pressure stress is greater than that induced by the weight of the tower above the section. However, for a few cases, the stress induced by the weight is greater than that induced by internal pressure for low-pressure thick-walled applications. The thick walls at low pressure could be for controlling tower deflections due to wind. For these limited cases the allowable stress is that determined by UG-23 (b) of the ASME vessel code, which is known as the B factor. The B factor is commonly associated with external pressure, because the case of the weight stress exceeding the internal pressure stress is rare, but it must be emphasized that the B factor is the allowable value of stress for longitudinal compressive loading like that encountered in towers. Thus, the B factor is more comprehensive than its external pressure application would indicate. Therefore, if Equation 4-71 is reversed and

$$|\sigma_{wt}| > |\sigma_p| \tag{4-72}$$

such that the leeward side governs, then

$$|\sigma_p - \sigma_w - \sigma_{wt}| < \text{B factor}$$

If Equations 4-70 and 4-71 hold,

$$|\sigma_w + \sigma_p - \sigma_{wt}| > \sigma_{all}E$$

where σ_{all} = allowable stress in *tension* for a given material at a given temperature and pressure
E = weld efficiency

Another form of Equation 4-71 may be determined by rewriting the equation as

$$\sigma_p > \sigma_{wt}$$

or

$$\frac{P(R_o - 0.4t)}{E_t} > \frac{W}{\pi(R_o^2 - R_i^2)}$$

in which $R_o > \left[\dfrac{3\,W - 1.2\pi\,PR_i^2}{1.8\,P\pi}\right]$ (4-73)

Equation 4-73 is another form of Equation 4-71, in terms of the vessel dimensions, where W is the total weight of the tower above the section being analyzed.

2. *Combination of wind (or earthquake) load, external pressure and the weight of the tower.*
On windward side

$$\sigma_{WS} = \sigma_w - \sigma_p - \sigma_{wt}$$

On leeward side

$$\sigma_{LS} = -\sigma_w - \sigma_p - \sigma_{wt}$$

For most applications with external pressure we have

$$\sigma_{LS} > \sigma_{WS}$$

$$|-\sigma_w - \sigma_p - \sigma_{wt}| > |\sigma_w - \sigma_p - \sigma_{wt}|$$

$$\sigma_p \neq 0$$

Since the value of σ_p is for external pressure, we must apply the B factor in Equation 4-67. After these criteria are satisfied, we turn our attention to the determination of wind loads that induce σ_w.

WIND DESIGN SPEEDS

The procedures for determining wind design speeds for structures, towers, and stacks vary from country to country, depending on how well records have been kept. The wind velocity is a function of the temperature gradient and terrain roughness. The first representation of a mean wind velocity profile in horizontally homogeneous terrain is the power law, first proposed in 1916. This law states

$$V_Z = V_o\left(\frac{Z}{Z_o}\right)^{\alpha} \tag{4-74}$$

where V_o = mean wind speed at a reference height Z
Z_o = reference height (normally 33 ft or 10 m)
α = a constant dependent upon roughness of terrain
Z = height above ground

Other proposals have been made to determine wind speeds. Simiu [9] states that the logarithmic law is a superior representation of strong wind profiles in the lower atmosphere. What is relevant to the reader is to be familiar with whatever standard is used. The discussions and examples presented in this text are slanted toward those standards in the United States. However, the techniques and base principles of engineering are applicable to all countries.

In the United States there are four basic codes governing wind—ANSI-A58.1-1982, the Uniform, the Basic, and the Standard Building Codes [10]. The ANSI-A58.1-1982 differs from the ANSI-A58.1-1972 in that three optional methods of determining wind design loads on a structure are given [11]. These options are as follows:

1. Choose a design wind speed (50-year mean recurrence interval) off the U.S. map provided on the document. The national map is a graphic display of isopleths (lines of equal wind speed) of the maximum values of the mean speed for which records have been kept, i.e., basic wind speeds that can be expected to occur within a particular period. This "particular period" is called the return period. The problem with a national map consisting of isopleths is that localized wind speeds can vary as much as 30 mph over the speed shown on the isopleth (particularly in mountainous regions).

   Hurricanes are fairly well accounted for on these maps. Tornadoes are considered to be nonexistent, because it is not economically feasible to design an entire building for tornado wind speeds. The reason for this is that the probability of a structure being hit by one is extremely small; however, nuclear sites are designed to withstand tornado winds.
2. Using site and structure factors calculate the design wind speed. The factors on the ANSI 1982 test used are as follows:
   (a) Importance Coefficient, I, a hospital or nuclear plant would be designed more conservatively than a barn on a farm.
   (b) Variation of wind speed with building height and surrounding terrain.
   (c) Gust response factor.
   (d) Velocity pressure coefficient, K_Z.
3. Test a model of the tower and its surroundings in a wind tunnel. Even though the 1972 ANSI standard does not mention this, the 1982 version specifies certain requirements for wind tunnels.

These three options are new to both the ANSI-A58.1 standard and to the three building codes—the Uniform, Basic, and the Standard. The latter three codes do not automatically adopt newly revised ANSI standards, thus making for inconsistency in wind code provisions in the United States.

The basic wind pressure in the ANSI-A58.1-1972 is

$$q_{30} = \rho V^2/2 = (0.5)(0.00238)(5{,}280/3{,}600)^2 V_{30}^2 = 0.00256\,V_{30}^2 \tag{4-75}$$

where q_{30} = basic wind pressure at 30 ft, above grade lb/ft^2
V_{30} = basic wind speed, mph

The effective velocity pressures of winds for buildings and structures, q_F, is

$$q_F = K_Z G_F q_{30} \tag{4-76}$$

where K_Z = velocity pressure coefficient that depends upon the type of exposure and height Z above the ground
G_F = dynamic gust response factor

In the 1982 ANSI-A58.1 Code the effective velocity pressure for wind is partially a combination of Equations 4-75 and 4-76,

$$q_Z = 0.00256\,K_Z(IV)^2 \tag{4-77}$$

V = basic wind speed, mph
I = importance factor = 1

A value of V can be approximated for the United States from the isopleths shown in Figure 4-19.

One of the major differences between the ANSI A58.1-1972 and 1982 is how the velocity pressure coefficient, K_Z, is determined. In the 1972 Code the value of

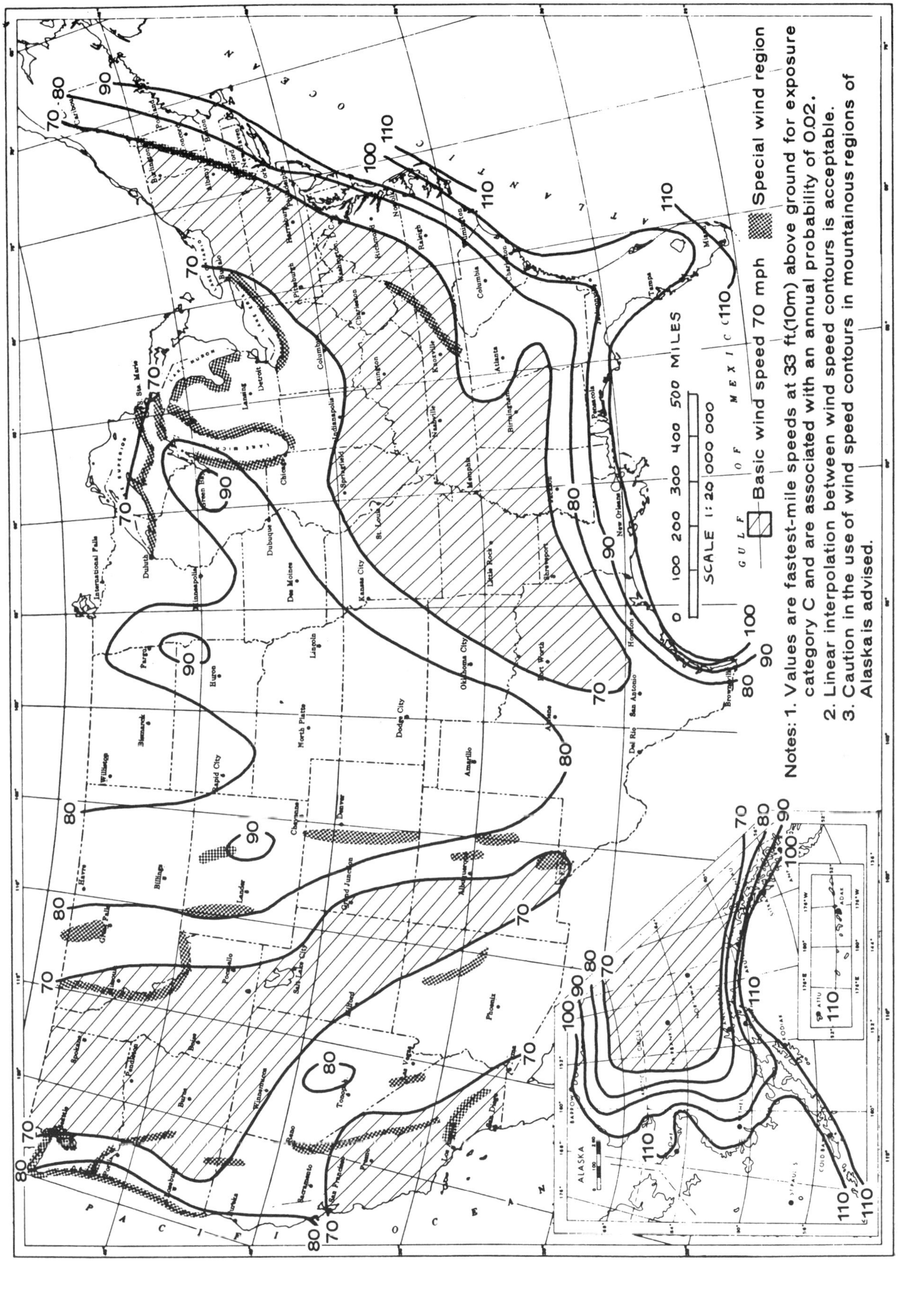

Figure 4-19. UBC and ANSI A58.1-1982 basic wind speed chart [10].

Table 4-9
Velocity Pressure Exposure Coefficient, K_Z [11]

Height above Ground Level, Z (feet)	K_Z			
	Exposure A	Exposure B	Exposure C	Exposure D
0–15	0.12	0.37	0.80	1.20
20	0.15	0.42	0.87	1.27
25	0.17	0.46	0.93	1.32
30	0.19	0.50	0.98	1.37
40	0.23	0.57	1.06	1.46
50	0.27	0.63	1.13	1.52
60	0.30	0.68	1.19	1.58
70	0.33	0.73	1.24	1.63
80	0.37	0.77	1.29	1.67
90	0.40	0.82	1.34	1.71
100	0.42	0.86	1.38	1.75
120	0.48	0.93	1.45	1.81
140	0.53	0.99	1.52	1.87
160	0.58	1.05	1.58	1.92
180	0.63	1.11	1.63	1.97
200	0.67	1.16	1.68	2.01
250	0.78	1.28	1.79	2.10
300	0.88	1.39	1.88	2.18
350	0.98	1.49	1.97	2.25
400	1.07	1.58	2.05	2.31
450	1.16	1.67	2.12	2.36
500	1.24	1.75	2.18	2.41

K_Z is a linear function of the height Z from heights of thirty to nine hundred feet. This results in a triangular wind distribution on the tower. In the 1982 Code the value of K_Z is a parabolic function (can be approximated with a step function) for wind loading depicted in Table 4-9 and for dynamic gust response, K_Z is governed by the power law, Equation 4-74.

$$K_Z = \begin{cases} 2.58\left(\dfrac{Z}{Z_g}\right)^{2/\alpha} & \text{for } Z \geqslant 15 \text{ feet} \\ \\ 2.58\left(\dfrac{15}{Z_g}\right)^{2/\alpha} & \text{for } Z < 15 \text{ feet} \end{cases} \tag{4-78}$$

where values of Z_g and α are given in Table 4-10. The parabolic function is a reflection of the old classical approach used in the ASA 58.1-1955, but is a more refined distribution. The treatment of K_Z in the dynamic gust response analysis is a new development in U.S. codes.

The force exerted on a tower immersed in a moving fluid is a function of the properties of the tower shape and properties of the fluid. The fluid properties of importance are the viscosity, density, and elasticity. Writing this relationship in functional form we have

$$F = f(\rho, V, \ell, \mu, a)$$

where a = velocity of sound = 0 in our case, because wind speeds are extremely low compared to sonic speeds

This equation shows that there is a relationship dictated by the dimensions of the parameters involved. Applying dimensional analysis makes the equation

$$C\left\{\frac{\rho V \ell}{\mu}, \frac{F}{\rho(V^2\ell^2)}\right\} = 0$$

where each of the two components is a dimensionless parameter. The equation can be solved for the first dimensionless combination by

$$\frac{F}{\rho V^2 \ell^2} = C_1\left(\frac{\rho V \ell}{\mu}\right) \tag{4-79}$$

Equation 4-79 implies that the parameters $F/(\rho V^2\ell^2)$ and $(\rho V\ell/\mu)$ have certain definite values that will be equal if a geometrically similar body with the same orientation is moved through the same fluid or another fluid for which $\rho V\ell/\mu$ has the same value as the first body. Two such bodies are said to be dynamically similar and dynamic similarity is the key to wind tunnel tests. Assuming that μ has no influence on the force F, we can deduce from

Equation 4-79 (see any basic fluid mechanics text) and obtain

$$C_F \frac{F}{\frac{\rho}{2} V^2 \ell^2} = 0$$

$$\text{or } F = \frac{C_F \rho V^2 \ell^2}{2} \qquad (4\text{-}80)$$

where C_F is a dimensionless empirical constant. Equation 4-80 states that, for a body of given orientation and shape that is immersed in a moving fluid, the force experienced is proportional to the kinetic energy per unit volume of the motion of the fluid $(\rho/2)V^2$ and a characteristic area ℓ^2. C_F is a dimensionless quantity that characterizes the force and is called the *force coefficient.* Two bodies that are immersed in moving fluids are said to be similar (geometric similarity) if their Reynolds numbers are equal. Then the flows are dynamically similar and have equal force coefficients. The Reynolds number $\rho V \ell/\mu$ is called a *similarity parameter.* Figure 4-20 shows the influence of the Reynolds number, corner radius, and surface roughness on the force coefficient on various bodies. The values of C_F are determined empirically and are shown in the figure. Sometimes this coefficient is referred to as the drag or pressure coefficient.

Kuethe and Schetzer [12], use the Kutta-Joukowski theorem to show that the force per unit length acting on a right cylinder of any cross section whatever is equal to $\rho V \Gamma$ and acts perpendicular to V. The symbol Γ is circulation flow about the cylinder and $\Gamma = \pi DV$. The Kutta-Joukowski principle is exemplified in Figure 4-21[13]. Here the pressure distribution around the cylinder is maximum ninety degrees to the air flow. Depending upon the relative stiffness of the tower sections and mass distribution, this perpendicular force vector can cause a phenomenon known as ovaling, which will be discussed later.

In computing the wind forces on a tower, Equation 4-80 takes the following form in using ANSI A58.1-1982:

$$F = q_Z \bar{G} C_F A_f \qquad (4\text{-}81)$$

where q_Z = wind pressure at height Z, Equation 4-77, lb/ft^2
$\bar{G}$ = gust response factor for main wind-force resisting systems of flexible structures
C_F = force coefficient
A_f = cross-sectional area of tower and other attachments, ft^2

The gust response factor, $\bar{G}$, when multiplied by the mean wind load, produces an equivalent static wind load

Table 4-10
Exposure Category Constants [11]

Exposure Category	α	z_g	D_o
A	3.0	1500	0.025
B	4.5	1200	0.010
C	7.0	900	0.005
D	10.0	700	0.003

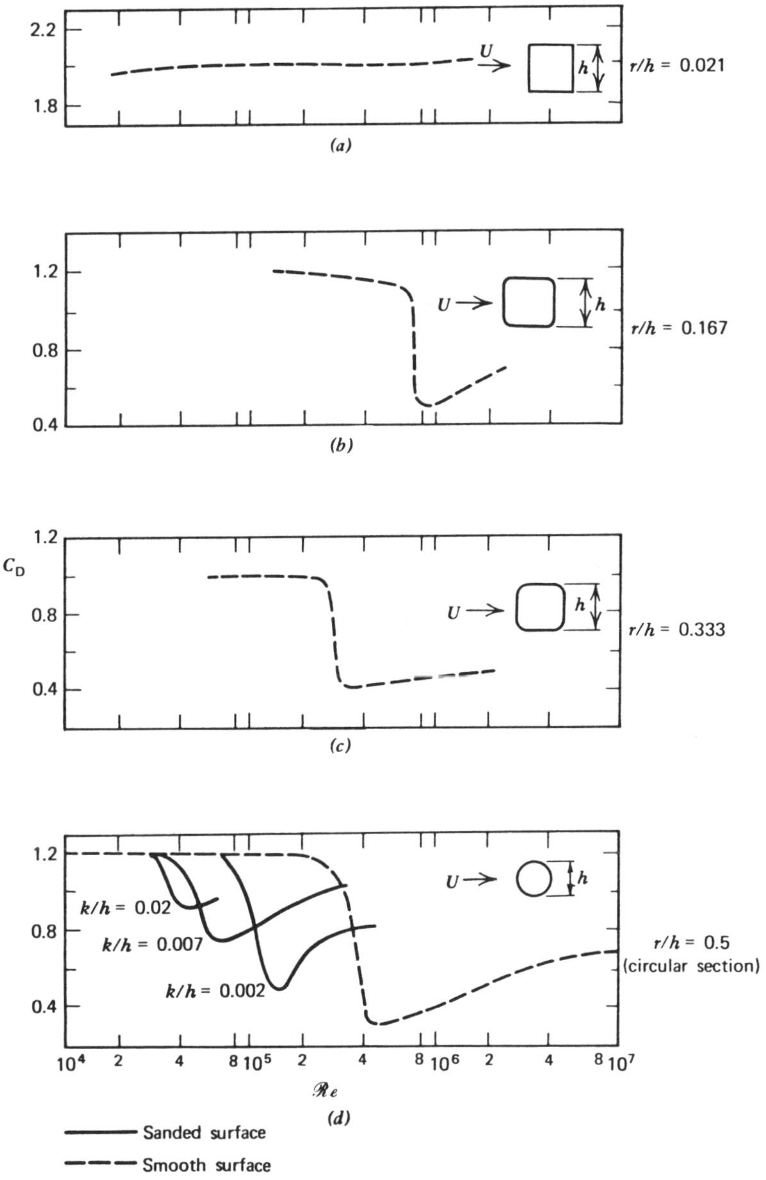

Figure 4-20. The curves depict the influence of the Reynolds number, corner radius, and surface roughness on the drag coefficient, square to circular cylinders; r is the corner radius and K is the sand grain size [9].

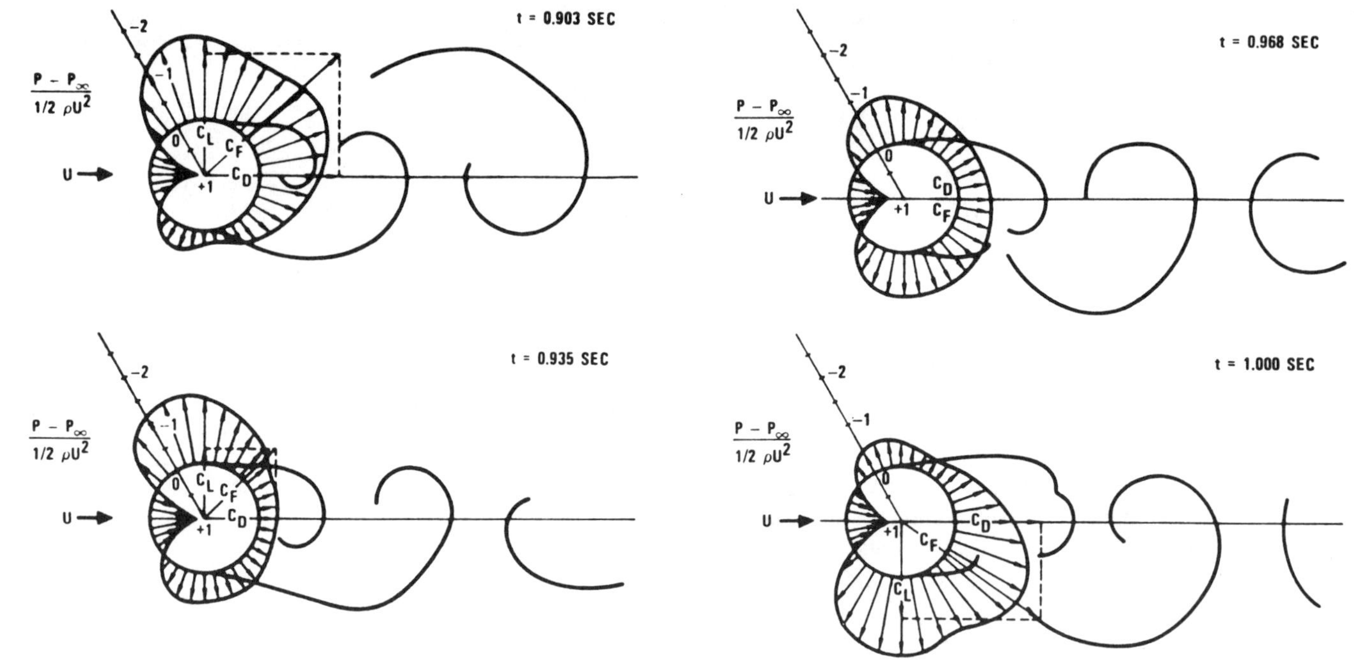

Figure 4-21. A sequence of pressure fields forming around a cylinder at $N_{RE} = 112{,}000$ for approximately one third of one cycle of vortex shedding. (*Flow-Induced Vibration* by R. Blevins. ©1977 by Van Nostrand Reinhold Company, Inc. Reprinted by permission.)

that would induce deflections equal to those of a gusty wind. MacDonald [14] refers to this approach as a quasi-static loading analysis. Quasi-static means that at any instant the stress and deflection induced in the tower are the same as if the instantaneous mean wind load were applied as a static load. Thus, the significant factor is identifying the single highest peak value of instantaneous mean wind speed, or that is, predicting the future worst peak value. Baker and others found at the end of the nineteenth century that there is a simple relationship between the gust frontal area and gust duration. This relationship provides a means of determining the size of the gust, and is illustrated in Figure 4-22. The figure indicates that the worst wind condition for a particular tower is not necessarily the maximum value of the wind velocity, but rather the highest wind speed of the particular size of gust capable of totally enveloping it. To compensate for this in a simple quasi-static analysis, ANSI A58.1-1982 gives the gust factor as

$$\bar{G} = 0.65 + \left(\frac{P}{\beta} + \frac{(3.32T_1)^2 S}{1 + 0.002C}\right)^{1/2} \qquad (4\text{-}82)$$

where β = structural damping coefficient (percentage of critical damping). For normal working stress conditions, $0.01 \leqslant \beta \leqslant 0.02$ for towers.

Table 4-11
Probability of Exceeding Wind Design Speed
$$P_a = 1 - (1 - P_a)^N$$

Annual Probability	Design Life of Structure in N Years						
P_a	1	5	10	15	25	50	100
0.10	0.100	0.410	0.651	0.794	0.928	0.995	0.999
0.05	0.050	0.226	0.401	0.537	0.723	0.923	0.994
0.01	0.010	0.049	0.096	0.140	0.222	0.395	0.634
0.005	0.005	0.025	0.049	0.072	0.118	0.222	0.394

P = probability of exceeding design wind speed during n years, where $P = 1 - (1 - P_a)^n$
P_a = annual probability of wind speed exceeding a given magnitude (Table 4-11)
T_1 = exposure factor evaluated at two-thirds the mean height of the structure

$$T_Z = \frac{2.35(C_L)^{0.5}}{(Z/30)^{1/\alpha}}$$

S = structure size factor (Figure 4-23)
C = average horizontal dimension of the building or structure in a direction normal to the wind, ft (see Example 4-2)

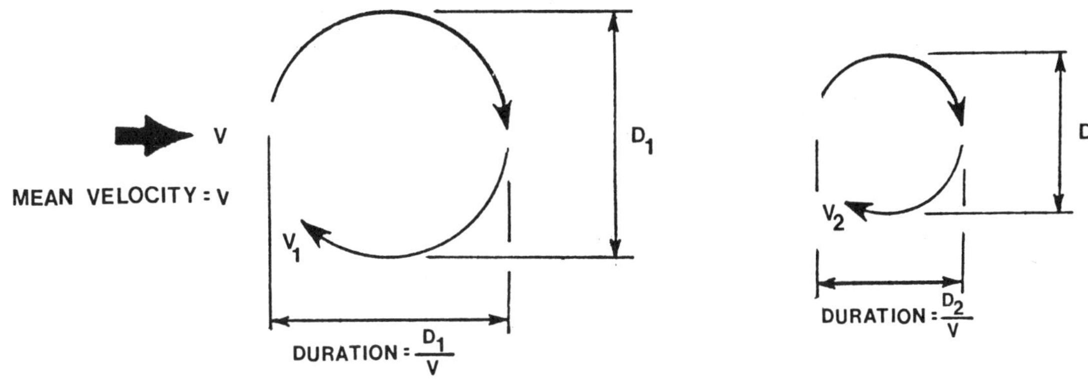

GUST DURATION	EFFECTIVE GUST DIAMETER
3 seconds	0 - 66 ft [20 m]
5 seconds	66 - 165 ft [50 m]
15 seconds	>165 ft

Figure 4-22. Diagram of relationship between gust duration and gust diameter.

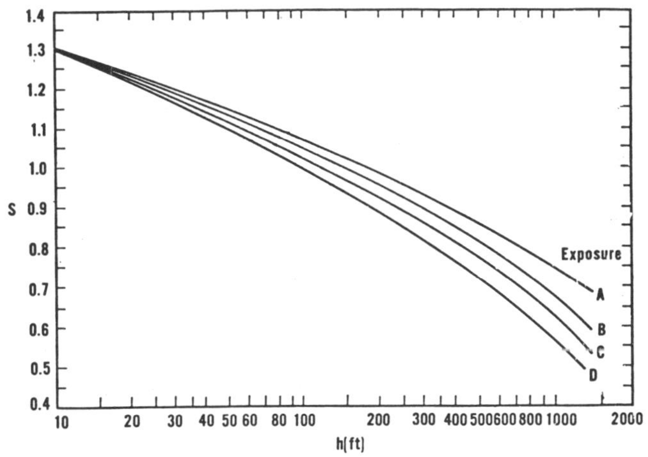

Figure 4-23. Structure size factor, S [11].

For a tower with many obstructions, such as piping, ladders, platforms, and clips that are comparable in size to the vessel, the gust response factor can be determined by:

$$\bar{G} = 0.65\left(\frac{1.25P}{\beta} + \frac{(3.32T_1)^2S}{1 + 0.001C}\right)^{1/2} \qquad (4\text{-}83)$$

The gust response factors given in Equations 4-82 and 4-83 are for flexible structures, such as towers, where the height exceeds the minimum horizontal dimension at least by five to one or the structure exhibits a natural frequency less than one. The fact that the tower may have a natural frequency less than one is significant.

Simiu and Scanlan [9] point out that for natural frequencies greater than one, the response spectra are dependent on the structure's height. However, for natural frequencies less than one, the spectra distribution has little influence on structural response, and the magnitude of turbulent fluctuation components, such as wind gusts, at or near the natural frequency of the tower could significantly affect the structural response. For this reason Equation 4-82 should be used for towers with particularly low natural frequencies.

Figure 4-24 shows a plot of wind gust velocity versus the structural response of a structure. The cyclic loading

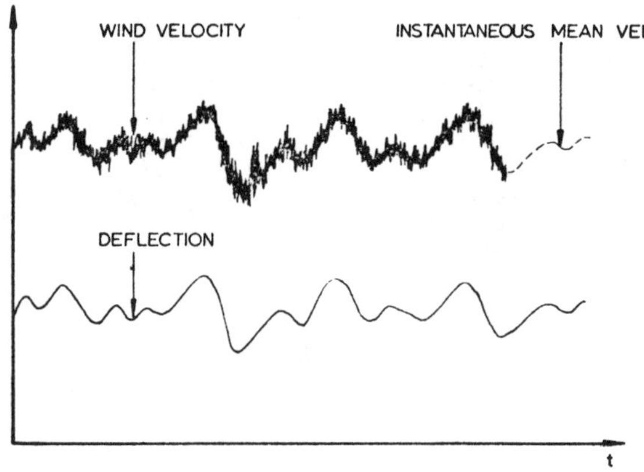

Figure 4-24. Quasi-static structural response spectra versus wind velocity [14].

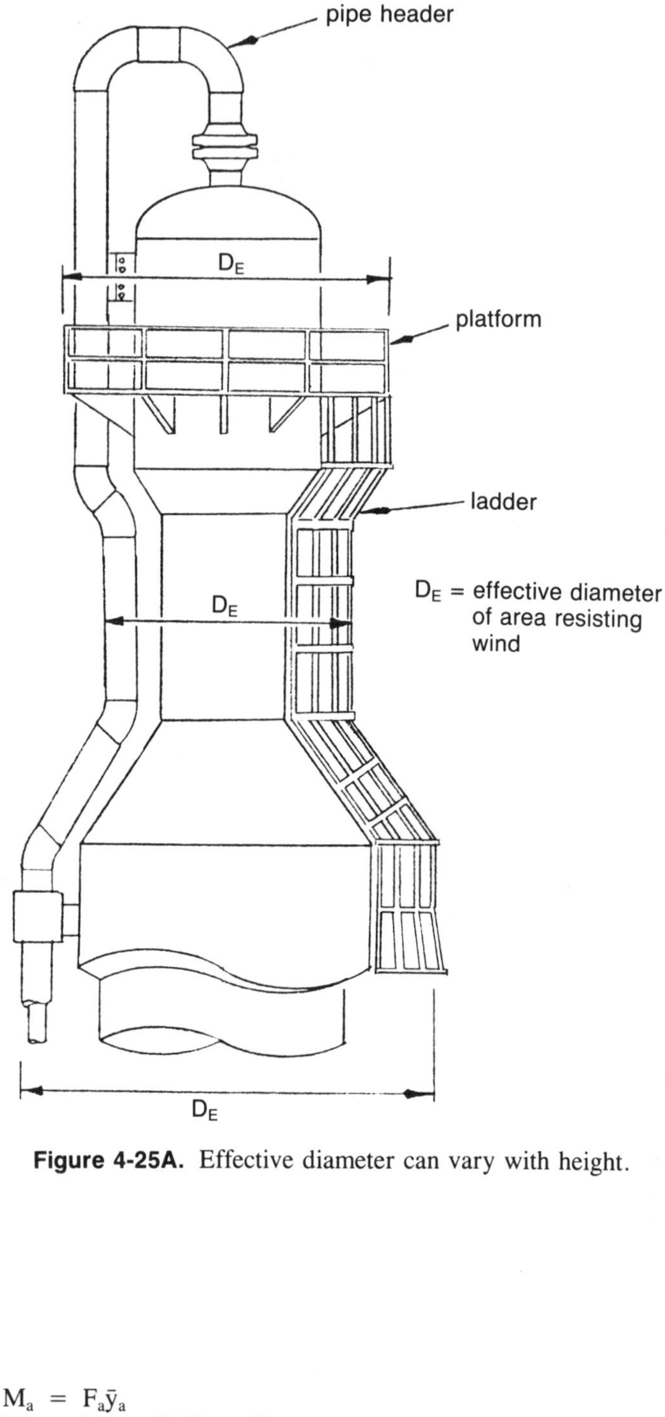

Figure 4-25A. Effective diameter can vary with height.

induced in the tower can result in fatigue failure of various vessel components.

Equation 4-81 contains the last parameter that must be defined, A_f, the total cross-sectional area of the tower and attachments that are perpendicular to the wind. This area is computed by first determining the equivalent diameter of the area facing the wind. This can be expressed as

$$\begin{aligned} D_e = {} & (\text{vessel OD}) + 2(\text{vessel insulation thickness}) \\ & + (\text{pipe OD}) + 2(\text{pipe insulation thickness}) \\ & + (\text{platform projection}) \\ & + (\text{ladder projection}) \end{aligned} \quad (4\text{-}84)$$

Equation 4-84 does not consider extraneous equipment attached to a tower, such as reboilers. The engineer must add the OD of the reboiler, plus twice the insulation thickness, plus any other equipment diameters to Equation 4-84. Doing this and multiplying by a length over which D_e is effective determines A_f. Figure 4-25 shows the effective or equivalent diameter.

WIND-INDUCED MOMENTS

After the wind pressure distribution is obtained from Equation 4-77, the distribution of section force vectors is obtained from Equation 4-81. The force vectors, shown in Figure 4-26, act through the centroids of the pressure distribution sections. Referring to Figure 4-26, we see that the wind moment distribution is obtained from the wind force vectors through the following relationships:

$$\begin{aligned} M_a &= F_a\bar{y}_a \\ M_b &= M_a + F_a(Z_a - Z_b) + F_b\bar{y}_b \\ M_c &= M_b + (F_a + F_b)(Z_b - Z_c) + F_c\bar{y}_c \\ M_d &= M_c + (F_a + F_b + F_c)(Z_c - Z_d) + F_d\bar{y}_d \\ M_e &= M_d + (F_a + F_b + F_c + F_d)(Z_d - Z_e) + F_e\bar{y}_e \end{aligned}$$

or in a general equation,

$$M_n = M_{n-1} + (Z_{n-1} - Z_{n-2}) \sum_{i=1}^{n-1} F_i + F_n\bar{y}_n \quad (4\text{-}85)$$

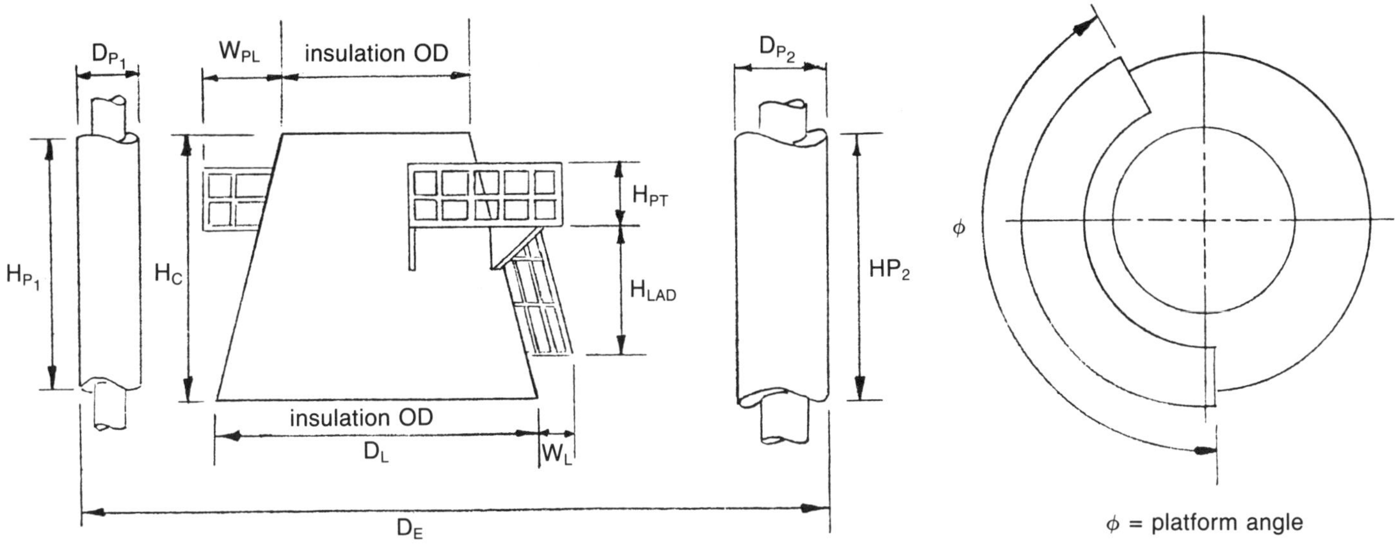

Figure 4-25B. Wind area and force calculations for conical sections.

WIND-INDUCED DEFLECTIONS OF TOWERS

Tall process towers and stacks are treated like cantilever beams in computing deflections induced by wind. Like a cantilever beam, when the tower deflects it translates and rotates at the same time. These translations and rotations are most expediently computed by the method of superposition. The three cases to consider in the superposition are a cantilever beam with a uniform load, an end load, and an end couple. These three cases and their accompanying equations are shown in Table 4-12.

The first case of the uniform load represents the wind load on the side of the tower, the second case of the edge load represents the wind shear at the various shell sections, and the third case of the end couple represents the case of couples produced at the shell section junctures by the translation and rotation of the upper sections. This combined loading is shown in Figure 4-26.

Adding the three cases we obtain the following:

$$\delta_i = \frac{\ell_i^2}{EI}\left(\frac{W_i\ell_i}{8} + \frac{Q_i\ell_i}{3} + \frac{M_i}{2}\right) \tag{4-86}$$

where δ_i = lateral translational deflection of section i, in.
ℓ_i = length of section i
W_i = concentrated wind load ($w_i\ell_i$), lb
w_i = wind profile, lb/ft
ℓ_i = section length, ft
Q_i = wind shear at each section juncture
M_i = moment induced by wind profile, in.-lb

For rotation we have

$$\Delta_{1-i} = \frac{\left(\sum_{i=1}^{n-1}\ell_i\right)\ell_i}{EI_i}\left(\frac{W_i\ell_i}{6} + \frac{Q_i\ell_i}{2} + M_{i-1}\right) \tag{4-87}$$

Total deflection = y

$$y = \sum_{i=1}^{n}\delta_i + \sum_{i=2}^{n}\Delta_{1-i} \tag{4-88}$$

WIND-INDUCED VIBRATIONS ON TALL TOWERS

Chapter 2 discussed the phenomenon of vortex shedding inducing vibrations in piping systems. This chapter focuses on the nature and techniques of analyzing vortex shedding.

Over the years many researchers have made wind tunnel tests, proposed various analytical procedures, and conducted field tests of various structures subjected to wind loads. Wind-induced vibration was first noticed on

Table 4-12
Cantilever Beam Formulas

	Beam	Formula	
Uniform Load	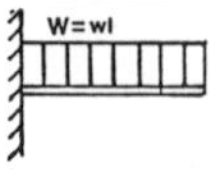	$\theta = \dfrac{W\ell^2}{6EI}$	$\delta = \dfrac{W\ell^3}{8EI}$
End Load	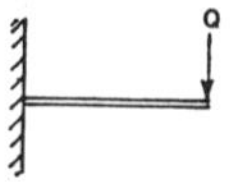	$\theta = \dfrac{Q\ell^2}{2EI}$	$\delta = \dfrac{Q\ell^3}{3EI}$
End Couple	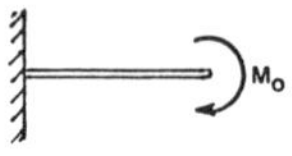	$\theta = \dfrac{M_0\ell}{EI}$	$\delta = \dfrac{M_0\ell^2}{2EI}$

$$\theta = \frac{W\ell^2}{6EI} + \frac{W\ell^2}{2EI} + \frac{M_0\ell}{EI}, \qquad \theta = \frac{\ell}{EI}\left(\frac{W\ell}{6} + \frac{W\ell}{2} + M\right)$$

$$\Delta = \theta\ell, \qquad \Delta = \frac{\left(\sum_{i=1}^{n-1} \ell_i\right)\ell_i}{EI_i}\left(\frac{W_i\ell_i}{6} + \frac{W_i\ell_i}{2} + M_i\right)$$

tall stacks by Baker at the turn of the century. Since then, many advances have been made in the field of aerodynamics allowing designers to adequately design tall structures. This chapter discusses tall process towers and Chapter 5 discusses tall stacks. The differences between the two will become more clear in the following discussion. Staley and Graven [15] summarized the state of the art of wind vibrations. Their studies indicate that even though vortex excitation of higher modes has been obtained in wind tunnel tests, existing free-standing stacks have always been observed to vibrate during vortex excitation at a frequency and with a mode shape associated with the fundamental mode. Furthermore, the shape of the dynamic force amplitude or existence of nearly constant frequency over the height of the stack (or "lock-in") implies that dynamic response will almost entirely be induced by the first mode. Staley and Graven concluded that all higher modes should be neglected in the dynamic analysis and that the frequency and associated critical wind velocity of the fundamental mode should be considered. For this reason the Rayleigh method is the industrially accepted method because it is used to determine an approximate value for the lowest natural frequency of a conservative system based on an assumed configuration of the first mode.

What is clear in wind tunnel tests and field observations is that at low Reynolds numbers the tower is dynamically stable, vulnerable only to forced vibrations and at higher Reynolds numbers a possibility of self-excited vibration will be present. From many field observations it can be concluded that the first peak vibration amplitudes occur at the critical wind velocity V_1, which corresponds to a Strouhal number of 0.2 with the forced vibration as the basic source of excitation. Thus, it is significant that the peak amplitudes of vibration determined by forced vibration theory are in very good agreement with field observations. This will be seen later in this chapter in Example 4-4.

Even though the Rayleigh method is the industrially accepted method for the present, there are other methods used to describe the vibration phenomena of tall process towers and stacks. One such method was devised by N. O. Myklestad, a great pioneer in the theory of vibrations. The Myklestad method used in cantilever beams is essentially a Holzer procedure applied to the beam problem. Its strong point is utilizing field and point transfer matrices to obtain relations that govern the flexural motion and vibrations of lumped-mass massless elastic-beam systems. This method is used in such applications as aircraft wings where the structural component is subjected to high Reynolds numbers. Since we have already delineated the difference between cylinders subjected to high and low Reynolds numbers and the fact that modes higher than the fundamental mode can be neglected, the Myklestad method has lost favor to the Rayleigh method. We are primarily interested in forced vibration peak amplitudes of relatively low natural frequencies. Although the Myklestad analysis is excellent for relatively clean aerodynamic surfaces such as wings and missiles, its practical use in process towers with attached ladders, platforms, and piping is questionable. Even for stacks. low Reynolds numbers allow for the fundamental mode to dictate.

Before the Rayleigh method is applied to our analysis, let us summarize some basic precepts. Equation 4-80 calculated the pressure force exerted on a cylinder by a static wind. When dynamic effects settle in maximum actual amplitudes, these amplitudes often exceed those under static conditions. The net result is to multiply Equation 4-80 by a magnification factor. To understand the magnification factor we must consider some basic principles.

Consider Figure 4-27 in which a system with a single degree of freedom is subjected to viscous damping and an externally imposed harmonic force. The spring is denoted by stiffness k, the friction coefficient by c, mass by m, displacement by x, impressed force as F sin ωt so, we have

$$m\dot{X} + cX + k\dot{X} = F \sin \omega t \qquad (4\text{-}89)$$

From the theory of differential equations we know that the solution of Equation 4-89 is the superposition of the general or complementary solution of the homogeneous Equation 4-89 and the particular solution of the same relation. Writing this in equation form we have

$$X = X_c + X_p$$

where X_c is the complementary function and X_p is the particular solution. This classical differential equation is

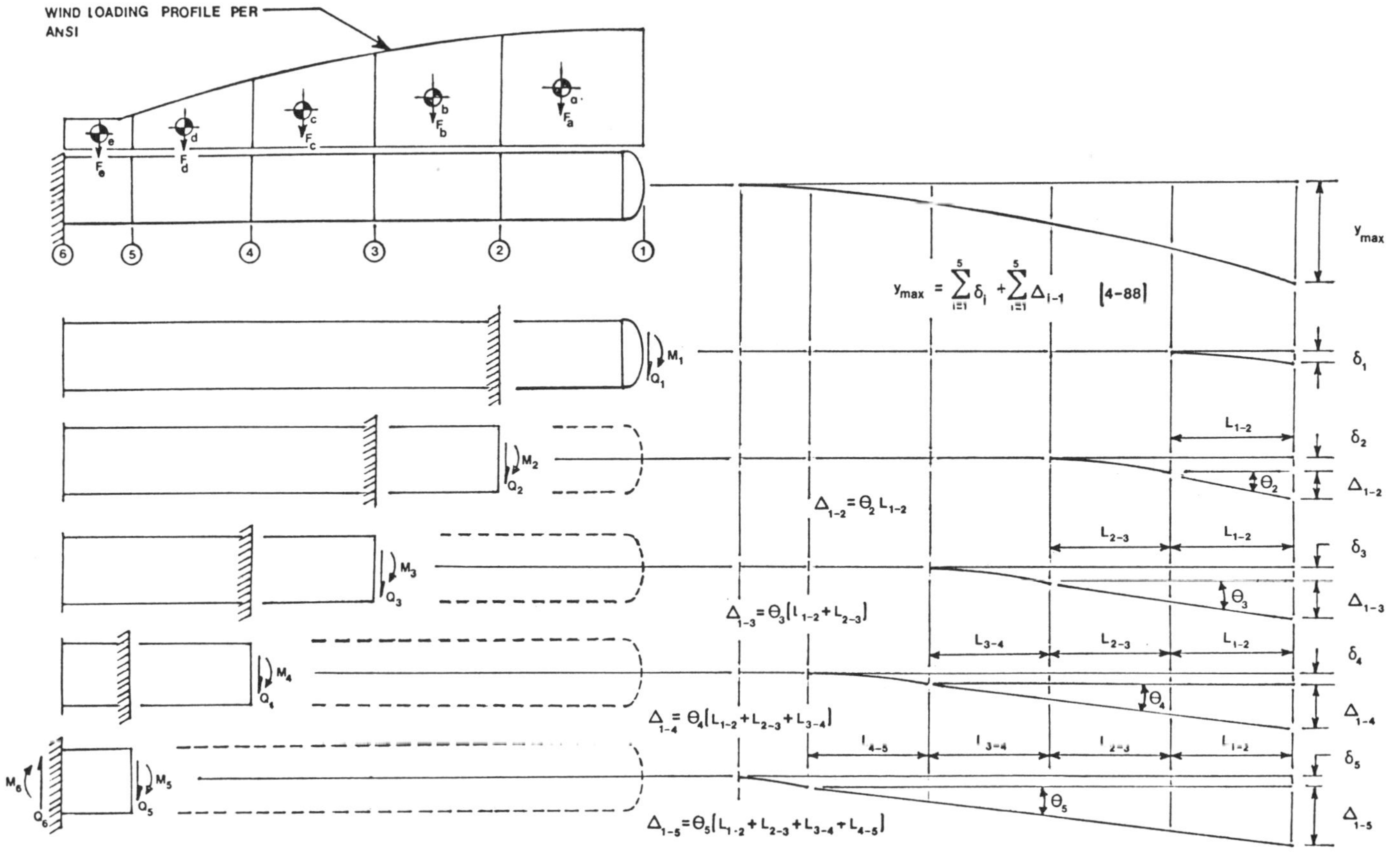

Figure 4-26. Schematic diagram of wind loadings and deflections of a tower.

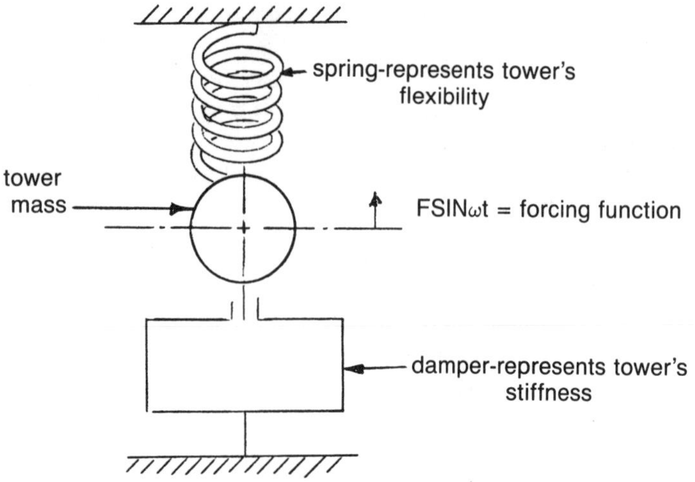

Figure 4-27. The vibration of a tower is modeled as a single degree of freedom, which is exposed to an externally imposed harmonic force and subjected to viscous damping.

solved in numerous sources and will not be delved into here. See Vierck [16] for a complete discussion of the solution. The final solution takes the form of the following:

$$X(t) = e^{-\xi\omega t}(A \cos \omega_D t + B \sin \omega_D t) + \frac{X_{st} \sin (\bar{\omega} t - \theta)}{\sqrt{(1 - r^2)^2 + (2r\xi)^2}} \quad (4\text{-}90)$$

$\xi = c/c_r$

$c_r = 2(mk)^{1/2}$ is the critical damping factor that is the criteria for critical damping such that

$c/c_r > 1$ = nonvibrating motion = overdamping

$c/c_r < 1$ = harmonic vibration = underdamping

c = a few percent of c_r for a tall, slender structure such as a tower

X_{st} = static deflection of the spring acted upon by the force F/K

$r = \bar{\omega}/\omega$ = frequency ratio of forced vibration frequency to free vibration frequency

$$\omega = \sqrt{\frac{K}{M}}$$

Letting $X = \dfrac{X_{st}}{\sqrt{(1 - r^2)^2 + (2r\xi)^2}}$

and $\tan \theta = \dfrac{2\xi r}{1 - r^2}$

we have

$$\frac{X}{X_{st}} = \frac{1}{\sqrt{(1 - r^2)^2 + (2r\xi)^2}} \quad (4\text{-}91)$$

The maximum actual amplitude X of forced vibration is obtained by multiplying the static deflection X_{st} by fraction X/X_{st}.

The fraction or ratio X/X_{st} is called the dynamic magnification factor, D. These formulations indicate that the nondimensional amplitude X/X_{st} and the phase angle, θ, are functions of the frequency ratio r and the damping factor ξ and are plotted in Figure 4-28. These curves indicate that the damping factor has a large influence on the amplitude and phase angle in the frequency region near resonance. From Equation 4-91 we see that at resonance the dynamic magnification factor, D, is inversely proportional to the damping ratio, or

$$D = \frac{1}{2\xi}$$

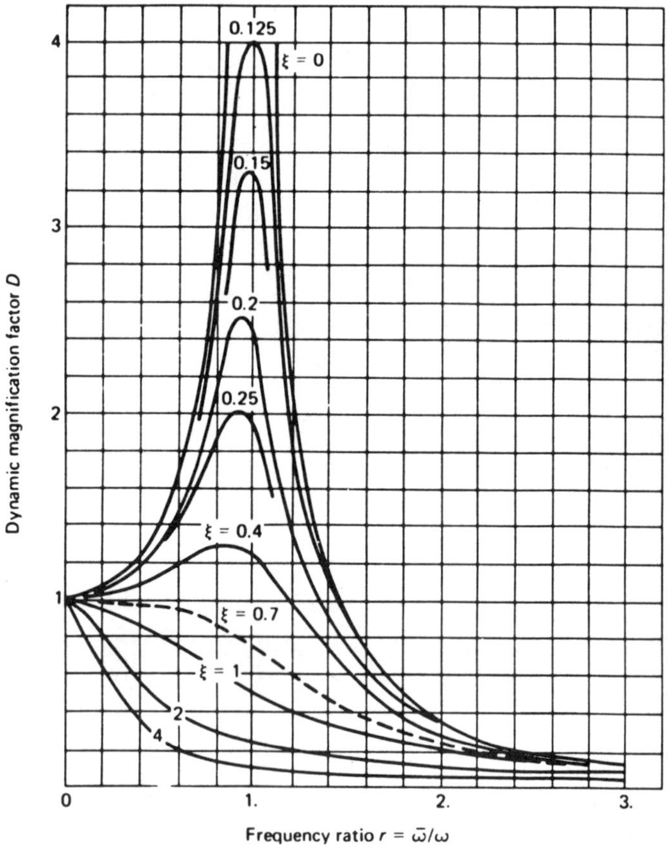

Figure 4-28. The dynamic magnification factor versus the frequency ratio for various amounts of damping. (From *Structural Dynamics* by M. Paz. ©1980 by Van Nostrand Reinhold Company, Inc. Reprinted by permission.)

The damping ratio, ξ, is not known and extremely difficult to measure at best. A practical method for experimentally determining the damping coefficient of a system is to initiate free vibration, and measure through decreasing amplitudes of oscillatory motion, as shown in Figure 4-29. This decrease or decay is termed the logarithmic decrement, δ, and is defined as the natural logarithm of the ratio of any two successive peak amplitudes, X_1 and X_2 in free vibration. Expressing this in equation form we have

$$\delta = \ln \frac{X_1}{X_2}$$

The evaluation of damping from the logarithmic decrement is given analytically by

$$X(t) = Ce^{-\xi\omega t} \cos(\omega_D t - \alpha)$$

It can be shown [17] that the dynamic magnification factor, D, and the logarithmic decrement, δ, are related using the previous expression as

$$\delta = \frac{\pi}{D} \qquad (4\text{-}92)$$

Most research data available for practical use are presented in terms of the logarithmic decrement, δ. Table 4-13 provides values of δ versus D for various structures. These values are acceptable for use in actual design of process towers and stacks.

Applying the dynamic magnification factor to Equation 4-80 we have

$$F = \frac{C_F D \rho V^2 \ell^2}{2} \qquad (4\text{-}93)$$

The force coefficient can be readily obtained from Figure 4-29. Equation 4-92 yields the maximum transverse force per unit area of the projected surface of a cylinder at resonance.

Equation 4-93 may be rewritten with the velocity in miles per hour as

$$F = 0.00086(C_F D)(H)V_1^2, \text{ for air at } 50°F \qquad (4\text{-}94a)$$

and

$$F = 0.01113\rho C_F D V_1^2(dH) \qquad (4\text{-}94b)$$

These equations apply when the top third of the tower is the controlling length. Often, the top fourth of the stack may be best to use as the controlling length. An example of this would be a section on top of the tower that is one fourth the total tower height and is significantly greater in diameter than the section below (see Example 4-4). Thus, for the top fourth of the tower Equation 4-93 becomes

$$F = 0.00065(C_F D)(d)(H)V_1^2, \text{ for air at } 50°F \qquad (4\text{-}94c)$$

and

$$F = 0.07728\rho C_F D V_1^2(dH) \qquad (4\text{-}94d)$$

where d = outside diameter of either upper 1/3 or 1/4 of tower, ft
H = total height of tower, ft
V_1 = first critical wind velocity,
V_1 = 3.40d/T, mph
T = first period of vibration, Hz
ρ = density of air at any specified temperature, lb/ft

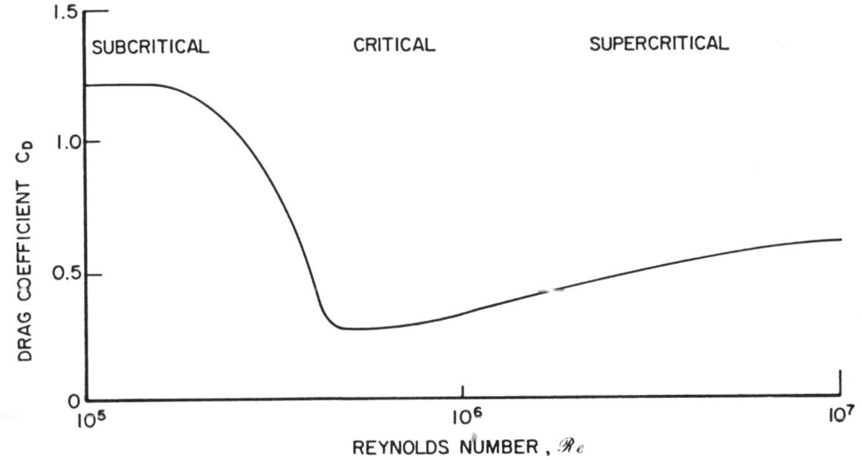

Figure 4-29. The Reynolds number versus the drag coefficient for a circular cylinder [9].

Table 4-13
Conservative Values for Logarithmic Decrement and Dynamic Magnification Factor for Tall Process Towers

	Logarithmic Decrement δ	Dynamic Magnification Factor D
Low damping: rocky-stiff soil, low-stressed pile support, or structural frame support	0.052	60
Average damping: moderately stiff soil, normal spread footing or pile support	0.080	40
High damping: soft soil, foundation on highly stressed friction piles	0.126	25

Suggested Structural Damping Coefficients [18]

A = Steel frame
B = Reinforced or prestressed concrete
Low stress levels
(a) $0.005 \leqslant c \leqslant 0.010$
(b) $0.005 \leqslant c \leqslant 0.010$
Working stress
(a) $0.01 \leqslant c \leqslant 0.02$
(b) $0.015 \leqslant c \leqslant 0.03$
Near yield
(a) $0.04 \leqslant c \leqslant 0.06$
(b) $0.05 \leqslant c \leqslant 0.10$

C_c = critical damping factor

$$C_c = 2(Mk)^{0.5} = 2\left(\frac{Wk}{386}\right)^{0.5} \frac{lb_f\text{-sec}}{in.}$$

C = damping factor = tower stiffness, lb_f/in.

$$\delta = 2\pi\left(\frac{C}{C_c}\right)$$

$$D_{max} = \pi/\delta$$

$$M = \text{tower mass} = \frac{W}{g}$$

k = tower stiffness, lb_f/in.

W = total tower weight, lb_f

For tall, slender towers of constant diameter, the first period of vibration is given by the expression

$$T = (1/0.51)(WH^4/gEI)^{0.5} \tag{4-95}$$

where g = 32.2 ft/sec
H = total height of tower, ft

The Rayleigh method applies only to undamped systems, but is found to be sufficiently accurate for computing the fundamental frequency of process towers, even though towers have varying shell thicknesses down the length that result in unevenly distributed mass and stiffness. The Rayleigh method is basically the conservation of energy, i.e., the total kinetic energy of the system is zero at the maximum displacement but is a maximum at the static equilibrium point. For the potential energy of the system, the reverse is true. Thus,

$$(K.E.)_{max} = (P.E.)_{max} = \text{total energy of the system}$$

The resulting equation will readily yield the natural frequency of the system.

To estimate the period of vibration using the Rayleigh method the tower is considered as a series of lumped masses. These lumped masses are determined by considering the weights of

1. Shell and heads
2. Trays and internals
3. Manways and nozzles
4. Insulation and fire proofing

These are summed for each section and the overall tower is considered as lumped masses at the centroid of each section along its entire length. The assumption is made that the stiffness is constant along the entire length of the tower; an assumption that greatly simplifies the computations for the various deflections of the section centroids. The more sections, the greater the overall accuracy achieved. Such a beam with lumped masses is shown in Figure 4-30. For such a simple, one-mass, vibrating system, Timoshenko et al. [19] have shown that the angular natural undamped frequency (rad/sec) for such a system is

$$\omega = \left(\frac{gWy}{Wy^2}\right)^{0.5} \tag{4-96}$$

Integrating Equation 4-96 numerically across the section centroids of the tower results in

$$\omega = [g(W_1y_a + W_2y_b + \ldots + W_ny_n)/(W_1y_a^2 + W_2y_b^2 + \ldots + W_ny_n^2)]^{0.5} \tag{4-97}$$

or

$$T = 2\pi\left[\left(\sum_{i=1}^{n} W_iy^2\right)\Big/\left(g\sum_{i=1}^{n} W_iy\right)\right]^{0.5} \tag{4-98}$$

The section weights, W_i, are computed by using cumulative weights down the tower. Summing moments about the base in Figure 4-30 we obtain the moment distribution in the tower as follows:

$$M_1 = 0$$

$$M_2 = W_1\left(\frac{R_1 + R_2}{2}\right)$$

$$M_3 = W_1\left(\frac{R_1}{2} + R_2 + \frac{R_3}{2}\right) + W_2\left(\frac{R_2 + R_3}{2}\right)$$

$$M_4 = W_1\left(\frac{R_1}{2} + R_2 + R_3\right) + W_2\left(\frac{R_2}{2} + R_3\right) + W_3\left(\frac{R_3}{2}\right) \tag{4-99}$$

The moments obtained are used to determine the deflections induced by vortex shedding. The method of deflection computation is based on the area-moment (conjugate beam) method applied to a cantilever beam. In this method the slope of the elastic bending curve of the actual beam is equal to the shear at the same point on the conjugate beam, which is an idealized beam corresponding to the actual member. The deflection y of the actual beam (or tower) at any point relative to its original position is equal to the bending moment at the corresponding point on the conjugate beam that has the same M/EI area of the actual beam. Figure 4-31 shows weights of the vessel sections distributed about the section centroids along with beam lengths used in the analysis. The conjugate beam method of computing deflection is demonstrated in Table 4-14. For an indepth analysis of the method the reader is referred to Higdon et al. [20]. The examples presented at the end of this chapter will clarify this approach.

OVALING

Ovaling is a resonance phenomenon more common in stacks rather than process towers. However, towers exhibit this phenomenon mostly during construction, before insulation and appurtenances are added to the vessel. To avoid ovaling, the designer should consider the following guidelines.

The cylinder is considered as a ring that has a natural frequency of

$$f_r = \frac{7.58t_s\sqrt{E}}{60D^2} \tag{4-100}$$

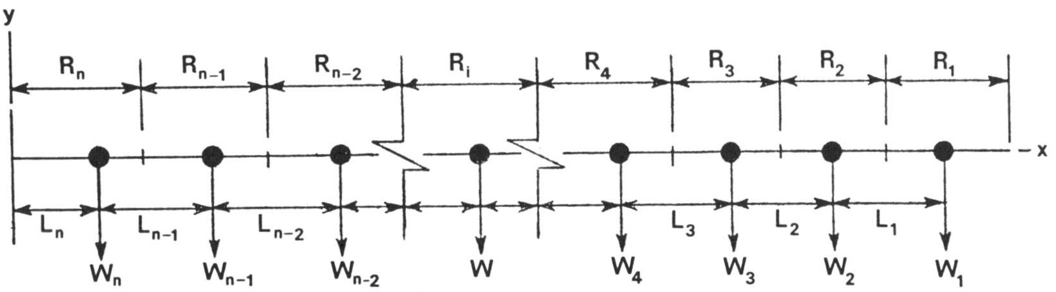

Figure 4-30. A tower modeled as a sectionless beam with distributed lumped masses.

Table 4-14A
Vibration Deflections Based on the Conjugate Beam Method

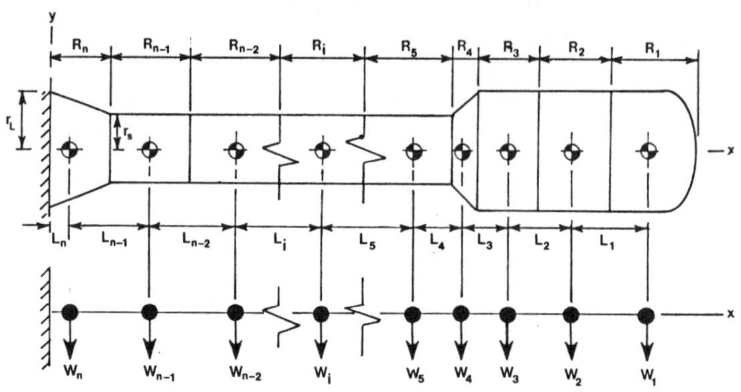

M_i (ft-lb$_f$)		I_i	M_i/E_iI_i	S_i	$\sum S_i = A_i = M_i dx/E_iI_i$	P_i	$\mu_i = M_i d^2x/E_iI_i$
$M_1 = 0$		I_1	$\frac{M_1}{E_1I_1}$	$\frac{M_2}{2E_2I_2} L_1 = S_1$	$S_1 + A_2 = A_1$	$\left(\frac{A_1 + A_2}{2}\right) L_1 = P_1$	$P_1 + \mu_2 = \mu_1$
$M_2 = W_1L_1$	$L_1 = \frac{R_1 + R_2}{2}$	I_2	$\frac{M_2}{E_2I_2}$	$\frac{\left(\frac{M_3}{E_3I_3} + \frac{M_2}{E_2I_2}\right)}{2} \times L_2 = S_2$	$S_2 + A_3 = A_2$	$\left(\frac{A_2 + A_3}{2}\right) \times L_2 = P_2$	$P_2 + \mu_3 = \mu_2$
$M_3 = W_1(L_1 + L_2) + W_2L_2$	$L_2 = \frac{R_2 + R_3}{2}$	I_3	$\frac{M_3}{E_3I_3}$	$\frac{\left(\frac{M_1}{E_4I_4} + \frac{M_3}{E_3I_3}\right)}{2} \times L_3 = S_3$	$S_3 + A_4 = A_3$	$\left(\frac{A_3 + A_4}{2}\right) \times L_3 = P_3$	$P_3 + \mu_4 = \mu_3$
$M_4 = W_1(L_1 + L_2 + L_3) + W_2 (L_2 + L_3) + W_3L_3$	$L_4 = \frac{R_5}{2} + R_4 - \bar{y}$	I_4	$\frac{M_4}{E_4I_4}$	$\frac{\left(\frac{M_5}{E_5I_5} + \frac{M_4}{E_4I_4}\right)}{2} \times L_4 = S_4$	$S_4 + A_5 = A_4$	$\left(\frac{A_4 + A_5}{2}\right) \times L_4 = P_4$	$P_4 + \mu_5 = \mu_4$
$M_i = M_{i-1} + L_{i-1} \sum_{i=1}^{i} W_{i-1}$	For cylindrical shells, $L_i = \frac{R_i + R_{i+1}}{2}$	I_i	$\frac{M_i}{E_iI_i}$	$\frac{\left(\frac{M_{i+1}}{E_{i+1}I_{i+1}} + \frac{M_i}{E_iI_i}\right)}{2} \times L_i = S_i$	$S_i + \sum_{j=n}^{i+1} A_{i+1} S_j = A_i$	$\left(\frac{A_i + A_{i+1}}{2}\right) \times L_i = P_i$	$P_i + \mu_{i+1} = \mu_i$
$M_n = M_{n-1} + L_{n-1} \times \sum_{i=1}^{n} W_{i-1}$	$L_n = R_n - \bar{y}$	I_n	$\frac{M_n}{E_nI_n}$	$\frac{\left(\frac{M_{n+1}}{E_nI_n} + \frac{M_n}{E_nI_n}\right)}{2} \times L_n = S_n$	$S_n = A_n$	$\frac{A_n}{2} L_n = P_n$	$P_n = \mu_n$
$M_{n+1} = M_n + L_n \times \sum_{i=1}^{n} W_i$							

Computation of lateral y deflections. For formulas of $\bar{y}$ see Table 4-16C.
deflection of real beam = bending moment diagram of conjugate beam
slope of real beam elastic curve = shear of conjugate beam
moment diagram of real beam = load diagram of conjugate beam
$y_i = (\mu_i)(12)$, in.

Table 4-14B
Vibration Deflections Based on the Conjugate Beam Method—Section Break Method

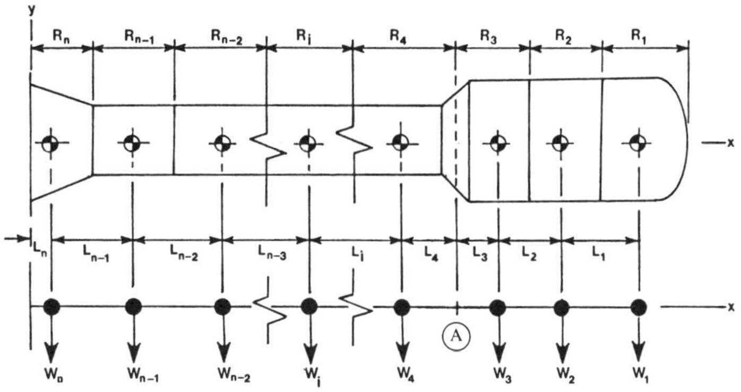

M_i	I_i	M_i/E_iI_i	S_i	$\sum S_i = A_i = M_i dx/E_iI_i$	P_i	$\mu_i = M_i d^2x/E_iI_i$
$M_1 = 0$	I_1	$\frac{M_1}{E_1I_1}$	$\frac{M_2}{2E_2I_2} L_1 = S_1$	$S_1 + A_2 = A_1$	$\left(\frac{A_1 + A_2}{2}\right) \times L_1 = P_1$	$P_1 + \mu_2 = \mu_1$
$M_2 = W_1\left(\frac{R_1 + R_2}{2}\right) = W_1L_1$	I_2	$\frac{M_2}{E_2I_2}$	$\frac{\left(\frac{M_3}{E_3I_3} + \frac{M_2}{E_2I_2}\right)}{2} \times L_2 = \frac{M_2}{2I_2} \times L_2 = S_2$	$S_2 + A_3 = A_2$	$\left(\frac{A_2 + A_3}{2}\right) \times L_2 = P_2$	$P_2 + \mu_3 = \mu_2$
$M_3 = W_1\left(\frac{R_1}{2} + R_2 + \frac{R_3}{2}\right) + W_2\left(\frac{R_2 + R_3}{2}\right) = W_1(L_1 + L_2) + W_2L_2$	I_3	$\frac{M_3}{E_3I_3}$	$\frac{\left(\frac{M_3}{E_4I_4} + \frac{M_3}{E_3I_3}\right)}{2} \times L_3 = S_3$	$S_3 + A_4 = A_3$	$\left(\frac{A_3 + A_4}{2}\right) \times L_3 = P_3$	$P_3 + \mu_4 = \mu_3$
$M_3 = W_1\left(\frac{R_1}{2} + R_2 + \frac{R_3}{2}\right) + W_2\left(\frac{R_2 + R_3}{2}\right) = W_1(L_1 + L_2 + L_3) + W_2L_2$	I_4	$\frac{M_3}{E_4I_4}$	$\frac{\left(\frac{M_4}{E_4I_4} + \frac{M_3}{E_4I_4}\right)}{2} \times L_4 = S_4$	$S_4 + A_5 = A_4$	$\left(\frac{A_4 + A_5}{2}\right) \times L_4 = P_4$	$P_4 + \mu_5 = \mu_4$
$M_4 = W_1\left(\frac{R_1}{2} + R_2 + R_3 + \frac{R_4}{2}\right) + W_2\left(\frac{R_2}{2} + R_3 + \frac{R_4}{2}\right) + W_3\left(\frac{R_3 + R_4}{2}\right)$	I_5	$\frac{M_4}{E_5I_5}$	$\frac{\left(\frac{M_5}{E_5I_5} + \frac{M_4}{E_4I_4}\right)}{2} \times L_5 = S_5$	A_6 $S_5 + \sum_{i=k}^{6} S_i = A_5$	$\left(\frac{A_5 + A_6}{2}\right) \times L_5 = P_5$	$P_5 + \mu_6 = \mu_5$
$M_n = M_{n-1} + L_{n-1}\sum_{i=1}^{n} W_{i-1}$	I_n	$\frac{M_n}{I_k}$	$\frac{\left(\frac{M_{n+1}}{E_nI_n} + \frac{M_n}{E_nI_n}\right)}{2} \times L_n = S_n$	$S_n = A_n$	$\frac{A_k}{2} L_n = P_n$	$P_n = \mu_n$
$M_{n+1} = M_n + L_n \times \sum_{i=1}^{n} W_i$		$y_i = (\mu_i)(12)$ ft				

(A) = Abrupt section break
$k = n +$ (number of abrupt section breaks)
In figure above, $k = n + 1$

Table 4-14C
Centroids of Shell Volumes

Ellipsoidal Head	Centroid of Shell Volume

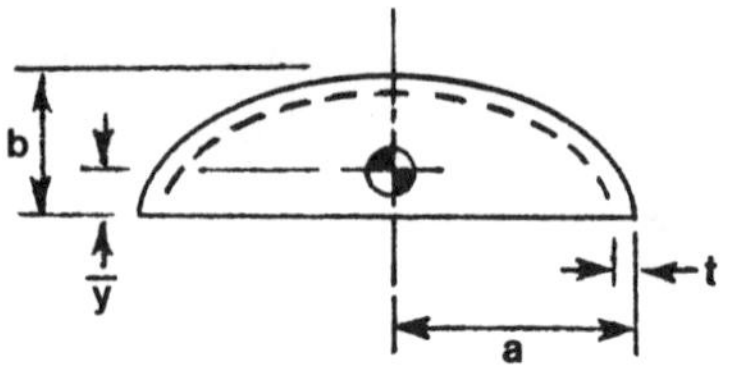

$$\bar{y} = \frac{b(4a - 3t)}{6\left(\frac{\pi a}{2} - t\right)}$$

Conical Section

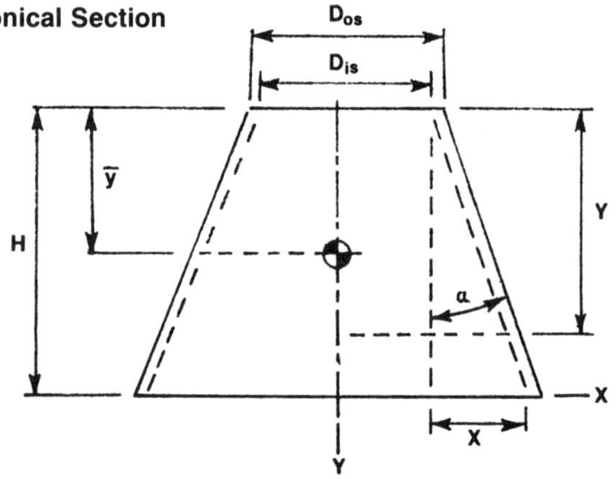

$$\bar{y} = \frac{3H(D_{o_s}^2 - D_{i_s}^2) + 8H^2(D_{o_s} - D_{i_s})\tan\alpha}{6[(D_{o_s}^2 - D_{i_s}^2) + 2H(D_{o_s} - D_{i_s})\tan\alpha]}$$

The vortex shedding frequency is given by

$$f_v = \frac{0.2V}{D} \tag{4-101}$$

where v = 45 mph or 66 fps

If for any section of the tower $f_r < 2f_v$, ovaling vibration is imminent. The resonance wind velocity that would theoretically induce ovaling is

$$V_o = \frac{60\, f_r D}{2s} \tag{4-102}$$

where s = Strouhal number = 0.2 for this application

To counter ovaling vibration, ovaling rings or helical strakes are added. These normally are not practical for process columns, because these vessels usually have many external, attached appurtenances. What is more commonly done with towers is to stiffen up shell sections to offset ovaling resonance. See Chapter 5 for more information on ovaling.

CRITERIA FOR VIBRATION ANALYSIS

While there is no absolute parameter available for determining whether a vibration analysis is required, there are certain guidelines for designing towers.

1. If the critical wind velocity, V_1, exceeds 60 mph then a vibration analysis is not required. Very few cases of severe vortex excitation have been observed for wind velocities in this range.

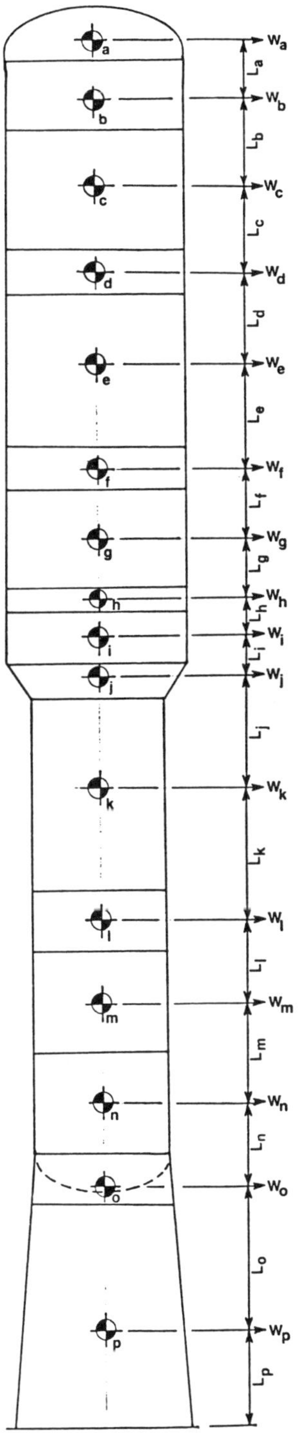

Figure 4-31. The vibration ensemble in which each section weight is located at the section centroid.

2. If the first critical wind velocity, V_1, is greater than the wind design speed, a vibration analysis is not required.
3. The limiting minimum height-to-diameter ratios H/d are as follows:

$$\left.\begin{array}{l} H/d \geqslant 13 \text{ unlined stacks} \\ H/d \geqslant 15 \text{ lined stacks} \\ H/d \geqslant 15 \text{ process columns} \end{array}\right\} \quad (4\text{-}103)$$

4. The Zorilla criterion for vibration analysis is as follows:

$$\begin{array}{ll} \dfrac{W}{LD^2} \leqslant 20 & \text{vibration analysis must be performed} \\ 20 \leqslant \dfrac{W}{LD^2} \leqslant 25 & \text{vibration analysis should be performed} \\ 25 < \dfrac{W}{LD^2} & \text{vibration analysis need not be performed} \end{array} \quad (4\text{-}104)$$

5. If the total force on the tower induced by the first critical wind velocity V_1 does not exceed $^1/_{15}$ of the operating (corroded) weight W or

$$\frac{\frac{1}{2}\rho V_1^2 Hd}{W} \leqslant \frac{1}{15} \quad (4\text{-}105)$$

Further guidelines and procedures for stacks are discussed in Chapter 5.

SEISMIC DESIGN OF TALL TOWERS

There are several ways to analyze earthquake forces imposed on a structure. The procedures outlined in the Uniform Building Code [10] are the simplest and most straightforward, but do not account for all of the significant dynamic properties of structures. Large, complex structures, such as 50-story buildings, nuclear power plants, large dams, and long suspension bridges, require a more thorough dynamic analysis. Fortunately, the UBC method is accurate enough for most tall, process tower/stack design problems and is presented here.

In seismic analysis the design spectrum is not a specification of a particular earthquake ground motion; it is a specification of the strengths of structures. For this reason the tower must be ductile enough to absorb energy without ultimate yield. This implies that for the structure to absorb energy that exceeds maximum design conditions the overall structure deformation will be ductile rather than brittle. The result is that while more rigorous analyses are very helpful in determining design criteria, practical design procedures are simplifications of the complex dynamic phenomenon used as "quasi" static criteria applied with elastic stress limits.

The Uniform Building Code 1982 [10] requires that all freestanding structures in seismic zones to be designed and constructed to withstand a total lateral force (base shear) given by

$$V = ZIKCSW \tag{4-106}$$

where Z = seismic zone factor (see Figure 4-32)
I = occupancy importance factor = 1 for all process towers and stacks
K = structure type coefficient
C = structure period response factor
S = site structure interaction factor
W = total operating weight of tower above ground

The structure type coefficient, K, is as follows:

K = 2.0 for vertical vessels on skirt supports
K = 2.5 for vertical vessels on skirts when $t_{shell} > 1.5\, t_{skirt}$

The structure period response factor, C, is determined by

$$C = \frac{1}{15\sqrt{T}} \text{ sec} \tag{4-107}$$

where T = structure period of vibration, sec, with $C_{max} = 0.12$

For short, stiff structures, such as horizontal vessel supports, in lieu of making a period calculation, the response factor C may be taken as equal to C_{max}.

For most industrially accepted design methods, the effects of the soil-structure interaction are considered. This is done in the Uniform Building Code by using the ratio of the fundamental elastic period of vibration of the tower, T, to the characteristic site period, T_s.

Formulations used to determine the fundamental natural period of vibration for seismic response vary as to the type of structural cross-section considered. The generally accepted equation for towers of uniform cross-section is

$$T = 1.79\left(\frac{wH^4}{EIg}\right)^{0.5} \tag{4-108}$$

where T = fundamental period, sec
H = total height, ft
w = weight per unit of height, lb/ft
t = shell thickness, in.
E = modulus of elasticity, psi
I = moment of inertia, ft^4
g = 32.2 ft/sec^2

When Equation 4-108 is applied to steel with a value of $E = 30 \times 10^6$ psi we have

$$T = (7.659 \times 10^{-6})\left(\frac{H}{D_m}\right)^2\sqrt{\frac{12wD_m}{t}} \tag{4-109}$$

and for $E = 29 \times 10^6$ psi,

$$T = (7.7896 \times 10^{-6})\left(\frac{H}{D_m}\right)^2\sqrt{\frac{12wD_m}{t}} \tag{4-110}$$

where D_m = mean diameter of tower, ft

For a tower with uniform cross section and tapered (conical) skirt the following relationship can be used in computing the fundamental period:

$$T = 2\pi\left(\frac{0.80\delta}{g}\right)^{0.5} \tag{4-111}$$

where δ = the calculated deflection at top of tower induced by 100% of its weight applied as a lateral load

With towers of varying cross sections and attaching equipment, a method used to determine the fundamental frequency was developed by Warren W. Mitchell in an unpublished work [21]. The solution is based on the Rayleigh method of equating potential and kinetic energies in a vibrating system. The resulting formulation is readily useful in computing fundamental periods of cylindrical, tapered-cylindrical, and step-tapered-cylindrical structures common to the petrochemical industry (CPI). The formulation is as follows:

$$T = \left(\frac{H}{100}\right)^2 \frac{\sqrt{\sum w\Delta\alpha = \left(\frac{1}{H}\right)\sum W\beta}}{\sum \bar{E}D^3 + \Delta\gamma} \tag{4-112}$$

where
T = period, sec
H = overall height of tower, ft
w = distributed weight (lb/ft) of each section
W = concentrated loads attached to the tower at any level, that add mass but do not contribute to the stiffness of the tower
$\bar{E}$ = modulus of elasticity (10^6 psi) for each section
α, β, γ = coefficients for a given elevation depending on the ratio of the height of the elevation above grade to the overall height of the tower (h_x/H)

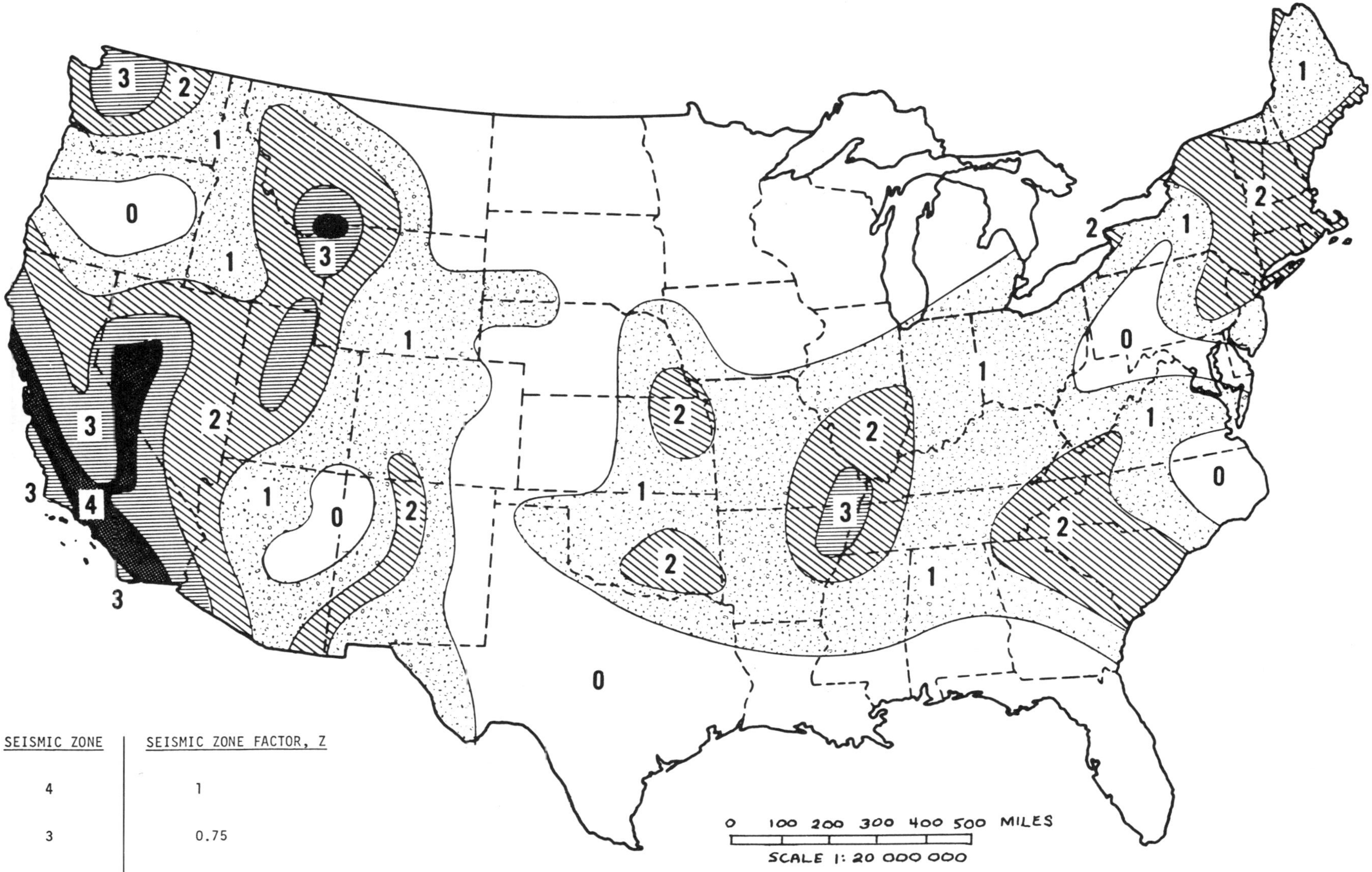

SEISMIC ZONE	SEISMIC ZONE FACTOR, Z
4	1
3	0.75
2	0.375
1	0.1875
0	0 (no earthquake design required)

Figure 4-32. Uniform Building Code map for seismic zones [10].

$\Delta\alpha, \Delta\gamma$ = differentials in the values of α and γ, from the top to the bottom of each section of uniform weight, diameter, and thickness. β is determined from each concentrated mass. Values of α, β, and γ are shown in Table 4-15.

In applying Equation 4-112 the following factors should be considered:

- If a tower's lower section is several times wider in diameter and shorter than the upper sections, then the tower's period can be more accurately determined by computing the upper section's period, assuming that the tower is fixed as to translational and rotational displacement. If a tower's shell diameter or thickness is significantly larger than that of the supporting skirt, the period calculated by Equation 4-112 may be overly conservative for earthquake design and a more accurate method may be desirable.
- For conical tower sections the Mitchell equation cannot be used because of lack of data for the coefficients α, β, and γ. The Rayleigh equation (Equation 4-97) is more comprehensive and ubiquitous in application.

Once the fundamental period of vibration is determined, the numerical coefficient for the site structure interaction (seismic site-structure resonance coefficient), S, can be determined. As previously stated, the soil-structure interaction is considered in most industrially accepted methods. The value of S is determined by the following formulas:

For $T/T_s \leqslant 1.0$,

$$S = 1.0 + \frac{T}{T_s} - 0.5\left(\frac{T}{T_s}\right)^2 \quad (4\text{-}113a)$$

For $T/T_s > 1.0$,

$$S = 1.2 + 0.6\frac{T}{T_s} - 0.3\left(\frac{T}{T_s}\right)^2 \quad (4\text{-}113b)$$

$$S \geqslant 1.0$$
$$(C)(S) \leqslant 0.14$$

The characteristic site period, T_s, falls into the following time:

$$0.5 \leqslant T_s \leqslant 2.5 \text{ sec}$$

When T_s is not properly established, S is taken as 1.5, except when T exceeds 2.5 seconds, S can be determined by assuming a value of 2.5 seconds for T_s.

VERTICAL DISTRIBUTION OF SHEAR FORCES

For towers having an overall height-to-base-width ratio greater than 3.0, a portion of the total earthquake force, V, shall be applied to the top of the tower according to the following relationships:

For,

$$\frac{h}{D} \leqslant 3.0, \; F_t = 0$$

$$3.0 \leqslant \frac{h}{D} < 6.12, \; F_t = 0.07\,TV \quad (4\text{-}114)$$

$$\frac{h}{D} \geqslant 6.12, \; F_t = 0.15V$$

where F_t = total force applied at top of structure
h = overall height of tower, ft
D = diameter of tower, ft
V = total base shear from Equation 4-106

The remainder of the total seismic force is distributed and applied to the mass distribution in the structure according to the following equation:

$$F_x = (V - F_t)\frac{W_x h_x}{\sum_{i=1}^{n} W_i h_i} \quad (4\text{-}115)$$

where F_x = lateral force applied to a mass at level x, lb
W_x = weight of mass at level x, lb
h_x = height of level x above the base (normally measured from bottom of the base plate of the tower), ft
Σwh = the sum of the products of w_x and h_x for all the masses within the structure, ft-lb

The seismic moments are computed from the following expression:

$$M = V_{x_i} L_{y_{i-1}} + F_{x_i} C_i \quad (4\text{-}116)$$

where $L_{y_{i-1}}$ = length of section below shear force, ft
C_i = $L_{y_i}/2$ for a cylinder

$$C_i = \left(\frac{L}{4}\right)\left[\frac{r_L^2 + 2r_L r_s + 3r_s^2}{r_L^2 + r_L r_s + r_s^2}\right] \text{ for a cone (see Figure 4-33)}$$

For an illustration of seismic analysis, see Example 4-3.

Table 4-15
Coefficients for Determining Period of Vibration of Free-Standing Cylindrical Shells Having Varying Cross Sections and Mass Distribution*

$\frac{h_x}{H}$	α	β	γ	$\frac{h_x}{H}$	α	β	γ
1.00	2.103	8.347	1.000000	0.50	0.1094	0.9863	0.95573
0.99	2.021	8.121	1.000000	0.49	0.0998	0.9210	0.95143
0.98	1.941	7.898	1.000000	0.48	0.0909	0.8584	0.94683
0.97	1.863	7.678	1.000000	0.47	0.0826	0.7987	0.94189
0.96	1.787	7.461	1.000000	0.46	0.0749	0.7418	0.93661
0.95	1.714	7.248	0.999999	0.45	0.0678	0.6876	0.93097
0.94	1.642	7.037	0.999998	0.44	0.0612	0.6361	0.92495
0.93	1.573	6.830	0.999997	0.43	0.0551	0.5872	0.91854
0.92	1.506	6.626	0.999994	0.42	0.0494	0.5409	0.91173
0.91	1.440	6.425	0.999989	0.41	0.0442	0.4971	0.90448
0.90	1.377	6.227	0.999982	0.40	0.0395	0.4557	0.89679
0.89	1.316	6.032	0.999971	0.39	0.0351	0.4167	0.88864
0.88	1.256	5.840	0.999956	0.38	0.0311	0.3801	0.88001
0.87	1.199	5.652	0.999934	0.37	0.0275	0.3456	0.87088
0.86	1.143	5.467	0.999905	0.36	0.0242	0.3134	0.86123
0.85	1.090	5.285	0.999867	0.35	0.0212	0.2833	0.85105
0.84	1.038	5.106	0.999817	0.34	0.0185	0.2552	0.84032
0.83	0.988	4.930	0.999754	0.33	0.0161	0.2291	0.82901
0.82	0.939	4.758	0.999674	0.32	0.0140	0.2050	0.81710
0.81	0.892	4.589	0.999576	0.31	0.0120	0.1826	0.80459
0.80	0.847	4.424	0.999455	0.30	0.010293	0.16200	0.7914
0.79	0.804	4.261	0.999309	0.29	0.008769	0.14308	0.7776
0.78	0.762	4.102	0.999133	0.28	0.007426	0.12576	0.7632
0.77	0.722	3.946	0.998923	0.27	0.006249	0.10997	0.7480
0.76	0.683	3.794	0.998676	0.26	0.005222	0.09564	0.7321
0.75	0.646	3.645	0.998385	0.25	0.004332	0.08267	0.7155
0.74	0.610	3.499	0.998047	0.24	0.003564	0.07101	0.6981
0.73	0.576	3.356	0.997656	0.23	0.002907	0.06056	0.6800
0.72	0.543	3.217	0.997205	0.22	0.002349	0.05126	0.6610
0.71	0.512	3.081	0.996689	0.21	0.001878	0.04303	0.6413
0.70	0.481	2.949	0.996101	0.20	0.001485	0.03579	0.6207
0.69	0.453	2.820	0.995434	0.19	0.001159	0.02948	0.5992
0.68	0.425	2.694	0.994681	0.18	0.000893	0.02400	0.5769
0.67	0.399	2.571	0.993834	0.17	0.000677	0.01931	0.5536
0.66	0.374	2.452	0.992885	0.16	0.000504	0.01531	0.5295
0.65	0.3497	2.3365	0.99183	0.15	0.000368	0.01196	0.5044
0.64	0.3269	2.2240	0.99065	0.14	0.000263	0.00917	0.4783
0.63	0.3052	2.1148	0.98934	0.13	0.000183	0.00689	0.4512
0.62	0.2846	2.0089	0.98789	0.12	0.000124	0.00506	0.4231
0.61	0.2650	1.9062	0.98630	0.11	0.000081	0.00361	0.3940
0.60	0.2464	1.8068	0.98455	0.10	0.000051	0.00249	0.3639
0.59	0.2288	1.7107	0.98262	0.09	0.000030	0.00165	0.3327
0.58	0.2122	1.6177	0.98052	0.08	0.000017	0.00104	0.3003
0.57	0.1965	1.5279	0.97823	0.07	0.000009	0.00062	0.2669
0.56	0.1816	1.4413	0.97573	0.06	0.000004	0.00034	0.2323
0.55	0.1676	1.3579	0.97301	0.05	0.000002	0.00016	0.1966
0.54	1.1545	1.2775	0.97007	0.04	0.000001	0.00007	0.1597
0.53	0.1421	1.2002	0.96688	0.03	0.000000	0.00002	0.1216
0.52	0.1305	1.1259	0.96344	0.02	0.000000	0.00000	0.0823
0.51	0.1196	1.0547	0.95973	0.01	0.000000	0.00000	0.0418
				0.	0.	0.	0.

* Mitchell formula: $T = \left(\frac{H}{100}\right)^2 \sqrt{\frac{\sum W\Delta\alpha + (1/H)\sum P\beta}{\sum \bar{E}D^3 f\Delta\gamma}}$

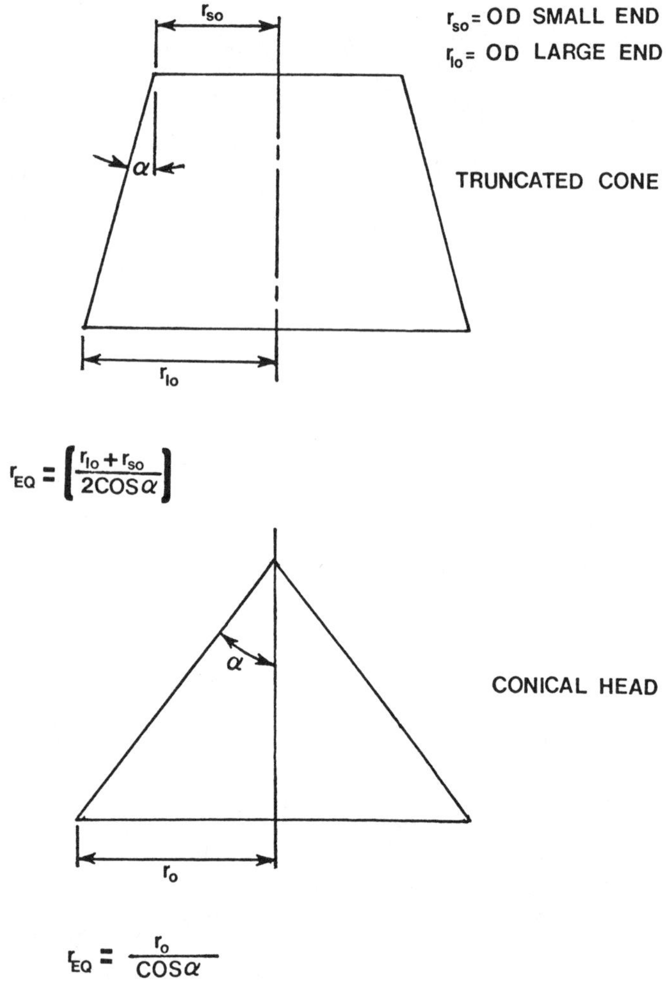

Figure 4-33. The equivalent radius for cones.

TOWER SHELL DISCONTINUITIES AND CONICAL SECTIONS

Most vessel codes do not discuss the analytical computation of tower shell discontinuity stresses, which are prevented by welding stiffening rings to the outside shell of conical sections. In addition, most codes do not consider discontinuity stresses on cylindrical shell sections. The ASME Section VIII Division I uses a safety factor of four to one to compensate for not computing these stresses.

Conical sections can be treated quite simply by utilizing the equivalent circle technique. Bednar [22] shows that the cone-to-cylinder stresses computed by the equivalent circle method are very close in magnitude to those computed by more exact methods. Because of its close approximate answers and simplicity, the equivalent circle method is normally the method used for treating conical sections in towers. The method will only be outlined here, as others [22] have already derived it.

Figure 4-33 shows how the sections of a truncated cone and a conical head are approximated by an equivalent circle, which is used to compute the section modulus and moment of inertia. These formulations are used in tower design and are demonstrated in the examples that follow.

Conical shells used in tower sections have a half apex angle $\alpha \leqslant 30$ degrees. When $h_c \leqslant 0.10H$, the cone can be approximated by considering two cylinders shown with dotted lines (Figure 4-34). In practice, stiffening rings must be used when required by the vessel code.

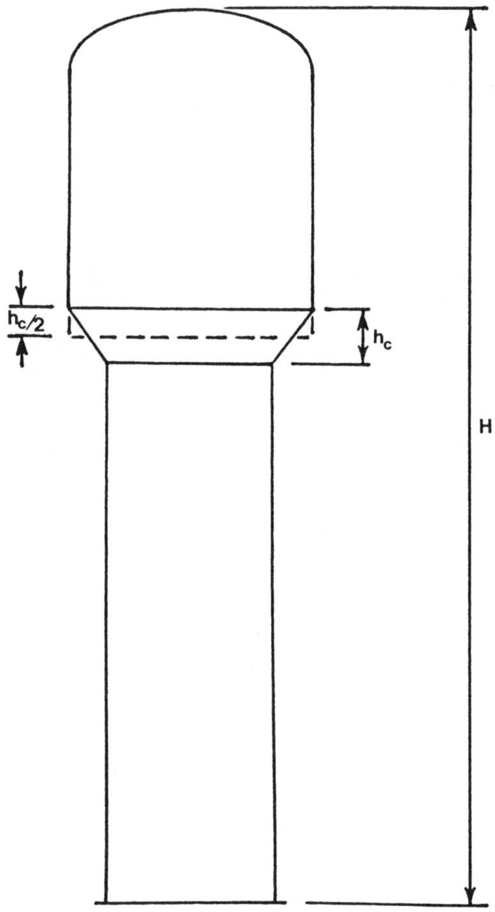

Figure 4-34. When $h_c \leqslant 0.1$ H, the cone can be approximated by considering the two cylinders shown with dotted lines.

EXAMPLE 4-1: WEAR PLATE REQUIREMENT ANALYSIS

A horizontal vessel containing hot oil is to be completely analyzed using the Zick method to determine wear plate requirements (Figure 4-35).

Vessel material = SA-516-70		Q = 7,828.981 lb	
Saddle material = SA-36		CA = 0	
Temp = 300°F			
Design pressure = 671 psi			
θ = 120°			

$$\beta = \left(180 - \frac{\theta}{2}\right) = 120°$$

$$\Delta = \frac{\pi}{180}\left(\frac{5\theta}{12} + 30\right) = 1.396 \text{ rad} = 80.0°$$

From Equation 4-1 at the saddle,

$$\sigma_1 = \frac{3(7{,}828.981)(10)}{\pi(21.0)^2(1.0)}\left[\frac{4(1.5)}{(10)}\left(1 - 0.703\right)\right]\left[\frac{0.297}{0.178}\right]$$

$$\sigma_1 = 50.501 \text{ psi}$$

$$\sigma_p = \frac{PD}{4\ (t - CA)} = \frac{(671)(10.0)}{4(1.0)} = 1{,}677.50$$

$$\sigma_{1_T} = \sigma_1 + \sigma_p = 1{,}728.00 \text{ psi}$$

At Midspan

From Equation 4-3

$$\sigma_2 = \frac{3(7{,}828.981)(10.0)}{\pi(21.0)^2(1.0)}\left[\frac{1 + 0.006}{1 + 0.222} - 0.600\right]$$

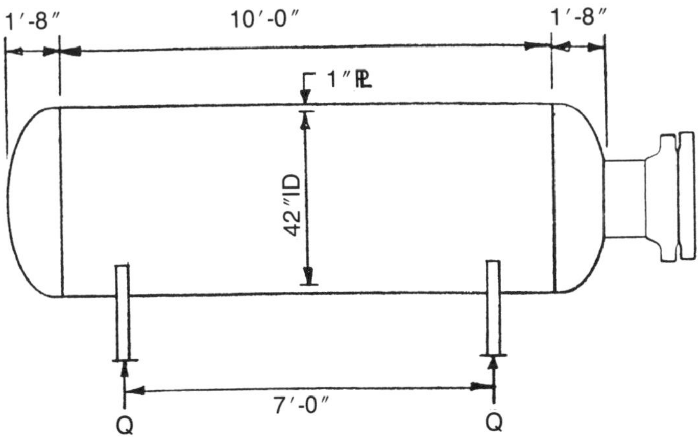

Figure 4-35. Horizontal pressure vessel containing hot oil.

$$\sigma_2 = 37.845 \text{ psi}$$

$$\sigma_{2_T} = \sigma_2 + \sigma_p = 1{,}715.345 \text{ psi}$$

σ_7 = circumferential stress at horn of saddle

$$8R = 8(1.750) = 14.00 \Rightarrow L < 8R$$

$$\frac{A}{R} = \frac{1.500}{1.750} = 0.857 \Rightarrow k = 0.05$$

$$\sigma_7 = \frac{-7{,}828.981}{4(0.941)[0.375 + 1.56[(21)(0.94)]^{0.5}]} - \frac{12(0.05)(7{,}828.981)(1.750)}{(10)(0.941)^2}$$

$$\sigma_7 = -284.547 - 928.358$$

$$\sigma_7 = 1{,}212.905 \text{ psi} << 1.25\ \sigma_{All} = 21{,}875$$

σ_9 = ring compressive stress in shell over saddle

$$\sigma_9 = \frac{7{,}828.981}{(0.941)[0.375 + 1.56[(21)(0.941)]^{0.5}]} \times \left[\frac{1 + \cos(114)}{\pi - 1.990 + \sin(114)\cos(114)}\right]$$

$$\sigma_9 = 865.678 \text{ psi} << 19{,}000 \text{ psi} = 0.5\ \sigma_y$$

Since the ring compressive stress and the circumferential stress at the saddle are less than one half of yield stress and the allowable stress, respectively, wear plates are not required.

EXAMPLE 4-2: MECHANICAL DESIGN OF A PROCESS COLUMN

A detailed mechanical design is required for the process column shown in Figure 4-36. The design criteria are as follows:

Design temperature: −150°F
Design wind speed = 100 mph
Internal pressure at top head: 150 psig
Internal pressure at bottom head: 162 psig
Shell material: 304 SS
Skirt material: A36
External pressure requirement: None
PWHT: Yes
Radiograph: Full
Ambient temperature = 10°F min, 100°F max

Wind distribution is to be computed from ASNI–A58.1–1982:

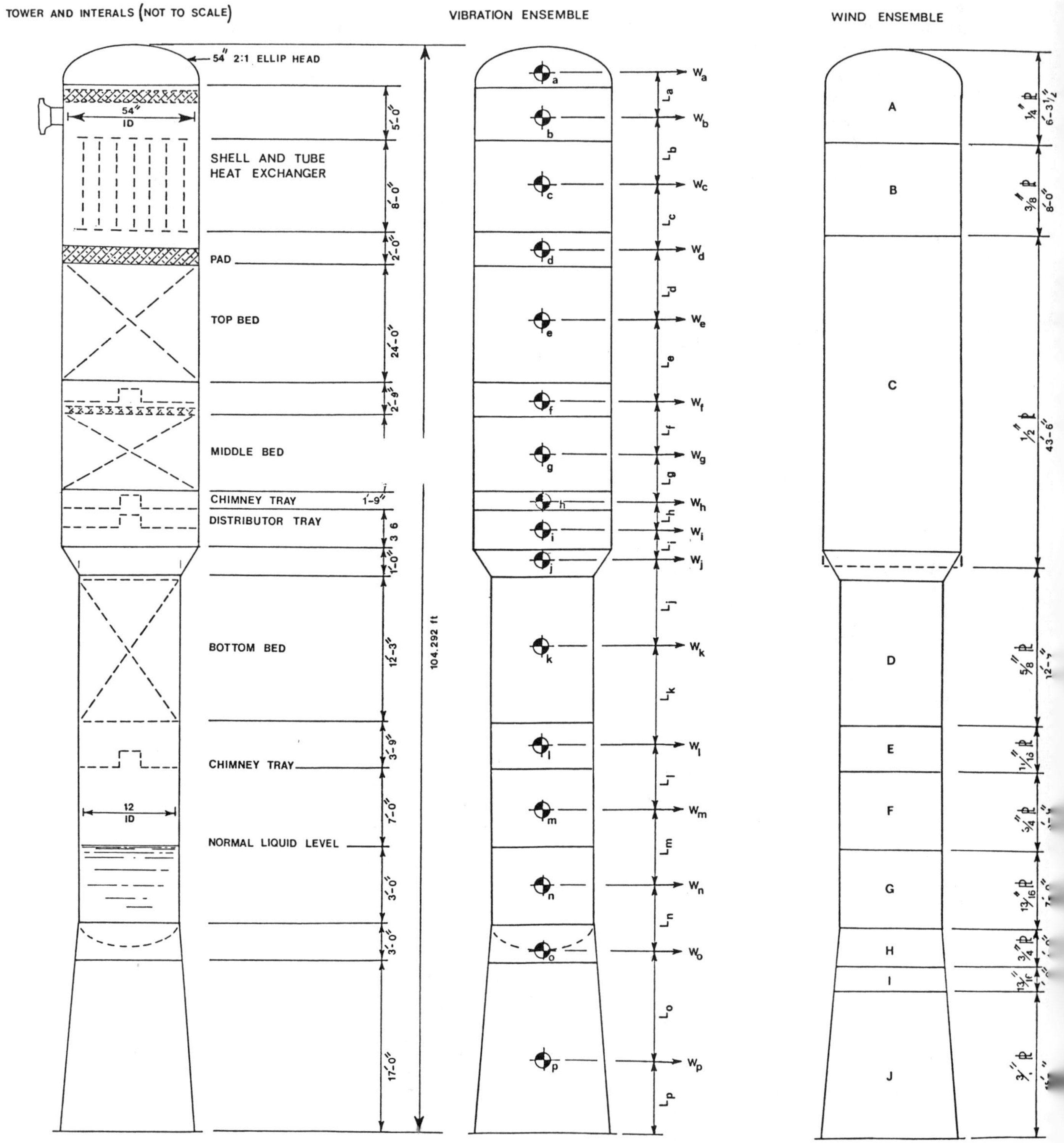

Figure 4-36. Tower analyses ensembles.

$$q_Z = 0.00256\, K_Z\, (IV)^2;\ V = 100 \text{ mph}$$

$$K_Z = 2.58 \left(\frac{Z}{Z_g}\right)^{2/\alpha} \quad \text{for } Z \geqslant 15 \text{ ft}$$

$$K_Z = 2.58 \left(\frac{15}{Z_g}\right)^{2/\alpha} \quad \text{for } Z < 15 \text{ ft}$$

Tower is in Exposure Category C, for which

$\alpha = 7.0,\ Z_g = 900,\ D_o = 0.005$

At 15 ft,

$$K_Z = 2.58 \left(\frac{15}{900}\right)^{2/7} = 0.801$$

For $Z \geqslant 15$ ft,

$$K_Z = 2.58 \left(\frac{Z}{900}\right)^{0.286}$$

For 15 ft and under,

$$q_Z = 0.00256(0.801)(IV)^2$$

From Tables 1 and 5 of A58.1–1982,

$I = 1.0$

$$q_Z = 0.00256(0.801)(100)^2 = 20.506 \text{ lb/ft}^2,\ Z \leqslant 15 \text{ ft}$$

$$q_Z = 0.00256(2.58)\left(\frac{Z}{900}\right)^{0.286}(100)^2$$

$$= 66.048\left(\frac{Z}{900}\right)^{0.286},\ Z > 15 \text{ ft}$$

Effective Cross-sectional Area (Figure 4-37)

Top portion of tower,

$$D_e = 66.5 \text{ in.} + \underbrace{(12.75 + 12) \text{ in.}}_{\text{12 in. pipe plus insulation}} + \underbrace{(12.75 + 11) \text{ in.}}_{\text{12 in. pipe plus insulation}}$$

$$+ \underbrace{12 \text{ in.}}_{\text{ladders and platforms}} = 127.00 \text{ in.}$$

Bottom portion,

$$D_e = 47.0 \text{ in.} + (12.75 + 12) \text{ in.}$$

$$+ (12.75 + 11) \text{ in.} + 12 \text{ in.} = 107.50 \text{ in.}$$

Gust Response Factor

$V = 100$ mph, $h = 104.292$ ft, $f = 0.981$ Hz

Structural damping coefficient $= 0.01$

$$c = \left(\frac{57.792}{104.292}\right)(127.00)$$

$$+ \left(\frac{46.500}{104.292}\right)(107.50) = 118.305$$

From A58.1 Table A9, $s = 1.00$

$$f = \frac{10.5\, fh}{sV} = \frac{(10.5)(0.981)(104.292)}{(1.00)(100)} = 10.743$$

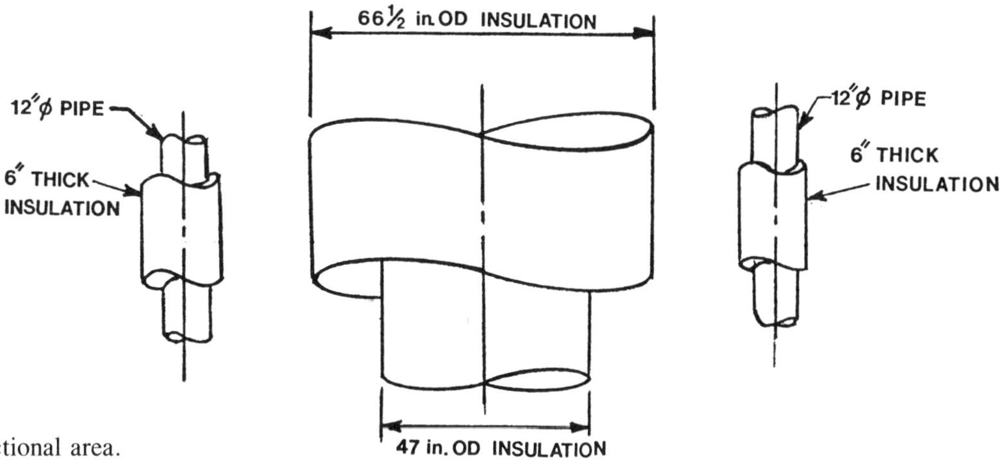

Figure 4-37. Effective cross sectional area.

$$\frac{c}{h} = \frac{\left(\frac{118.305}{12}\right)}{104.292} = 0.095\ ;\ c = \frac{118.305}{12} = 9.859$$

$$\gamma = \frac{0.23}{104.292} = 0.0022$$

$$J = 0.0055,\ y = 0.145$$

$$P = \bar{f}Jy = (10.743)(0.0055)(0.145) = 0.009$$

$$T_1 = \frac{2.35(D_o)^{0.5}}{\left(\frac{Z}{30}\right)^{1/\alpha}} = \frac{2.35(0.005)^{0.5}}{\left(\frac{69.528}{30}\right)^{1/7}} = 0.147$$

$$s = 1.1$$

For a tower with many attachments and connecting piping,

$$G = 0.65 + \left[\frac{0.009}{0.01} + \frac{[(3.32)(0.147)]^2(1.1)}{1 + (0.002)(9.859)}\right]^{0.5} \quad (4\text{-}83)$$

$$G = 0.65 + 1.076 = 1.726$$

From Simiu[9],

$$q = q_Z G$$

Figure 4-38 shows the wind pressure distribution q plotted along the tower length.

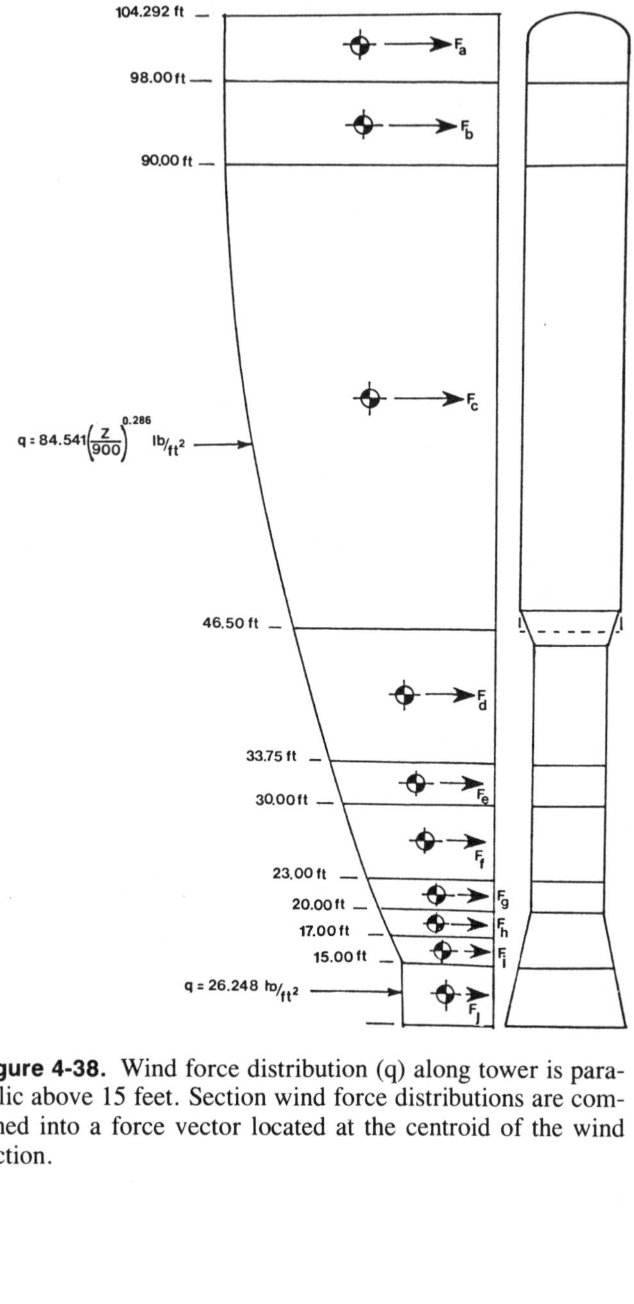

Figure 4-38. Wind force distribution (q) along tower is parabolic above 15 feet. Section wind force distributions are combined into a force vector located at the centroid of the wind section.

Centroid of a Spandrel Segment

The centroid of a parabolic segment is shown in Figure 4-39. Applying the general equation, Equation 4-117, to our case we obtain

$$\bar{Z} = \frac{(n + \alpha)}{2(n + 2\alpha)}\left(\frac{2\alpha q_n + nq_{n-1}}{\alpha q_n + nq_{n-1}}\right)(Z_n - Z_{n-1}) \quad (4\text{-}117)$$

From Figure 4-39 we obtain a general expression for the composite area,

$$A_f = \frac{(Z_n - Z_{n-1})}{(n + \alpha)}(\alpha q_n + nq_{n-1}),\ \text{lb/ft} \quad (4\text{-}118)$$

Solving for the wind resultant force acting through each centroid we obtain,

$$F = A_f D_e C_f$$

Using this equation we compute the wind force distribution. From Figure 4-20, $C_f \approx 0.6$. Solving for section properties we have the following:

$$q_Z = 66.048\left(\frac{Z}{900}\right)^{2/7}$$

Section A

$$Z = 104.292\ \text{ft}$$

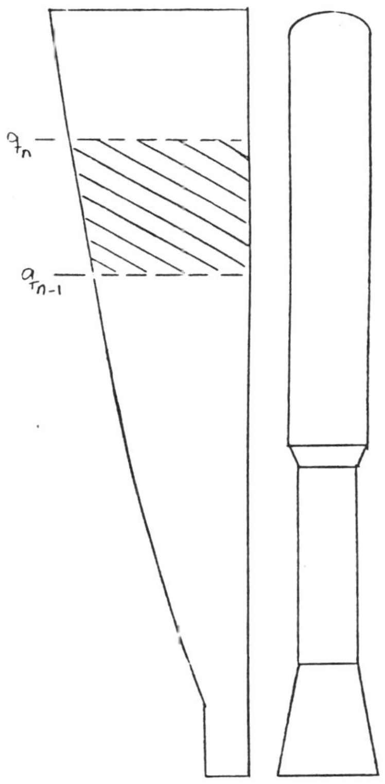

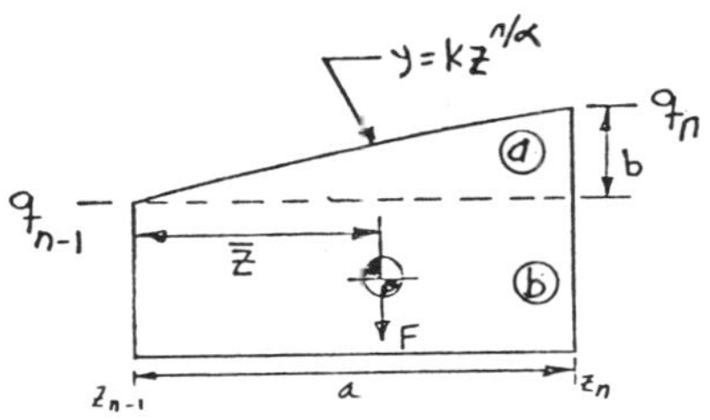

AREA ⓐ—Parabolic Area

$$y = kZ^{n/\alpha}\,;\; K = \frac{b}{a^{n/\alpha}}$$

$$dA = y\,dz = KZ^{n/\alpha}\,dz$$

$$A = \int_0^a dA = K\int_0^a Z^{n/\alpha}\,dz = \frac{\left(\frac{b}{a^{n/\alpha}}\right)a^{\left(\frac{n+\alpha}{\alpha}\right)}}{\frac{n+\alpha}{\alpha}}$$

$$A_a = \frac{\alpha\, ba}{n+\alpha}$$

$$a = Z_n - Z_{n-1};\; b = q_n - q_{n-1}$$

$$A_a = \frac{\alpha(q_n - q_{n-1})(Z_1 - Z_{n-1})}{n+\alpha}$$

Centroid of Parabolic Area ⓐ

$$\int \bar{Z}dA = \int \bar{Z}ydZ = \int_0^a \left(\frac{b}{a^{n/\alpha}}\right) Z^{1+n/\alpha}dZ = \frac{ba^2}{\left(2+\frac{n}{\alpha}\right)}$$

$$\bar{Z}A = \int \bar{Z}dA \Rightarrow \frac{\bar{Z}\alpha(q_n - q_{n-1})(Z_n - Z_{n-1})}{(n+\alpha)} = \frac{(q_n - q_{n-1})(Z_n - Z_{n-1})^2}{\left(2-\frac{n}{\alpha}\right)}$$

$$\Rightarrow \bar{Z} = \frac{(Z_n - Z_{n-1})(n+\alpha)}{(n+2\alpha)}$$

Centroid of Total of Areas ⓐ & ⓑ

Component	Area	$\bar{Z}$	$\bar{Z}A$
Rectangle	$q_{n-1}(Z_n - Z_{n-1})$	$\frac{(Z_n - Z_{n-1})}{2}$	$\frac{q_{n-1}(Z_n - Z_{n-1})^2}{2}$
Parabolic Area	$\frac{\alpha(q_n - q_{n-1})(Z_n - Z_{n-1})}{(n+\alpha)}$	$\frac{(Z_n - Z_{n-1})(n+\alpha)}{(n+2\alpha)}$	$\frac{\alpha(q_n - q_{n-1})(Z_n - Z_{n-1})^2}{(n+2\alpha)}$

$$\sum A = \frac{(Z_n - Z_{n-1})}{(n+\alpha)}(\alpha q_n + nq_{n-1}) \quad \text{lb/ft} \qquad (4\text{-}117)$$

$$\sum \bar{Z}A = \frac{(Z_n - Z_{n-1})^2(nq_{n-1} + 2\alpha q_n)}{2(n+2\alpha)}$$

$$\bar{Z}\sum A = \sum \bar{Z}A$$

$$\bar{Z} = \frac{(n+\alpha)(nq_{n-1} + 2\alpha q_n)(Z_n - Z_{n-1})}{2(n+2\alpha)(\alpha q_n + nq_{n-1})} \qquad (4\text{-}118)$$

$$F = A_f D_e C_f$$

$$M_a = F_a \bar{Z}_a$$

$$M_b = M_a + F_a(Z_a - Z_b) + F_b \bar{Z}_b$$

$$M_c = M_b + (F_a + F_b)(Z_b - Z_c) + F_c \bar{Z}_c$$

$$M_d = M_c + (F_a + F_b + F_c)(Z_c - Z_d) + F_d \bar{Z}_d$$

$$M_n = M_{n-1} + (Z_{n-2} - Z_{n-1})\sum_{i=1}^{n-1} F_i + F_n \bar{Z}_n \qquad (4\text{-}85)$$

Figure 4-39. Wind area section properties.

$$q_Z = 66.048\left(\frac{104.292}{900}\right)^{0.286} = 35.659$$

$$q_n = (1.726)(35.659) = 61.547$$

$$q_{n-1} = (1.726)\left(\frac{98}{900}\right)^{0.286}(66.048) = 60.461$$

$$A_f = \frac{(6.292)[(7)(61.547) + (2)(60.461)]}{2 + 7}$$

$$= 385.735 \text{ lb/ft}$$

$$\overline{Z} = \frac{(9)}{2(2 + 14)}\left[\frac{(2)(7)(61.547) + (2)(60.461)}{(7)(61.547) + (2)(60.461)}\right](6.292)$$

$$= 3.151 \text{ ft}$$

Section B

$$q_n = 60.461$$

$$q_{n-1} = (1.726)\left(\frac{90}{900}\right)^{0.286}(66.048) = 59.007$$

$$A_f = \frac{(8.00)[(7)(60.461) + (2)(59.007)]}{9}$$

$$= 481.103 \text{ lb/ft}$$

$$\overline{Z} = \frac{9}{32}\left[\frac{(2)(7)(60.461) + (2)(59.007)}{(7)(60.461) + (2)(59.007)}\right](8.00)$$

$$= 4.009 \text{ ft}$$

Section C

$$q_n = 59.007$$

$$q_{n-1} = (1.726)\left(\frac{46.50}{900}\right)^{0.286}(66.048) = 48.852$$

$$A_f = \frac{(43.50)[(7)(59.007) + (2)(48.852)]}{9}$$

$$= 2{,}468.640 \text{ lb/ft}$$

$$\overline{Z} = \frac{9}{32}\left[\frac{(2)(7)(59.007) + (2)(48.852)}{(7)(59.007) + (2)(48.852)}\right](43.50)$$

$$= 22.128 \text{ ft}$$

Section D

$$q_n = 48.852$$

$$q_{n-1} = (1.726)\left(\frac{33.75}{900}\right)^{0.286}(66.048) = 44.573$$

$$A_f = \frac{(12.75)[(7)(48.852) + (2)(44.573)]}{9}$$

$$= 610.739$$

$$\overline{Z} = \frac{9}{32}\left[\frac{(2)(7)(48.852) + (2)(44.573)}{(7)(48.852) + (2)(44.573)}\right](12.75)$$

$$= 6.430 \text{ ft}$$

Section E

$$q_n = 44.573$$

$$q_{n-1} = (1.726)\left(\frac{30.00}{900}\right)^{0.286}(66.048) = 43.097$$

$$A_f = \frac{(3.75)[(7)(44.573) + (2)(43.097)]}{9}$$

$$= 165.919$$

$$\overline{Z} = \frac{9}{32}\left[\frac{(2)(7)(44.573) + (2)(43.097)}{(7)(44.573) + (2)(43.097)}\right](3.75)$$

$$= 1.881 \text{ ft}$$

Section F

$$q_n = 43.097$$

$$q_{n-1} = (1.726)\left(\frac{23.00}{900}\right)^{0.286}(66.048) = 39.943$$

$$A_f = \frac{(7.00)[(7)(43.097) + (2)(39.943)]}{9}$$

$$= 296.773 \text{ lb/ft}$$

$$\overline{Z} = \frac{9}{32}\left[\frac{(2)(7)(43.097) + (2)(39.943)}{(7)(43.097) + (2)(39.943)}\right](7.00)$$

$$= 3.525 \text{ ft}$$

Section G

$$q_n = 39.943$$

$$q_{n-1} = (1.726)\left(\frac{20.00}{900}\right)^{0.286}(66.048) = 38.378$$

$$A_f = \frac{(3.00)[(7)(39.943) + (2)(38.378)]}{9}$$

$$= 118.786 \text{ lb/ft}$$

$$\bar{Z} = \frac{9}{32}\left[\frac{(2)(7)(39.943) + (2)(38.378)}{(7)(39.943) + (2)(38.378)}\right](3.00)$$

$$= 1.506 \text{ ft}$$

Section H

$$q_n = 38.378$$

$$q_{n-1} = (1.726)\left(\frac{17.00}{900}\right)^{0.286}(66.048) = 36.635$$

$$A_f = \frac{(3.00)[(7)(38.378) + (2)(36.635)]}{9}$$

$$= 113.972 \text{ lb/ft}$$

$$\bar{Z} = \frac{9}{32}\left[\frac{(2)(7)(38.378) + (2)(36.635)}{(7)(38.378) + (2)(36.635)}\right](3.00)$$

$$= 1.507 \text{ ft}$$

Section I

$$q_n = 36.635$$

$$q_{n-1} = (1.726)\left(\frac{15.00}{900}\right)^{0.286}(66.048) = 35.347$$

$$A_f = \frac{(2.00)[(7)(36.635) + (2)(35.347)]}{9}$$

$$= 72.697 \text{ lb/ft}$$

$$\bar{Z} = \frac{9}{32}\left[\frac{(2)(7)(36.635) + (2)(35.347)}{(7)(36.635) + (2)(35.347)}\right](2.00)$$

$$= 1.003 \text{ ft}$$

Section J

$$q_n = 35.347$$

$$q_{n-1} = 35.347$$

$$A_f = q(Z_n - Z_{n-1}) = (35.347)(15.0)$$
$$= 530.205 \text{ lb/ft}$$

$$\bar{Z} = \frac{15}{2} = 7.50 \text{ ft}$$

Now solving for the section forces we have

$$F = A_f D_e C_f$$

$$F_A = (385.735)\left(\frac{127.00}{12}\right)(0.6) = 2{,}449.417 \text{ lb}$$

$$F_B = (481.103)\left(\frac{127.00}{12}\right)(0.6) = 3{,}055.004 \text{ lb}$$

$$F_C = (2{,}468.640)\left(\frac{127.00}{12}\right)(0.6) = 15{,}675.864 \text{ lb}$$

$$F_D = (610.739)\left(\frac{107.50}{12}\right)(0.6) = 3{,}282.722 \text{ lb}$$

$$F_E = (165.919)\left(\frac{107.50}{12}\right)(0.6) = 891.815 \text{ lb}$$

$$F_F = (296.773)\left(\frac{107.50}{12}\right)(0.6) = 1{,}595.155 \text{ lb}$$

$$F_G = (118.786)\left(\frac{107.50}{12}\right)(0.6) = 638.475 \text{ lb}$$

$$F_H = (113.972)\left(\frac{107.50}{12}\right)(0.6) = 612.600 \text{ lb}$$

$$F_I = (72.697)\left(\frac{107.50}{12}\right)(0.6) = 390.746 \text{ lb}$$

$$F_J = (530.205)\left(\frac{107.50}{12}\right)(0.6) = 2{,}849.852 \text{ lb}$$

$$\sum_{i=A}^{J} F_i = 31{,}441.650 \text{ lb}$$

Solving for section moments we use the following expression:

$$M_n = M_{n-1} + (Z_{n-z} - Z_{n-1}) \sum_{i=1}^{n-1} F_{i-1} + F_n\bar{Z}_n \quad (4\text{-}85)$$

$$M_A = (2449.417)(3.151) = 7{,}718.113 \text{ ft-lb}$$

$$M_B = 7{,}718.113 + (2449.417)(8.00) + (3{,}055.004)(4.009) = 39{,}560.960 \text{ ft-lb}$$

$$M_C = 39{,}560.960 + (5504.421)(43.5) + (15{,}675.864)(22.128) = 625{,}878.792 \text{ ft-lb}$$

$$M_D = 625{,}878.792 + (21{,}180.285)(12.75) + (3{,}282.722)(6.430) = 917{,}035.328 \text{ ft-lb}$$

$$M_E = 917{,}035.328 + (24{,}463.007)(3.75) + (891.815)(1.881) = 1{,}010{,}449.109 \text{ ft-lb}$$

$$M_F = 1{,}010{,}449.109 + (25{,}354.822)(7.00) + (1{,}595.155)(3.525) = 1{,}193{,}555.784 \text{ ft-lb}$$

$$M_G = 1{,}193{,}555.784 + (26{,}949.977)(3.00) + (638.475)(1.506) = 1{,}275{,}367.258 \text{ ft-lb}$$

$$M_H = 1{,}275{,}367.258 + (27{,}588.452)(3.00) + (612.600)(1.491) = 1{,}359{,}046.001 \text{ ft-lb}$$

$$M_I = 1{,}359{,}046.001 + (28{,}201.052)(2.00) + (390.746)(1.003) = 1{,}415{,}840.023 \text{ ft-lb}$$

$$M_J = 1{,}415{,}840.023 + (28{,}591.798)(15.0) + (2{,}849.852)(7.50) = 1{,}866{,}090.883 \text{ ft-lb}$$

Section Moments of Inertia

Section a—1/4-in. PL

$$I_a = \frac{\pi}{64}\left[\left(\frac{54.50}{12}\right)^4 - \left(\frac{54.50}{12}\right)^4\right] = 0.756 \text{ ft}^4$$

Section b—3/8 in. PL

The shell and tube heat exchanger section moment of inertia is approximated by a set of concentric circles of tubes. The concentric circle pattern approximates that of the exchanger tube sheet. Using the parallel-axis theorem, we arrive at the section moment of inertia. Referring to Figure 4-40 we analyze the exchanger as follows:

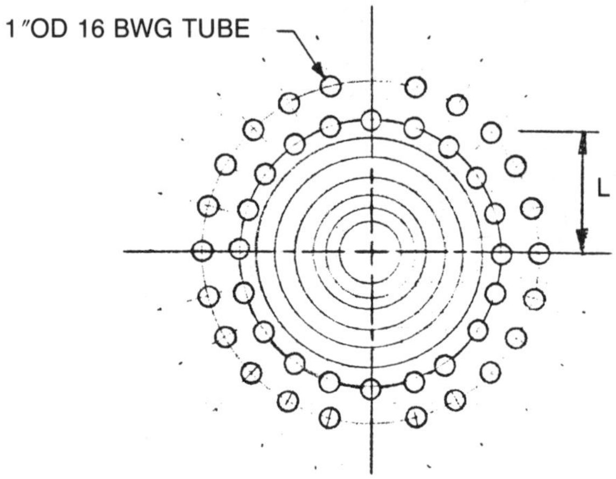

Figure 4-40. The tube bundle is modeled in banks of concentric circles used to approximate the section moment of inertia. The tube bundle enhances the section stiffness.

For 1-in. OD 16 BWG tube,

$\bar{I} = 0.0210 \text{ in.}^4$ = metal cross section

$A = 0.191 \text{ in.}^2$ = metal area

K = number of tubes per circle

n = number of circles

Thus, from the parallel-axis theorem the composite moment of inertia is

$$I = \sum_{i=1}^{n} K_i(\bar{I} + AL^2)$$

Values of I are tabulated in Table 4-16.

For enclosing shell, 3/8-in. PL,

$$I = \frac{\pi}{64}[(54.75)^4 - (54.00)^4] = 23{,}676.070 \text{ in.}^4$$

$$\sum I = 114{,}735.441 \text{ in.}^4 + 23{,}676.070 \text{ in.}^4 = 138{,}411.511 \text{ in.}^4$$

or for the total cross-sectional area,

$$I_b = 6.675 \text{ ft}^4$$

Table 4-16
Values of I for Tube Bundle

Circle	L (in.)	L^2 (in.^{2})	A (in.^{2})	$\bar{I}$ (in.^{4})	K	I (in.^{4})	$\sum$I (in.^{4})
1	25.75	663.063	0.191	0.021	129	16,364.299	16,364.299
2	24.50	600.250	0.191	0.021	123	14,127.503	30,491.802
3	23.25	540.563	0.191	0.021	117	12,104.531	42,596.333
4	22.00	484.000	0.191	0.021	111	10,284.531	52,880.927
5	20.75	430.563	0.191	0.021	104	8,574.543	61,455.470
6	19.50	380.250	0.191	0.021	98	7,138.100	68,593.570
7	18.25	333.063	0.191	0.021	92	5,871.903	74,465.473
8	17.00	289.000	0.191	0.021	86	4,765.174	79,230.647
9	15.75	248.063	0.191	0.021	79	3,759.613	82,990.260
10	14.50	210.250	0.191	0.021	73	2,946.846	85,937.106
11	13.25	175.563	0.191	0.021	67	2,260.750	88,197.856
12	12.00	144.000	0.191	0.021	60	1,662.840	89,860.696
13	10.75	115.563	0.191	0.021	54	1,203.257	91,063.953
14	9.50	90.250	0.191	0.021	48	837.492	91,901.445
15	8.25	68.063	0.191	0.021	42	554.821	92,456.266
16	7.00	49.000	0.191	0.021	35	334.915	92,791.181
17	5.75	33.063	0.191	0.021	29	189.226	92,980.407
18	4.50	20.250	0.191	0.021	23	93.788	93,074.195
19	3.25	10.563	0.191	0.021	16	35.641	93,109.836
20	2.00	4.000	0.191	0.021	10	9.740	93,119.576
21	0.75	0.563	0.191	0.021	4	1.270	93,120.846
22	27.00	729.000	0.191	0.021	155	21,614.595	114,735.441

Section c—1/2-in. ℄

$$I = \frac{\pi}{64}\left[\left(\frac{55.00}{12}\right)^4 - \left(\frac{54.00}{12}\right)^4\right]$$

$$= 1.533 \text{ ft}^4$$

Section d—5/8-in. ℄

$$I = \frac{\pi}{64}\left[\left(\frac{43.25}{12}\right)^4 - \left(\frac{42.00}{12}\right)^4\right]$$

$$= 0.917 \text{ ft}^4$$

Section e—11/16-in. ℄

$$I = \frac{\pi}{64}\left[\left(\frac{43.375}{12}\right)^4 - \left(\frac{42.00}{12}\right)^4\right]$$

$$= 1.013 \text{ ft}^4$$

Section f—3/4-in. ℄

$$I = \frac{\pi}{64}\left[\left(\frac{43.50}{12}\right)^4 - \left(\frac{42.00}{12}\right)^4\right]$$

$$= 1.110 \text{ ft}^4$$

Section g—13/16-in. ℄

$$I = \frac{\pi}{64}\left[\left(\frac{43.625}{12}\right)^4 - \left(\frac{42.00}{12}\right)^4\right]$$

$$= 1.208 \text{ ft}^4$$

Referring to Figures 4-36 and 4-41 we calculate the section moments of inertia as follows:

Section h

$$r_{eq} = \left(\frac{26.193 + 21.8125}{2 \cos 6.934°}\right) = 24.180 \text{ in.}$$

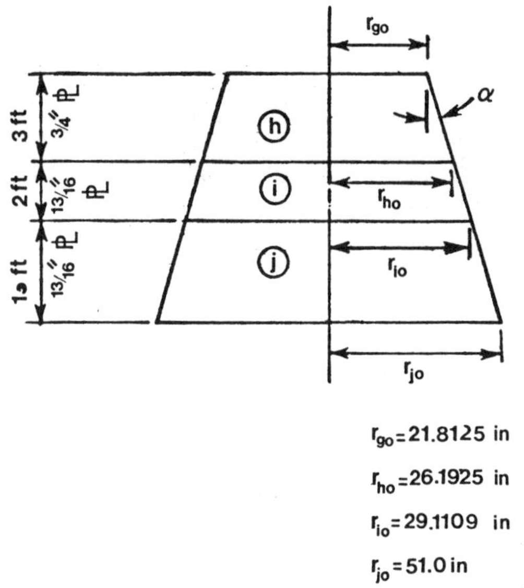

Figure 4-41. The equivalent radii for the skirt sections.

$$I_h = \frac{\pi}{64}\left[\left(\frac{48.36}{12}\right)^4 - \left(\frac{46.86}{12}\right)^4\right] = 1.533 \text{ ft}^4$$

Section i

$$r_{eq} = \left(\frac{29.111 + 26.193}{2 \cos 6.934°}\right) = 27.856 \text{ in.}$$

$$I_i = \frac{\pi}{64}\left[\left(\frac{55.712}{12}\right)^4 - \left(\frac{54.087}{12}\right)^4\right] = 2.547 \text{ ft}^4$$

Section j

$$r_{eq} = \left(\frac{51.00 + 29.111}{2 \cos 6.934°}\right) = 40.351 \text{ in.}$$

$$I_j = \frac{\pi}{64}\left[\left(\frac{80.701}{12}\right)^4 - \left(\frac{79.076}{12}\right)^4\right] = 7.846 \text{ ft}^4$$

Values of each section's wind force, shear, bending moment, and moment of inertia are summarized in Table 4-17 for the entire tower.

Deflections

$$\delta_1 = \frac{(6.292)^2}{(3.157 \times 10^9)}\left[\frac{(2{,}449.417)(6.292)}{8}\right]$$

$$= 2.416 \times 10^{-5} \text{ ft}$$

$$\delta_2 = \frac{(8.00)^2}{(2.787 \times 10^{10})}\left[\frac{(3{,}055.004)(8.00)}{8} + \frac{(2{,}449.417)(8.00)}{3} + \frac{(7{,}718.113)}{2}\right]$$

$$= 3.088 \times 10^{-5} \text{ ft}$$

$$\delta_3 = \frac{(43.50)^2}{(6.402 \times 10^9)}\left[\frac{(15{,}675.864)(43.5)}{8} + \frac{(5{,}504.421)(43.5)}{3} + \frac{(39{,}560.960)}{2}\right]$$

$$= 0.055 \text{ ft}$$

$$\delta_4 = \frac{(12.75)^2}{(3.829 \times 10^9)}\left[\frac{(3{,}282.722)(12.75)}{8} + \frac{(21{,}180.285)(12.75)}{3} + \frac{(625{,}878.792)}{2}\right]$$

$$= 0.017 \text{ ft}$$

$$\delta_5 = \frac{(3.75)^2}{(4.230 \times 10^9)}\left[\frac{(891.815)(3.75)}{8} + \frac{(24{,}463.007)(3.75)}{3} + \frac{(917{,}035.328)}{2}\right]$$

$$= 0.002$$

$$\delta_6 = \frac{(7.00)^2}{(4.635 \times 10^9)}\left[\frac{(1{,}595.155)(7.00)}{8} + \frac{(25{,}354.822)(7.00)}{3} + \frac{(1{,}010{,}449.109)}{2}\right]$$

$$= 0.006 \text{ ft}$$

Table 4-17
Summation of Section Wind Force, Shear, Bending Moment and Moment of Inertia

Section	F_i (lb)	Q_i (lb)	M_i (ft-lb)	I_i (ft^4)	EI (lb-ft^2)
a	2,449.417	2,449.417	7,718.113	0.756	3.157×10^9
b	3,055.004	5,504.421	39,560.960	6.675	2.787×10^{10}
c	15,675.864	21,180.285	625,878.792	1.533	6.402×10^9
d	3,282.722	24,463.007	917,035.328	0.917	3.829×10^9
e	891.815	25,354.822	1,010,449.109	1.013	4.230×10^9
f	1,595.155	26,949.977	1,193,555.784	1.110	4.635×10^9
g	638.475	27,588.452	1,275,367.258	1.208	5.045×10^9
h	612.600	28,201.052	1,359,046.001	1.533	6.402×10^9
i	390.746	28,591.798	1,415,840.023	2.547	1.064×10^{10}
j	2,849.852	31,441.650	1,866,090.883	7.846	3.276×10^{10}

$$\delta_7 = \frac{(3.00)^2}{(5.045 \times 10^9)}\left[\frac{(638.475)(3.00)}{8} + \frac{(26{,}949.977)(3.0)}{3} + \frac{(1{,}193{,}555.784)}{2}\right]$$

$$= 0.001 \text{ ft}$$

$$\delta_8 = \frac{(3.00)^2}{(6.402 \times 10^9)}\left[\frac{(612.00)(3.00)}{8} + \frac{(27{,}588.452)(3.0)}{3} + \frac{(1{,}275{,}367.258)}{2}\right]$$

$$= 0.001 \text{ ft}$$

$$\delta_9 = \frac{(2.00)^2}{(1.064 \times 10^{10})}\left[\frac{(390.746)(2.00)}{8} + \frac{(28{,}201.052)(2.00)}{3} + \frac{(1{,}359{,}046.001)}{2}\right]$$

$$= 2.626 \times 10^{-4} \text{ ft}$$

$$\delta_{10} = \frac{(15.00)^2}{(3.276 \times 10^{10})}\left[\frac{(2{,}849.852)(15.00)}{8} + \frac{(28{,}591.798)(15.00)}{3} + \frac{(1{,}415{,}840.023)}{2}\right]$$

$$= 0.006 \text{ ft}$$

$$\Delta_{1\text{-}2} = \frac{(6.292)(8.00)}{(2.787 \times 10^{10})}\left[\frac{(3{,}055.004)(8.00)}{6} + \frac{(2{,}449.417)(8.00)}{2} + 7{,}718.113\right]$$

$$= 3.899 \times 10^{-5} \text{ ft}$$

$$\Delta_{1\text{-}3} = \frac{(14.292)(43.50)}{(6.402 \times 10^9)}\left[\frac{(15{,}675.864)(43.5)}{6} + \frac{(5{,}504.421)(43.5)}{2} + 39{,}560.960\right]$$

$$= 0.027 \text{ ft}$$

$$\Delta_{1\text{-}4} = \frac{(57.792)(12.75)}{(3.829 \times 10^9)}\left[\frac{(3{,}282.722)(12.75)}{6} + \frac{(21{,}180.285)(12.75)}{2} + 625{,}878.792\right]$$

$$= 0.148 \text{ ft}$$

$$\Delta_{1\text{-}5} = \frac{(70.542)(3.75)}{(4.230 \times 10^9)}\left[\frac{(891.815)(3.75)}{6} + \frac{(24{,}463.007)(3.75)}{2} + 917{,}035.328\right]$$

$$= 0.060 \text{ ft}$$

$$\Delta_{1\text{-}6} = \frac{(74.292)(7.00)}{(4.635 \times 10^9)}\left[\frac{(1{,}595.155)(7.00)}{6} + \frac{(25{,}354.822)(7.00)}{2} + 1{,}010{,}449.109\right]$$

$$= 0.124 \text{ ft}$$

$$\Delta_{1\text{-}7} = \frac{(81.292)(3.00)}{(5.045 \times 10^9)}\left[\frac{(638.475)(3.00)}{6} + \frac{(26{,}949.977)(3.00)}{2} + 1{,}193{,}555.784\right]$$

$$= 0.060 \text{ ft}$$

$$\Delta_{1\text{-}8} = \frac{(84.292)(3.00)}{(6.402 \times 10^9)}\left[\frac{(612.60)(3.00)}{6} + \frac{(27{,}588.452)(3.0)}{2} + 1{,}275{,}367.258\right]$$

$$= 0.052 \text{ ft}$$

$$\Delta_{1\text{-}9} = \frac{(87.292)(2.00)}{(1.064 \times 10^{10})}\left[\frac{(390.746)(2.00)}{6} + \frac{(28{,}591.798)(2.00)}{2} + 1{,}359{,}046.001\right]$$

$$= 0.023 \text{ ft}$$

$$\Delta_{1\text{-}10} = \frac{(89.292)(15.00)}{(3.276 \times 10^{10})}\left[\frac{(2{,}849.852)(15.00)}{6} + \frac{(28{,}591.798)(15.00)}{2} + 1{,}415{,}840.023\right]$$

$$= 0.067 \text{ ft}$$

$$y_i = \sum \delta_i + \sum \Delta_i$$

$$= 0.088 \text{ ft} + 0.561 \text{ ft} = 0.649 \text{ ft}$$
$$= 7.787 \text{ in.}$$

Tower Section Stress Calculations

Section a—1/4-in. ℄

For tension on the windward side, using Equation 4-31,

$$\sigma = \left(\frac{PD}{4t}\right) \pm \left(\frac{16D_oM}{\pi t(D_o + D_i)(D_o^2 + D_i^2)}\right) - \left(\frac{2W}{\pi t(D_o + D_i)}\right)$$

$P = 150$ psi, $E = 1.0$, $t = 0.25$ in., $T = -100°F$,

$W = 0.423$ kips

$D_i = 54.0$ in., $D_o = 54.50$ in.

$$\sigma = \left[\frac{(150.0)(54.0)}{(4)(0.25)}\right] \pm \left[\frac{16(54.50)(12)(7{,}718.113)}{\pi(0.25)(108.50)(5{,}886.25)}\right] - \left[\frac{2(1{,}814)}{\pi(0.25)(108.50)}\right]$$

$$\sigma = 8{,}100 \pm 161.009 - 42.57$$

$\sigma = 8{,}218.43$ psi tension on windward side

$\sigma = 7{,}896.42$ psi compression on leeward side

Internal pressure circumferential stress,

$$\sigma_p = \frac{PD}{2t} = \frac{(150.0)(54.0)}{2(0.25)(1.0)}$$

$$= 16{,}200 \text{ psi} < 18{,}800 = \sigma_a$$

The circumferential stress governs in this section and is less than the allowable stress.

Section b— 3/8-in. ℄

$$\sigma = \left[\frac{(150.0)(54.0)}{(4)(0.375)}\right] \pm \left[\frac{16(54.75)(39{,}560.960)(12)}{\pi(0.375)(108.50)(5{,}913.563)}\right] - \left[\frac{2(15{,}201.00)}{\pi(0.375)(108.75)}\right]$$

$$\sigma = 5{,}400 \pm 550.16 - 237.30$$

$$\sigma = 5{,}712.86 \text{ psi tension}$$

$$\sigma = 4{,}612.54 \text{ psi compression}$$

$$\sigma_p = \frac{(150.0)(54.0)}{2(0.375)(1.0)} = 10{,}800 \text{ psi}$$

$$< 18{,}800 = \sigma_a$$

Section c—1/2-in. PL

$$\sigma = \left[\frac{(150.0)(54.0)}{(4)(0.5)}\right]$$

$$\pm \left[\frac{16(55.00)(625{,}878.792)(12)}{\pi(0.50)(109.00)(5{,}941.00)}\right]$$

$$- \left[\frac{2(37{,}723.0)}{\pi(0.50)(109)}\right]$$

$$\sigma = 4{,}050.00 \pm 6{,}497.53 - 440.646$$

$$\sigma = 10{,}106.884 \text{ psi tension} < 18{,}800 = \sigma_a$$

$$\sigma = -2{,}888.176 \text{ psi compression}$$

$$\sigma_p = \frac{(150.0)(54.0)}{2(0.50)(1.0)} = 8{,}100 \text{ psi}$$

Section d—5/8-in. PL

$$\sigma = \left[\frac{(150.0)(42.0)}{(4)(0.625)}\right]$$

$$\pm \left[\frac{(16)(43.25)(917{,}035.328)(12)}{\pi(0.688)(85.375)(3{,}645.391)}\right]$$

$$- \left[\frac{2(43{,}418.60)}{\pi(0.688)(85.375)}\right]$$

$$\sigma = 2{,}520.00 \pm 11{,}320.361 - 470.584$$

$$\sigma = 13{,}369.777 \text{ psi tension} < 18{,}800 \text{ psi} = \sigma_a$$

$$\sigma = -9{,}290.944 \text{ psi compression}$$

$$\sigma_p = \frac{(150.0)(42.0)}{2(0.625)} = 5{,}040.00 \text{ psi}$$

Section e—11/16-in. PL

$$\sigma = \left[\frac{(150.0)(42.0)}{(4)(0.688)}\right]$$

$$\pm \left[\frac{16(43.375)(1{,}010{,}449.109)(12)}{\pi(0.688)(85.375)(3{,}645.391)}\right]$$

$$- \left[\frac{2(44{,}561.927)}{\pi(0.688)(85.375)}\right]$$

$$\sigma = 2{,}289.24 \pm 12{,}509.56 - 482.98$$

$$\sigma = 14{,}315.82 \text{ psi tension} < 18{,}800 \text{ psi} = \sigma_a$$

$$\sigma = -10{,}703.30 \text{ psi compression}$$

$$\sigma_p = \frac{(150.00)(42.00)}{2(0.688)} = 4{,}578.488 \text{ psi}$$

Section f—3/4-in. PL

$$\sigma = \left[\frac{(150.0)(42.0)}{(4)(0.75)}\right]$$

$$\pm \left[\frac{16(43.50)(1{,}193{,}555.784)(12)}{\pi(0.75)(85.50)(3{,}656.250)}\right]$$

$$- \left[\frac{2(46{,}958.955)}{\pi(0.75)(85.50)}\right]$$

$$\sigma = 2{,}100.00 \pm 13{,}533.81 - 466.199$$

$$\sigma = 15{,}167.61 \text{ psi tension} < 18{,}800 \text{ psi} = \sigma_a$$

$$\sigma = -11{,}900.01 \text{ psi compression}$$

$$\sigma_p = \frac{(150.00)(42.00)}{2(0.75)} = 4{,}200.00 \text{ psi}$$

Section g—13/16-in. PL

$$\sigma = \left[\frac{(150.0)(42.0)}{(4)(0.813)}\right]$$

$$\pm \left[\frac{16(43.625)(1{,}275{,}367.258)(12)}{\pi(0.813)(85.625)(3{,}667.141)}\right]$$

$$- \left[\frac{2(48{,}073.488)}{\pi(0.813)(85.625)}\right]$$

$\sigma = 1{,}937.27 \pm 13{,}319.972 - 439.64$

$\sigma = 14{,}817.602$ psi tension $< 18{,}800$ psi $= \sigma_a$

$\sigma = -11{,}822.342$ psi compression

All section stresses are less than the allowable stress of 18,800 psi. Thus, the tower thicknesses are acceptable.

SKIRT AND BASE PLATE DESIGN

First we determine the size and number of anchor bolts required.

$$A_b = \left[\frac{4M}{BC} - W\right] \Big/ N\sigma_a \tag{4-42}$$

Using an A-193-B7 high-strength bolt with an allowable stress of 40,000 psi per AISC and assuming a bolt circle of 107 in., the required bolt area, A_b, is

$$A_b = \frac{\left[\dfrac{(4)(12)(1{,}866{,}090.883)}{107} - 63{,}815.727\right]}{(12)(40{,}000)}$$

$$= 1.611 \text{ in.}^2$$

We select a 1⅝-in. ϕ bolt of 8-thread series with a minimum root area of 1.680 in.² Thus, using Figure 4-15, the new bolt circle becomes

$BC = 102 \text{ in.} + 2(2.375) = 106.75 \text{ in.}$

The new required area becomes

$$A_b = \frac{\left[\dfrac{(48)(1{,}866{,}090.883)}{106.75} - 63{,}815.727\right]}{(12)(40{,}000)}$$

$$= 1.615 \text{ in.}^2 < 1.680 \text{ in.}^2$$

$$\text{Bolt spacing} = \frac{\pi(106.75)}{12}$$

$$= 27.947 \text{ in.} > 18 \text{ in. minimum}$$

Maximum bearing pressure on contact area:

D_o = OD of base plate $= 106.75 + 2(2.375)$
$= 111.50$ in.

D_i = ID of base plate $= 111.50 - 2(6.75)$
$= 98.00$ in.

$$A_c = \frac{\pi(D_o^2 - D_i^2)}{4} = 2221.302 \text{ in.}^2$$

$$I_c = \frac{\pi(D_o^4 - D_i^4)}{64} = 3{,}059{,}323.380$$

$$\sigma_c = \left(\frac{NF_i}{A_c}\right) + \left(\frac{W}{A_c}\right) + \left(\frac{MD_o}{2I_c}\right) \tag{4-40}$$

$$F_i = \frac{8MD_o}{N(D_o^2 + D_i^2)} - \frac{W_E}{N} \tag{4-41}$$

$$F_i = \frac{(8)(12)(1{,}866{,}090.883)(111.50)}{(12)(22{,}036.250)} - \frac{63{,}815.727}{12}$$

$$= 70{,}219.061 \text{ lb}$$

$$\sigma_c = \frac{(12)(70{,}219.061)}{2{,}221.302} + \frac{63{,}815.727}{2{,}221.302} + \frac{(1{,}866{,}090.883)(12)(111.50)}{2(3{,}059{,}323.380)}$$

$$\sigma_c = 379.340 + 28.729 + 408.069$$

$$\sigma_c = 816.138 \text{ psi} < F_b = 1.33(900) = 1{,}197 \text{ psi}$$

The concrete bearing strength criteria are met, so we can continue to the base plate design.

$$t_3 = \frac{D_o - D_i}{2} = \frac{111.50 - 98.00}{2}$$

= 6.75 = Base ℄ width = BP ℄ W

$$k = \frac{1}{1 + \dfrac{\sigma_s}{n\sigma_c}} \tag{4-45}$$

$\sigma_s = 20{,}000$ = allowable working stress for steel, psi

$\sigma_c = 1{,}000$ psi

$n = \sigma_s/\sigma_c = 10$

$k = 0.333$

With $k = 0.333$, $C_c = 1.588$, $C_t = 2.376$, $Z = 0.431$, $j = 0.782$, and six iterations of Equation 4-45 using Table 4-7, we obtain a value of k of 0.186.

For $k = 0.186$,
$\sigma_c = 655.834$ psi and $\sigma_s = 26{,}850.892$ psi
The allowable stress in the base ring = 36,000 psi
The allowable strength in the concrete = 3,000 psi

Using Equation 4-54 and solving for the maximum induced stress in the concrete,

$$\sigma_{c(max)} = (655.834)\left[\frac{2(0.183)(106.75) + 6.75}{2(0.183)(106.75)}\right]$$

$$\sigma_{c(max)} = 769.139 \text{ psi}$$

$$\text{BTHK} = L\left[\frac{3 \quad \sigma_{c(max)}}{\sigma_{all}}\right]^{0.5} \qquad (4\text{-}56)$$

$$L = \frac{111.50 - 102.00}{2} = 4.750 \text{ in.}$$

$$\text{BTHK} = (4.75)\left[\frac{(3)(769.139)}{20{,}000}\right]^{0.5} = 1.613 \text{ in.}$$

Make base ℙ 1⅝ in. thick

Solving for the compression ring thickness using Equation 4-63 we have

$$t_{CR} = \left[\frac{(64{,}605.803)(2.375)}{4(20{,}000)[4.75 - (1.25)(1.625)]}\right] = 0.705 \text{ in.}$$

Make compression ring ¾ in. thick

Using Equation 4-64 and Figure 4-16 in calculating the compression ring thickness, we have

$$18{,}000\, G_W t_G^3 - (F_i)t_G^2 - \frac{G_H^2(F_i)}{1{,}500} = 0 \qquad (4\text{-}64)$$

$F_i = 64{,}605.803$

$G_H = 9.00$ in.; $G_W = 4.25$ in.

Equation 4-64 becomes

$$t_G^3 - 0.845\, t_G^2 - 0.046 = 0$$

$$t_G = 0.901 \text{ in.}$$

Make gusset plates 15/16 in. thick

Calculating the minimum skirt-to-base-plate weld size we have from Equation 4-65

$$\xi = \left(\frac{4M}{\pi D_{SK}^2}\right) + \left(\frac{W}{\pi D_{SK}}\right) \text{ lb/lin in.}$$

$$\xi = \left[\frac{4(12)(1{,}866{,}090.883)}{\pi(102.00)^2}\right] + \frac{63{,}815.727}{\pi(102.00)}$$

$$= 2{,}939.611$$

For wind or earthquake,

$$\sigma_w = 1.33\, \sigma_a(0.55) = 1.33(12{,}700)(0.55)$$
$$= 9{,}290.05 \text{ psi}$$

$$\text{weld size} = w = \frac{2{,}939.611}{(2)(9290.05)} = 0.158$$

Use at least a 3/16-in. weld on each side of skirt.

Anchor bolt torque is determined by Equation 4-66. For lubricated bolts with Fel-Pro C-5A,

$C = 0.15$

$T = (0.15)(1.625)(64{,}605.803) = 15{,}747.664$ in.-lb

or

$T = 1{,}313$ ft-lb with torque wrench

Checking the skirt thickness for reaction of the bolting ring against the skirt we have from Equation 4-39,

$$t = 1.76\left(\frac{F_i B}{m(GH)\sigma_{all}}\right)^{2/3} r^{1/3} \qquad (4\text{-}39)$$

$$t = 1.76\left[\frac{(70{,}219.061)(2.375)}{(27{,}947)(9.00)(20{,}000)}\right]^{2/3}\left(\frac{102}{2}\right)^{1/3}$$

$t = 0.672$ in. < 13/16 in. skirt thickness at chair
Skirt thickness meets chair ring reaction criteria.

A sketch of the skirt and chair design is shown in Figure 4-42.

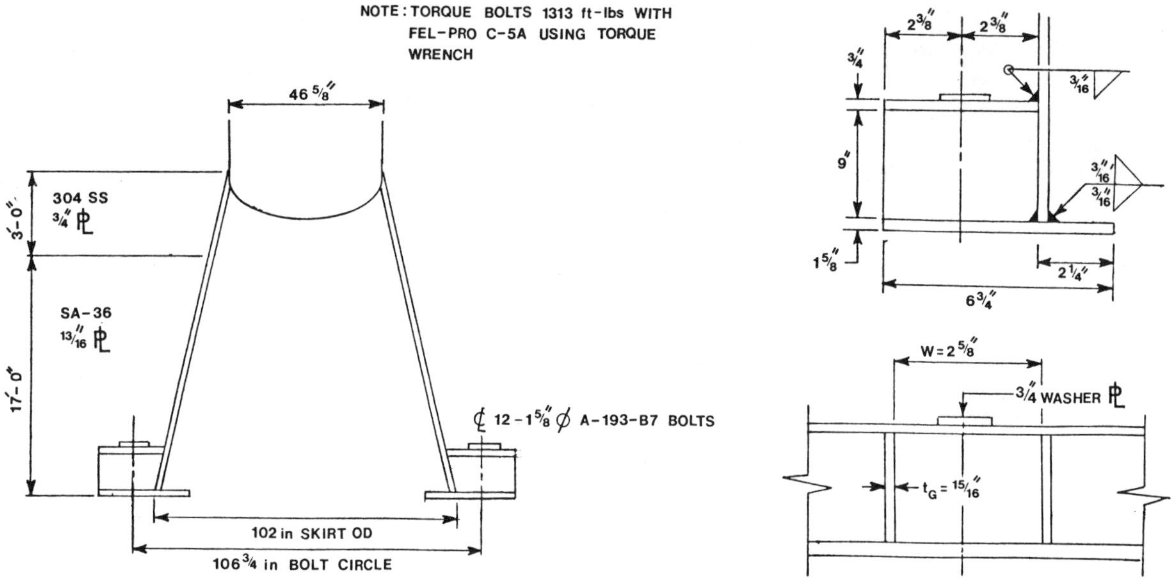

Figure 4-42. The skirt detail.

Section Centroids

Referring to Figure 4-21 we have the following:

Section a

$$\bar{y}_a = \frac{b(4a - 3t)}{6\left(\frac{\pi a}{2} - t\right)}$$

$$= \frac{\frac{54.50}{4}[4(27.25) - 3(0.25)]}{6\left(\frac{\pi(27.25)}{2} - 0.25\right)}$$

$$\bar{y}_a = 5.857 \text{ in.} = 0.488 \text{ ft}$$

$$L_1 = 0.488 \text{ ft} + \frac{5.0}{2} = 2.988 \text{ ft}$$

Section b

$$\bar{y}_b = \frac{5.00}{2} = 2.50 \text{ ft}$$

Section c

$$\bar{y}_c = \frac{8.00}{2} = 4.00 \text{ ft}$$

$$L_2 = 2.50 + 4.00 = 6.50 \text{ ft}$$

Section d

$$\bar{y}_d = \frac{2.00}{2} = 1.00 \text{ ft}$$

$$L_3 = 4.00 + 1.00 = 5.00 \text{ ft}$$

Section e

$$\bar{y}_e = \frac{24.00}{2} = 12.00 \text{ ft}$$

$$L_4 = 1.00 + 12.00 = 13.00 \text{ ft}$$

Section f

$$\bar{y}_f = \frac{2.75}{2} = 1.375 \text{ ft}$$

$$L_5 = 12.00 + 1.375 = 13.375 \text{ ft}$$

Section g

$$\bar{y}_g = \frac{8.00}{2} = 4.00 \text{ ft}$$

$$L_6 = 1.375 + 4.00 = 5.375 \text{ ft}$$

Section h

$$\bar{y}_h = \frac{1.75}{2} = 0.875 \text{ ft}$$

$$L_7 = 4.00 + 0.875 = 4.875 \text{ ft}$$

Section i

$$\bar{y}_i = \frac{3.50}{2} = 1.750 \text{ ft}$$

$$L_8 = 0.875 + 1.75 = 2.625 \text{ ft}$$

Section j

$$\bar{y}_j = \frac{3H(D_{o_s}^2 - D_{i_s}^2) + 8(D_{o_s} - D_{i_s}) \tan \alpha}{6[(D_{o_s}^2 - D_{i_s}^2) + 2H(D_{o_s} - D_{i_s}) \tan \alpha]}$$

$$D_{o_s} - D_{i_s} = 43.25 - 42.00 = 1.25$$

$$D_{o_s}^2 - D_{i_s}^2 = 1{,}870.563 - 1{,}764.00 = 106.563$$

$$\alpha = \arctan\left(\frac{6}{12}\right) - 26.565°$$

$$\bar{y}_j = \frac{3(12)(106.563) + 8(12)^2(1.25) \tan (26.565)}{6[106.563 + 2(12)(1.25) \tan (26.565)]}$$

$$= 6.247 \text{ in.}$$

$$\bar{y}_j = 0.521 \text{ ft}$$

$$L_9 = 1.750 + (1.0 - 0.521) = 2.229 \text{ ft}$$

Section k

$$\bar{y}_k = \frac{12.25}{2} = 6.125 \text{ ft}$$

$$L_{10} = 0.521 + 6.125 = 6.646 \text{ ft}$$

Section l

$$\bar{y}_l = \frac{3.75}{2} = 1.875 \text{ ft}$$

$$L_{11} = 6.125 + 1.875 = 8.00 \text{ ft}$$

Section m

$$\bar{y}_m = \frac{7.00}{2} = 3.50 \text{ ft}$$

$$L_{12} = 1.875 + 3.50 = 5.375 \text{ ft}$$

Section n

$$\bar{y}_n = \frac{3.00}{2} = 1.50 \text{ ft}$$

$$L_{13} = 3.50 + 1.50 = 5.00 \text{ ft}$$

Section o

$$D_{o_s} = 43.625,\ D_{i_s} = 42.125,\ D_{o_s} - D_{i_s} = 1.500 \text{ in.}$$

$$D_{o_s}^2 = 1{,}903.141,\ D_{i_s}^2 = 1{,}774.516,\ D_{o_s}^2 - D_{i_s}^2 = 128.625$$

$$\alpha = \arctan\left(\frac{102.00 - 43.625}{2(240.00)}\right) = 6.934$$

$$\bar{y}_o = \frac{3(36)(128.625) + 8(36)^2(1.50) \tan (6.934)}{6[128.625 + 2(36)(1.50) \tan (6.934)]}$$

$$= 18.556 \text{ in.}$$

$$\bar{y}_o = 1.546 \text{ ft}$$

$$L_{14} = 1.50 + 1.546 = 3.046 \text{ ft}$$

Section p

$$D_{o_s} = 52.381 \text{ in.},\ D_{i_s} = 50.756 \text{ in.},\ D_{o_s} - D_{i_s} = 1.625$$

$$D_{o_s}^2 = 2{,}743.795,\ D_{i_s}^2 = 2{,}576.197,$$

$$D_{o_s}^2 - D_{i_s}^2 = 167.598 \text{ in.}$$

$$\bar{y}_p = \frac{3(204)(167.598) + 8(204)^2(1.625) \tan (6.934)}{6[167.598 + 2(204)(1.625) \tan (6.934)]}$$

$$= 113.044 \text{ in.}$$

$$\bar{y}_p = 9{,}420 \text{ ft}$$

$$L_{15} = (3.00 - 1.546) + 9.420 = 10.874 \text{ ft}$$

$$L_{16} = 17.00 - 9.420 = 7.580 \text{ ft}$$

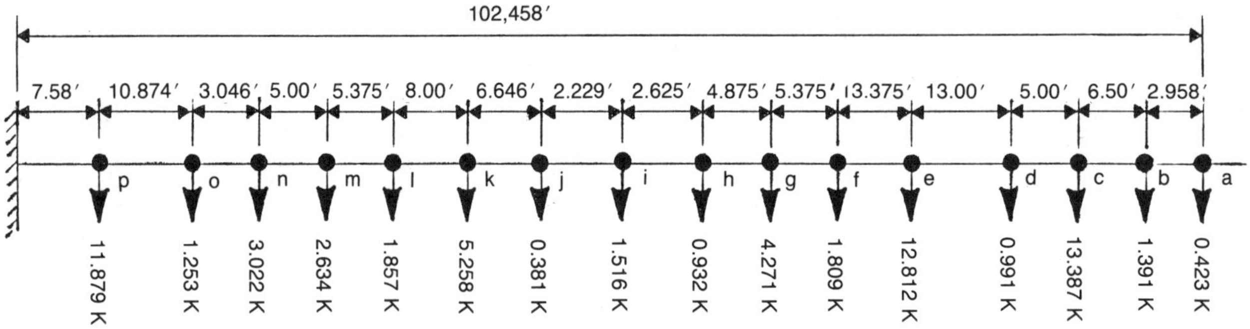

Figure 4-43. The vibration ensemble of lumped masses.

Vortex-Induced Vibration

Referring to Figure 4-43 we have the following:

$M_a = 0$
$M_b = (0.423)(2.958) = 1.251$ kip-ft
$M_c = 1.251 + (1.814)(6.50) = 13.042$ kip-ft
$M_d = 13.042 + (15.201)(5.00) = 89.047$ kip-ft
$M_e = 89.047 + (16.192)(13.00) = 299.543$ kip-ft
$M_f = 299.538 + (29.004)(13.375) = 687.472$ kip-ft
$M_g = 687.467 + (30.813)(5.375) = 853.091$ kip-ft
$M_h = 853.086 + (35.084)(4.875) = 1{,}024.126$ kip-ft
$M_i = 1{,}024.121 + (36.016)(2.625) = 1{,}118.668$ kip-ft
$M_j = 1{,}118.663 + (37.532)(2.229) = 1{,}202.322$ kip-ft
$M_k = 1{,}202.322 + (37.913)(6.646) = 1{,}454.292$ kip-ft
$M_l = 1{,}454.292 + (43.171)(800) = 1{,}799.660$ kip-ft
$M_m = 1{,}799.660 + (45.028)(5.375) = 2{,}041.685$ kip-ft
$M_n = 2{,}041.685 + (47.662)(5.00) = 2{,}279.995$ kip-ft
$M_o = 2{,}279.995 + (50.684)(3{,}046) = 2{,}434.379$ kip-ft
$M_p = 2{,}434.379 + (51.937)(10.874) = 2{,}999.142$ kip-ft
$M_{Base} = 2{,}999.142 + (63.816)(7.58) = 3{,}482.867$ kip-ft

$$\xi_i = \frac{M_i}{E_i I_i}, \quad (30 \times 10^6)\text{lb/in.}^2(144)\text{in.}^6/\text{ft}^2$$

$$= 4.320 \times 10^9 \text{ lb/ft}^2$$

$$\xi_2 = \frac{1{,}251}{(4.32 \times 10^9)(0.756)} = 3.830 \times 10^{-7}$$

$$\xi_3 = \frac{13{,}042}{(4.32 \times 10^9)(6.675)} = 4.523 \times 10^{-7}$$

$$\xi_4 = \frac{89{,}047}{(4.32 \times 10^9)(1.533)} = 1.345 \times 10^{-5}$$

$$\xi_5 = \frac{299{,}543}{(4.32 \times 10^9)(1.533)} = 4.523 \times 10^{-5}$$

$$\xi_6 = \frac{687{,}472}{(4.32 \times 10^9)(1.533)} = 1.038 \times 10^{-4}$$

$$\xi_7 = \frac{853{,}091}{(4.32 \times 10^9)(1.533)} = 1.288 \times 10^{-4}$$

$$\xi_8 = \frac{1{,}024{,}126}{(4.32 \times 10^9)(1.533)} = 1.546 \times 10^{-4}$$

$$\xi_9 = \frac{1{,}118{,}668}{(4.32 \times 10^9)(1.533)} = 1.689 \times 10^{-4}$$

$$\xi_{10} = \frac{1{,}202{,}322}{(4.32 \times 10^9)(1.910)} = 1.457 \times 10^{-4}$$

$$\xi_{11} = \frac{1{,}454{,}292}{(4.32 \times 10^9)(0.917)} = 3.671 \times 10^{-4}$$

$$\xi_{12} = \frac{1{,}799{,}660}{(4.32 \times 10^9)(1.013)} = 4.112 \times 10^{-4}$$

$$\xi_{13} = \frac{2{,}041{,}685}{(4.32 \times 10^9)(1.110)} = 4.258 \times 10^{-4}$$

$$\xi_{14} = \frac{2{,}279{,}995}{(4.32 \times 10^9)(1.208)} = 4.369 \times 10^{-4}$$

$$\xi_{15} = \frac{2{,}434{,}379}{(4.32 \times 10^9)(1.532)} = 3.678 \times 10^{-4}$$

$$\xi_{16} = \frac{2{,}999{,}142}{(4.32 \times 10^9)(7.011)} = 9.902 \times 10^{-5}$$

$$\xi_{17} = \frac{3{,}482{,}867}{(4.32 \times 10^9)(7.011)} = 1.150 \times 10^{-4}$$

$$\boxed{S_i = \left(\frac{\xi_{i+1} + \xi_i}{2}\right) L_i}$$

$$S_1 = \frac{(3.83 \times 10^{-7})(2.958)}{2} = 5.665 \times 10^{-7}$$

$$S_2 = \frac{[(3.83 \times 10^{-7}) + (4.523 \times 10^{-7})]}{2}(6.50)$$

$$= 2.715 \times 10^{-6}$$

$$S_3 = \frac{[(4.523 \times 10^{-7}) + (1.345 \times 10^{-5})]}{2}(5.00)$$

$$= 3.476 \times 10^{-5}$$

$$S_4 = \frac{[(1.345 \times 10^{-5}) + (4.523 \times 10^{-5})]}{2}(13.00)$$

$$= 3.814 \times 10^{-4}$$

$$S_5 = \frac{[(4.523 \times 10^{-4}) + (1.038 \times 10^{-4})]}{2}(13.375)$$

$$= 9.966 \times 10^{-4}$$

$$S_6 = \frac{(1.038 + 1.288)(10^{-4})}{2}(5.375) = 6.251 \times 10^{-4}$$

$$S_7 = \frac{(1.288 + 1.546)(10^{-4})}{2}(4.875) = 6.908 \times 10^{-4}$$

$$S_8 = \frac{(1.546 + 1.689)(10^{-4})}{2}(2.625) = 4.246 \times 10^{-4}$$

$$S_9 = \frac{(1.689 + 1.457)(10^{-4})}{2}(2.229) = 3.506 \times 10^{-4}$$

$$S_{10} = \frac{(1.457 + 3.671)(10^{-4})}{2}(6.646) = 1.704 \times 10^{-3}$$

$$S_{11} = \frac{(3.671 + 4.112)(10^{-4})}{2}(8.00) = 3.113 \times 10^{-3}$$

$$S_{12} = \frac{(4.112 + 4.258)(10^{-4})}{2}(5.375) = 2.249 \times 10^{-3}$$

$$S_{13} = \frac{(4.258 + 4.369)(10^{-4})}{2}(5.00) = 2.157 \times 10^{-3}$$

$$S_{14} = \frac{(4.369 + 3.678)(10^{-4})}{2}(3.046) = 1.226 \times 10^{-3}$$

$$S_{15} = \frac{(3.678 \times 10^{-4} + 9.902 \times 10^{-5})}{2}(10.874)$$

$$= 2.538 \times 10^{-3}$$

$$S_{16} = \frac{(9.902 \times 10^{-5} + 1.150 \times 10^{-4})}{2}(7.58)$$

$$= 8.111 \times 10^{-4}$$

$$\boxed{A_i = S_i + \sum_{j=n}^{i+1} S_j = S_i + A_{i+1}}$$

$$\begin{aligned}
A_1 &= 8.111 \times 10^{-4} \\
A_2 &= 8.111 \times 10^{-4} + 2.538 \times 10^{-3} = 3.349 \times 10^{-3} \\
A_3 &= 3.349 \times 10^{-3} + 1.226 \times 10^{-3} = 4.575 \times 10^{-3} \\
A_4 &= 4.575 \times 10^{-3} + 2.157 \times 10^{-3} = 6.732 \times 10^{-3} \\
A_5 &= 6.732 \times 10^{-3} + 2.249 \times 10^{-3} = 8.981 \times 10^{-3} \\
A_6 &= 8.981 \times 10^{-3} + 3.113 \times 10^{-3} = 1.209 \times 10^{-2} \\
A_7 &= 1.209 \times 10^{-2} + 1.704 \times 10^{-3} = 1.379 \times 10^{-2} \\
A_8 &= 1.379 \times 10^{-2} + 3.506 \times 10^{-4} = 1.414 \times 10^{-2} \\
A_9 &= 1.414 \times 10^{-2} + 4.246 \times 10^{-4} = 1.457 \times 10^{-2} \\
A_{10} &= 1.457 \times 10^{-2} + 6.908 \times 10^{-4} = 1.526 \times 10^{-2} \\
A_{11} &= 1.526 \times 10^{-2} + 6.251 \times 10^{-4} = 1.589 \times 10^{-2} \\
A_{12} &= 1.589 \times 10^{-2} + 9.966 \times 10^{-4} = 1.688 \times 10^{-2} \\
A_{13} &= 1.688 \times 10^{-2} + 3.814 \times 10^{-4} = 1.726 \times 10^{-2} \\
A_{14} &= 1.726 \times 10^{-2} + 3.476 \times 10^{-5} = 1.730 \times 10^{-2} \\
A_{15} &= 1.730 \times 10^{-2} + 2.715 \times 10^{-6} = 1.730 \times 10^{-2} \\
A_{16} &= 1.730 \times 10^{-2} + 5.665 \times 10^{-7} = 1.730 \times 10^{-2}
\end{aligned}$$

$$P_i = \left(\frac{A_i + A_{i+1}}{2}\right) L_i$$

$$P_{16} = \left(\frac{8.111 \times 10^{-4}}{2}\right)(7.58) = 3.074 \times 10^{-3}$$

$$P_{15} = \left(\frac{8.111 \times 10^{-4} + 3.349 \times 10^{-3}}{2}\right)(10.874)$$

$$= 2.262 \times 10^{-2}$$

$$P_{14} = \left(\frac{3.349 \times 10^{-3} + 4.575 \times 10^{-3}}{2}\right)(3.046)$$

$$= 1.207 \times 10^{-2}$$

$$P_{13} = \left(\frac{4.575 \times 10^{-3} + 6.732 \times 10^{-3}}{2}\right)(5.00)$$

$$= 2.827 \times 10^{-2}$$

$$P_{12} = \left(\frac{6.732 \times 10^{-3} + 8.981 \times 10^{-3}}{2}\right)(5.375)$$

$$= 4.223 \times 10^{-2}$$

$$P_{11} = \left(\frac{8.981 \times 10^{-3} + 1.209 \times 10^{-2}}{2}\right)(8.00)$$

$$= 8.428 \times 10^{-2}$$

$$P_{10} = \left(\frac{1.379 \times 10^{-2} + 1.414 \times 10^{-2}}{2}\right)(6.646)$$

$$= 9.281 \times 10^{-2}$$

$$P_{9} = \left(\frac{1.414 \times 10^{-2} + 1.457 \times 10^{-2}}{2}\right)(2.229)$$

$$= 3.200 \times 10^{-2}$$

$$P_{8} = \left(\frac{1.457 \times 10^{-2} + 1.526 \times 10^{-2}}{2}\right)(2.625)$$

$$= 3.915 \times 10^{-2}$$

$$P_{7} = \left(\frac{1.526 \times 10^{-2} + 1.589 \times 10^{-2}}{2}\right)(4.875)$$

$$= 7.593 \times 10^{-2}$$

$$P_{6} = \left(\frac{1.589 \times 10^{-2} + 1.688 \times 10^{-2}}{2}\right)(5.375)$$

$$= 8.807 \times 10^{-2}$$

$$P_{5} = \left(\frac{1.688 \times 10^{-2} + 1.726 \times 10^{-2}}{2}\right)(13.375)$$

$$= 2.283 \times 10^{-1}$$

$$P_{4} = \left(\frac{1.726 \times 10^{-2} + 1.730 \times 10^{-2}}{2}\right)(13.00)$$

$$= 2.246 \times 10^{-1}$$

$$P_{3} = \left(\frac{1.730 \times 10^{-2} + 1.730 \times 10^{-2}}{2}\right)(5.00)$$

$$= 8.650 \times 10^{-2}$$

$$P_{2} = \left(\frac{1.730 \times 10^{-2} + 1.730 \times 10^{-2}}{2}\right)(6.50)$$

$$= 1.125 \times 10^{-1}$$

$$P_{1} = \left(\frac{1.730 \times 10^{-2} + 1.730 \times 10^{-2}}{2}\right)(2.958)$$

$$= 5.117 \times 10^{-2}$$

$$\begin{aligned}
\mu_{16} &= 3.074 \times 10^{-3} \text{ ft} = y(16) = 0.037 \text{ in.} \\
\mu_{15} &= 3.074 \times 10^{-3} + 2.262 \times 10^{-2} \\
&= 2.569 \times 10^{-2} \text{ ft} = y(15) = 0.308 \text{ in.} \\
\mu_{14} &= 2.569 \times 10^{-2} + 1.207 \times 10^{-2} \\
&= 3.776 \times 10^{-2} \text{ ft} = y(14) = 0.453 \text{ in.} \\
\mu_{13} &= 3.776 \times 10^{-2} + 2.827 \times 10^{-2} \\
&= 6.603 \times 10^{-2} \text{ ft} = y(13) = 0.792 \text{ in.} \\
\mu_{12} &= 6.603 \times 10^{-2} + 4.223 \times 10^{-2} \\
&= 1.083 \times 10^{-1} \text{ ft} = y(12) = 1.300 \text{ in.} \\
\mu_{11} &= 1.083 \times 10^{-1} + 8.428 \times 10^{-2} \\
&= 1.925 \times 10^{-1} \text{ ft} = y(11) = 2.310 \text{ in.} \\
\mu_{10} &= 1.925 \times 10^{-1} + 9.281 \times 10^{-2} \\
&= 2.854 \times 10^{-1} \text{ ft} = y(10) = 3.425 \text{ in.} \\
\mu_{9} &= 2.854 \times 10^{-1} + 3.200 \times 10^{-2} \\
&= 3.174 \times 10^{-1} \text{ ft} = y(9) = 3.809 \text{ in.} \\
\mu_{8} &= 3.174 \times 10^{-1} + 3.915 \times 10^{-2} \\
&= 3.565 \times 10^{-1} \text{ ft} = y(8) = 4.278 \text{ in.} \\
\mu_{7} &= 3.565 \times 10^{-1} + 7.593 \times 10^{-2} \\
&= 4.324 \times 10^{-1} \text{ ft} = y(7) = 5.189 \text{ in.} \\
\mu_{6} &= 4.324 \times 10^{-1} + 8.807 \times 10^{-2} \\
&= 5.205 \times 10^{-1} \text{ ft} = y(6) = 6.246 \text{ in.} \\
\mu_{5} &= 5.205 \times 10^{-1} + 2.283 \times 10^{-1} \\
&= 7.488 \times 10^{-1} \text{ ft} = y(5) = 8.986 \text{ in.} \\
\mu_{4} &= 7.488 \times 10^{-1} + 2.246 \times 10^{-1} \\
&= 9.734 \times 10^{-1} \text{ ft} = y(4) = 11.681 \text{ in.} \\
\mu_{3} &= 9.734 \times 10^{-1} + 8.650 \times 10^{-2} \\
&= 1.060 \times 10^{-1} \text{ ft} = y(3) = 12.720 \text{ in.} \\
\mu_{2} &= 1.060 + 1.125 \times 10^{-1} \\
&= 1.172 \text{ ft} = y(2) = 14.064 \text{ in.} \\
\mu_{1} &= 1.172 + 5.117 \times 10^{-2} \\
&= 1.224 \text{ ft} = y(1) = 14.683 \text{ in.}
\end{aligned}$$

Section weights and displacements for computing the tower's period of vibration are listed in Table 4-18.

The first period of vibration, T, is determined as follows:

$$T = 2\pi \sqrt{\frac{\sum Wy^2}{g \sum Wy}} \tag{4-98}$$

Table 4-18
Tower Vibration Deflections

Section	Deflection y (in.)	W (lb)	Wy (in.-lb)	Wy^2 (lb-in.^{2})
a	14.683	423	6,210.909	91,194.777
b	14.064	1,391	19,563.024	275,134.370
c	12.720	13,387	170,282.640	2,165,995.181
d	11.681	911	11,575.871	135,217.749
e	8.986	12,812	115,128.632	1,034,545.887
f	6.246	1,809	11,299.014	70,573.641
g	5.189	4,271	22,162.219	114,999.754
h	4.278	932	3,987.096	17,056.797
i	3.809	1,516	5,774.444	21,994.857
j	3.425	381	1,304.925	4,469.368
k	2.310	5,258	12,145.980	28,057.214
l	1.300	1,857	2,414.100	3,138.330
m	0.792	2,634	2,086.128	1,652.213
n	0.453	3,022	1,368.966	620.142
o	0.308	1,253	385.924	118.865
p	0.037	20,493	439.523	16.262
			$\sum Wy = 386{,}129.395$	$\sum Wy^2 = 3{,}964{,}785.407$

$$T = 2\pi\sqrt{\frac{3{,}964{,}785.407}{(386.4)(386{,}129.395)}} = 1.024 \text{ sec/cycle}$$

$f = 1/T = 0.976$ Hz ≈ 0.981 Hz = the initial assumed value

Considering the upper third of the tower as being the effective length for vortex shedding the first critical wind velocity is as follows:

$$L = \frac{104.292}{3} = 34.764 \text{ ft} \Rightarrow d$$

$$= \frac{54.00 + 1.00}{12} = 4.583 \text{ ft}$$

$$V_1 = \frac{3.40\ d}{T} = \frac{(3.40)(4.583)\text{ ft}}{1.024\ \frac{\text{sec}}{\text{cycle}}}$$

$$= 15.217\ \frac{\text{ft}}{\text{sec}} = 10.375 \text{ mph}$$

The second critical wind velocity, V_2, is

$$V_2 = 6.25\ V_1 = 65.163 \text{ mph}$$

From Equation 4-94a and Table 4-13 we have

$$F_v = 0.00086(0.6)(60)(4.583)(104.292)(10.375)^2$$
$$= 1{,}592.930 \text{ lb}$$

To determine if we have a problem with vortex-induced vibrations, we must compare the force amplitude of 1,592.93 lb against the corresponding maximum wind force amplitude for the same region (either top 1/3 or 1/4 of tower—in this case, the top 1/3). Using Figure 4-39 we have the following:

L = 34.764 ft, from above

104.292 ft–34.764 ft = 69.528 ft

$$q_n = (1.726)(35.659) = 61.547 \text{ lb/ft}^2$$

$$q_{n-1} = (1.726)\left(\frac{69.528}{900}\right)^{0.286}(66.048)$$

$$q_{n-1} = 54.808 \text{ lb/ft}^2$$

$$A_f = \frac{(34.764)[(7)(61.547) + (2)(54.808)]}{(2 + 7)} = 2{,}087.559$$

$$F = (2087.559)\left(\frac{127.00}{12}\right)(0.6) = 13{,}255.999 \text{ lb}_f$$

or

$$F_{wind} = 13{,}256 \text{ lb}_f > 1{,}608.56 \text{ lb}_f = F_v$$

Since $F_{wind} > F_{vibration}$, the wind stresses are greater than those at resonance vibration, so no further vibration analysis is required.

If the vibration amplitude force had been greater than the maximum wind force, further investigation would have been required. Dynamic stresses at the critical wind velocity can be approximated by taking the ratio of the vibration force amplitude, F_v, to the maximum wind force amplitude, F_W, and multiplying this ratio by the bending stress term in Equation 4-29. The pressure stress, which is a primary stress, and the weight load stress in tension are unaffected. Shear is not considered in the equation because it is almost always negligible. Defining the ratio of vibration force amplitude to the maximum wind force amplitude as R, Equation 4-29 becomes

$$\sigma_D = \pm\left(\frac{PD}{4t}\right) \pm R\left(\frac{16\,D_oM}{\pi t(D_o + D_i)(D_o^2 + D_i^2)}\right) \pm \left(\frac{2W}{\pi t(D_o + D_i)}\right)$$

where σ_D = dynamic stress, psi
σ_D < static allowable stress

Staley and Graven [15] state that when dynamic stresses are combined with axial compressive stress, the result can be compared to the allowable static stress. The same is generally true for tension, but one must be cognizant of discontinuity stresses at the locations or irregular changes of geometry, such as welds. The latter can be avoided by using stiffening rings. Certainly a more accurate and detailed analysis, such as the octahedral shear stress theory of yield, can be used, but such a detailed analysis can be avoided in most tower designs. A detailed fatigue analysis is mandatory in many applications and should always be used in case of doubt. Weaver [24] discovered in wind-tunnel tests that vortex shedding cannot presently be analyzed as a response spectra because of its random nature and unpredictable motion. This greatly complicates the study of vortex excitation by use of finite element methods, but efforts are being made.

Finally, if the Reynolds number is greater than approximately 350,000, a vibration analysis is not required, because in these regions the vortices break-up. In our case,

$$N_{Re} = \frac{D_E V \rho}{\mu}$$

where D_E = effective wind diameter at top 1/3 or 1/4 of tower = 127.0 in.
V = critical wind velocity = 15.292 ft/sec
$\rho \cong 0.071$ lb_m/ft^3
$\mu \cong 1.285 \times 10^{-5}$ lb_m/ft-sec

$$N_{Re} = \frac{\left(\frac{127.00}{12}\right)\text{ft}\,(15.292)\left(\frac{\text{ft}}{\text{sec}}\right)(0.071)\frac{\text{lb}_m}{\text{ft}^3}}{(1.285 \times 10^{-5})\frac{\text{lb}_m}{\text{ft-sec}}} = 894{,}215$$

Since $N_{Re} = 894{,}215 > 350{,}000$, a vibration analysis is not required since we are outside the range of vortex formation.

Vortex formation has been observed at $N_{Re} \geq 3.5 \times 10^6$, but wind velocities encountered would not cause Reynolds numbers that high.

Equivalent Diameter Approach versus ANSI–A58.1–1982

In determining wind loadings in this example we used the formulation to compute wind forces:

$$F = q_z\bar{G}C_fA_f \tag{4-81}$$

in which A_f is computed using the total width of the tower, insulation, ladders, platforms, and attached piping as an equivalent or effective diameter of a cylinder, called the effective or equivalent cylinder. This equivalent cylinder represents the total wind area. Such an analysis is called a quasi-analysis, because it is not exact. The equivalent cylinder concept used for conical sections is similar when compared to the exact analysis of a cone.

The ANSI–A58.1–1982 uses a more refined and equally complex analysis to determine the wind loadings. The relationship used for wind force is as follows:

$$F = q_ZG_ZC_fA_{Cf} + q_Z\bar{G}C_fA_{Tf}$$

where G_Z = gust response factor for cladding and components calculated at height Z
A_{Cf} = area of insulation (cladding) of tower, and all external attachments such as platforms, ladders, and piping that resist wind
A_{Tf} = area of the tower shell itself that resists wind

The term G_Z is given in Table 8 of ANSI–A58.1–1982 that is determined by the following expressions:

$$G_Z = 0.65 + 3.65\ T_Z$$

$$T_Z = \frac{2.35\ (D_o)^{0.5}}{(Z/30)^{1/\alpha}}$$

For category A,

$$D_o = 0.025,\ \alpha = 3.0,\ Z_g = 1{,}500$$

For category B,

$$D_o = 0.010,\ \alpha = 4.5,\ Z_g = 1{,}200$$

For category C,

$$D_o = 0.005,\ \alpha = 7.0,\ Z_g = 900$$

For category D,

$$D_o = 0.003,\ \alpha = 10.0,\ Z_g = 700$$

G_Z, which is used for cladding and components, varies with height and is a parabolic distribution. The term $\bar{G}$ is used with the tower shell and only is constant along the height of the tower.

Comparing the two methods we set Equation 4-81 equal to the comparable expression given by ANSI–A58.1–1982.

$$F = q_Z\bar{G}C_fA_f = q_ZG_ZC_fA_f + q_Z\bar{G}C_fA_f$$

We now define the following variables:

$$\frac{A_{Cf}}{A_f} = x \text{ and } \frac{A_{Tf}}{A_f} = y$$

from which

$$q_Z\bar{G}C_fA_f = q_ZC_fA_f(xG_Z + y\bar{G})$$
$$\bar{G} = xG_Z + y\bar{G}$$
$$\bar{G}(1-y) > xG_Z$$

Now, for many, if not almost all cases,

$$\bar{G} > G_Z$$

This is certainly true as one moves up the tower in computing G_Z. It can be safely said that

$$\bar{G} > G_{Z(avg)}$$

where $G_{Z(avg)}$ = average value of G_Z across the height of the tower.

Also, rarely is y as great as 0.5, making the inequality above more credible.

After applying the real numbers for several cases, it is seen that the equivalent diameter method using the ANSI–A58.1–1982 gust factor for flexible structures, $\bar{G}$, is more conservative than the ANSI method of using the two gust response factors G_Z and $\bar{G}$. Thus, being more conservative than ANSI–A58.1–1982, one meets the minimum requirements of the standard, as it is stated in the title, "minimum design loads for buildings and other structures." Certainly, using the formula for lattice structures, Equation 4-83, is a conservative approach.

For designing a tower without a computer software package, the equivalent diameter method is recommended. In such a design, one is faced with numerous calculations, which leads to a greater possibility of error. Also, the use of two gust factors with one varying in height adds considerable complexity to the problem. When using a high-speed electronic computer the use of two gust factors would be a very good method to use, although cumbersome to verify. Certainly, some could argue that with less conservatism a cheaper vessel is produced. Such a consideration must be analyzed in each separate circumstance. For some, the additional man-hours may offset the economics of the vessel or time may be the ruling criterion.

EXAMPLE 4-3: SEISMIC ANALYSIS OF A VERTICAL TOWER

A client has a vertical tower that is to be moved from a plant in Jackass Flats, Nevada to a location northeast of Los Angeles, California. The vessel must be analyzed for seismic zone 4 to determine if it can be moved. This result is to be compared to a wind analysis for an 80-mph wind.

Seismic Analysis

$$V = ZIKCSW \tag{4-106}$$

For zone 4, Z = 1, I = 1, K = 2.0, W = 15,571 lb

Since the tower is not of uniform thickness, equation 4-108 cannot be used. Either the Rayleigh equation (Equation 4-97) or a modified form of the Rayleigh Equation, the Mitchell Equation (Equation 4-112), can only be used. For illustration purposes the Mitchell Equation will be applied and then compared to the more accurate Rayleigh method.

Using values in Table 4-15 we determine the values to be used in Equation 4-112. Connecting piping exerts a concentrated load of 2.7 kips at the support point midway in Section ②–③, using the values in Table 4-15 we construct Table 4-19,

$$\text{where } A = \sum (W\Delta\alpha + w\beta/H)$$

$$B = \sum \bar{E}\left(\frac{D}{10}\right)^3 t\Delta\gamma$$

Using Table 4-18 we have

$$T = \left(\frac{72}{100}\right)^2 \sqrt{\frac{0.440}{0.261}} = 0.673 \text{ sec/cycle}$$

Using the more accurate Rayleigh method, Equation 4-97, the value of T is

$$T = 0.734 \text{ sec/cycle}$$

in which the Mitchell Equation is in 8.3% error (which is quite normal). For application of the Rayleigh Equation see Examples 4-2 and 4-4.

Now, we must solve for the bending moments induced by the seismic forces. First we find the base shear using Equation 4-106. To accomplish this we have the following:

$$\frac{t_{vessel}}{t_{skirt}} = \frac{1.125}{1.0} = 1.125 < 1.5 \therefore K = 2.0$$

$$\text{Flexibility factor} = C = \frac{1}{15(T)^{0.5}} = 0.078$$

The characteristic site period, T_s, is determined by a soils consultant to fall within the following range:

$$0.5 \leqslant T_s \leqslant 0.55$$

To be safe, we will use the lower value of 0.5. Now,

$$\frac{T}{T_s} = \frac{0.734}{0.50} = 1.468 > 1.0 \quad \text{in which Equation 4-113b}$$

applies. Thus, we have

$$S = 1.2 + 0.6\left(\frac{T}{T_s}\right) - 0.3\left(\frac{T}{T_s}\right)^2$$

$$S = 1.2 + 0.6(1.468) - 0.3(1.468)^2$$

$$S = 1.434$$

Solving for V we have

$$V = (1)(1)(2.0)(0.078)(1.434)(15{,}571) = 3{,}484.0 \text{ lb}$$

Using Equation 4-114 to find F_t, we have the following:

$$\frac{h}{D} = \frac{72}{2} = 36 \Rightarrow F_t = 0.15V$$

$$F_t = (0.15)(3{,}484.0) = 522.60$$

From Equation 4-115 we obtain

$$F_x = (V - F_t)\frac{W_y h_y}{\sum_{i=1}^{n} W_i h_i} = (3{,}484 - 522.60)\frac{W_y h_y}{515{,}380}$$

$$= 0.0057\ W_h h_y$$

To solve this equation we must set up the table shown in Table 4-20. After determining the values for W, h_y, $W_y h_y$, F_x, and V_x we solve for the seismic moments using Equation 4-116:

$M = V_{x_i} L_{y_{i-1}} + F_x C_i$	$\sum M$
$M_2 = (0.30)(5.083) + (0.49)(2.542) = 2.770$	2.770
$M_3 = (0.49)(7.50) + (1.19)(3.75) = 8.138$	10.908
$M_4 = (1.68)(5.417) + (0.18)(2.708) = 9.588$	20.496
$M_5 = (1.86)(2.00) + (0.09)(1.00) = 3.810$	24.306
$M_6 = (1.95)(8.00) + (0.28)(4.00) = 16.720$	41.026
$M_7 = (2.23)(4.00) + (0.16)(2.00) = 9.240$	50.266
$M_8 = (2.39)(10.00) + (0.32)(5.00) = 25.500$	75.766
$M_9 = (2.71)(10.00) + (0.22)(5.00) = 28.200$	103.966
$M_{10} = (2.93)(9.083) + (0.23)(4.542) = 27.658$	131.624
$M_{11} = (3.16)(8.00) + (0.10)(4.00) = 25.680$	157.304
$M_{12} = (3.26)(2.917) + (0.01)(1.458) = 9.524$	166.828

The wind moment for an 80-mph wind was calculated to be 106,716 ft-lb. Since 166,828 ft-lb > 106,716 ft-lb seismic phenomena govern.

The skirt and base plate analysis is identical for seismic and wind analyses. Just as in Example 4-2, the seismic forces and moments are used instead of the wind forces and moments. In the case of this tower a thicker base plate was welded on, the number of gusset plates were doubled, and anchor bolts of a high strength alloy were used to meet the seismic criteria. In an earthquake zone other than zero, a comparison of seismic to wind should always be made.

Table 4-19
Numeric Integration of Period of Vibration, T sec/cycle

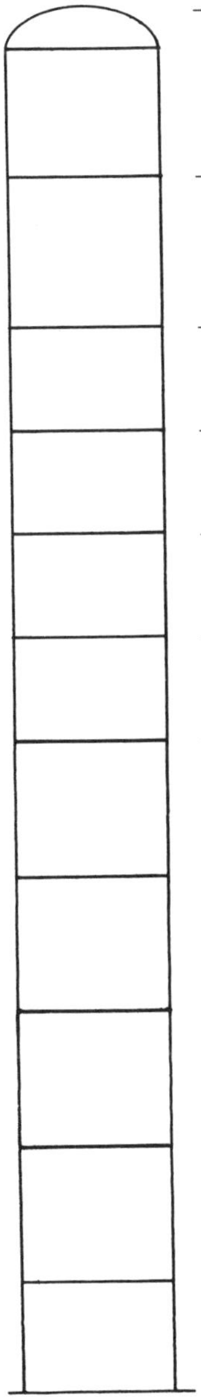

W kips/ft	$\frac{h_y}{H}$	α	$\Delta\alpha$	β	$W\Delta\alpha$ + $W\Delta\beta/H$	γ	$\Delta\gamma$	$\bar{E}\left(\frac{D}{10}\right)^3 t\Delta\gamma$
	1.00	2.103				1.000	1.000	
0.091			0.597		0.054			0
	0.923	1.506				1.000	1.000	
2.7	0.878		0.567	5.840	0.219			0
0.079					0.045			
	0.820	0.939				1.000	1.000	
0.100			0.329		0.033			1.234×10^{-6}
	0.745	0.610				0.998	0.998	
0.142			0.067		0.010			9.541×10^{-5}
	0.717	0.543				0.997	0.997	
0.125			0.278		0.035			0.0010
	0.607	0.265				0.986	0.986	
0.161			0.097		0.016			0.0015
	0.552	0.168				0.973	0.973	
0.161			0.124		0.020			0.0080
	0.414	0.044				0.904	0.904	
0.155			0.037		0.006			0.0211
	0.276	0.007				0.763	0.763	
0.285			0.007		0.002			0.0649
	0.151	0.0004				0.504	0.504	
0.312			0.0004		0.0001			0.1175
	0.040	0				0.160	0.160	
0.380			0		0			0.0472
	0.000	0				0	0	
					A = 0.440			B = 0.261

Table 4-20
Wind Load Distribution

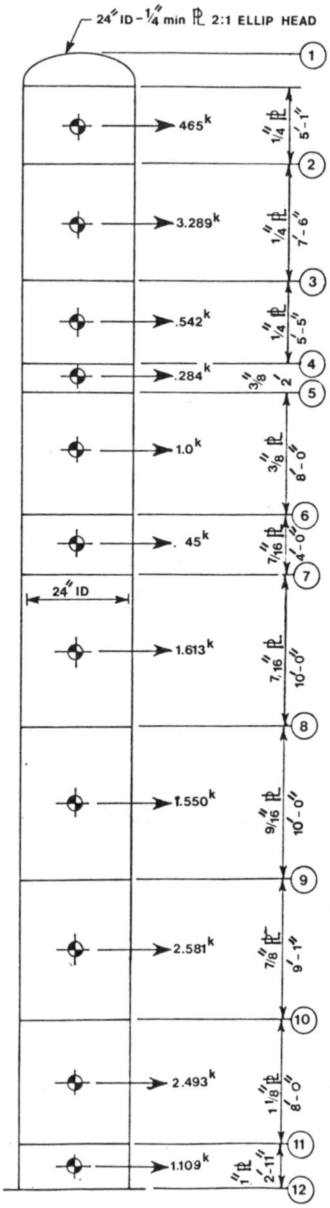

W kips	h_x	$W_x h_x$	F_x	V_x	M kip-ft	$\sum M_y$ kip-ft
	72		0.30	0.30		
0.465	69.5	32.32	0.19			
				0.49	2.770	2.770
3.289	63.2	207.86	1.19			
				1.68	8.138	10.908
0.542	56.7	30.73	0.18			
				1.86	9.588	20.496
0.284	53.0	15.05	0.09			
				1.95	3.810	24.306
1.0	48.0	48.00	0.28			
				2.23	16.720	41.026
0.645	42.0	27.09	0.16			
				2.39	9.240	41.026
1.613	35.0	56.46	0.32			
				2.71	25.500	75.766
1.550	25.0	38.75	0.22			
				2.93	28.200	103.966
2.581	15.5	40.01	0.23			
				3.16	27.658	131.624
2.493	7.0	17.45	0.10			
				3.26	25.680	157.304
1.109	1.5	1.66	0.01	–	9.524	166.828
15.571		515.38	3.270	3.27		

EXAMPLE 4-4: VIBRATION ANALYSIS FOR TOWER WITH LARGE VORTEX-INDUCED DISPLACEMENTS

A phone call from a plant manager reveals that an existing tower needs to be analyzed for wind vibrations. The tower was designed, built and installed overseas and is vibrating so badly all the natives drove off the plant site in fear of the tower falling over.

The tower with the appropriate wind load distribution is shown in Figure 4-44. The tower is divided into wind zones at 30 ft, 40 ft, and 75 ft and according to shell diameter and thickness. The variation of wind zones based on the shell diameter and thickness is necessary since the tower's section moment of inertia will vary.

To begin the analysis we start with defining the effective diameter of each section as illustrated in Figure 4-45. Thus we have the following:

Zone 1—Sections 7, 8, and 9

$$D_e = [32 \text{ in.} + 2\,(4) \text{ in.}] + [6.625 \text{ in.} + 2(3.5) \text{ in.}] + [2.375 \text{ in.} + 2(3.0) \text{ in.}] + [4.5 \text{ in.} + 2(3) \text{ in.}]$$

$$D_e = 40.00 + 13.625 + 8.375 + 10.50$$

$$D_e = 72.5 \text{ in.} = 6.042 \text{ ft}$$

Zone 2—Section 6

$$D_e = [32 \text{ in.} + 8 \text{ in.}] + [4.5 \text{ in.} + 2(2) \text{ in.}] = 48.5 \text{ in.} = 4.042 \text{ ft}$$

Zone 3—Section 5

$$D_e = [25.25 \text{ in.} + 2(2.5) \text{ in.}] + [4.5 \text{ in.} + 2(2.5) \text{ in.}] + [3.5 \text{ in.} + 2(3.5) \text{ in.}]$$

$$D_e = 30.25 \text{ in.} + 9.50 \text{ in.} + 10.50 \text{ in.}$$

$$D_e = 50.25 \text{ in.} = 4.188 \text{ ft}$$

Zone 4—Section 4

$$D_e = 25.25 \text{ in.} + 2(2.5) \text{ in.} = 30.25 \text{ in.}$$

$$D_e = 2.521 \text{ ft}$$

Zone 5—Section 3

$$D_e = [25.25 \text{ in.} + 2(4.5) \text{ in.}] + [4.5 \text{ in.} + 2(2.5) \text{ in.}] + [3.5 \text{ in.} + 2(3.5) \text{ in.}]$$

$$D_e = 34.25 \text{ in.} + 9.50 \text{ in.} + 10.50 \text{ in.}$$

$$D_e = 54.25 \text{ in.} = 4.521 \text{ ft}$$

Zone 6—Section 2

$$D_e = [24.50 \text{ in.} + 2(4.5) \text{ in.}] + [3.50 \text{ in.} + 2(4) \text{ in.}]$$

$$D_e = 33.50 \text{ in.} + 11.50 \text{ in.}$$

$$D_e = 45.00 \text{ in.} = 3.75 \text{ ft}$$

Zone 7—Section 1

$$D_e = [36.625 \text{ in.} + 2(4.5) \text{ in.}] + [6.625 \text{ in.} + 2(5) \text{ in.}] + [3.50 \text{ in.} + 2(4.5) \text{ in.}] + [6.625 \text{ in.} + 2(5) \text{ in.}]$$

$$D_e = 46.625 \text{ in.} + 16.625 \text{ in.} + 12.50 \text{ in.} + 16.625 \text{ in.}$$

$$D_e = 104.875 \text{ in.} = 8.740 \text{ ft}$$

Moments of Inertia

$$I = \frac{\pi}{64}(D_o^4 - D_i^4)$$

$$I_1 = \frac{\pi}{64}[(36.625)^4 - (36.000)^4] = 5{,}876.389 \text{ in.}^4$$

$$= 0.283 \text{ ft}^4$$

Transition Piece—Section 2

Referring to Figure 4-45,

$$r_{eq} = \left(\frac{18.375 + 12.375}{2 \cos 26.565°}\right)$$

$$r_{eq} = 17.190 \Rightarrow D_{eq} = 34.380 \text{ in.}$$

$$I_2 = \frac{\pi}{64}[(34.380)^4 - (33.630)^4]$$

$$I_2 = 5791.250 \text{ in.}^4 = 0.279 \text{ ft}^4$$

$$I_3 = \frac{\pi}{64}[(24.50)^4 - (24.00)^4] = 1{,}400.172 \text{ in.}^4$$

$$= 0.068 \text{ ft}^4$$

$$I_4 = \frac{\pi}{64}[(24.75)^4 - (24.00)^4] = 2{,}133.181 \text{ in.}^4$$

$$= 0.103 \text{ ft}^4$$

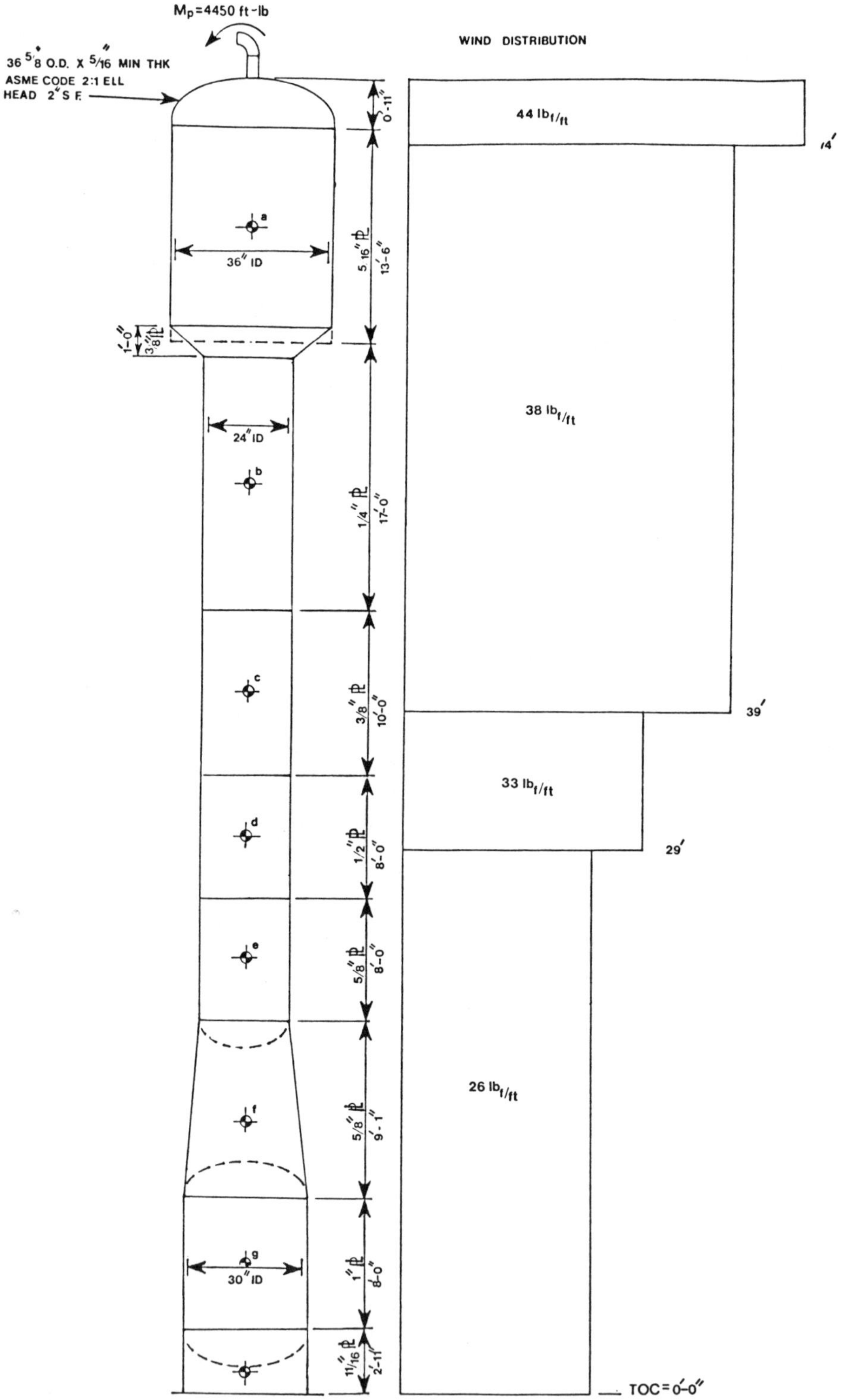

Figure 4-44. Tower wind ensemble.

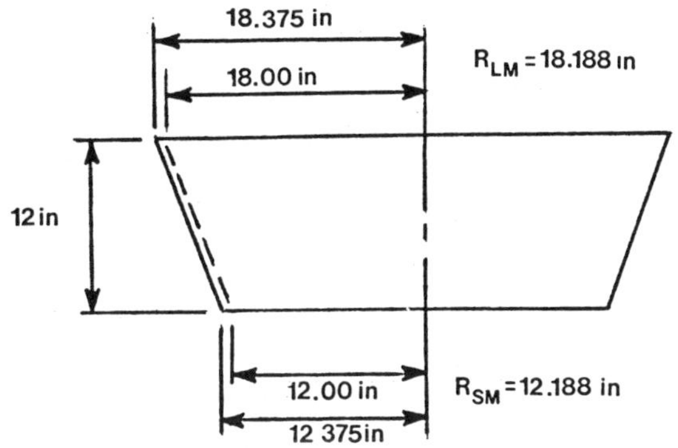

Figure 4-45. Transition piece of section 2 of Figure 4-44.

$$I_5 = \frac{\pi}{64}[(25.00)^4 - (24.00)^4] = 2{,}888.744 \text{ in.}^4$$

$$= 0.139 \text{ ft}^4$$

$$I_6 = \frac{\pi}{64}[(25.25)^4 - (24.00)^4] = 3{,}667.316 \text{ in.}^4$$

$$= 0.177 \text{ ft}^4$$

Section 7

Referring to Figure 4-46

$$r_{eq} = \left(\frac{15.625 + 12.625}{2 \cos 4.764°}\right) = 14.174 \text{ in.}$$

$$\Rightarrow D_{eq} = 28.348 \text{ in.}$$

$$I_7 = \frac{\pi}{64}[(28.348)^4 - (27.723)^4]$$

$$I_7 = 2{,}704.843 \text{ in.}^4 = 0.130 \text{ ft}^4$$

$$I_8 = \frac{\pi}{64}[(32.00)^4 - (30.00)^4] = 11{,}711.072 \text{ in.}^4$$

$$= 0.565 \text{ ft}^4$$

$$I_9 = \frac{\pi}{64}[(32.00)^4 - (30.625)^4] = 8{,}292.684 \text{ in.}^4$$

$$= 0.400 \text{ ft}^4$$

Wind Moment Calculations

Sections 1 and 2

$$M_2 = (788.425)\left(\frac{3.417}{2} + 13.5\right) + (2{,}690.172)\left(\frac{13.5}{2}\right) + 4{,}450$$

$$M_2 = 11{,}990.762 + 18{,}158.661 + 4{,}450.00$$
$$M_2 = 34{,}599.423 \text{ ft-lb}$$

Sections 2 and 3

$$M_3 = (788.425)(15.209 + 17.00) + (2{,}690.172)(6.75 + 17.00) + 4{,}450 + (1{,}453.50)\left(\frac{17}{2}\right)$$

$$M_3 = 25{,}394.381 + 63{,}891.585 + 4{,}450 + 12{,}354.75$$
$$M_3 = 106{,}090.716 \text{ ft-lb}$$

Section 3 and 4

$$M_4 = (788.425)(32.209 + 10.00) + (2{,}690.172)(23.75 + 10.00) + (1{,}453.50)(10.00) + (721.552)\left(\frac{10}{2}\right) + (268.547)\left(\frac{3.0}{2}\right)$$

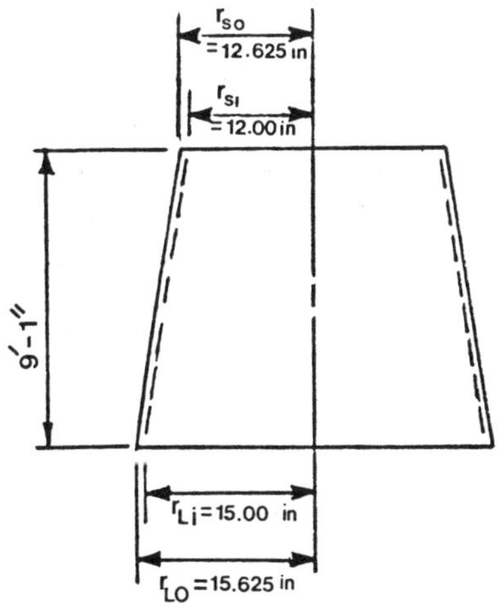

Figure 4-46. Section 7 of Figure 4-44.

$$M_4 = 33{,}278.631 + 90{,}793.305 + 4{,}450 + 14{,}535.00 + 3{,}607.76 + 402.821$$

$$M_4 = 147{,}067.517 \text{ ft-lb}$$

Sections 4 and 5

$$M_5 = (788.425)(42.209 + 8.0) + (2{,}690.172)(43.75 + 8.0) + 4{,}450 + (1{,}453.50)(10 + 8.0) + (721.552)(5 + 8) + (268.547)(1.5 + 8) + (349.411)\left(\frac{8}{2}\right) + (39.328)\left(\frac{1}{2}\right)$$

$$M_5 = 39{,}586.031 + 139{,}216.401 + 4{,}450 + 26{,}163.00 + 9{,}380.176 + 2{,}551.197 + 1{,}397.644 + 19.664$$

$$M_5 = 222{,}764.113 \text{ ft-lb}$$

Section 5 and 6

$$M_6 = (788.425)(50.209 + 8.0) + (2{,}690.172)(51.75 + 8.0) + 4{,}450 + (1{,}453.50)(18.0 + 8.0) + (721.552)(13 + 8) + (268.547)(9.5 + 8) + (349.411)(4 + 8) + (39.328)(0.5 + 8) + (522.662)\left(\frac{8}{2}\right)$$

$$M_6 = 45{,}893.431 + 160{,}737.777 + 4{,}450 + 37{,}791.00 + 15{,}152.592 + 4{,}699.573 + 4{,}192.932 + 334.288 + 2{,}090.648$$

$$M_6 = 270{,}892.241 \text{ ft-lb}$$

Sections 6 and 7

$$M_7 = (788.425)(58.209 + 9.083) + (2{,}690.172)(59.75 + 9.083) + 4{,}450 + (1{,}453.50)(26.0 + 9.083) + (721.552)(21 + 9.083) + (268.547)(17.5 + 9.083) + (349.411)(12 + 9.083) + (39.328)(8.5 + 9.083) + (522.662)\left(\frac{9.083}{2}\right)$$

$$M_7 = 53{,}054.695 + 185{,}172.609 + 4{,}450 + 50{,}993.141 + 21{,}706.449 + 7{,}138.785 + 7{,}366.632 + 691.504 + 2{,}373.669$$

$$M_7 = 332{,}947.484 \text{ ft-lb}$$

Sections 7 and 8

$$M_8 = (788.425)(67.292 + 8) + (2{,}690.172)(68.833 + 8) + 4{,}450 + (1{,}453.50)(35.083 + 8) + (721.552)(30.083 + 8) + (268.547)(26.583 + 8) + (349.411)(21.083 + 8) + (39.328)(17.583 + 8) + (522.662)(4.542 + 8) + (754.042)\left(\frac{8}{2}\right)$$

$$M_8 = 59{,}362.095 + 206{,}693.985 + 4{,}450 + 62{,}621.141 + 27{,}478.865 + 9{,}287.161 + 10{,}161.920 + 1{,}006.128 + 6{,}555.227 + 3{,}016.168$$

$$M_8 = 390{,}632.690 \text{ ft-lb}$$

Sections 8 and 9

$$M_9 = (788.425)(75.292 + 2.917) + (2{,}690.172)(76.833 + 2.917) + 4{,}450 + (1{,}453.50)(43.083 + 2.917) + (721.552)(38.083 + 2.917) + (268.547)(34.583 + 2.917) + (349.411)(29.083 + 2.917) + (39.328)(25.583 + 2.917) + (522.662)(12.542 + 2.917) + (754.042)(4 + 2.917) + (274.942)\left(\frac{2.917}{2}\right)$$

$$M_9 = 61{,}661.931 + 214{,}541.217 + 4{,}450 + 66{,}861.00 + 29{,}583.632 + 10{,}070.513 + 11{,}181.152 + 1{,}120.848 + 8{,}079.832 + 5{,}215.709 + 401.003$$

$$M_9 = 413{,}166.837 \text{ ft-lb}$$

Wind Deflections

$$y_1 = \frac{(13)^2}{(4.176 \times 10^9)(0.283)}\left[\frac{(3{,}478.597)(13)}{8} + \frac{4{,}450}{2}\right]$$

$$= 0.00113 \text{ ft}$$

$$y_2 = \frac{(17)^2}{(4.176 \times 10^9)(0.068)}\left[\frac{(3{,}478.597)(17.0)}{3} + \frac{(1{,}453.50)(17.0)}{8} + \frac{34{,}599.42}{2}\right] = 0.04081 \text{ ft}$$

$$y_3 = \frac{(10)^2}{(4.176 \times 10^9)(0.103)}\left[\frac{(4{,}932.097)(10)}{3}\right.$$

$$+ \frac{(990.099)(10)}{8} + \frac{106{,}090.72}{2}\Bigg] = 0.01658 \text{ ft}$$

$$y_4 = \frac{(8)^2}{(4.176 \times 10^9)(0.139)}\Bigg[\frac{(5{,}922.196)(8)}{3}$$

$$+ \frac{(388{,}738)(8)}{8} + \frac{147{,}067.52}{2}\Bigg] = 0.00989 \text{ ft}$$

$$y_5 = \frac{(8)^2}{(4.176 \times 10^9)(0.177)}\Bigg[\frac{(6{,}310.934)(8)}{3}$$

$$+ \frac{(522.662)(8)}{8} + \frac{222{,}764.113}{2}\Bigg] = 0.01115 \text{ ft}$$

$$y_6 = \frac{(9.083)^2}{(4.176 \times 10^9)(0.260)}\Bigg[\frac{(6{,}833.596)(9.083)}{3}$$

$$+ \frac{(572.730)(9.083)}{8} + \frac{270{,}892.241}{2}\Bigg] = 0.012 \text{ ft}$$

$$y_7 = \frac{(8)^2}{(4.176 \times 10^9)(0.565)}\Bigg[\frac{(7{,}406.326)(8)}{3}$$

$$+ \frac{(754.042)(8)}{8} + \frac{332{,}947.48}{2}\Bigg] = 0.00507 \text{ ft}$$

$$y_8 = \frac{(2.91667)^2}{(4.176 \times 10^9)(0.400)}\Bigg[\frac{(8{,}160.368)(2.91667)}{3}$$

$$+ \frac{(274.042)(2.91667)}{8} + \frac{390{,}632.69}{2}\Bigg] = 0.00203 \text{ ft}$$

$$\Delta_{1\text{-}2} = \frac{(13.5)(17)}{(4.176 \times 10^9)(0.068)}\Bigg[\frac{(3{,}478.597)(17)}{2}$$

$$+ \frac{(1{,}453.50)(17)}{6} + 34{,}599.42\Bigg] = 0.05519 \text{ ft}$$

$$\Delta_{1\text{-}3} = \frac{(30.50)(10)}{(4.176 \times 10^9)(0.103)}\Bigg[\frac{(4{,}932.097)(10)}{2}$$

$$+ \frac{(990.099)(10)}{6} + 106{,}090.72\Bigg] = 0.07709 \text{ ft}$$

$$\Delta_{1\text{-}4} = \frac{(40.50)(8)}{(4.176 \times 10^9)(0.139)}\Bigg[\frac{(5{,}922.196)(8)}{2}$$

$$+ \frac{(388.738)(8)}{6} + 147{,}067.52\Bigg] = 0.09560 \text{ ft}$$

$$\Delta_{1\text{-}5} = \frac{(48.50)(8)}{(4.176 \times 10^9)(0.177)}\Bigg[\frac{(6{,}310.934)(8)}{2}$$

$$+ \frac{(522.662)(8)}{6} + 222{,}764.113\Bigg] = 0.13055 \text{ ft}$$

$$\Delta_{1\text{-}6} = \frac{(56.50)(9.083)}{(4.176 \times 10^9)(0.260)}\Bigg[\frac{(6{,}833.596)(9.083)}{2}$$

$$+ \frac{(572.730)(9.083)}{6} + 270{,}892.241\Bigg] = 0.143 \text{ ft}$$

$$\Delta_{1\text{-}7} = \frac{(73.583)(8)}{(4.176 \times 10^9)(0.565)}\Bigg[\frac{(7{,}406.321)(8)}{2}$$

$$+ \frac{(754.042)(8)}{6} + 332{,}947.48\Bigg] = 0.09071 \text{ ft}$$

$$\Delta_{1\text{-}8} = \frac{(73.583)(2.9167)}{(4.176 \times 10^9)(0.400)}\Bigg[\frac{(8{,}160.368)(2.9167)}{2}$$

$$+ \frac{(274.942)(2.9167)}{6} + 390{,}632.69\Bigg] = 0.05174 \text{ ft}$$

δ = total deflection at top of vessel

$$= \sum y_i + \sum \Delta_{i-1}$$

$= 0.743$ ft

$= 8.910$ in. at top for static gust wind

Referring to Figure 4-47, we determine

$M_1 = 0$

$M_2 = (4.71)(6.961) = 32.786$ kip-ft

$M_3 = 32.786 + (4.823)(8.789) = 75.175$ kip-ft

$M_4 = 75.175 + (7.533)(13.25) = 174.987$ kip-ft

$M_5 = 174.987 + (10.013)(9.00) = 265.104$ kip-ft

$M_6 = 265.104 + (12.023)(8.00) = 361.288$ kip-ft

$M_7 = 361.288 + (14.253)(8.862) = 487.598$ kip-ft

$M_8 = 487.598 + (17.693)(8.221) = 633.032$ kip-ft

$M_9 = 633.032 + (21.233)(5.458) = 748.922$ kip-ft

$M_{10} = 748.922 + (23.143)(1.458) = 782.664$ kip-ft

$T = M/I$

$$T_2 = \frac{M_2}{I_2} = \frac{32{,}786}{0.279} = 117{,}512.54$$

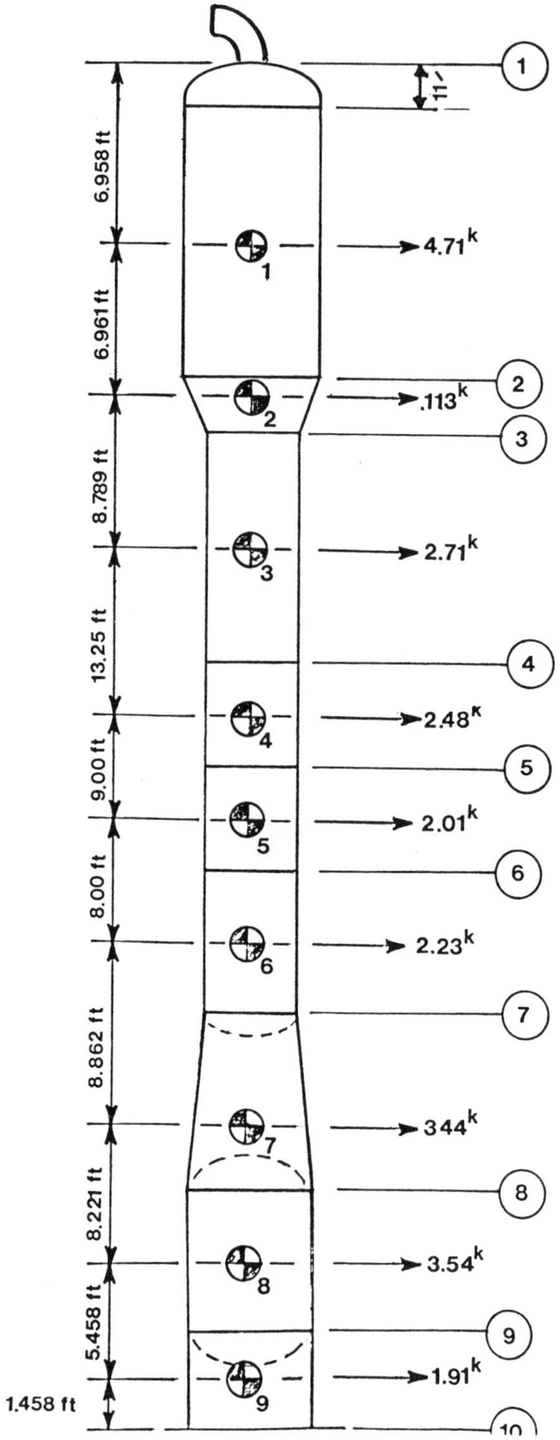

Figure 4-47. Tower vibration ensemble.

$$T_3 = \frac{M_3}{I_3} = \frac{75{,}175}{0.068} = 1{,}105{,}514.71$$

$$T_4 = \frac{M_4}{I_4} = \frac{174{,}987}{0.103} = 1{,}698{,}902.91$$

$$T_5 = \frac{M_5}{I_5} = \frac{265{,}104}{0.139} = 1{,}907{,}223.02$$

$$T_6 = \frac{M_6}{I_6} = \frac{361{,}288}{0.177} = 2{,}041{,}175.14$$

$$T_7 = \frac{M_7}{I_7} = \frac{487{,}598}{0.130} = 3{,}750{,}753.85$$

$$T_8 = \frac{M_8}{I_8} = \frac{633{,}032}{0.565} = 1{,}120{,}410.62$$

$$T_9 = \frac{M_9}{I_9} = \frac{748{,}922}{0.400} = 1{,}872{,}305.00$$

$$T_{10} = \frac{M_{10}}{I_9} = \frac{782{,}644}{0.400} = 1{,}956{,}660.00$$

$$S_i = M\,dx/I$$

$$S_{10} = \frac{(1{,}956{,}660.00 + 1{,}872{,}305.00)}{2}(1.458)$$

$$= 2{,}791{,}315.49$$

$$S_9 = \frac{(1{,}872{,}305.00 + 1{,}120{,}410.62)}{2}(5.458)$$

$$= 8{,}167{,}120.93$$

$$S_8 = \frac{(1{,}120{,}410.62 + 3{,}750{,}753.85)}{2}(8.221)$$

$$= 20{,}022{,}921.55$$

$$S_7 = \frac{(3{,}750{,}753.85 + 2{,}041{,}175.14)}{2}(8.862)$$

$$= 25{,}664{,}037.35$$

$$S_6 = \frac{(2{,}041{,}175.14 + 1{,}907{,}223.02)}{2}(8.00)$$

$$= 15{,}793{,}592.64$$

$$S_5 = \frac{(1{,}907{,}223.02 + 1{,}698{,}902.91)}{2}(9.00)$$

$$= 16{,}227{,}566.69$$

$$S_4 = \frac{(1{,}698{,}902.91 + 1{,}105{,}514.71)}{2}(13.25)$$

$$= 18{,}579{,}266.73$$

$$S_3 = \frac{(1{,}105{,}514.71 + 117{,}512.54)}{2}(8.789)$$

$$= 5{,}374{,}593.25$$

$$S_2 = \frac{(117{,}512.54)}{2}(6.961)$$

$$= 409{,}002.40$$

$$A_i = \sum S_i = (M_i\, dx)/I_i$$

$$A_{10} = 2{,}791{,}315.49$$

$$A_9 = 2{,}791{,}315.49 + 8{,}167{,}120.93 = 10{,}958{,}436.42$$

$$A_8 = 10{,}958{,}436.42 + 20{,}022{,}921.55 = 30{,}981{,}357.97$$

$$A_7 = 30{,}981{,}357.97 + 25{,}664{,}037.35 = 56{,}645{,}395.32$$

$$A_6 = 56{,}645{,}395.32 + 15{,}793{,}592.64 = 72{,}438{,}987.96$$

$$A_5 = 72{,}438{,}987.96 + 16{,}227{,}566.69 = 88{,}666{,}554.65$$

$$A_4 = 88{,}666{,}554.65 + 18{,}579{,}266.73 = 107{,}245{,}821.4$$

$$A_3 = 107{,}245{,}821.4 + 5{,}374{,}593.25 = 112{,}620{,}414.7$$

$$A_2 = 112{,}620{,}414.7 + 409{,}002.40 = 113{,}029{,}417.1$$

$$P_i = \left(\frac{A_i + A_{i+1}}{2}\right) L_i$$

$$P_{10} = \frac{2{,}791{,}315.49}{2}(1.458)$$

$$= 2{,}034{,}868.99$$

$$P_9 = \frac{(2{,}791{,}315.49 + 10{,}958{,}436.42)}{2}(5.458)$$

$$= 37{,}523{,}072.96$$

$$P_8 = \frac{(10{,}958{,}436.42 + 30{,}981{,}357.97)}{2}(8.221)$$

$$= 172{,}393{,}524.9$$

$$P_7 = \frac{(30{,}981{,}357.97 + 56{,}645{,}395.32)}{2}(8.862)$$

$$= 388{,}274{,}143.9$$

$$P_6 = \frac{(56{,}645{,}395.32 + 72{,}438{,}987.96)}{2}(8.00)$$

$$= 516{,}337{,}533.2$$

$$P_5 = \frac{(72{,}438{,}987.96 + 88{,}666{,}554.65)}{2}(9.00)$$

$$= 724{,}974{,}941.7$$

$$P_4 = \frac{(88{,}666{,}554.65 + 107{,}245{,}821.4)}{2}(13.25)$$

$$= 1{,}297{,}919{,}492$$

$$P_3 = \frac{(107{,}245{,}821.4 + 112{,}620{,}414.7)}{2}(8.789)$$

$$= 966{,}202{,}175.0$$

$$P_2 = \frac{(112{,}620{,}414.7 + 113{,}029{,}417.1)}{2}(6.961)$$

$$= 785{,}374{,}239.6$$

$$\mu_{10} = 2{,}034{,}868.99$$

$$\mu_9 = 2{,}034{,}868.99 + 37{,}523{,}072.96 = 39{,}557{,}941.95$$

$$\mu_8 = 39{,}557{,}941.95 + 172{,}393{,}524.9 = 211{,}951{,}466.9$$

$$\mu_7 = 211{,}951{,}466.9 + 388{,}274{,}143.9 = 600{,}225{,}610.8$$

$$\mu_6 = 600{,}225{,}610.8 + 516{,}337{,}533.2 = 1{,}116{,}563{,}144$$

$$\mu_5 = 1{,}116{,}563{,}144 + 724{,}974{,}941.7 = 1{,}841{,}538{,}086$$

$$\mu_4 = 1{,}841{,}538{,}086 + 1{,}297{,}919{,}492 = 3{,}139{,}457{,}578$$

$$\mu_3 = 3{,}139{,}457{,}578 + 966{,}202{,}175.0 = 4{,}105{,}659{,}753$$

$$\mu_2 = 4{,}105{,}659{,}753 + 785{,}374{,}239.6$$
$$= 4{,}891{,}033{,}993$$

$$y_i = \frac{\mu_i}{144E_i} \text{ ft}$$

$$y_2 = \frac{4{,}891{,}033{,}993}{144(30 \times 10^6)} = 1.13 \text{ ft} = 13.59 \text{ in.}$$

$$y_3 = \frac{4{,}105{,}659{,}753}{(4.32 \times 10^9)} = 0.950 \text{ ft} = 11.405 \text{ in.}$$

$$y_4 = \frac{3{,}139{,}457{,}578}{(4.32 \times 10^9)} = 0.727 \text{ ft} = 8.721 \text{ in.}$$

$$y_5 = \frac{1{,}841{,}538{,}086}{(4.32 \times 10^9)} = 0.426 \text{ ft} = 5.115 \text{ in.}$$

$$y_6 = \frac{1{,}116{,}563{,}144}{(4.32 \times 10^9)} = 0.258 \text{ ft} = 3.102 \text{ in.}$$

$$y_7 = \frac{600{,}225{,}610.8}{(4.32 \times 10^9)} = 0.139 \text{ ft} = 1.667 \text{ in.}$$

$$y_8 = \frac{211{,}951{,}466.9}{(4.32 \times 10^9)} = 0.049 \text{ ft} = 0.589 \text{ in.}$$

$$y_9 = \frac{39{,}557{,}941.95}{(4.32 \times 10^9)} = 0.009 \text{ ft} = 0.110 \text{ in.}$$

$$y_{10} = \frac{2{,}034{,}868.99}{(4.32 \times 10^9)} = 4.710 \times 10^{-4} \text{ ft} = 0.006 \text{ in.}$$

The tower section weights and displacements are combined in Table 4-21 to determine the period of vibration of the tower.

First critical wind velocity, V_1

$$V_1 = \frac{3.40 \text{ d}}{T}$$

$$L = \frac{76.96}{4} = 19.24 \text{ ft}$$

$$d = \left(\frac{13.0}{19.24}\right)(8.740) + \left(\frac{6.24}{19.24}\right)(3.750) = 7.122$$

From Equation 4-101, at resonance

f_V = vortex shedding frequency = natural frequency

$$V_1 = \frac{f_V D}{s} = \frac{(0.97)(7.122)}{0.2} = 34.540 \frac{\text{ft}}{\text{sec}}$$

$$= 23.550 \text{ mph}$$

Considering the top portion (Section 1) we have

$$V_1 = \frac{(0.97)(8.74)}{0.2} = 42.39 \frac{\text{ft}}{\text{sec}} = 28.90 \text{ mph}$$

Table 4-21
Values for Determining Tower's Period of Vibration

Section	Deflection y (in.)	W (lb_f)	Wy (in.-lb_f)	Wy^2 (in.^{2}-lb_f)
1	13.59	4,710	64,008.90	869,880.95
2	11.41	113	1,289.33	14,711.26
3	8.72	2,710	23,631.20	206,064.06
4	5.12	2,480	12,697.60	65,011.71
5	3.10	2,010	6,231.00	19,316.10
6	1.67	2,230	3,724.10	6,219.25
7	0.59	3,440	2,029.60	1,197.46
8	0.11	3,540	389.40	42.83
9	0.01	1,910	19.10	0.19
			$\sum Wy$ = 114,029.23	$\sum Wy^2$ = 1,182,443.81

First Period of Vibration, T

$$T = 2\pi\sqrt{\frac{\sum Wy^2}{g\sum Wy}} = 2\pi\sqrt{\frac{(1{,}182{,}443.81)}{(386.4)(114{,}020.23)}} = 1.03 \text{ sec/cycle} \Rightarrow f = 0.97 \text{ Hz}$$

Since the field measurements indicated an air velocity at resonance to be 30 mph and a stack deflection of 13 inches, this analysis agrees with empirical results. From the calculations for the first critical wind velocity, it appears that the larger diameter of Section 1 has a larger influence on this deflection. For this reason we use the top 1/4 of the tower rather than the top 1/3. Now,

$$V_2 = 6.25;\ V_1 = (6.25)(28.90) = 180.63 \text{ mph}$$

A tower that has been fabricated and installed in the field is beyond design changes. Unlike stacks (see Chapter 5), vortex strakes are difficult to install on many towers and impossible on others. Shortening the tower height is impractical, since the tower's internals are necessary (unlike a stack). Consequently, the only resolution is to mount guy wires to the tower's upper section (normally 2/3 the height). Except for special applications, guy wires are to be avoided in practice. They use a lot of space and plant maintenance people sometimes must temporarily remove one or two to gain access to an area for equipment installation or some other reason. Problems then may arise in keeping the tower from falling over during this temporary time interval, remembering to reconnect the guy wire(s), and making sure the wires are properly tensioned once they are reconnected. Despite these disadvantages, guy wires were essential in this application.

EXAMPLE 4-5: SADDLE PLATE ANALYSIS OF A HORIZONTAL VESSEL

A proposed horizontal vessel design shown in Figure 4-48 is fully loaded with corn syrup used by a confectionery manufacturing plant in Fayetteville, Arkansas. The corn syrup has a specific gravity of $\gamma = 1.4$ and is at 90°F. The thickness of the head and shell is 1/2 in. since the corn syrup is at 90°F, there is practically no thermal expansion of the vessel, so only uniform compression is considered in evaluating the saddles. Even though a Zick analysis indicates that the vessel is grossly overstressed, the saddle in Figure 4-48 is to be evaluated.

To analyze the saddle plate, refer to Figure 4-48 c. Each section of the saddle plate, A–B, B–C, C–D, is considered separately. Each section supports a portion of the vessel weight indicated by the dotted lines. Sections A–B and C–D support equal weights.

Section A–B and C–D

$$x = 4.27 \text{ ft} = 51.24 \text{ in.},\ R_i = 6.0 \text{ ft} = 72 \text{ in.}$$

From Appendix A, Equation A-8, the fluid volume in vessel above Section A–B is as follows:

$$V = \frac{2\pi(36)(72)^2}{6} - \frac{\pi(72)^2}{2}\left[51.24 - \frac{(51.24)^3}{3(36)^2}\right]$$

$$= 59{,}948.76 \text{ in.}^3 = 259.52 \text{ gal in one head}$$

From Equation A-1 in Appendix A the partial volume of liquid in the cylindrical portion is calculated.

$$V = \frac{(72)^2(150)(12)}{2}\left[\frac{\pi(146.48)}{180} - 0.55\right]$$

$$= 9{,}351{,}647.46 \text{ in.}^3$$

$$= 40{,}483.32 \text{ gal}$$

Total fluid volume above Section A–B is

$$V = 40{,}483.32 \text{ gal} + 2(259.52) \text{ gal} = 41{,}002.36 \text{ gal}$$
$$= 5{,}481.22 \text{ ft}^3$$

The total fluid weight is then

$$W_{FAB} = W_{FCD} = (5{,}481.22) \text{ ft}^3 (62.4) \frac{\text{lb}}{\text{ft}^3} (1.4)$$

$$= 478{,}839.22 \text{ lb}$$

Metal Weight Above Each Section, A–B and C–D

For outside surface on heads that are 1/2 in. thick,

$$V = \frac{2\pi(36.25)(72.5)^2}{6} - \frac{\pi(72.5)^2}{2}\left[51.74 - \frac{(51.74)^3}{3(36.25)^2}\right]$$

$$= 62{,}434.25 \text{ in.}^3$$

The inside volume in the head was determined in computing the fluid volume as being 59,948.76 in^3. The metal volume in one head is then

$$V_M = 62{,}434.25 \text{ in.}^3 - 59{,}948.76 \text{ in.}^3 = 2{,}485.49 \text{ in.}^3$$

For two heads,

$$V_M = 2(2{,}485.49) = 4{,}970.98 \text{ in.}^3$$

The metal volume in the cylinder portion above Section A–B is determined as follows:
For outside surface,

$$V = \frac{(72.5)^2(150)(12)}{2}\left[\frac{\pi(146.72)}{180} - 0.55\right]$$

$$= 9{,}512{,}090.41 \text{ in.}^3$$

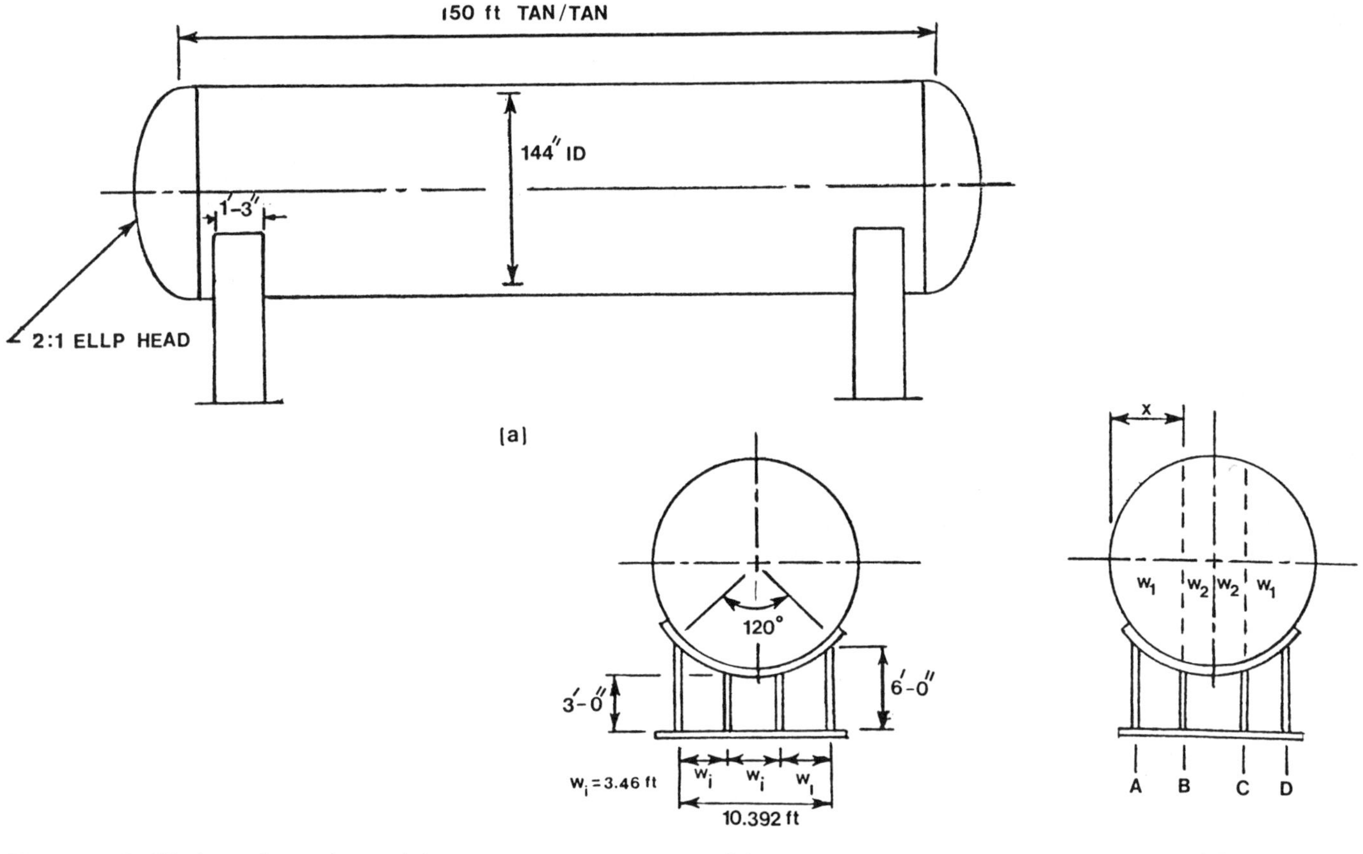

Figure 4-48. Horizontal vessel containing corn syrup.

The inside volume was determined in computing the fluid volume as being 9,351,647.46 in^3. The metal volume in the cylinder is then

$$V = 9{,}512{,}090.41 \text{ in.}^3 - 9{,}351{,}647.46 \text{ in.}^3 = 160{,}442.95 \text{ in.}^3$$

The total metal volume above Section A–B is

$$V_M = 4{,}970.98 \text{ in.}^3 + 160{,}442.95 \text{ in.}^3 = 165{,}413.93 \text{ in.}^3$$

The metal weight is

$$W_M = (165{,}413.93) \text{ in.}^3 (0.283) \frac{\text{lb}}{\text{in.}^3} = 46{,}812.14 \text{ lb}$$

Combining with fluid weight the total weight,

$$W = 46{,}812.14 \text{ lb} + 478{,}839.22 \text{ lb} = 525{,}651.36 \text{ lb}$$

For each saddle,

$$W_{AB} = W_{CD} = \frac{525{,}651.36 \text{ lb}}{2} = 262{,}825.68 \text{ lb} = Q$$

Section B–C

Similarly to Section A–B, for the head, the liquid volume is determined from Equation A-7 in Appendix A.

$$V = \frac{\pi(72)^2}{2}\left[51.24 - \frac{(51.24)^3}{3(36)^2}\right] = 135{,}483.43 \text{ in.}^3$$

For total volume,

$$V = 2(135{,}483.43) = 270{,}966.86 \text{ in.}^3 \text{ for one head}$$

For both heads,

$$V = 2(270{,}966.86) = 541{,}933.73 \text{ in.}^3$$

Liquid volume for cylinder portion is

$$V = \pi(72)^2(150)(12) \text{ in.}^3 - 2(9{,}351{,}647.46) \text{ in.}^3 = 10{,}611{,}534.46 \text{ in.}^3$$

The total liquid volume above Section B-C is

$$V_F = 10{,}611{,}534.46\ \text{in.}^3 + 541{,}933.73\ \text{in.}^3 = 11{,}153{,}468.19\ \text{in.}^3$$

Metal Volume Above Section B-C

For outside surface on a single head, using Equation A-7

$$V = \frac{\pi(72.5)^2}{2}\left[51.74 - \frac{(51.74)^3}{3(36.25)^2}\right] = 137{,}097.79\ \text{in.}^3$$

The inside volume was determined from calculating the liquid volume as being 135,483.43 in.^3 Thus the metal volume for a single head is

$$V_M = 137{,}097.43\ \text{in.}^3 - 135{,}483.43\ \text{in.}^3 = 1{,}614.35\ \text{in.}^3$$

For two heads,

$$V_M = 2(1{,}614.36)\text{in.}^3 = 3{,}228.72\ \text{in.}^3$$

The metal volume for the cylindrical portion is determined using Equation A-1 and the total volume of a cylinder as follows:

$$V = \pi(72.0)^2(150)(12) - \frac{(72)^2(150)(12)}{2}\left[\frac{\pi(146.48°)}{180} - 0.55\right]$$

$$V = 19{,}963{,}181.93\ \text{in.}^3 \text{ for inside volume}$$

For outside volume,

$$V = \pi(72.5)^2(150)(12) - \frac{(72.5)^2(150)(12)}{2} \times \left[\frac{\pi(146.72°)}{180} - 0.55\right]$$

$$V = 20{,}205{,}196.86\ \text{in.}^3$$

The metal volume is

$$V_M = 20{,}205{,}196.86\ \text{in.}^3 - 19{,}963{,}181.93\ \text{in.}^3 = 242{,}014.93\ \text{in.}^3$$

For both sides of centerline in Figure 4-48 c,

$$V_M = 2(242{,}014.93)\text{in.}^3 = 484{,}029.86\ \text{in.}^3$$

Combining both the cylindrical and head metal weights we have

$$V_M = 484{,}029.86\ \text{in.}^3 + 3228.72\ \text{in.}^3 = 487{,}258.58\ \text{in.}^3$$

The total metal weight above Section B–C is

$$W_M = (487{,}258.58)\text{in.}^3\ (0.2833)\ \frac{\text{lb}}{\text{ft}^3} = 137{,}894.18\ \text{lb}$$

The total liquid volume above Section B-C is

$$W_L = \frac{(11{,}153{,}468.19)\text{in.}^3}{1{,}728\ \text{in.}^3}\ (62.4)\ \frac{\text{lb}}{\text{ft}^3}\ (1.4)\text{ft}^3 = 563{,}869.78\ \text{lb}$$

The total weight above Section B–C is

$$W_T = 137{,}894.18\ \text{lb} + 563{,}869.78\ \text{lb} = 701{,}763.96\ \text{lb}$$

For each saddle,

$$W_{BC} = Q = \frac{701{,}763.96\ \text{lb}}{2} = 350{,}881.98\ \text{lb}$$

Saddle Plate Buckling Analysis

The critical buckling stress for a plate is determined from Equation 4-17a.

$$\sigma_{cr} = \frac{K_i \pi^2 E}{12(1 - \nu^2)\left(\frac{b_e}{t_s}\right)^2} \tag{4-17a}$$

$$\text{where } b_e = \frac{d_i\ t_s}{[d_i\ t_s + 2t_w(b - 1)]}$$

$$t_s = t_w = 0.50\ \text{in.}$$

$$b_e = \frac{(12)(3.46)(0.50)}{[(12)(3.46)(0.5) + 2(0.5)(15 - 1)]} = 0.597\ \text{in.}$$

Also,

$$b_e = Kt_s \tag{4-15}$$

$$b_e = (1.28)(0.5) = 0.64\ \text{in. (use 0.597 in.)}$$

Adding more length to web plate will not increase the local buckling strength for pure compression. The same also holds for bending and shear. Substituting the value of b_e above into Equation 4-17a we have

$$\sigma_{cr} = \frac{(1.28)\pi^2 (29 \times 10^6)}{12\left(1 - \frac{1}{9}\right)\left(\frac{(3.46)(12)}{0.5}\right)^2} = 4{,}980.860 \text{ psi}$$

Substituting this value into Equation 4-18 we determine the buckling load for compressive loading as follows:

$$F_B = n(A_s + 2b_e t_s)\sigma_r \qquad (4\text{-}18)$$

$$F_B = (4)[7.5 + 2(0.597)(0.5)] = 161{,}321.389 \text{ lb}$$

Since 161,321.4 lb < 262,825.7 lb < 350,881.98 lb

we must use more stiffening plates if we are to use a 1/2-in. saddle plate.
Now

$$F_{B_{min}} = 351{,}000 \text{ lb}$$

From Equation 4-18 we have

$$351{,}000 = n[7.5 + 2(0.597)(0.5)]\sigma_{cr}$$

$$\sigma_{cr} = \frac{351{,}000}{n}$$

The effective plate width normal to the web plate axis is

$$d_e = d_i (0.25 + 0.91\lambda^2)$$

$$\text{where } \lambda = \left(\frac{t_s}{w_i}\right)\left(\frac{E}{\sigma_{cr}}\right)^{0.5}$$

$$\lambda = \left(\frac{0.5}{(3.46)(12)}\right)\left(\frac{29 \times 10^6}{25{,}000}\right)^{0.5} = 0.41$$

$$w_e = (3.46)(12)(0.25 + 0.91\,(0.41)^2) = 16.74 \text{ in.}$$

$$n = \frac{10.392(12)}{16.74} = 7.449 \rightarrow \text{use 8 stiffeners}$$

$$\sigma_{cr} = \frac{351{,}000}{8} = 43{,}875.0 \text{ psi}$$

$$F_B = (8)[7.5 + 2(0.597)(0.5)](43{,}875.0) = 2{,}842{,}047.0 \text{ lb}$$

Since 2,842,047 lb >> 351,000 lb, eight stiffeners are sufficient.

Horizontal Reaction Force on Saddle

From Equation 4-19 the horizontal reaction is

$$F = Q\left[\frac{1 + \cos\beta - \frac{1}{2}\sin^2\beta}{\pi - \beta + \sin\beta\cos\beta}\right]$$

$$\beta = \left(180 - \frac{\theta}{2}\right) = 180 - \frac{120}{2} = 120°$$

$$Q = \frac{2(262{,}825.68) + 350{,}881.98}{2} = 438{,}266.67 \text{ lb}$$

$$F = (438{,}266.67)\left[\frac{1 - 0.50 - 0.38}{\pi - 2.09 + (0.87)(-0.50)}\right]$$

$$= 85{,}294.56 \text{ lb}$$

The effective area resisting this force is

$$A_E = \left(\frac{R}{3}\right) t_s = \left(\frac{72.5}{3}\right)(0.50) = 12.08 \text{ in.}^2$$

This results in a stress of

$$\sigma = \frac{85{,}294.56 \text{ lb}}{12.08 \text{ in.}^2} = 7{,}058.86 \text{ psi}$$

Referring to Table 4-6, the allowable stress for A-36 is $0.60\ \sigma_y = 22{,}000$ psi. Since 7,058.86 psi < 22,000 psi, the saddle is sufficient for the horizontal reaction.

NOTATION

- A = dimension from saddle centerline to tangent of head (Figure 4-2) ft, in.
- A_c = effective area of concrete, ft²
- b = plate width (Equation 4-15) in.
- BP = bearing pressure, psi
- BPW = base plate thickness, in.
- C = constant for bolt torque (Equation 4-66), dimensionless; friction coefficient (Equation 4-89) dimensionless; structure period response factor (Equation 4-106) dimensionless
- CA = corrosion allowance, in.
- C_r = critical damping factor (Equation 4-90), dimensionless
- C_s = compressive strength of concrete (Table 4-7), psi

D = diameter (Equation 4-27), in.; dynamic magnification factor (Equation 4-91), dimensionless
D_E = effective wind diameter (Figure 4-22), in.
D_o = outside diameter, in.
D_i = inside diameter, in.
E = welding joint efficiency (Table 4-2), dimensionless; modulus of elasticity, psi
F = wind force (Equation 4-94)
F_i = bold uplift force (Equation 4-39), lb_f
f_r = natural frequency of a ring (Equation 4-100), Hz
f_v = vortex shedding frequency Equation 4-101, Hz
G_F = dynamic gust response factor, dimensionless
G_H = gusset plate height (Equations 4-39 and 4-63), in.
G_W = gusset plate width (Equation 4-63), in.
H = depth of vessel head (Figure 4-2), in.
I = moment of inertia (Equation 4-24), in^4; occupancy importance factor (Equation 4-106), dimensionless
I_c = moment of inertia of effective area of concrete, in^4
K = coefficient of buckling for shear (Equation 4-15 and Figure 4-3), dimensionless
k = dimensionless parameter for concrete (Table 4-7)
K_i = plate buckling coefficient (Equation 4-15), dimensionless
K_Z = velocity pressure coefficient (Table 4-9 and Equation 4-78)
L = length of a horizontal vessel from seam to seam (Figurc 4-2), ft, in.
L_e = effective column length (Equation 4-19), in.
M = bending moment, in.-lb, ft-lb
m = bolt spacing (Equation 4-39), in.
M_C = compressive bending moment in the shell of a horizontal vessel (Figure 4-2), ft-lb
M_T = tensional bending moment in the shell of a horizontal vessel (Figure 4-2), ft-lb
N = number of anchor bolts (Equation 4-40), dimensionless
P = buckling load for compressive loading (Equation 4-18), lb_f; probability of exceeding wind design speed during n years (Table 4-11) and Appendix A), dimensionless
P_a = annual probability of wind speed exceeding a given magnitude—see (Appendix A), dimensionless
R = mean radius of shell (Figure 4-2), ft, in.
R_i = inside vessel radius (Equation 4-73), in.
R_o = outside vessel radius (Equation 4-73), in.
r = inside radius of vessel (Figure 4-2), ft
Q = reaction at saddle (w/2), lb_f
q_F = velocity pressure of wind on structures (Equation 4-76), lb_f/ft^2
q_{30} = basic wind pressure at 30 ft, lb_f/ft^2
S = Strouhal number used (Equation 4-102), dimensionless; structure size factor (Equation 4-82)
T = bolt torque as defined (Equation 4-66), in.-lb
T_1 = exposure factor for wind (Table 4-11), dimensionless
t_{CR} = compression plate thickness (Equation 4-63), in.
t_g = gusset plate thickness, in.
t_h = head thickness (Equation 4-7), in.
t_s = shell thickness (Equation 4-1), in.
v_o = theoretical ovaling velocity (Equation 4-102), mph or ft/sec
v_1 = first critical wind velocity (Equation 4-94), mph
v_{30} = basic wind speed at thirty feet used as design wind speed (Equation 4-75), mph
W = vessel weight (Equation 4-40), lb_f
x_{st} = static deflection of a spring acted upon by a force (Equation 4-90), in.
$x(t)$ = displacement as a function of time (Equation 4-90), in.
y = total lateral displacement of tower (Equation 4-88, Figure 4-21), in., ft
Z = elevation or height above a reference point, such as the ground (Equation 4-74), ft
Z = reference height in which basic wind speed is considered (30 ft or 10 m), ft

Greek Symbols

α = $\pi - (\pi/180)(\theta/2 + \beta/20)$ (Equation 4-6), degrees
β = $(180 - \theta/2)$, degrees
Δ = $(\pi/180)(5\theta/12 + 30)$, degrees
δ_i = lateral translational deflection of tower, (Equation 4-88 and Figure 4-26), in.
θ = angle of contact of saddle with shell (Figure 4-1), degrees, radians; rotational displacement of tower (Figure 4-26), degrees
λ = $(t/b)(E/\sigma_{CR})$ used in Equation 4-18, dimensionless
ρ = radius of gyration = $(I/A)^{0.5}$
σ = general term for stress, psi
σ_a = allowable stress values (Table 4-3) psi
σ_c = allowable stress induced on concrete (Equation 4-40), psi; general term for compressive stress (Equation 4-16), psi
σ_{CR} = critical stress in a flat plate defined in Equation 4-15, psi

σ_c' = elastic buckling stress (Equation 4-16), psi; 28-day ultimate compressive strength of concrete (Table 4-7), psi
σ_g = stress due to weight, lb_f
σ_p = pressure stress induced by either internal or external pressure, psi; longitudinal stress in Equation 4-67, psi
σ_s = tensile stress in steel, psi
σ_w = stress induced by wind or earthquake response spectra, psi
σ_y = minimum yield stress for a ductile material, psi
ν = Poisson ratio for a given material, dimensionless
ϕ = concrete bearing parameter (Equation 4-20), dimensionless

REFERENCES

1. *ASME Boiler and Pressure Vessel Code,* Section VIII Division 1, American Society of Mechanical Engineers, New York.
2. Zick, L. P., "Stresses in Large Horizontal Cylindrical Pressure Vessels on Two Saddle Supports," *Welding Research Journal* Supplement, 1971.
3. Brownell, L. E. and Young, E. H., *Process Equipment Design,* John Wiley and Sons, New York, 1959.
4. U.S. Steel, *Steel Design Manual,* U.S. Steel, Pittsburgh, Pennsylvania, 1981.
5. American Institute of Steel Construction, *Manual of Steel Construction,* Eighth Edition, AISC, Chicago, Illinois, 1980.
6. Timoshenko, S., *Theory of Plates and Shells,* McGraw-Hill Book Co., New York, 1959.
7. Bickford, J. H., *An Introduction to the Design and Behavior of Bolted Joints,* Marcel Dekker, Inc., New York, 1981.
8. Faires, V. M., *Design of Machine Elements,* The Macmillan Co., New York, 1962.
9. Simiu, E. and Scanlan, R. H., *Wind Effects on Structures,* John Wiley and Sons, New York, 1978.
10. International Conference of Building Officials, *Uniform Building Code,* Whittier, California, 1982.
11. American National Standards Institute, Inc., "ANSI A58.1-Minimum Design Loads for Buildings and Other Structures-1982," New York.
12. Kuethe, A. M. and Schetzer, J. D., *Foundations of Aerodynamics,* John Wiley and Sons, New York, 1959.
13. Blevins, R. D., *Flow-Induced Vibration,* Van Nostrand Rheinhold Co., New York, 1977.
14. Macdonald, A. J., *Wind Loading on Buildings,* Applied Science Publishers, Ltd., London, England, 1980.
15. Staley, C. M. and Graven, G. G., *The Static and Dynamic Wind Design of Steel Stacks,* ASME 72-Pet-30, New York.
16. Vierck, R. K., *Vibration Analysis,* Harper and Row, New York, 1979.
17. Paz, M., *Structural Dynamics,* Van Nostrand Rheinhold Co. New York, 1980.
18. Australian Standard 1170, Part 2–1983 SAA Loading Code, Part 2–Wind Forces, p. 55.
19. Timoshenko, S., Young, D. H., Weaver, W., *Vibration Problems In Engineering,* John Wiley and Sons, New York, 1974.
20. Higdon, A., Olsen, E. H., Stiles, W. B., Weese, J. A., and Riley, W. F., *Mechanics of Materials,* John Wiley and Sons, New York, 1976.
21. Mitchell, Warren W., "Determination of the Period of Vibration of Multi-Diameter Columns by the Method Used on Rayleigh's Principle," an unpublished work prepared for the Engineering Department of the Standard Oil Company of California, San Francisco, California, 1962.
22. Bednar, H. H., *Pressure Vessel Design Handbook,* Van Nostrand Rheinhold Co., New York, 1981.
23. Boardman, H. C., "Stresses at Junction of Cone and Cylinder in Tanks With Cone Bottoms or Ends," *Pressure Vessel and Piping Design,* collected papers, ASME, N.Y., 1960.
24. Weaver, William, Jr., "Wind-Induced Vibrations in Antenna Members," American Society of Civil Engineers, Paper No. 3336, Vol. 127, Part 1, N.Y., N.Y., 1962.

Appendix A

Partial Volumes and Pressure Vessel Calculations

PARTIAL VOLUME OF A CYLINDER

$$V_P = \frac{R_i^2 L}{2}\left(\frac{\pi\alpha^\circ}{180^\circ} - \sin\alpha\right) = \text{partial volume shown in shaded region} \qquad \text{(A-1)}$$

L = length of cylinder

R = inside radius of cylinder

Example (Figure A-1)

For a cylinder with 144-in. ID find the partial volume of a fluid head of 60 in., if L = 100 ft:

$$\frac{\alpha}{2} = 80.41^\circ$$

$$V_P = \frac{(72)^2(1200)}{2}\left[\frac{\pi(160.81)}{180} - \sin(160.81^\circ)\right]$$

$$V_P = 7{,}707{,}650.2 \text{ in.}^3 = 33{,}366.5 \text{ gal}$$

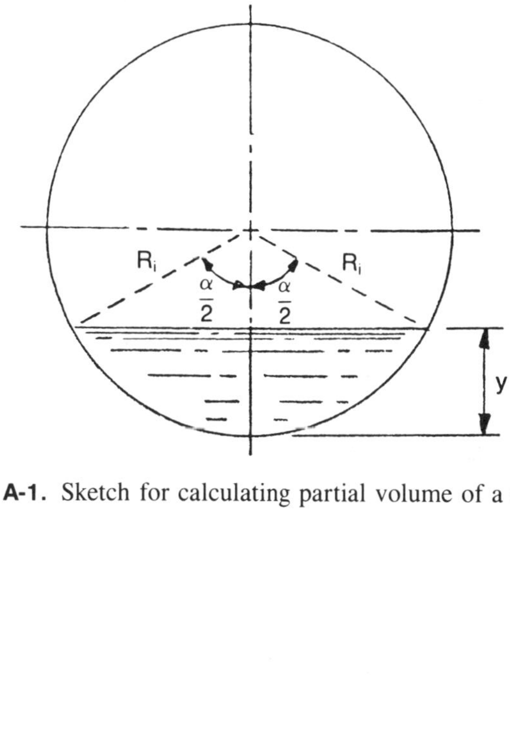

Figure A-1. Sketch for calculating partial volume of a cylinder.

PARTIAL VOLUME OF A HEMISPHERICAL HEAD

$$V_P = \frac{\pi y^2(3R_i - D)}{3} \qquad \text{(A-2)}$$

V_P = partial volume shown in shaded region

Example

For vertical volume in Figure A-2a find partial volume for a head with R_i = 50 in. and y = 35 in.:

$$V_P = \frac{\pi(35)^2[3(50) - 100]}{3} = 64{,}140.85 \text{ in.}^3 = 277.7 \text{ gal}$$

Example

For horizontal volume in Figure A-2b find partial volume for a head with R_i = 50 in. and y = 35 in.:

$$V_P = \frac{277.7}{2} = 138.85 \text{ gal}$$

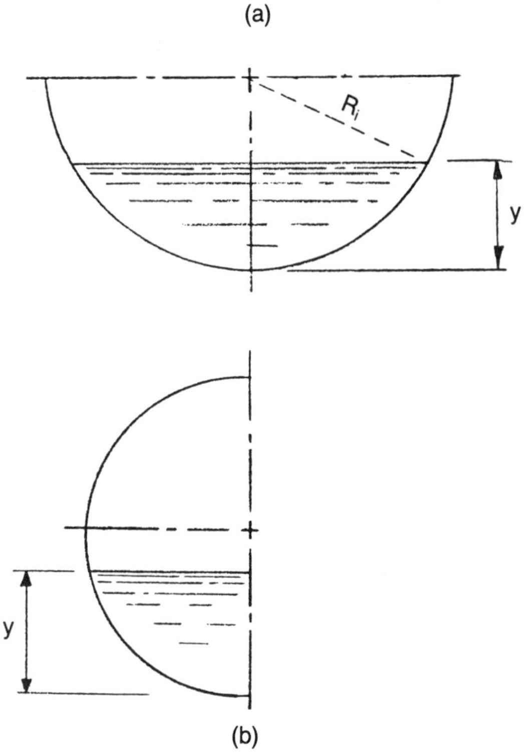

Figure A-2. Partial volume of vertical hemispherical head. (B) Partial volume of horizontal hemispherical head.

PARTIAL VOLUMES OF SPHERICALLY DISHED HEADS

Horizontal Head

The partial volume of a horizontal head (Figure A-3) is

$$V = \alpha \left| \frac{\sqrt{(\rho^2 - y^2)^3} - \sqrt{(\rho^2 - R_i^2)^3}}{3} - \frac{L(R_i^2 - y^2)}{2} \right| \quad \text{(A-3)}$$

Vertical Head

The partial volume of a vertical head (Figure A-4) is

$$V = \frac{\pi y(3x^2 + y^2)}{6} \quad \text{(A-4)}$$

or

$$V = \frac{\pi y^2(3\rho - y)}{3} \quad \text{(A-5)}$$

Example—Spherically Dished Horizontal Head

A spherically dished head with a 114-in. ϕ OD is spun from 1-in. plate. Determine the partial volume of 10 in. of liquid. From vessel head manufacturer's catalog we determine the following:

IDD = 16.786 in. (Figure A-5), ρ = 108 in.

$$R_i = \frac{114 - 2(1.0)}{2} = 56.0 \text{ in.}$$

$$\alpha = 159.43° = 2.78$$

$$L = 108 - 16.786 = 91.21 \text{ in.}$$

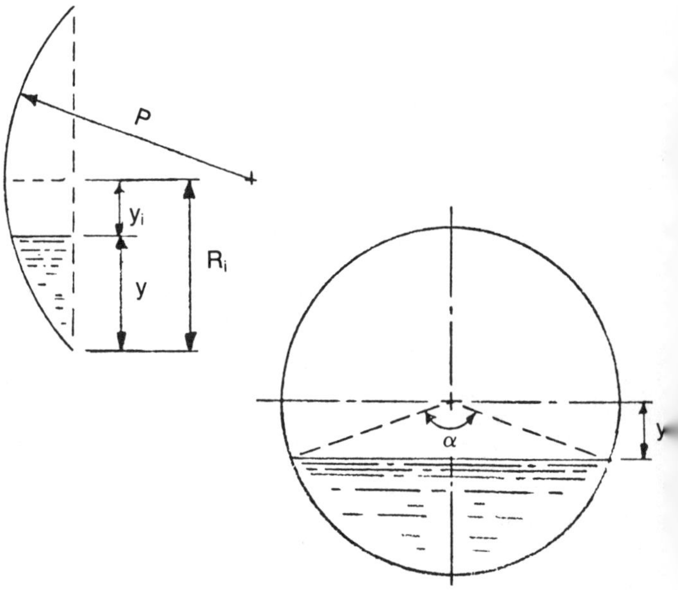

Figure A-3. Partial volume of spherically dished horizontal heads.

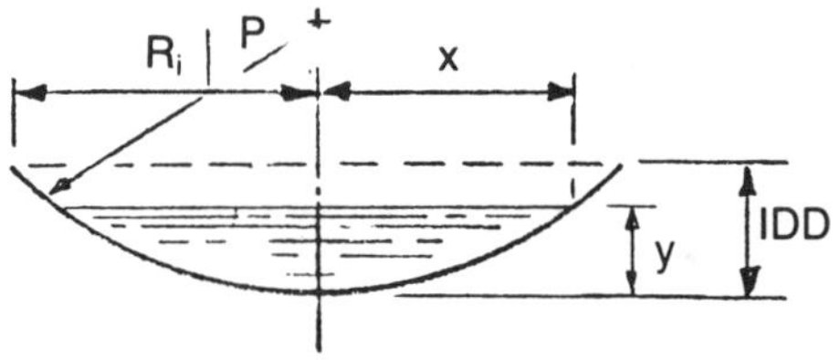

Figure A-4. Partial volume of spherically dished vertical heads.

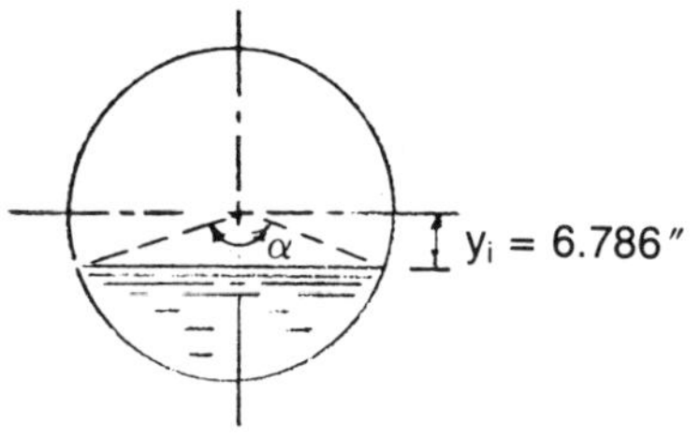

Figure A-5.

$$V = (2.78)\left|\frac{\sqrt{(108^2 - 6.786^2)^3} - \sqrt{(108^2 - 56^2)^3}}{3} - \frac{(91.21)(56^2 - 6.786^2)}{2}\right|$$

$$V = 38{,}893.21 \text{ in.}^3 = 168.37 \text{ gal}$$

Example— Spherically Dished Vertical Head

For the same head above, determine the partial volume of a head of liquid of 9 in.

$$x = 55.456 \text{ in.}$$

$$V = \frac{\pi(9)[3(55.456)^2 + 9^2]}{6} = 14{,}874 \text{ in.}^3 = 64.4 \text{ gal}$$

PARTIAL VOLUMES OF ELLIPTICAL HEADS

The exact partial volume of a horizontal elliptical head (Figure A-6) is as follows:

$$V = \frac{(IDD)\alpha}{6R_i}\sqrt{(R_i^2 - y_i^2)^3} \qquad \text{(A-6)}$$

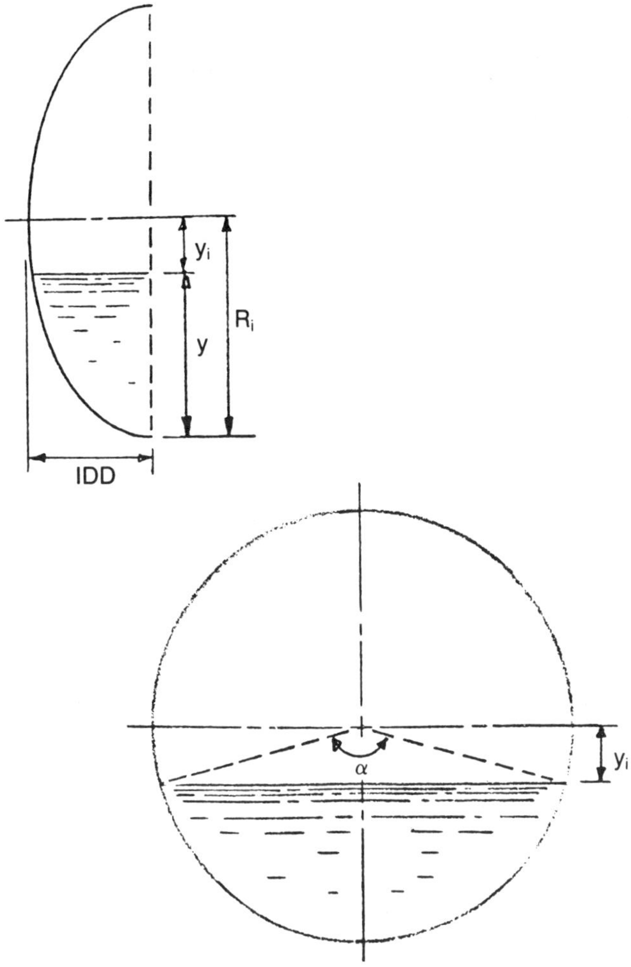

Figure A-6. Partial volume of horizontal elliptical head.

Vertical Elliptical Heads

Volume of top portion ② of Figure A-7 is

$$V_{②} = \frac{\pi R_i^2}{2}\left[y - \frac{y^3}{3(IDD)^2}\right] \qquad \text{(A-7)}$$

Volume of bottom portion ① is

$$V_{①} = \frac{2\pi(IDD)R_i^2}{6} - \frac{\pi R_i^2}{2}\left[y - \frac{y^3}{3(IDD)^2}\right] \qquad \text{(A-8)}$$

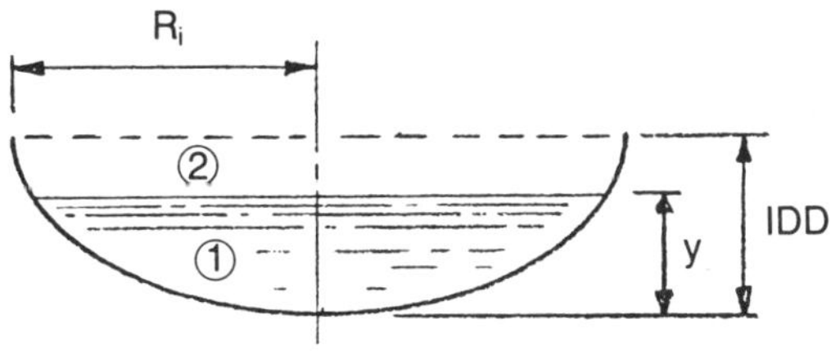

Figure A-7. Partial volume of vertical elliptical head.

Horizontal Head Example

Find the partial volume of a 2:1 (R_i/IDD = 2) elliptical head that is 108-in. OD. The level of the liquid is 35 in., and the head is spun from 1-in. plate.

$$IDD = \frac{108 - 2(1.0)}{4} = 26.50 \text{ in.}$$

From Equation A-6 and Figure A-8 we have the following:

$$V = \frac{(IDD)\,\alpha}{6R_i}\sqrt{(R_i^2 - y_i^2)^3}$$

$$\alpha = 138.80° = 2.42$$

$$V = \frac{(19.0)(2.42)}{6(53)}\sqrt{((53)^2 - (19)^2)^3}$$

$$V = 17{,}512.94 \text{ in.}^3 = 75.81 \text{ gal}$$

Vertical Head Example

For some head above, determine the partial volume for a vertical head with 19 in. of liquid. Using Equation A-8 we have the following:

$$V = \frac{2\pi(IDD)R_i^2}{6} - \frac{\pi R_i}{2}\left[y_i - \frac{y_i^3}{3(IDD)^2}\right]$$

$$V = \frac{2\pi(26.50)(53.0)^2}{6} - \frac{\pi(53.0)}{2}\left[19.0 - \frac{(19.0)^3}{3(26.50)^2}\right]$$

$$V = 77{,}951.81 \text{ in.}^3 - 1310.75 \text{ in.}^3$$

$$V = 76{,}641.06 \text{ in.}^3 = 331.78 \text{ gal}$$

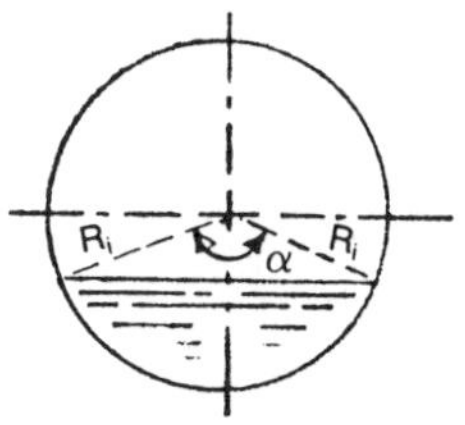

Figure A-8.

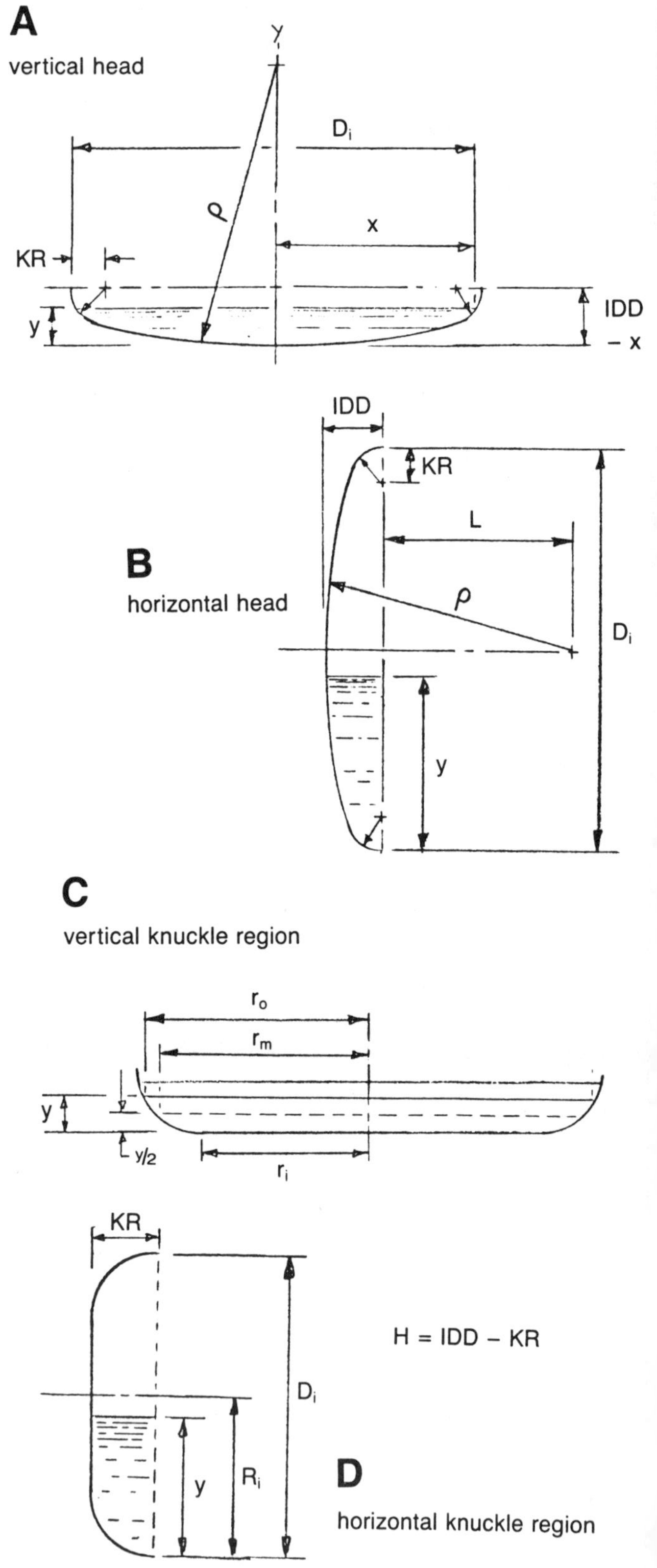

Figure A-9. Partial volumes of torispherical heads: (A) vertical, (B) horizontal, (C) vertical knuckle region, (D) horizontal knuckle region.

PARTIAL VOLUMES OF TORISPHERICAL HEADS

For Figures A-9 and A-10,

V_k = knuckle volume
V_D = dish volume
KR = knuckle radius
y = height of liquid
IDD = inside depth of dish
ρ = inside dish radius

For vertical heads (Figure A-9c) the knuckle-cylinder partial volume is

$$V_k = \frac{\pi y}{6}(r_o^2 + 4r_M^2 + r_i^2) \tag{A-9}$$

The partial volume of the dish region of a vertical head is

$$V_D = \frac{\pi y(3x^2 + y^2)}{6} \tag{A-10}$$

The total partial volume in a vertical head is

$$V_V = \frac{\pi H}{6}(r_o^2 + 4r_M^2 + r_i^2) + \frac{\pi y(3x^2 + y^2)}{6} \tag{A-11}$$

where y_d = IDD − KR

Horizontal Torispherical Heads

Partial Volume of Dish ① (Figure A-11)

$$V_{①} = \alpha\left|\frac{\sqrt{(\rho^2 - y_i^2)^3} - \sqrt{(\rho^2 - R_i^2)^3}}{3} - \frac{L(R_i^2 - y_i^2)}{2}\right| \tag{A-12}$$

Volume of Knuck-Cylinder Region ② (Figure A-12)

$$V_{②} = \alpha\left[\frac{4(KR)}{3\pi} + (R_i - KR) + (R_i - KR)^2\right] \tag{A-13}$$

The total partial volume for a horizontal torispherical head is as follows:

$$V_T = V_{①} + V_{②}$$

$$V_T = \alpha\left|\frac{\sqrt{(\rho^2 - y_i^2)^3} - \sqrt{(\rho^2 - R_i^2)^3}}{3} - \frac{L(R_i^2 - y_i^2)}{2}\right| + \alpha\left[\frac{4(KR)}{3\pi} + (R_i - KR) + (R_i - KR)^2\right] \tag{A-14}$$

where L = ρ − IDD

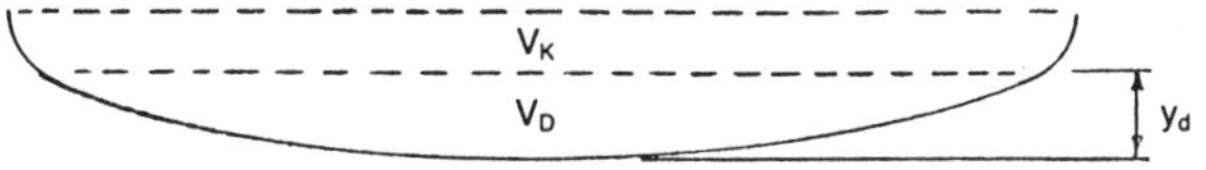

Figure A-10.

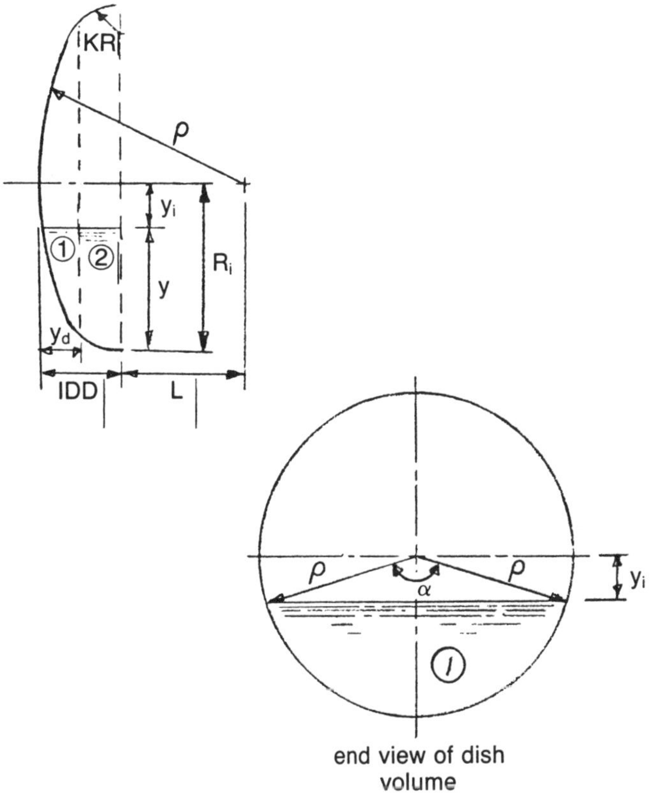

Figure A-11. Sketch for example partial volume calculation of horizontal torispherical head.

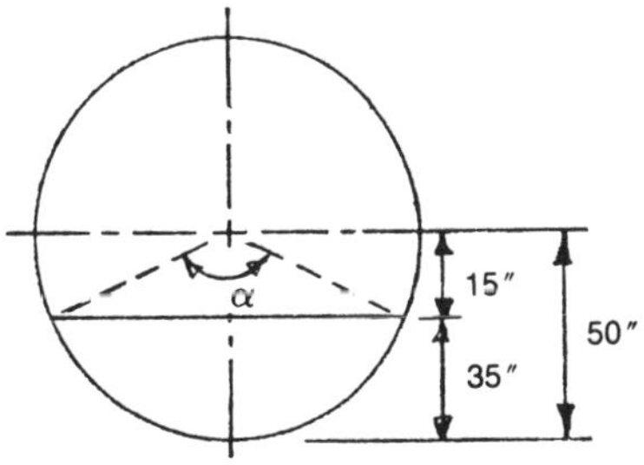

Figure A-12.

Horizontal Head Example

A 102-in. ϕ OD flanged and dished (torispherical) head made to ASME specifications ($KR \geqslant 0.60\rho$ and $KR > 3t_h$, t_h = head thickness) is spun from 1-in. plate. The head is horizontal and the liquid level is 35-in. determine the partial volume.

From the vessel head manufacturer's catalog and Figure A-12 we determine the following:

ρ = 96 in., KR = 6.125 in., IDD = 17.562 in.

$$R_i = \frac{100}{2} = 50 \text{ in.}, \; L = 96.0 - 17.562 = 78.438 \text{ in.}$$

From Equation A-14 we have

$$V_T = (2.532) \left| \frac{\sqrt{(96^2 - 15^2)^3} - \sqrt{(96^2 - 50^2)^3}}{3} - \frac{(78.438)(50^2 - 15^2)}{2} \right| + (2.532) \left[\frac{4(6.125)}{3\pi} + (50.00 - 6.125) + (50.00 - 6.125)^2 \right]$$

V_T = 34,093.44 in.^{3} = 147.59 gal

Vertical Head Example

A 138-in. ϕ OD F&D (flanged and dished) head *not* made to ASME specifications is spun from 1½-in. plate. The head is vertical and the liquid level is 18-in. Determine the partial volume.

From the vessel head manufacturer's catalog we determine the following:

ρ = 132 in., KR = 3 in., IDD = 20.283 in.

$$R_i = \frac{138 - 2(1.5)}{2} = 67.50 \text{ in.};$$

$$x = 67.50 - [(3)^2 - H^2]^{0.5} = 66.446 \text{ in.}$$

For knuckle-cylinder region,

$r_o = R_i = 67.50$; $r_i \cong R_i - KR = 67.50 - 3.00 = 64.50$ in.

$$r_m = \frac{67.50 + 64.50}{2} = 66.0;$$

$$h = |20.283 - (3.0 + 15.0)| = 2.283 \text{ in.}$$

$$V_V = \frac{\pi(2.283)}{6}[(67.50)^2 + 4(66.0)^2 + (64.50)^2] + \frac{\pi(17.283)[3(64.500)^2 + (17.283)^2]}{6}$$

V_V = 31,247.726 in.^{3} + 115,645.832 in.^{3}

V_V = 146,893.558 in.^{3} = 635.903 gal

INTERNAL PRESSURE ASME FORMULATIONS WITH OUTSIDE DIMENSIONS

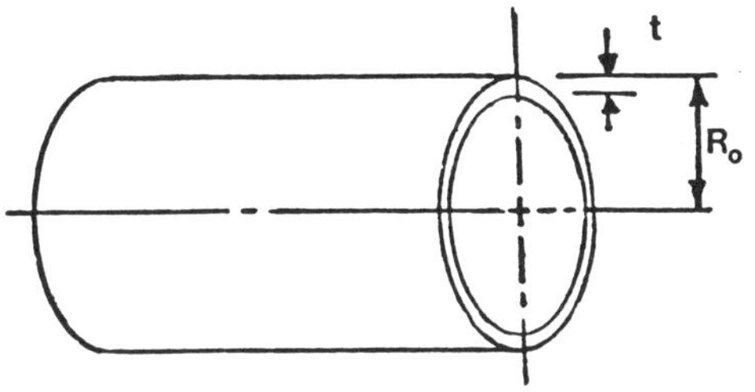

Cylindrical Shell

Longitudinal Joint

$$t = \frac{PR}{\sigma E + 0.4P} \qquad P = \frac{\sigma Et}{R - 0.4t}$$

Circumferential Joint

$$t = \frac{PR_o}{2\sigma E + 1.4P} \qquad P = \frac{2\sigma Et}{R_o - 1.4t}$$

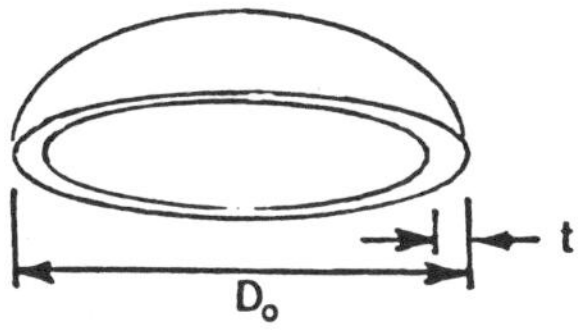

2:1 Ellipsoidal Head

$$t = \frac{PD_o}{2\sigma E + 1.8P} \qquad P = \frac{2\sigma Et}{D_o - 1.8t}$$

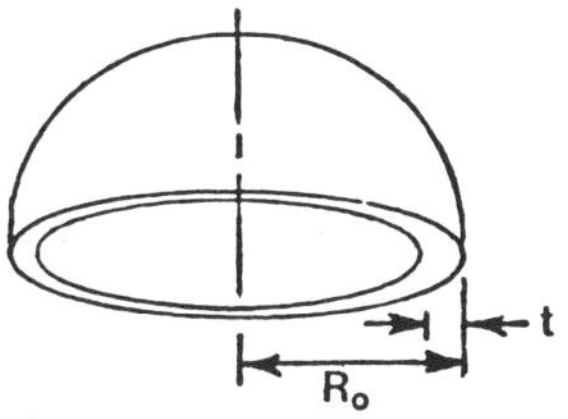

Sphere and Hemispherical Head

$$t = \frac{PR_o}{2\sigma E + 0.8P} \qquad P = \frac{2\sigma Et}{R_o - 0.8t}$$

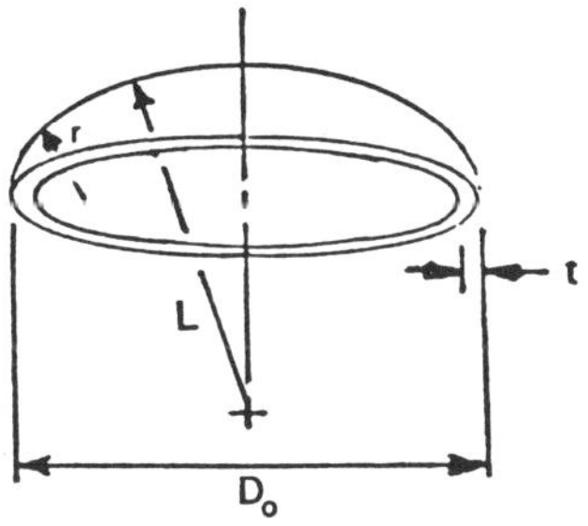

ASME Flanged and Dished Head when $L/R = 16\frac{2}{3}$

$$t = \frac{0.885PL}{\sigma E + 0.8P} \qquad P = \frac{\sigma Et}{0.885L - 0.8t}$$

When $L/R < 16\frac{2}{3}$

$$t = \frac{PLM}{2\sigma E + P(M - 0.2)} \qquad P = \frac{2\sigma Et}{ML - t(M - 0.2)}$$

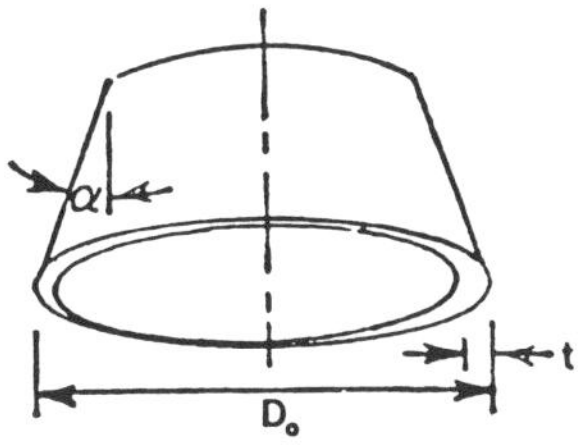

Conical Section

$$t = \frac{PD_o}{2\cos\alpha(\sigma E + 0.4P)} \qquad P = \frac{2SEt\cos\alpha}{D_o - 0.8t\cos\alpha}$$

INTERNAL PRESSURE ASME FORMULATIONS WITH INSIDE DIMENSIONS

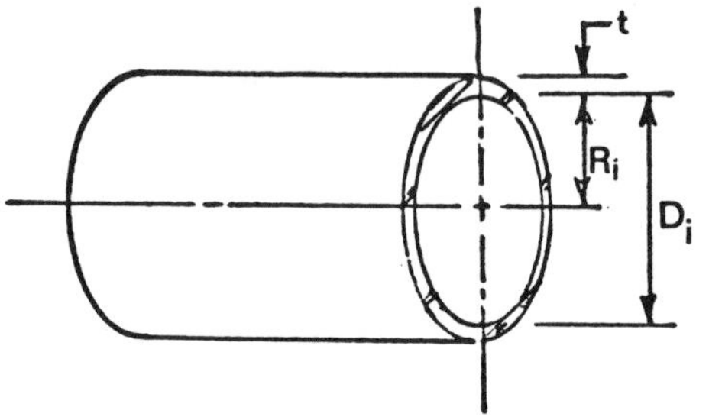

Cylindrical Shell

Longitudinal Joint

$$t = \frac{PR_i}{\sigma E - 0.6P} \qquad P = \frac{\sigma Et}{R_i + 0.6t}$$

Circumferential Joint

$$t = \frac{PR_i}{2\sigma E + 0.4P} \qquad P = \frac{2\sigma Et}{R_i - 0.4t}$$

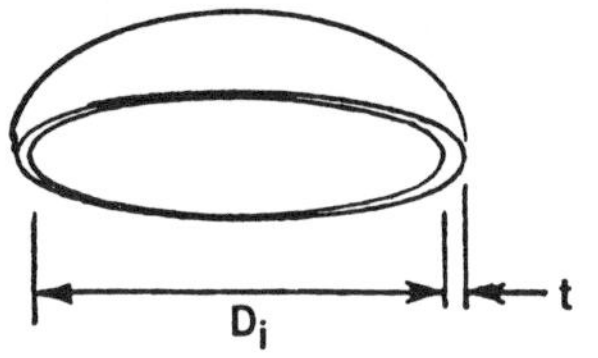

2:1 Ellipsoidal Head

$$t = \frac{PD_i}{2\sigma E - 0.2P} \qquad P = \frac{2\sigma Et}{D_i + 0.2t}$$

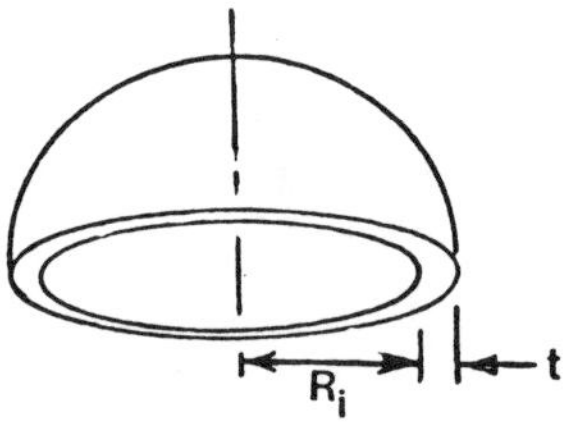

Sphere and Hemispherical Head

$$t = \frac{PR_i}{2SE - 0.2P} \qquad P = \frac{2\sigma Et}{R + 0.2t}$$

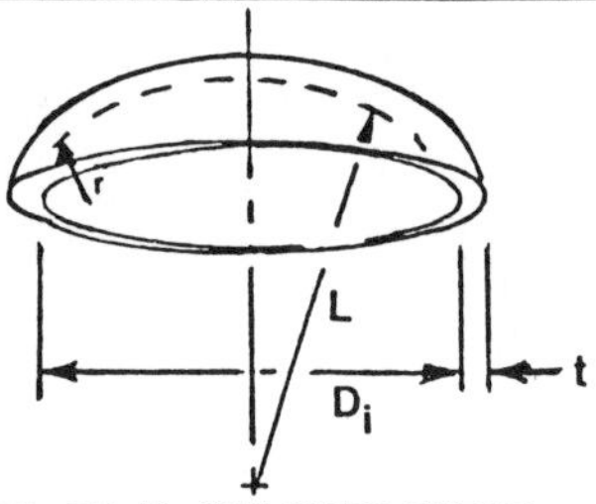

FOR VALUES OF M SEE SUPPLEMENT

ASME Flanged and Dished Head when L/R = 16⅔

$$t = \frac{0.885PL}{\sigma E - 0.1P} \qquad P = \frac{\sigma Et}{0.885L + 0.1t}$$

When L/R < 16⅔

$$t = \frac{PLM}{2\sigma E - 0.2P} \qquad P = \frac{2\sigma Et}{LM + 0.2t}$$

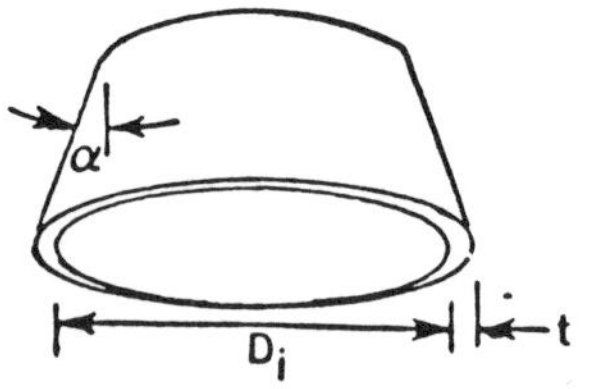

Conical Section

$$t = \frac{PD_i}{2\cos\alpha(\sigma E - 0.6P)} \qquad P = \frac{2\sigma Et\cos\alpha}{D_i + 1.2t\cos\alpha}$$

Supplement for ASME Formulations

1. For a cylindrical shell, when the wall thickness exceeds one half the inside radius or $P > 0.385\sigma E$, the formulas in ASME Code Appendix 1-2 shall be used.
2. For hemispherical heads without a straight flange, the efficiency of the head-to-shell joint is to be used if it is less than the efficiency of the seams in the head.
3. For ellipsoidal heads, where the ratio of the major axis is other than 2:1, refer to ASME Code Appendix 1-4(c).
4. To use the formulations for a conical section in the table, the half apex angle, α, shall not exceed 30°. If $\alpha > 30°$, then a special analysis is required per ASME Code Appendix 1-5(e).
5. For an ASME flanged and dished head (torispherical head) when $L/r < 16\frac{2}{3}$ the following values of M shall be used:

Values of Factor M

L/r	1.00	1.25	1.50	1.75	2.00	2.25	2.50	2.75	3.00	3.25	3.50	4.00	4.50	5.00	5.50	6.00	6.50
M	1.00	1.03	1.06	1.08	1.10	1.13	1.15	1.17	1.18	1.20	1.22	1.25	1.28	1.31	1.34	1.36	1.39
L/r	7.00	7.50	8.00	8.50	9.00	9.50	10.0	10.5	11.0	11.5	12.0	13.0	14.0	15.0	16.0	16⅔	*
M	1.41	1.44	1.46	1.48	1.50	1.52	1.54	1.56	1.58	1.60	1.62	1.65	1.69	1.72	1.75	1.77	

* The maximum allowed ratio: $L - t = D$. When $L/r > 16\frac{2}{3}$ (non-ASME Code construction), the values of M may be calculated by

$$M = \frac{1}{4\left(3 + \sqrt{\frac{L}{r}}\right)}$$

Appendix B

National Wind Design Standards

A standard is a collection of current practices, past experiences, and research knowledge. Standards that are developed by consensus groups (e.g., ASTM, ANSI), trade associations (e.g., AISC, ACI), or government groups (e.g., HUD, CPSC) carry more authority than other standards because they reflect wider ranges of materials.

The ANSI A58.1–1982 is a collection of information that is considered to be the state-of-the-art in the design of buildings and other structures. Local and regional building codes adopt portions of the ANSI standard for their own use. These local and regional codes are developed to represent the needs and interests of their respective areas and are written in legal language to be incorporated into state and local laws. Because these building codes are regional or local in scope, they often do not include everything in the ANSI standard, which is national in perspective. For this reason, one must be certain that a local code written for one area is applicable to the site being considered.

The ANSI standard does not have as much authority as the ASME vessel codes, and, unfortunately, does not have a referral committee or group to officially interpret the document. Therefore, one must make decisions based on past experience and accepted methods of design. The ANSI standard (Paragraph 6.6, p. 16) states that in determining the value for the gust response factor a rational analysis can be used. A note below the paragraph states that *one* such procedure for determining the gust response factor is in the standard's appendix. The note at the top of the appendix (p. 52) states clearly that it is not a part of the ANSI A58.1 miminum design standard. What all this implies is that one may follow the guide of the ANSI standard's appendix or use another rational analysis, which includes another wind standard. Thus, one *can* use another standard for design purposes. One of the most widely accepted international standards is the Australian Standard 1170, Part 2–1983, SAA Loading Code Part 2—Wind Forces.

The Australian Standard 1170 is more applicable to the process industries because in it are shape factors for geometries that are more common in that industry, e.g., circular shapes. However, before applying the shape factors of the Australian standard to the ANSI or any other national standard, one must be very careful to correctly convert the factors. This is because the codes have different basis upon which these factors are determined, and a direct application of other parameters is not possible. This is discussed later after we discuss the basis for the various standards.

CRITERIA FOR DETERMINING WIND SPEED

Wind is caused by differential heating of air masses by the sun. These masses of air at approximately one mile above the ground circulate air around their centers of pressure. At this altitude, the velocity and direction of the wind is almost entirely determined by macro-scale forces caused by large scale weather systems. Below this gradient height, the wind is modified by surface roughness, which reduces its velocity and changes its direction and turbulence. A secondary criterion, except for extreme wind conditions, is the temperature gradient, which affects the vertical mobility of turbulent eddies and therefore influences the surface velocity and the gradient height. Therefore, the exact nature of the surface wind at any point depends, first, on the general weather situation, which determines the gradient wind and the temperature gradient, and, second, on the surrounding topography and ground roughness which, together with the temperature gradient, modify the gradient wind to the surface wind.

Wind motion is further complicated by the rotation of the earth, which induces additional forces that cause the air moving across the earth's surface to be subjected to a force at right angles to the wind velocity vector. These additional forces are known as Coriolis forces.

Each country has adopted its own standard for measuring wind velocity. The U.S. National Weather Service and U.S. codes use the fastest-mile wind speed, which is defined as the average speed of one mile of air passing an anemometer. Thus, a fastest-mile wind speed of 120 mph means that a "mile" of wind passed the anemometer during a 30-second period. Other nations, namely Australia and Great Britain, use the two-second gust speed. This is based on the worst 2-second mean as measured by a cup anemometer. The mean gust speeds are recorded over a period of time such that a mean recurrence interval is determined. The mean recurrence interval is the reciprocal of the probability of exceeding a wind speed of a given magnitude at a particular location in one year. The risk, or probability, R, that the design wind speed will be equaled or surpassed at least once in the life of the tower is given by the expression

$$R = 1 - (1 - P_a)^n$$

where P_a = annual probability of exceedance (reciprocal of the mean recurrence interval)

n = life of the tower or stack

The risk that a given wind speed of specified magnitude will be equaled or exceeded increases with the period of time that the tower is exposed to the wind. Values of risk of exceeding design wind speed for a designated annual probability and a given design life of the structure are shown in Table B-1.

For example, if the design wind speed for a tower is based on an annual probability of 0.02 (mean recurrence interval of 50 years) and the projected tower life is 25 years, there is a 0.40 probability that the design wind speed will be exceeded during the life of the structure. The United States and Australian wind codes use the 50-year recurrence interval.

Table B-1
Probability of Exceeding Wind Design Speed
$P_A = 1 - (1 - P_A)^N$

Annual Probability	Design Life of Structure in N Years						
P_A	1	5	10	15	25	50	100
0.10	0.100	0.410	0.651	0.794	0.928	0.995	0.999
0.05	0.050	0.226	0.401	0.537	0.723	0.923	0.994
0.01	0.010	0.049	0.096	0.140	0.222	0.395	0.634
0.005	0.005	0.025	0.049	0.072	0.118	0.222	0.394

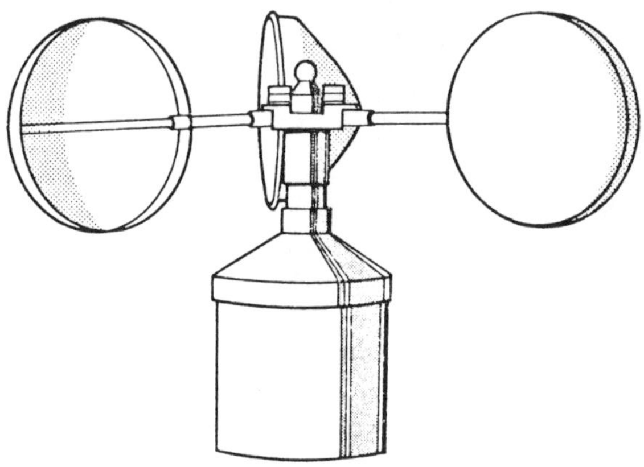

Figure B-1. Cup generator anemometer.

The instrument for measuring the wind in the United States, Great Britain, and Australia is the cup-generator anemometer shown in Figure B-1. This device is operated by the wind striking the cups, which drive a small permanent alternator. The indicator, which incorporates a rectifier, is simply a voltmeter calibrated in miles per hour. In most recent cup-generator models the generator output is used to activate a pen-chart recorder which provides a record of continuous wind speed.

WIND SPEED RELATIONSHIPS

As stated previously, another method can be substituted for the appendix in ANSI A58.1. What this means is that another code could be used instead of the appendix. To do this one must be careful to utilize the correct conversion factors between standards. To accomplish this we refer to Figure B-2. For a 100-mph fastest mile wind speed in ANSI A58.1 we wish to determine the equivalent fastest mile wind speed for a 2-second gust using either the Australian or British code. From Figure B-2 we read from the ordinate 1.54 for 2 sec. Knowing that one mile of wind moving at 100 mph will pass the anemometer in 36 sec, we read 36 sec on the curve and arrive at $V_t/V_{3600} = 1.30$. Thus, the equivalent fastest mile wind speed is

$$V = \left(\frac{1.54}{1.30}\right)(100) \text{ mph} = 118.4 \text{ mph}$$

for a 2-sec gust. For 110 mph, the values becomes

$$V = (1.18)(110) \text{ mph} = 129.8 \text{ mph}$$

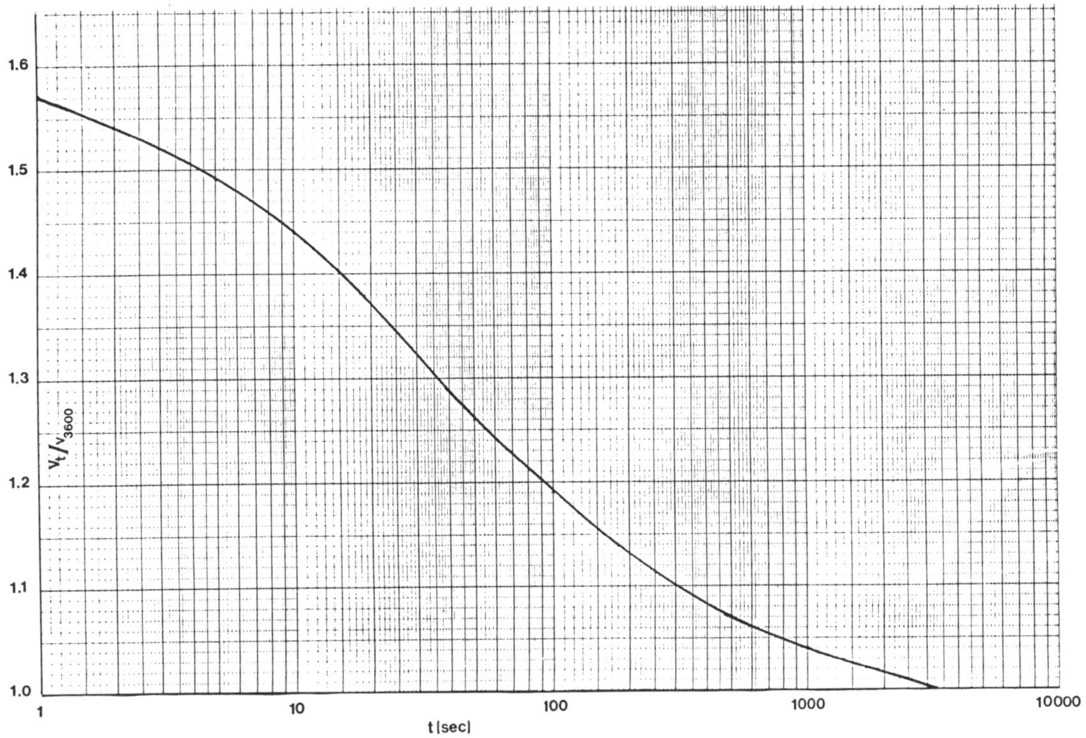

Figure B-2. Ratio of probable maximum wind speed averaged over t seconds to hourly mean speed.

Thus, the 1.18 factor would have to be used in the 2-sec gust code if that code were to be substituted for Appendix A of ANSI A58.1-1982.

Similarly, the Canadian code we must convert to obtain an equivalent fastest mile wind speed from the mean hourly. The mean hourly implies that the wind moves an average of 100 mph across the anemometer in a period of 3,600 sec. Reading Figure B-2 we have $V_t/V_{3600} = 1.0$. Thus

$$\frac{1.0}{1.3} = 0.769$$

which yields an equivalent velocity of 76.9 mph. With the Canadian code one must use 0.769 in use of shape constants and the various other parameters when using with ANSI A58.1. A comparison of the major wind codes is given in Tables B-2, B-3, B-4, and B-5.

ANSI A58.1-1982 WIND CATEGORIES

In the ANSI A58.1-1982 there are four wind categories—A, B, C, and D. The categories are described as follows:

Category A—A very restricted category in which the wind speed is drastically reduced. Most petrochemical and power facilities do not fall within this category. The wind force is reduced because the structure is considered to be among many tall structures. One example would be a ten-story building in downtown Manhattan, New York, where the taller buildings would block the stronger air currents.

Category B—A classification that encompasses some tall structures, but not enough to block the majority of wind gusts. An example of this category would be a tower in the midst of a large petrochemical facility where there were other towers that would block some of the wind force. A forest surrounding a tower is another example.

Category C—The most common classification for petrochemical applications. This category is open terrain where the tower would receive full impact from the wind with minimum ground resistance to the wind. An example of this category would be an open field or an airport.

Category D—A classification for wind moving over water. A beachhead, in which there is flat beach up to a row of buildings would be in Category D. Miami beach, from the ocean front up to the facade of hotels, is a good example. Behind the hotel fronts would be Category C. Another example of this classification would be a tall vertical vessel on an offshore structure.

Table B-2
Major U.S. and Foreign Building Codes and Standards Used in Wind Design

Code or Standard	Edition	Organization	Address
Australian Standard 1170, Part 2-Wind Forces	1983	Standards Association of Australia	Standards House 80 Arthur Street/North Sydney, N.S.W. Australia
British Code of Basic Data for Design of Buildings (CP3)	1972	British Standards Institution	British Standards Institution 2 Park Street London, W1A 285, England
Wind Loading Handbook (commentary on CP3)	1974	Building Research Establishment	Building Research Station Garston, Watford, WD2 7JR, England
National Building Code of Canada (NRCC No. 17303)	1980	National Research Council of Canada	National Research Council of Canada
The Supplement to the National Building Code of Canada (NRCC 17724)	1980	National Research Council of Canada	Ottawa, Ontario K1A OR6 Canada
ANSI A58.1-1982	1982	American National Standards Institute	1430 Broadway New York, New York 10018
Uniform Building Code	1982	International Conference of Building Officials	5360 South Workman Mill Road Whittier, California 90601
Standard Building Code	1982 with 1983 rev.	Southern Building Code Congress International	900 Montclair Road Birmingham, Alabama 35213
Basic Building Code	1984	Building Officials and Code Administrators International, Inc.	17926 South Halsted Street Homewood, Illinois 60430

Table B-3
Reference Wind Speed

Reference	Australian (SAA, 1983)	British (BSI, 1982)	Canadian (NRCC, 1980)	United States (ANSI, 1982)
Averaging time	2–3 second gust speed	2-second gust speed	Mean hourly	Fastest mile
Equivalent reference wind speed to fastest mile 100 mph	118.4	118.4	76.9	100

Table B-4
Parameters Used in the Major National Standards

Parameter	Australian (SAA, 1983)	British (BSI, 1972)	Canadian (NRCC, 1980)	United States (ANSI, 1982)
Wind Speed				
Terrain roughness	4	4	3	4
Local terrain	Yes	Yes	None	None
Height variation	Yes	Yes	Yes	Yes
Ref. speed	2-sec gusts	2-sec gusts	Mean hourly	Fastest mile
Wind Pressure				
Pressure coefficients	Tables in appendix includes figures	Tables, includes figures	Figures and tables in commentaries	Tables, figures and notes
Gusts				
Magnitude	Gust speed	Gust speed	Gust effect factor	Gust response factor
Spatial correlation	Reduction for large area	None	Gust effect factor	Area averaging
Gust frequency	Dynamic consideration for h/b > 5	Dynamic consideration not included	Dynamic consideration for h/b > 4 in. or for h > 400 ft	Dynamic consideration for h/b > 5
Analysis procedure	This standard is considered by many the best for use in the process industries. Figures and tables are easy to read. The standard actually provides the user with equations to curves. The analysis procedure is straight-forward.	Overall a very good code, its weakest part is the lack of dynamic consideration.	An excellent wind standard. The analysis procedure is straight-forward and the documents—code and supplement contain tables and figures easy to read.	Although the appendix is technically not considered a part of the standard, it contains figures difficult to read, namely Figure 6. For many structures the data extend beyond the limits of the curves in Figures 6 and 7. In the method in the appendix, one must assume an initial natural frequency, resulting in an iterative process. This method is extremely difficult in designing petrochemical towers without the use of a computer.

Table B-5
Limitations of Codes and Standards

Code or Standard	Statement of Limitation	Location
Australian Standard 1170, Part 2—1983	"Minimum Design Loads on Structures"	Title
National Building Code of Canada (NRCC, 1980)	". . . Essentially a Set of Minimum Regulations . . ."	Guide to the Use of the Code
British CP3	". . . Does Not Apply to Buildings . . . That Are of Unusual Shape or Location For Which Special Investigations May Be Necessary . . ."	Section 1 (Scope)
United States ANSI A58.1	"Minimum Design Loads . . ." "Specific Guidelines Are Given For . . . Wind Tunnel Investigations . . . For Buildings . . . Having Irregular Shapes . . ."	Title Paragraph 6.1
Uniform Building Code	"The purpose . . . is to provide minimum standards . . ."	Section 102
Basic Building Code (BOCA, 1984)	"The Basic Minimum Wind Speeds Are Shown in Figure 912.1 . . ."	Section 912.1
Standard Building Code, 1982 (SBCCI, 1982)	"The Purpose of This Code is to Provide Minimum Requirements . . ." "The Building Official May Require Evidence to Support the Design Pressures Used in the Design of Structures Not Included in This Section."	Preface Article 1205.2(a)

Appendix C

Properties of Pipe

PROPERTIES OF PIPE *

The following formulas are used in the computation of the values shown in the table:

† weight of pipe per foot (pounds)	=	$10.6802t(D-t)$
weight of water per foot (pounds)	=	$0.3405d^2$
square feet outside surface per foot	=	$0.2618D$
square feet inside surface per foot	=	$0.2618d$
inside area (square inches)	=	$0.785d^2$
area of metal (square inches)	=	$0.785(D^2-d^2)$
moment of inertia (inches⁴)	=	$0.0491(D^4-d^4)$
	=	$A_m R_g^2$
section modulus (inches³)	=	$\frac{0.0982(D^4-d^4)}{D}$
radius of gyration (inches)	=	$0.25\sqrt{D^2+d^2}$

A_m = area of metal (square inches)
d = inside diameter (inches)
D = outside diameter (inches)
R_g = radius of gyration (inches)
t = pipe wall thickness (inches)

† The ferritic steels may be about 5% less, and the austenitic stainless steels about 2% greater than the values shown in this table which are based on weights for carbon steel.

*** schedule numbers**

Standard weight pipe and schedule 40 are the same in all sizes through 10-inch; from 12-inch through 24-inch, standard weight pipe has a wall thickness of ⅜-inch.

Extra strong weight pipe and schedule 80 are the same in all sizes through 8-inch; from 8-inch through 24-inch, extra strong weight pipe has a wall thickness of ½-inch.

Double extra strong weight pipe has no corresponding schedule number.

a: ANSI B36.10 steel pipe schedule numbers

b: ANSI B36.10 steel pipe nominal wall thickness designation

c: ANSI B36.19 stainless steel pipe schedule numbers

nominal pipe size *outside diameter,* in.	schedule number* a	b	c	wall thickness, in.	inside diameter, in.	inside area, sq. in.	metal area, sq. in.	sq ft outside surface, per ft	sq ft inside surface, per ft	weight per ft, lb†	weight of water per ft, lb	moment of inertia, in.⁴	section modulus, in.³	radius gyration, in.
⅛ *0.405*	......	......	10S	0.049	0.307	0.0740	0.0548	0.106	0.0804	0.186	0.0321	0.00088	0.00437	0.1271
	40	Std	40S	0.068	0.269	0.0568	0.0720	0.106	0.0705	0.245	0.0246	0.00106	0.00525	0.1215
	80	XS	80S	0.095	0.215	0.0364	0.0925	0.106	0.0563	0.315	0.0157	0.00122	0.00600	0.1146
¼ *0.540*	......	......	10S	0.065	0.410	0.1320	0.0970	0.141	0.1073	0.330	0.0572	0.00279	0.01032	0.1694
	40	Std	40S	0.088	0.364	0.1041	0.1250	0.141	0.0955	0.425	0.0451	0.00331	0.01230	0.1628
	80	XS	80S	0.119	0.302	0.0716	0.1574	0.141	0.0794	0.535	0.0310	0.00378	0.01395	0.1547
⅜ *0.675*	......	......	SS	0.065	0.710	0.396	0.1582	0.220	0.1859	0.538	0.1716	0.01197	0.0285	0.2750
	......	......	10S	0.065	0.545	0.2333	0.1246	0.177	0.1427	0.423	0.1011	0.00586	0.01737	0.2169
	40	Std	40S	0.091	0.493	0.1910	0.1670	0.177	0.1295	0.568	0.0827	0.00730	0.02160	0.2090
	80	XS	80S	0.126	0.423	0.1405	0.2173	0.177	0.1106	0.739	0.0609	0.00862	0.02554	0.1991
½ *0.840*	......	......	5S	0.065	0.710	0.3959	0.1583	0.220	0.1859	0.538	0.171	0.0120	0.0285	0.2750
	......	......	10S	0.083	0.674	0.357	0.1974	0.220	0.1765	0.671	0.1547	0.01431	0.0341	0.2692
	40	Std	40S	0.109	0.622	0.304	0.2503	0.220	0.1628	0.851	0.1316	0.01710	0.0407	0.2613
	80	XS	80S	0.147	0.546	0.2340	0.320	0.220	0.1433	1.088	0.1013	0.02010	0.0478	0.2505
	160	......	......	0.187	0.466	0.1706	0.383	0.220	0.1220	1.304	0.0740	0.02213	0.0527	0.2402
	......	XXS	......	0.294	0.252	0.0499	0.504	0.220	0.0660	1.714	0.0216	0.02425	0.0577	0.2192
¾ *1.050*	......	......	5S	0.065	0.920	0.665	0.2011	0.275	0.2409	0.684	0.2882	0.02451	0.0467	0.349
	......	......	10S	0.083	0.884	0.614	0.2521	0.275	0.2314	0.857	0.2661	0.02970	0.0566	0.343
	40	Std	40S	0.113	0.824	0.533	0.333	0.275	0.2157	1.131	0.2301	0.0370	0.0706	0.334
	80	XS	80S	0.154	0.742	0.432	0.435	0.275	0.1943	1.474	0.1875	0.0448	0.0853	0.321
	160	......	......	0.218	0.614	0.2961	0.570	0.275	0.1607	1.937	0.1284	0.0527	0.1004	0.304
	......	XXS	......	0.308	0.434	0.1479	0.718	0.275	0.1137	2.441	0.0641	0.0579	0.1104	0.2840
1 *1.315*	......	......	5S	0.065	1.185	1.103	0.2553	0.344	0.310	0.868	0.478	0.0500	0.0760	0.443
	......	......	10S	0.109	1.097	0.945	0.413	0.344	0.2872	1.404	0.409	0.0757	0.1151	0.428
	40	Std	40S	0.133	1.049	0.864	0.494	0.344	0.2746	1.679	0.374	0.0874	0.1329	0.421
	80	XS	80S	0.179	0.957	0.719	0.639	0.344	0.2520	2.172	0.311	0.1056	0.1606	0.407
	160	......	......	0.250	0.815	0.522	0.836	0.344	0.2134	2.844	0.2261	0.1252	0.1903	0.387
	......	XXS	......	0.358	0.599	0.2818	1.076	0.344	0.1570	3.659	0.1221	0.1405	0.2137	0.361
1¼ *1.660*	......	......	5S	0.065	1.530	1.839	0.326	0.434	0.401	1.107	0.797	0.1038	0.1250	0.564
	......	......	10S	0.109	1.442	1.633	0.531	0.434	0.378	1.805	0.707	0.1605	0.1934	0.550
	40	Std	40S	0.140	1.380	1.496	0.669	0.434	0.361	2.273	0.648	0.1948	0.2346	0.540
	80	XS	80S	0.191	1.278	1.283	0.881	0.434	0.335	2.997	0.555	0.2418	0.2913	0.524
	160	......	......	0.250	1.160	1.057	1.107	0.434	0.304	3.765	0.458	0.2839	0.342	0.506
	......	XXS	......	0.382	0.896	0.631	1.534	0.434	0.2346	5.214	0.2732	0.341	0.411	0.472
1½ *1.900*	......	......	5S	0.065	1.770	2.461	0.375	0.497	0.463	1.274	1.067	0.1580	0.1663	0.649
	......	......	10S	0.109	1.682	2.222	0.613	0.497	0.440	2.085	0.962	0.2469	0.2599	0.634

*Courtesy of ITT Grinnell.

PROPERTIES OF PIPE (Continued)

nominal pipe size *outside diameter,* in.	schedule number* a	schedule number* b	schedule number* c	wall thick-ness, in.	inside diam-eter, in.	inside area, sq. in.	metal area, sq. in.	sq ft outside surface, per ft	sq ft inside surface, per ft	weight per ft, lb†	weight of water per ft, lb	moment of inertia, $in.^4$	section modu-lus, $in.^3$	radius gyra-tion, in.
1½ *1.900*	40	Std	40S	0.145	1.610	2.036	0.799	0.497	0.421	2.718	0.882	0.310	0.326	0.623
	80	XS	80S	0.200	1.500	1.767	1.068	0.497	0.393	3.631	0.765	0.391	0.412	0.605
	160	......	......	0.281	1.338	1.406	1.429	0.497	0.350	4.859	0.608	0.483	0.508	0.581
	......	XXS	......	0.400	1.100	0.950	1.885	0.497	0.288	6.408	0.412	0.568	0.598	0.549
	......	......	......	0.525	0.850	0.567	2.267	0.497	0.223	7.710	0.246	0.6140	0.6470	0.5200
	......	......	......	0.650	0.600	0.283	2.551	0.497	0.157	8.678	0.123	0.6340	0.6670	0.4980
2 *2.375*	......	......	5S	0.065	2.245	3.96	0.472	0.622	0.588	1.604	1.716	0.315	0.2652	0.817
	......	......	10S	0.109	2.157	3.65	0.776	0.622	0.565	2.638	1.582	0.499	0.420	0.802
	40	Std	40S	0.154	2.067	3.36	1.075	0.622	0.541	3.653	1.455	0.666	0.561	0.787
	80	XS	80S	0.218	1.939	2.953	1.477	0.622	0.508	5.022	1.280	0.868	0.731	0.766
	160	......	......	0.343	1.689	2.240	2.190	0.622	0.442	7.444	0.971	1.163	0.979	0.729
	......	XXS	......	0.436	1.503	1.774	2.656	0.622	0.393	9.029	0.769	1.312	1.104	0.703
	......	......	......	0.562	1.251	1.229	3.199	0.622	0.328	10.882	0.533	1.442	1.2140	0.6710
	......	......	......	0.687	1.001	0.787	3.641	0.622	0.262	12.385	0.341	1.5130	1.2740	0.6440
2½ *2.875*	......	......	5S	0.083	2.709	5.76	0.728	0.753	0.709	2.475	2.499	0.710	0.494	0.988
	......	......	10S	0.120	2.635	5.45	1.039	0.753	0.690	3.531	2.361	0.988	0.687	0.975
	40	Std	40S	0.203	2.469	4.79	1.704	0.753	0.646	5.793	2.076	1.530	1.064	0.947
	80	XS	80S	0.276	2.323	4.24	2.254	0.753	0.608	7.661	1.837	1.925	1.339	0.924
	160	......	......	0.375	2.125	3.55	2.945	0.753	0.556	10.01	1.535	2.353	1.637	0.894
	......	XXS	......	0.552	1.771	2.464	4.03	0.753	0.464	13.70	1.067	2.872	1.998	0.844
	......	......	......	0.675	1.525	1.826	4.663	0.753	0.399	15.860	0.792	3.0890	2.1490	0.8140
	......	......	......	0.800	1.275	1.276	5.212	0.753	0.334	17.729	0.554	3.2250	2.2430	0.7860
3 *3.500*	......	......	5S	0.083	3.334	8.73	0.891	0.916	0.873	3.03	3.78	1.301	0.744	1.208
	......	......	10S	0.120	3.260	8.35	1.274	0.916	0.853	4.33	3.61	1.822	1.041	1.196
	40	Std	40S	0.216	3.068	7.39	2.228	0.916	0.803	7.58	3.20	3.02	1.724	1.164
	80	XS	80S	0.300	2.900	6.61	3.02	0.916	0.759	10.25	2.864	3.90	2.226	1.136
	160	......	......	0.437	2.626	5.42	4.21	0.916	0.687	14.32	2.348	5.03	2.876	1.094
	......	XXS	......	0.600	2.300	4.15	5.47	0.916	0.602	18.58	1.801	5.99	3.43	1.047
	......	......	......	0.725	2.050	3.299	6.317	0.916	0.537	21.487	1.431	6.5010	3.7150	1.0140
	......	......	......	0.850	1.800	2.543	7.073	0.916	0.471	24.057	1.103	6.8530	3.9160	0.9840
3½ *4.000*	......	......	5S	0.083	3.834	11.55	1.021	1.047	1.004	3.47	5.01	1.960	0.980	1.385
	......	......	10S	0.120	3.760	11.10	1.463	1.047	0.984	4.97	4.81	2.756	1.378	1.372
	40	Std	40S	0.226	3.548	9.89	2.680	1.047	0.929	9.11	4.28	4.79	2.394	1.337
	80	XS	80S	0.318	3.364	8.89	3.68	1.047	0.881	12.51	3.85	6.28	3.14	1.307
	......	XXS	......	0.636	2.728	5.845	6.721	1.047	0.716	22.850	2.530	9.8480	4.9240	1.2100
4 *4.500*	......	......	5S	0.083	4.334	14.75	1.152	1.178	1.135	3.92	6.40	2.811	1.249	1.562
	......	......	10S	0.120	4.260	14.25	1.651	1.178	1.115	5.61	6.17	3.96	1.762	1.549
	......	......	......	0.188	4.124	13.357	2.547	1.178	1.082	8.560	5.800	5.8500	2.6000	1.5250
	40	Std	40S	0.237	4.026	12.73	3.17	1.178	1.054	10.79	5.51	7.23	3.21	1.510
	80	XS	80S	0.337	3.826	11.50	4.41	1.178	1.002	14.98	4.98	9.61	4.27	1.477
	120	......	......	0.437	3.626	10.33	5.58	1.178	0.949	18.96	4.48	11.65	5.18	1.445
	......	......	......	0.500	3.500	9.621	6.283	1.178	0.916	21.360	4.160	12.7710	5.6760	1.4250
	160	......	......	0.531	3.438	9.28	6.62	1.178	0.900	22.51	4.02	13.27	5.90	1.416
	......	XXS	......	0.674	3.152	7.80	8.10	1.178	0.825	27.54	3.38	15.29	6.79	1.374
	......	......	......	0.800	2.900	6.602	9.294	1.178	0.759	31.613	2.864	16.6610	7.4050	1.3380
	......	......	......	0.925	2.650	5.513	10.384	1.178	0.694	35.318	2.391	17.7130	7.8720	1.3060
5 *5.563*	......	......	5S	0.109	5.345	22.44	1.868	1.456	1.399	6.35	9.73	6.95	2.498	1.929
	......	......	10S	0.134	5.295	22.02	2.285	1.456	1.386	7.77	9.53	8.43	3.03	1.920
	40	Std	40S	0.258	5.047	20.01	4.30	1.456	1.321	14.62	8.66	15.17	5.45	1.878
	80	XS	80S	0.375	4.813	18.19	6.11	1.456	1.260	20.78	7.89	20.68	7.43	1.839
	120	......	......	0.500	4.563	16.35	7.95	1.456	1.195	27.04	7.09	25.74	9.25	1.799
	160	......	......	0.625	4.313	14.61	9.70	1.456	1.129	32.96	6.33	30.0	10.80	1.760
	......	XXS	......	0.750	4.063	12.97	11.34	1.456	1.064	38.55	5.62	33.6	12.10	1.722
	......	......	......	0.875	3.813	11.413	12.880	1.456	0.998	43.810	4.951	36.6450	13.1750	1.6860
	......	......	......	1.000	3.563	9.966	14.328	1.456	0.933	47.734	4.232	39.1110	14.0610	1.6520

PROPERTIES OF PIPE (Continued)

nominal pipe size *outside diameter,* in.	schedule number* a	b	c	wall thickness, in.	inside diameter, in.	inside area, sq. in.	metal area, sq. in.	sq ft outside surface, per ft	sq ft inside surface, per ft	weight per ft, lb†	weight of water per ft, lb	moment of inertia, in.[4]	section modulus, in.[3]	radius gyration, in.
	......	......	5S	0.109	6.407	32.2	2.231	1.734	1.677	5.37	13.98	11.85	3.58	2.304
	......	......	10S	0.134	6.357	31.7	2.733	1.734	1.664	9.29	13.74	14.40	4.35	2.295
	......	......	......	0.219	6.187	30.100	4.410	1.734	1.620	15.020	13.100	22.6600	6.8400	2.2700
	40	Std	40S	0.280	6.065	28.89	5.58	1.734	1.588	18.97	12.51	28.14	8.50	2.245
6	80	XS	80S	0.432	5.761	26.07	8.40	1.734	1.508	28.57	11.29	40.5	12.23	2.195
6.625	120	......	......	0.562	5.501	23.77	10.70	1.734	1.440	36.39	10.30	49.6	14.98	2.153
	160	......	......	0.718	5.189	21.15	13.33	1.734	1.358	45.30	9.16	59.0	17.81	2.104
	......	XXS	......	0.864	4.897	18.83	15.64	1.734	1.282	53.16	8.17	66.3	20.03	2.060
	......	......	......	1.000	4.625	16.792	17.662	1.734	1.211	60.076	7.284	72.1190	21.7720	2.0200
	......	......	......	1.125	4.375	15.025	19.429	1.734	1.145	66.084	6.517	76.5970	23.1240	1.9850
	......	......	5S	0.109	8.407	55.5	2.916	2.258	2.201	9.91	24.07	26.45	6.13	3.01
	......	......	10S	0.148	8.329	54.5	3.94	2.258	2.180	13.40	23.59	35.4	8.21	3.00
	......	......	......	0.219	8.187	52.630	5.800	2.258	2.150	19.640	22.900	51.3200	11.9000	2.9700
8	20	......	......	0.250	8.125	51.8	6.58	2.258	2.127	22.36	22.48	57.7	13.39	2.962
8.625	30	......	......	0.277	8.071	51.2	7.26	2.258	2.113	24.70	22.18	63.4	14.69	2.953
	40	Std	40S	0.322	7.981	50.0	8.40	2.258	2.089	28.55	21.69	72.5	16.81	2.938
	60	......	......	0.406	7.813	47.9	10.48	2.258	2.045	35.64	20.79	88.8	20.58	2.909
	80	XS	80S	0.500	7.625	45.7	12.76	2.258	1.996	43.39	19.80	105.7	24.52	2.878
	100	......	......	0.593	7.439	43.5	14.96	2.258	1.948	50.87	18.84	121.4	28.14	2.847
	120	......	......	0.718	7.189	40.6	17.84	2.258	1.882	60.63	17.60	140.6	32.6	2.807
8	140	......	......	0.812	7.001	38.5	19.93	2.258	1.833	67.76	16.69	153.8	35.7	2.777
8.625	160	......	......	0.906	6.813	36.5	21.97	2.258	1.784	74.69	15.80	165.9	38.5	2.748
	......	......	......	1.000	6.625	34.454	23.942	2.258	1.734	81.437	14.945	177.1320	41.0740	2.7190
	......	......	......	1.125	6.375	31.903	26.494	2.258	1.669	90.114	13.838	190.6210	44.2020	2.6810
	......	......	5S	0.134	10.482	86.3	4.52	2.815	2.744	15.15	37.4	63.7	11.85	3.75
	......	......	10S	0.165	10.420	85.3	5.49	2.815	2.728	18.70	36.9	76.9	14.30	3.74
	......	......	......	0.219	10.312	83.52	7.24	2.815	2.70	24.63	36.2	100.46	18.69	3.72
	20	......	......	0.250	10.250	82.5	8.26	2.815	2.683	28.04	35.8	113.7	21.16	3.71
	30	......	......	0.307	10.136	80.7	10.07	2.815	2.654	34.24	35.0	137.5	25.57	3.69
	40	Std	40S	0.365	10.020	78.9	11.91	2.815	2.623	40.48	34.1	160.8	29.90	3.67
10	60	XS	80S	0.500	9.750	74.7	16.10	2.815	2.553	54.74	32.3	212.0	39.4	3.63
10.750	80	......	......	0.593	9.564	71.8	18.92	2.815	2.504	64.33	31.1	244.9	45.6	3.60
	100	......	......	0.718	9.314	68.1	22.63	2.815	2.438	76.93	29.5	286.2	53.2	3.56
	120	......	......	0.843	9.064	64.5	26.24	2.815	2.373	89.20	28.0	324	60.3	3.52
	......	......	......	0.875	9.000	63.62	27.14	2.815	2.36	92.28	27.6	333.46	62.04	3.50
	140	......	......	1.000	8.750	60.1	30.6	2.815	2.291	104.13	26.1	368	68.4	3.47
	160	......	......	1.125	8.500	56.7	34.0	2.815	2.225	115.65	24.6	399	74.3	3.43
	......	......	......	1.250	8.250	53.45	37.31	2.815	2.16	126.82	23.2	428.17	79.66	3.39
	......	......	......	1.500	7.750	47.15	43.57	2.815	2.03	148.19	20.5	478.59	89.04	3.31
	......	......	5S	0.156	12.438	121.4	6.17	3.34	3.26	20.99	52.7	122.2	19.20	4.45
	......	......	10S	0.180	12.390	120.6	7.11	3.34	3.24	24.20	52.2	140.5	22.03	4.44
	20	......	......	0.250	12.250	117.9	9.84	3.34	3.21	33.38	51.1	191.9	30.1	4.42
	30	......	......	0.330	12.090	114.8	12.88	3.34	3.17	43.77	49.7	248.5	39.0	4.39
	......	Std	40S	0.375	12.000	113.1	14.58	3.34	3.14	49.56	49.0	279.3	43.8	4.38
	40	......	......	0.406	11.938	111.9	15.74	3.34	3.13	53.53	48.5	300	47.1	4.37
	......	XS	80S	0.500	11.750	108.4	19.24	3.34	3.08	65.42	47.0	362	56.7	4.33
12	60	......	......	0.562	11.626	106.2	21.52	3.34	3.04	73.16	46.0	401	62.8	4.31
12.750	80	......	......	0.687	11.376	101.6	26.04	3.34	2.978	88.51	44.0	475	74.5	4.27
	......	......	......	0.750	11.250	99.40	28.27	3.34	2.94	96.2	43.1	510.7	80.1	4.25
	100	......	......	0.843	11.064	96.1	31.5	3.34	2.897	107.20	41.6	562	88.1	4.22
	......	......	......	0.875	11.000	95.00	32.64	3.34	2.88	110.9	41.1	578.5	90.7	4.21
	120	......	......	1.000	10.750	90.8	36.9	3.34	2.814	125.49	39.3	642	100.7	4.17
	140	......	......	1.125	10.500	86.6	41.1	3.34	2.749	139.68	37.5	701	109.9	4.13
	......	......	......	1.250	10.250	82.50	45.16	3.34	2.68	153.6	35.8	755.5	118.5	4.09
	160	......	......	1.312	10.126	80.5	47.1	3.34	2.651	160.27	34.9	781	122.6	4.07

PROPERTIES OF PIPE (Continued)

nominal pipe size *outside diameter,* in.	schedule number*			wall thick-ness, in.	inside diam-eter, in.	inside area, sq. in.	metal area, sq. in.	sq ft outside surface, per ft	sq ft inside surface, per ft	weight per ft, lb†	weight of water per ft, lb	moment of inertia, in.^{4}	section modu-lus, in.^{3}	radius gyra-tion, in.
	a	b	c											
	......	......	5S	0.156	13.688	147.20	6.78	3.67	3.58	23.0	63.7	162.6	23.2	4.90
	......	......	10S	0.188	13.624	145.80	8.16	3.67	3.57	27.7	63.1	194.6	27.8	4.88
	......	......	......	0.210	13.580	144.80	9.10	3.67	3.55	30.9	62.8	216.2	30.9	4.87
	......	......	......	0.219	13.562	144.50	9.48	3.67	3.55	32.2	62.6	225.1	32.2	4.87
	10	......	......	0.250	13.500	143.1	10.80	3.67	3.53	36.71	62.1	255.4	36.5	4.86
	......	......	......	0.281	13.438	141.80	12.11	3.67	3.52	41.2	61.5	285.2	40.7	4.85
	20	......	......	0.312	13.376	140.5	13.42	3.67	3.50	45.68	60.9	314	44.9	4.84
	......	......	......	0.344	13.312	139.20	14.76	3.67	3.48	50.2	60.3	344.3	49.2	4.83
14	30	Std	......	0.375	13.250	137.9	16.05	3.67	3.47	54.57	59.7	373	53.3	4.82
14.000	40	......	......	0.437	13.126	135.3	18.62	3.67	3.44	63.37	58.7	429	61.2	4.80
	......	......	......	0.469	13.062	134.00	19.94	3.67	3.42	67.8	58.0	456.8	65.3	4.79
	......	XS	......	0.500	13.000	132.7	21.21	3.67	3.40	72.09	57.5	484	69.1	4.78
	60	......	......	0.593	12.814	129.0	24.98	3.67	3.35	84.91	55.9	562	80.3	4.74
	......	......	......	0.625	12.750	127.7	26.26	3.67	3.34	89.28	55.3	589	84.1	4.73
	80	......	......	0.750	12.500	122.7	31.2	3.67	3.27	106.13	53.2	687	98.2	4.69
	100	......	......	0.937	12.126	115.5	38.5	3.67	3.17	130.73	50.0	825	117.8	4.63
	120	......	......	1.093	11.814	109.6	44.3	3.67	3.09	150.67	47.5	930	132.8	4.58
	140	......	......	1.250	11.500	103.9	50.1	3.67	3.01	170.22	45.0	1127	146.8	4.53
	160	......	......	1.406	11.188	98.3	55.6	3.67	2.929	189.12	42.6	1017	159.6	4.48
	......	......	5S	0.165	15.670	192.90	8.21	4.19	4.10	28	83.5	257	32.2	5.60
	......	......	10S	0.188	15.624	191.70	9.34	4.19	4.09	32	83.0	292	36.5	5.59
	10	......	......	0.250	15.500	188.7	12.37	4.19	4.06	42.05	81.8	384	48.0	5.57
	20	......	......	0.312	15.376	185.7	15.38	4.19	4.03	52.36	80.5	473	59.2	5.55
	30	Std	......	0.375	15.250	182.6	18.41	4.19	3.99	62.58	79.1	562	70.3	5.53
16	40	XS	......	0.500	15.000	176.7	24.35	4.19	3.93	82.77	76.5	732	91.5	5.48
16.000	60	......	......	0.656	14.688	169.4	31.6	4.19	3.85	107.50	73.4	933	116.6	5.43
	80	......	......	0.843	14.314	160.9	40.1	4.19	3.75	136.46	69.7	1157	144.6	5.37
	100	......	......	1.031	13.938	152.6	48.5	4.19	3.65	164.83	66.1	1365	170.6	5.30
	120	......	......	1.218	13.564	144.5	56.6	4.19	3.55	192.29	62.6	1556	194.5	5.24
	140	......	......	1.437	13.126	135.3	65.7	4.19	3.44	223.64	58.6	1760	220.0	5.17
	160	......	......	1.593	12.814	129.0	72.1	4.19	3.35	245.11	55.9	1894	236.7	5.12
	......	......	5S	0.165	17.670	245.20	9.24	4.71	4.63	31	106.2	368	40.8	6.31
	......	......	10S	0.188	17.624	243.90	10.52	4.71	4.61	36	105.7	417	46.4	6.30
	10	......	......	0.250	17.500	240.5	13.94	4.71	4.58	47.39	104.3	549	61.0	6.28
	20	......	......	0.312	17.376	237.1	17.34	4.71	4.55	59.03	102.8	678	75.5	6.25
	......	Std	......	0.375	17.250	233.7	20.76	4.71	4.52	70.59	101.2	807	89.6	6.23
18	30	......	......	0.437	17.126	230.4	24.11	4.71	4.48	82.06	99.9	931	103.4	6.21
	......	XS	......	0.500	17.00	227.0	27.49	4.71	4.45	93.45	98.4	1053	117.0	6.19
18.000	40	......	......	0.562	16.876	223.7	30.8	4.71	4.42	104.75	97.0	1172	130.2	6.17
	60	......	......	0.750	16.500	213.8	40.6	4.71	4.32	138.17	92.7	1515	168.3	6.10
	80	......	......	0.937	16.126	204.2	50.2	4.71	4.22	170.75	88.5	1834	203.8	6.04
	100	......	......	1.156	15.688	193.3	61.2	4.71	4.11	207.96	83.7	2180	242.2	5.97
	120	......	......	1.375	15.250	182.6	71.8	4.71	3.99	244.14	79.2	2499	277.6	5.90
	140	......	......	1.562	14.876	173.8	80.7	4.71	3.89	274.23	75.3	2750	306	5.84
	160	......	......	1.781	14.438	163.7	90.7	4.71	3.78	308.51	71.0	3020	336	5.77
	......	......	5S	0.188	19.634	302.40	11.70	5.24	5.14	40	131.0	574	57.4	7.00
	......	......	10S	0.218	19.564	300.60	13.55	5.24	5.12	46	130.2	663	66.3	6.99
	10	......	......	0.250	19.500	298.6	15.51	5.24	5.11	52.73	129.5	757	75.7	6.98
	20	Std	......	0.375	19.250	291.0	23.12	5.24	5.04	78.60	126.0	1114	111.4	6.94
20	30	XS	......	0.500	19.000	283.5	30.6	5.24	4.97	104.13	122.8	1457	145.7	6.90
20.000	40	......	......	0.593	18.814	278.0	36.2	5.24	4.93	122.91	120.4	1704	170.4	6.86
	60	......	......	0.812	18.376	265.2	48.9	5.24	4.81	166.40	115.0	2257	225.7	6.79
	......	......	......	0.875	18.250	261.6	52.6	5.24	4.78	178.73	113.4	2409	240.9	6.77
	80	......	......	1.031	17.938	252.7	61.4	5.24	4.70	208.87	109.4	2772	277.2	6.72
	100	......	......	1.281	17.438	238.8	75.3	5.24	4.57	256.10	103.4	3320	332	6.63

PROPERTIES OF PIPE (Continued)

nominal pipe size *outside diameter,* in.	schedule number*			wall thickness, in.	inside diameter, in.	inside area, sq in.	metal area, sq in.	sq ft outside surface, per ft	sq ft inside surface, per ft	weight per ft, lb†	weight of water per ft, lb	moment of inertia, in.[4]	section modulus, in.[3]	radius gyration, in.
	a	b	c											
20 *20.000*	120	......	......	1.500	17.000	227.0	87.2	5.24	4.45	296.37	98.3	3760	376	6.56
	140	......	......	1.750	16.500	213.8	100.3	5.24	4.32	341.10	92.6	4220	422	6.48
	160	......	......	1.968	16.064	202.7	111.5	5.24	4.21	379.01	87.9	4590	459	6.41
22 *22.000*	......	......	5S	0.188	21.624	367.3	12.88	5.76	5.66	44	159.1	766	69.7	7.71
	......	......	10S	0.218	21.564	365.2	14.92	5.76	5.65	51	158.2	885	80.4	7.70
	10	......	......	0.250	21.500	363.1	17.18	5.76	5.63	58	157.4	1010	91.8	7.69
	20	Std	......	0.375	21.250	354.7	25.48	5.76	5.56	87	153.7	1490	135.4	7.65
	30	XS	......	0.500	21.000	346.4	33.77	5.76	5.50	115	150.2	1953	177.5	7.61
	......	......	......	0.625	20.750	338.2	41.97	5.76	5.43	143	146.6	2400	218.2	7.56
	......	......	......	0.750	20.500	330.1	50.07	5.76	5.37	170	143.1	2829	257.2	7.52
	60	......	......	0.875	20.250	322.1	58.07	5.76	5.30	197	139.6	3245	295.0	7.47
	80	......	......	1.125	19.750	306.4	73.78	5.76	5.17	251	132.8	4029	366.3	7.39
	100	......	......	1.375	19.250	291.0	89.09	5.76	5.04	303	126.2	4758	432.6	7.31
	120	......	......	1.625	18.750	276.1	104.02	5.76	4.91	354	119.6	5432	493.8	7.23
	140	......	......	1.875	18.250	261.6	118.55	5.76	4.78	403	113.3	6054	550.3	7.15
	160	......	......	2.125	17.750	247.4	132.68	5.76	4.65	451	107.2	6626	602.4	7.07
24 *24.000*	10	......	......	0.250	23.500	434	18.65	6.28	6.15	63.41	188.0	1316	109.6	8.40
	20	Std	......	0.375	23.250	425	27.83	6.28	6.09	94.62	183.8	1943	161.9	8.35
	......	XS	......	0.500	23.000	415	36.9	6.28	6.02	125.49	180.1	2550	212.5	8.31
	30	......	......	0.562	22.876	411	41.4	6.28	5.99	140.80	178.1	2840	237.0	8.29
	......	......	......	0.625	22.750	406	45.9	6.28	5.96	156.03	176.2	3140	261.4	8.27
	40	......	......	0.687	22.626	402	50.3	6.28	5.92	171.17	174.3	3420	285.2	8.25
	......	......	......	0.750	22.500	398	54.8	6.28	5.89	186.24	172.4	3710	309	8.22
	......	......	5S	0.218	23.564	436.1	16.29	6.28	6.17	55	188.9	1152	96.0	8.41
	......	......	......	0.875	22.250	388.6	63.54	6.28	5.83	216	168.6	4256	354.7	8.18
	60	......	......	0.968	22.064	382	70.0	6.28	5.78	238.11	165.8	4650	388	8.15
	80	......	......	1.218	21.564	365	87.2	6.28	5.65	296.36	158.3	5670	473	8.07
	100	......	......	1.531	20.938	344	108.1	6.28	5.48	367.40	149.3	6850	571	7.96
	120	......	......	1.812	20.376	326	126.3	6.28	5.33	429.39	141.4	7830	652	7.87
	140	......	......	2.062	19.876	310	142.1	6.28	5.20	483.13	134.5	8630	719	7.79
	160	......	......	2.343	19.314	293	159.4	6.28	5.06	541.94	127.0	9460	788	7.70
26 *26.000*	......	......	......	0.250	25.500	510.7	19.85	6.81	6.68	67	221.4	1646	126.6	9.10
	10	......	......	0.312	25.376	505.8	25.18	6.81	6.64	86	219.2	2076	159.7	9.08
	......	Std	......	0.375	25.250	500.7	30.19	6.81	6.61	103	217.1	2478	190.6	9.06
	20	XS	......	0.500	25.000	490.9	40.06	6.81	6.54	136	212.8	3259	250.7	9.02
	......	......	......	0.625	24.750	481.1	49.82	6.81	6.48	169	208.6	4013	308.7	8.98
	......	......	......	0.750	24.500	471.4	59.49	6.81	6.41	202	204.4	4744	364.9	8.93
	......	......	......	0.875	24.250	461.9	69.07	6.81	6.35	235	200.2	5458	419.9	8.89
	......	......	......	1.000	24.000	452.4	78.54	6.81	6.28	267	196.1	6149	473.0	8.85
	......	......	......	1.125	23.750	443.0	87.91	6.81	6.22	299	192.1	6813	524.1	8.80
28 *28.000*	......	......	......	0.250	27.500	594.0	21.80	7.33	7.20	74	257.3	2098	149.8	9.81
	10	......	......	0.312	27.376	588.6	27.14	7.33	7.17	92	255.0	2601	185.8	9.79
	......	Std	......	0.375	27.250	583.2	32.54	7.33	7.13	111	252.6	3105	221.8	9.77
	20	XS	......	0.500	27.000	572.6	43.20	7.33	7.07	147	248.0	4085	291.8	9.72
	30	......	......	0.625	26.750	562.0	53.75	7.33	7.00	183	243.4	5038	359.8	9.68
	......	......	......	0.750	26.500	551.6	64.21	7.33	6.94	218	238.9	5964	426.0	9.64
	......	......	......	0.875	26.250	541.2	74.56	7.33	6.87	253	234.4	6865	490.3	9.60
	......	......	......	1.000	26.000	530.9	84.82	7.33	6.81	288	230.0	7740	552.8	9.55
	......	......	......	1.125	25.750	520.8	94.98	7.33	6.74	323	225.6	8590	613.6	9.51
30 *30.000*	......	......	5S	0.250	29.500	683.4	23.37	7.85	7.72	79	296.3	2585	172.3	10.52
	10	......	10S	0.312	29.376	677.8	29.19	7.85	7.69	99	293.7	3201	213.4	10.50
	......	Std	......	0.375	29.250	672.0	34.90	7.85	7.66	119	291.2	3823	254.8	10.48
	20	XS	......	0.500	29.000	660.5	46.34	7.85	7.59	158	286.2	5033	335.5	10.43
	30	......	......	0.625	28.750	649.2	57.68	7.85	7.53	196	281.3	6213	414.2	10.39

PROPERTIES OF PIPE (Continued)

nominal pipe size *outside diameter,* in.	schedule number* a	schedule number* b	schedule number* c	wall thick-ness, in.	inside diam-eter, in.	inside area, sq. in.	metal area, sq. in.	sq ft outside surface, per ft	sq ft inside surface, per ft	weight per ft, lb†	weight of water per ft lb	moment of inertia, in.⁴	section modu-lus, in.³	radius gyra-tion, in.
	40	......	......	0.750	28.500	637.9	68.92	7.85	7.46	234	276.6	7371	491.4	10.34
30	......	......	......	0.875	28.250	620.7	80.06	7.85	7.39	272	271.8	8494	566.2	10.30
30.000	......	......	......	1.000	28.000	615.7	91.11	7.85	7.33	310	267.0	9591	639.4	10.26
	......	......	......	1.125	27.750	604.7	102.05	7.85	7.26	347	262.2	10653	710.2	10.22
	......	......	......	0.250	31.500	779.2	24.93	8.38	8.25	85	337.8	3141	196.3	11.22
	10	......	......	0.312	31.376	773.2	31.02	8.38	8.21	106	335.2	3891	243.2	11.20
	......	Std	......	0.375	31.250	766.9	37.25	8.38	8.18	127	332.5	4656	291.0	11.18
	20	XS	......	0.500	31.000	754.7	49.48	8.38	8.11	168	327.2	6140	383.8	11.14
32	30	......	......	0.625	30.750	742.5	61.59	8.38	8.05	209	321.9	7578	473.6	11.09
32.000	40	......	......	0.688	30.624	736.6	67.68	8.38	8.02	230	319.0	8298	518.6	11.07
	......	......	......	0.750	30.500	730.5	73.63	8.38	7.98	250	316.7	8990	561.9	11.05
	......	......	......	0.875	30.250	718.3	85.52	8.38	7.92	291	311.6	10372	648.2	11.01
	......	......	......	1.000	30.000	706.8	97.38	8.38	7.85	331	306.4	11680	730.0	10.95
	......	......	......	1.125	29.750	694.7	109.0	8.38	7.79	371	301.3	13023	814.0	10.92
	......	......	......	0.250	33.500	881.2	26.50	8.90	8.77	90	382.0	3773	221.9	11.93
	10	......	......	0.312	33.376	874.9	32.99	8.90	8.74	112	379.3	4680	275.3	11.91
	......	Std	......	0.375	33.250	867.8	39.61	8.90	8.70	135	376.2	5597	329.2	11.89
	20	XS	......	0.500	33.000	855.3	52.62	8.90	8.64	179	370.8	7385	434.4	11.85
34	30	......	......	0.625	32.750	841.9	65.53	8.90	8.57	223	365.0	9124	536.7	11.80
34.000	40	......	......	0.688	32.624	835.9	72.00	8.90	8.54	245	362.1	9992	587.8	11.78
	......	......	......	0.750	32.500	829.3	78.34	8.90	8.51	266	359.5	10829	637.0	11.76
	......	......	......	0.875	32.250	816.4	91.01	8.90	8.44	310	354.1	12501	735.4	11.72
	......	......	......	1.000	32.000	804.2	103.67	8.90	8.38	353	348.6	14114	830.2	11.67
	......	......	......	1.125	31.750	791.3	116.13	8.90	8.31	395	343.2	15719	924.7	11.63
	......	......	......	0.250	35.500	989.7	28.11	9.42	9.29	96	429.1	4491	249.5	12.64
	10	......	......	0.312	35.376	982.9	34.95	9.42	9.26	119	426.1	5565	309.1	12.62
	......	Std	......	0.375	35.250	975.8	42.01	9.42	9.23	143	423.1	6664	370.2	12.59
	20	XS	......	0.500	35.000	962.1	55.76	9.42	9.16	190	417.1	8785	488.1	12.55
36	30	......	......	0.625	34.750	948.3	69.50	9.42	9.10	236	411.1	10872	604.0	12.51
36.000	40	......	......	0.750	34.500	934.7	83.01	9.42	9.03	282	405.3	12898	716.5	12.46
	......	......	......	0.875	34.250	920.6	96.50	9.42	8.97	328	399.4	14903	827.9	12.42
	......	......	......	1.000	34.000	907.9	109.96	9.42	0.90	374	393.6	16851	936.2	12.38
	......	......	......	1.125	33.750	894.2	123.19	9.42	8.89	419	387.9	18763	1042.4	12.34
	......	......	......	0.250	41.500	1352.6	32.82	10.99	10.86	112	586.4	7126	339.3	14.73
	......	Std	......	0.375	41.250	1336.3	49.08	10.99	10.80	167	579.3	10627	506.1	14.71
	20	XS	......	0.500	41.000	1320.2	65.18	10.99	10.73	222	572.3	14037	668.4	14.67
42	30	......	......	0.625	40.750	1304.1	81.28	10.99	10.67	276	565.4	17373	827.3	14.62
42.000	40	......	......	0.750	40.500	1288.2	97.23	10.99	10.60	330	558.4	20689	985.2	14.59
	......	......	......	1.000	40.000	1256.6	128.81	10.99	10.47	438	544.8	27080	1289.5	14.50
	......	......	......	1.250	39.500	1225.3	160.03	10.99	10.34	544	531.2	33233	1582.5	14.41
	......	......	......	1.500	39.000	1194.5	190.85	10.99	10.21	649	517.9	39181	1865.7	14.33

INSULATION WEIGHT FACTORS

To determine the weight per foot of any piping insulation, use the pipe size and nominal insulation thickness to find the insulation weight factor F in the chart shown below. Then multiply F by the density of the insulation in pounds per cubic foot.

Example. For 4″ pipe with 4″ nominal thickness insulation, $F = .77$. If the insulation density is 12 pounds per cubic foot, then the insulation weight is $.77 \times 12 = 9.24$ lb/ft.

Nominal Pipe Size	Nominal Insulation Thickness										
	1″	1½″	2″	2½″	3″	3½″	4″	4½″	5″	5½″	6″
1	.057	.10	.16	.23	.31	.40					
1¼	.051	.12	.15	.22	.30	.39					
1½	.066	.11	.21	.29	.38	.48					
2	.080	.14	.21	.29	.37	.47	.59				
2½	.091	.19	.27	.36	.46	.58	.70	.83			
3	.10	.17	.25	.34	.44	.56	.68	.81			
3½	.15	.23	.31	.41	.54	.66	.78	..	.97		
4	.13	.21	.30	.39	.51	.63	.77	.96	1.10		
5	.15	.24	.34	.45	.58	.71	.88	1.04	1.20		
6	.17	.27	.38	.51	.64	.83	.97	1.13	1.34		
8	..	.34	.47	.66	.80	.97	1.17	1.36	1.56	1.75	
10	..	.43	.59	.75	.93	1.12	1.32	1.54	1.76	1.99	
12	..	.50	.68	.88	1.07	1.28	1.52	1.74	1.99	2.24	2.50
14	..	.51	.70	.90	1.11	1.34	1.57	1.81	2.07	2.34	2.62
16	..	.57	.78	1.01	1.24	1.49	1.74	2.01	2.29	2.58	2.88
18	..	.64	.87	1.12	1.37	1.64	1.92	2.21	2.51	2.82	3.14
20	..	.70	.96	1.23	1.50	1.79	2.09	2.40	2.73	3.06	3.40
24	..	.83	1.13	1.44	1.77	2.10	2.44	2.80	3.16	3.54	3.92

LOAD CARRYING CAPACITIES OF THREADED HOT ROLLED STEEL ROD CONFORMING TO ASTM A-36

Nominal Rod Diameter, in.	⅜	½	⅝	¾	⅞	1	1⅛	1¼	1½	1¾	2	2¼	2½	2¾	3	3¼	3½
Root Area of Thread, sq. in.	.068	.126	.202	.302	.419	.552	.693	.889	1.293	1.744	2.300	3.023	3.719	4.619	5.621	6.720	7.918
Max. Safe Load, lbs. at Rod Temp. of 650°F	610	1130	1810	2710	3770	4960	6230	8000	11630	15700	20700	27200	33500	41580	50580	60480	71280

WEIGHTS OF PIPING MATERIALS

1″ PIPE 1.313″ O.D.

PIPE	Schedule No.	40	80	160									
	Wall Designation	Std.	XS		XXS								
	Thickness—In.	.133	.179	.250	.358								
	Pipe—Lbs/Ft	**1.68**	**2.17**	**2.84**	**3.66**								
	Water—Lbs/Ft	**.37**	**.31**	**.23**	**.12**								
WELDING FITTINGS	L.R. 90° Elbow	**.3** .3	**.4** .3	**.6** .3	**.7** .3								
	S.R. 90° Elbow	**.2** .2											
	L.R. 45° Elbow	**.2** .2	**.3** .2	**.4** .2	**.4** .2								
	Tee	**.8** .4	**.9** .4	**1.1** .4	**1.3** .4								
	Lateral	**1.7** 1.1	**2.5** 1.1										
	Reducer	**.3** .2	**.4** .2	**.4** .2	**.5** .2								
	Cap	**.2** .3	**.3** .3	**.4** .3	**.4** .3								

	Temperature Range °F	100-199	200-299	300-399	400-499	500-599	600-699	700-799	800-899	900-999	1000-1099	1100-1200
INSULATION: 85% Magnesia Calcium Silicate	Nom. Thick., In.	1	1	1½	2	2	2½	2½	2½	3	3	3
	Lbs/Ft.	**.72**	**.72**	**1.23**	**1.94**	**1.94**	**2.76**	**2.76**	**2.76**	**3.70**	**3.70**	**3.70**
Combination	Nom. Thick., In.						2½	2½	2½	3	3	3
	Lbs/Ft						**3.30**	**3.30**	**3.30**	**4.70**	**4.70**	**4.70**
*Asbestos Fiber—Sodium Silicate	Nom. Thick., In.	1	1	1	1	1	1½	1½	2	2	3	3
	Lbs/Ft	**.91**	**.91**	**.91**	**.91**	**.91**	**1.61**	**1.61**	**2.54**	**2.54**	**4.91**	**4.91**

	Pressure Rating psi	Cast Iron		Steel						
		125	250	150	300	400	600	900	1500	2500
FLANGES	Screwed or Slip-On	**2.5** 1.5	**4** 1.5	**2.5** 1.5	**4** 1.5	**5** 1.5	**5** 1.5	**12** 1.5	**12** 1.5	**15** 1.5
	Welding Neck			**2.3** 1.5	**5** 1.5	**7** 1.5	**7** 1.5	**12** 1.5	**12** 1.5	**16** 1.5
	Lap Joint			**2.5** 1.5	**4** 1.5	**5** 1.5	**5** 1.5	**12** 1.5	**12** 1.5	**15** 1.5
	Blind	**2.5** 1.5	**4** 1.5	**2.5** 1.5	**5** 1.5	**5** 1.5	**5** 1.5	**12** 1.5	**12** 1.5	**15** 1.5
FLANGED FITTINGS	S.R. 90° Elbow	**6** 3.6					**15** 3.7		**28** 3.8	
	L.R. 90° Elbow	**8** 3.8								
	45° Elbow	**5** 3.2					**14** 3.4		**26** 3.6	
	Tee	**11** 5.4					**20** 5.6		**39** 5.7	
VALVES	Flanged Bonnet Gate				**20** 1.2		**25** 1.5		**80** 4.3	
	Flanged Bonnet Globe or Angle								**84** 3.5	
	Flanged Bonnet Check									
	Pressure Seal Bonnet—Gate						**31** 1.7	**31** 1.7		
	Pressure Seal Bonnet—Globe									

Boldface type is weight in pounds. Lightface type beneath weight is weight factor for insulation.

Insulation thicknesses and weights are based on average conditions and do not constitute a recommendation for specific thicknesses of materials. Insulation weights are based on 85% magnesia and hydrous calcium silicate at 11 lbs/cubic foot. The listed thicknesses and weights of combination covering are the sums of the inner layer of diatomaceous earth at 21 lbs/cubic foot and the outer layer at 11 lbs/cubic foot.

Insulation weights include allowances for wire, cement, canvas, bands and paint, but not special surface finishes.

To find the weight of covering on flanges, valves or fittings, multiply the weight factor by the weight per foot of covering used on straight pipe.

Valve weights are approximate. When possible, obtain weights from the manufacturer.

Cast iron valve weights are for flanged end valves; steel weights for welding end valves.

All flanged fitting, flanged valve and flange weights include the proportional weight of bolts or studs to make up all joints.

* 16 lb cu. ft. density.

1¼″ PIPE 1.660″ O.D. WEIGHTS OF PIPING MATERIALS

PIPE					
Schedule No.	40	80	160		
Wall Designation	Std.	XS		XXS	
Thickness—In.	.140	.191	.250	.382	
Pipe—Lbs/Ft	**2.27**	**3.00**	**3.77**	**5.22**	
Water—Lbs/Ft	.65	.56	.46	.27	

WELDING FITTINGS					
L.R. 90° Elbow	**.6** .3	**.8** .3	**.9** .3	**1.2** .3	
S.R. 90° Elbow	**.4** .2				
L.R. 45° Elbow	**.4** .2	**.5** .2	**.6** .2	**.8** .2	
Tee	**1.3** .5	**1.6** .5	**1.9** .5	**2.4** .5	
Lateral	**2.4** 1.2	**3.9** 1.2			
Reducer	**.5** .2	**.5** .2	**.6** .2	**.8** .2	
Cap	**.3** .3	**.4** .3	**.5** .3	**.6** .3	

INSULATION		Temperature Range °F 100-199	200-299	300-399	400-499	500-599	600-699	700-799	800-899	900-999	1000-1099	1100-1200
85% Magnesia Calcium Silicate	Nom. Thick., In.	1	1	1½	2	2	2½	2½	2½	3	3	3
	Lbs/Ft	**.65**	**.65**	**1.47**	**1.83**	**1.83**	**2.65**	**2.65**	**2.65**	**3.58**	**3.58**	**3.58**
Combination	Nom. Thick., In.						2½	2½	2½	3	3	3
	Lbs/Ft						**3.17**	**3.17**	**3.17**	**5.76**	**5.76**	**5.76**
*Asbestos Fiber—Sodium Silicate	Nom. Thick., In.	1	1	1	1½	1½	1½	2	2½	2½	3	3
	Lbs/Ft	**.82**	**.82**	**.82**	**1.93**	**1.93**	**1.93**	**2.45**	**3.58**	**3.58**	**4.82**	**4.82**

	Pressure Rating psi	Cast Iron 125	250	Steel 150	300	400	600	900	1500	2500
FLANGES	Screwed or Slip-On	**2.5** 1.5	**5** 1.5	**3.5** 1.5	**5** 1.5	**7** 1.5	**7** 1.5	**13** 1.5	**13** 1.5	**23** 1.5
	Welding Neck			**3.5** 1.5	**7** 1.5	**8** 1.5	**8** 1.5	**13** 1.5	**13** 1.5	**25** 1.5
	Lap Joint			**3.5** 1.5	**5** 1.5	**7** 1.5	**7** 1.5	**13** 1.5	**13** 1.5	**22** 1.5
	Blind	**3.5** 1.5	**5** 1.5	**3.5** 1.5	**7** 1.5	**7** 1.5	**7** 1.5	**13** 1.5	**13** 1.5	**23** 1.5
FLANGED FITTINGS	S.R. 90° Elbow	**8** 3.6			**17** 3.7		**18** 3.8		**33** 3.9	
	L.R. 90° Elbow	**10** 3.9			**18** 3.9					
	45° Elbow	**7** 3.3			**15** 3.4		**16** 3.5		**31** 3.7	
	Tee	**13** 5.4			**23** 5.6		**28** 5.7		**49** 5.9	
VALVES	Flanged Bonnet Gate				**40** 4		**60** 4.2		**97** 4.6	
	Flanged Bonnet Globe or Angle									
	Flanged Bonnet Check				**21** 4					
	Pressure Seal Bonnet—Gate							**38** 1.8	**38** 1.1	
	Pressure Seal Bonnet—Globe									

Boldface type is weight in pounds. Lightface type beneath weight is weight factor for insulation.

Insulation thicknesses and weights are based on average conditions and do not constitute a recommendation for specific thicknesses of materials. Insulation weights are based on 85% magnesia and hydrous calcium silicate at 11 lbs/cubic foot. The listed thicknesses and weights of combination covering are the sums of the inner layer of diatomaceous earth at 21 lbs/cubic foot and the outer layer at 11 lbs/cubic foot.

Insulation weights include allowances for wire, cement, canvas, bands and paint, but not special surface finishes.

To find the weight of covering on flanges, valves or fittings, multiply the weight factor by the weight per foot of covering used on straight pipe.

Valve weights are approximate. When possible, obtain weights from the manufacturer.

Cast iron valve weights are for flanged end valves; steel weights for welding end valves.

All flanged fitting, flanged valve and flange weights include the proportional weight of bolts or studs to make up all joints.

* 16 lb cu. ft. density.

WEIGHTS OF PIPING MATERIALS — 1.900″ O.D. $1\frac{1}{2}''$ PIPE

PIPE					
Schedule No.	40	80	160		
Wall Designation	Std.	XS		XXS	
Thickness—In.	.145	.200	.281	.400	
Pipe—Lbs/Ft	**2.72**	**3.63**	**4.86**	**6.41**	
Water—Lbs/Ft	.88	.77	.61	.41	

WELDING FITTINGS				
L.R. 90° Elbow	**.8** .4	**1.1** .4	**1.4** .4	**1.8** .4
S.R. 90° Elbow	**.6** .3	**.7** .3		
L.R. 45° Elbow	**.5** .2	**.7** .2	**.8** .2	**1** .2
Tee	**2** .6	**2.5** .6	**3.1** .6	**3.7** .6
Lateral	**3.3** 1.3	**5.4** 1.3		
Reducer	**.6** .2	**.7** .2	**.9** .2	**1.2** .2
Cap	**.4** .3	**.5** .3	**.7** .3	**.7** .3

INSULATION	Temperature Range °F	100-199	200-299	300-399	400-499	500-599	600-699	700-799	800-899	900-999	1000-1099	1100-1200
85% Magnesia Calcium Silicate	Nom. Thick., In.	1	1	1½	2	2	2½	2½	2½	3	3	3
	Lbs/Ft	**.84**	**.84**	**1.35**	**2.52**	**2.52**	**3.47**	**3.47**	**3.47**	**4.52**	**4.52**	**4.52**
Combination	Nom. Thick., In.						2½	2½	2½	3	3	3
	Lbs/Ft						**4.20**	**4.20**	**4.20**	**5.62**	**5.62**	**5.62**
*Asbestos Fiber—Sodium Silicate	Nom. Thick., In.	1	1	1	1½	1½	2	2	2½	2½	3	3
	Lbs/Ft	**1.07**	**1.07**	**1.07**	**1.85**	**1.85**	**3.50**	**3.50**	**4.76**	**4.76**	**6.16**	**6.16**

	Pressure Rating psi	Cast Iron		Steel						
		125	250	150	300	400	600	900	1500	2500
FLANGES	Screwed or Slip-On	**3.5** 1.5	**7** 1.5	**3.5** 1.5	**8** 1.5	**9** 1.5	**9** 1.5	**19** 1.5	**19** 1.5	**31** 1.5
	Welding Neck			**4** 1.5	**9** 1.5	**12** 1.5	**12** 1.5	**19** 1.5	**19** 1.5	**34** 1.5
	Lap Joint			**3.5** 1.5	**8** 1.5	**9** 1.5	**9** 1.5	**19** 1.5	**19** 1.5	**31** 1.5
	Blind	**3.5** 1.5	**7** 1.5	**3.5** 1.5	**9** 1.5	**10** 1.5	**10** 1.5	**19** 1.5	**19** 1.5	**31** 1.5
FLANGED FITTINGS	S.R. 90° Elbow	**10** 3.7		**12** 3.7	**23** 3.8		**26** 3.9		**46** 4	
	L.R. 90° Elbow	**12** 4		**13** 4	**24** 4					
	45° Elbow	**9** 3.4		**11** 3.4	**21** 3.5		**23** 3.5		**39** 3.7	
	Tee	**17** 5.6		**20** 5.6	**30** 5.7		**37** 5.8		**70** 6	
VALVES	Flanged Bonnet Gate	**27** 6.8			**55** 4.2		**70** 4.5		**125** 5	
	Flanged Bonnet Globe or Angle				**40** 4.2		**45** 4.2		**170** 5	
	Flanged Bonnet Check			**30** 4.1	**35** 4.1		**40** 4.2		**110** 4.5	
	Pressure Seal Bonnet—Gate							**42** 1.9	**42** 1.2	
	Pressure Seal Bonnet—Globe									

Boldface type is weight in pounds. Lightface type beneath weight is weight factor for insulation.

Insulation thicknesses and weights are based on average conditions and do not constitute a recommendation for specific thicknesses of materials. Insulation weights are based on 85% magnesia and hydrous calcium silicate at 11 lbs/cubic foot. The listed thicknesses and weights of combination covering are the sums of the inner layer of diatomaceous earth at 21 lbs/cubic foot and the outer layer at 11 lbs/cubic foot.

Insulation weights include allowances for wire, cement, canvas, bands and paint, but not special surface finishes.

To find the weight of covering on flanges, valves or fittings, multiply the weight factor by the weight per foot of covering used on straight pipe.

Valve weights are approximate. When possible, obtain weights from the manufacturer.

Cast iron valve weights are for flanged end valves; steel weights for welding end valves.

All flanged fitting, flanged valve and flange weights include the proportional weight of bolts or studs to make up all joints.

* 16 lb cu. ft. density.

2″ PIPE 2.375″ O.D. WEIGHTS OF PIPING MATERIALS

PIPE	Schedule No.	40	80	160	
	Wall Designation	Std.	XS		XXS
	Thickness—In.	.154	.218	.343	.436
	Pipe—Lbs/Ft	**3.65**	**5.02**	**7.44**	**9.03**
	Water—Lbs/Ft	**1.46**	**1.28**	**.97**	**.77**
WELDING FITTINGS	L.R. 90° Elbow	**1.5** .5	**2** .5	**2.9** .5	**3.5** .5
	S.R. 90° Elbow	**1** .3	**1.3** .3		
	L.R. 45° Elbow	**.8** .2	**1.1** .2	**1.6** .2	**1.8** .2
	Tee	**3** .6	**3.7** .6	**5** .6	**5.7** .6
	Lateral	**5** 1.4	**7.8** 1.4		
	Reducer	**.9** .3	**1.2** .3	**1.6** .3	**1.9** .3
	Cap	**.5** .4	**.7** .4	**1.2** .4	**1.2** .4

	Temperature Range °F	100-199	200-299	300-399	400-499	500-599	600-699	700-799	800-899	900-999	1000-1099	1100-1200
INSULATION — 85% Magnesia Calcium Silicate	Nom. Thick., In.	1	1	1½	2	2	2½	2½	3	3	3	3½
	Lbs/Ft	**1.01**	**1.01**	**1.71**	**2.53**	**2.53**	**3.48**	**3.48**	**4.42**	**4.42**	**4.42**	**5.59**
Combination	Nom. Thick., In.						2½	2½	3	3	3	3½
	Lbs/Ft						**4.28**	**4.28**	**5.93**	**5.93**	**5.93**	**7.80**
*Asbestos Fiber—Sodium Silicate	Nom. Thick., In.	1	1	1	1½	1½	2	2	2½	2½	3	3
	Lbs/Ft	**1.26**	**1.26**	**1.26**	**2.20**	**2.20**	**3.32**	**3.32**	**4.57**	**4.57**	**5.99**	**5.99**

	Pressure Rating psi	Cast Iron		Steel						
		125	250	150	300	400	600	900	1500	2500
FLANGES	Screwed or Slip-On	**6** 1.5	**9** 1.5	**6** 1.5	**9** 1.5	**11** 1.5	**11** 1.5	**32** 1.5	**32** 1.5	**48** 1.5
	Welding Neck			**6** 1.5	**10** 1.5	**13** 1.5	**13** 1.5	**31** 1.5	**31** 1.5	**48** 1.5
	Lap Joint			**6** 1.5	**9** 1.5	**12** 1.5	**12** 1.5	**32** 1.5	**32** 1.5	**48** 1.5
	Blind	**6** 1.5	**10** 1.5	**4.8** 1.5	**10** 1.5	**12** 1.5	**12** 1.5	**31** 1.5	**31** 1.5	**49** 1.5
FLANGED FITTINGS	S.R. 90° Elbow	**16** 3.8	**24** 3.8	**19** 3.8	**29** 3.8		**35** 4		**83** 4.2	
	L.R. 90° Elbow	**18** 4.1	**27** 4.1	**22** 4.1	**31** 4.1					
	45° Elbow	**14** 3.4	**22** 3.5	**16** 3.4	**24** 3.5		**33** 3.7		**73** 3.9	
	Tee	**23** 5.7	**37** 5.7	**27** 5.7	**41** 5.7		**52** 6		**129** 6.3	
VALVES	Flanged Bonnet Gate	**37** 6.9	**52** 7.1	**40** 4	**65** 4.2		**80** 4.5		**190** 5	
	Flanged Bonnet Globe or Angle	**30** 7	**64** 7.3	**30** 3.8	**45** 4		**85** 4.5		**235** 5.5	
	Flanged Bonnet Check	**26** 7	**51** 7.3	**35** 3.8	**40** 4		**60** 4.2		**300** 5.8	
	Pressure Seal Bonnet—Gate								**150** 2.5	
	Pressure Seal Bonnet—Globe								**165** 3	

Boldface type is weight in pounds. Lightface type beneath weight is weight factor for insulation.

Insulation thicknesses and weights are based on average conditions and do not constitute a recommendation for specific thicknesses of materials. Insulation weights are based on 85% magnesia and hydrous calcium silicate at 11 lbs/cubic foot. The listed thicknesses and weights of combination covering are the sums of the inner layer of diatomaceous earth at 21 lbs/cubic foot and the outer layer at 11 lbs/cubic foot.

Insulation weights include allowances for wire, cement, canvas, bands and paint, but not special surface finishes.

To find the weight of covering on flanges, valves or fittings, multiply the weight factor by the weight per foot of covering used on straight pipe.

Valve weights are approximate. When possible, obtain weights from the manufacturer.

Cast iron valve weights are for flanged end valves; steel weights for welding end valves.

All flanged fitting, flanged valve and flange weights include the proportional weight of bolts or studs to make up all joints.

* 16 lb cu. ft. density.

WEIGHTS OF PIPING MATERIALS   2.875″ O.D. 2½″ PIPE

PIPE	Schedule No.	40	80	160	
	Wall Designation	Std.	XS		XXS
	Thickness—In.	.203	.276	.375	.552
	Pipe—Lbs/Ft	**5.79**	**7.66**	**10.01**	**13.70**
	Water—Lbs/Ft	**2.08**	**1.84**	**1.54**	**1.07**
WELDING FITTINGS	L.R. 90° Elbow	**2.9** .6	**3.8** .6	**4.9** .6	**6.5** .6
	S.R. 90° Elbow	**1.9** .4	**2.5** .4		
	L.R. 45° Elbow	**1.6** .3	**2.1** .3	**2.7** .3	**3.5** .3
	Tee	**5.2** .8	**6.4** .8	**7.9** .8	**9.9** .8
	Lateral	9.2 1.5	14 1.5		
	Reducer	**1.6** .3	**2.1** .3	**2.7** .3	**3.4** .3
	Cap	**.8** .4	**1** .4	**2** .4	**2.1** .4

INSULATION	Temperature Range °F	100-199	200-299	300-399	400-499	500-599	600-699	700-799	800-899	900-999	1000-1099	1100-1200
85% Magnesia Calcium Silicate	Nom. Thick., In.	1	1	1½	2	2	2½	2½	3	3	3½	3½
	Lbs/Ft	**1.14**	**1.14**	**2.29**	**3.23**	**3.23**	**4.28**	**4.28**	**5.46**	**5.46**	**6.86**	**6.86**
Combination	Nom. Thick., In.						2½	2½	3	3	3½	3½
	Lbs/Ft						**5.20**	**5.20**	**7.36**	**7.36**	**9.58**	**9.58**
*Asbestos Fiber—Sodium Silicate	Nom. Thick., In.	1	1	1	1½	1½	2	2	2½	2½	3	3
	Lbs/Ft	**1.44**	**1.44**	**1.44**	**3.09**	**3.09**	**4.34**	**4.34**	**5.75**	**5.75**	**7.34**	**7.34**

	Pressure Rating psi	Cast Iron 125	Cast Iron 250	Steel 150	Steel 300	Steel 400	Steel 600	Steel 900	Steel 1500	Steel 2500
FLANGES	Screwed or Slip-On	**8** 1.5	**14** 1.5	**9** 1.5	**14** 1.5	**17** 1.5	**17** 1.5	**46** 1.5	**46** 1.5	**69** 1.5
	Welding Neck			**9** 1.5	**14** 1.5	**20** 1.5	**20** 1.5	**46** 1.5	**46** 1.5	**66** 1.5
	Lap Joint			**9** 1.5	**14** 1.5	**18** 1.5	**18** 1.5	**45** 1.5	**45** 1.5	**67** 1.5
	Blind	**8** 1.5	**15** 1.5	**9** 1.5	**16** 1.5	**19** 1.5	**19** 1.5	**45** 1.5	**45** 1.5	**70** 1.5
FLANGED FITTINGS	S.R. 90° Elbow	**21** 3.8	**36** 3.9	**27** 3.8	**42** 3.9		**50** 4.1		**114** 4.4	
	L.R. 90° Elbow	**25** 4.2	**40** 4.2	**30** 4.2	**47** 4.2					
	45° Elbow	**19** 3.5	**34** 3.6	**22** 3.5	**35** 3.6		**46** 3.8		**99** 3.9	
	Tee	**32** 5.7	**55** 5.8	**42** 5.7	**61** 5.9		**77** 6.2		**169** 6.6	
VALVES	Flanged Bonnet Gate	**50** 7	**82** 7.1	**60** 4	**100** 4.2		**105** 4.6		**275** 5.2	
	Flanged Bonnet Globe or Angle	**43** 7.1	**87** 7.4	**50** 4	**70** 4.1		**120** 4.6		**325** 5.5	
	Flanged Bonnet Check	**36** 7.1	**71** 7.4	**40** 4	**50** 4		**105** 4.6		**320** 5.5	
	Pressure Seal Bonnet—Gate								**215** 2.5	
	Pressure Seal Bonnet—Globe								**230** 2.8	

Boldface type is weight in pounds. Lightface type beneath weight is weight factor for insulation.

Insulation thicknesses and weights are based on average conditions and do not constitute a recommendation for specific thicknesses of materials. Insulation weights are based on 85% magnesia and hydrous calcium silicate at 11 lbs/cubic foot. The listed thicknesses and weights of combination covering are the sums of the inner layer of diatomaceous earth at 21 lbs/cubic foot and the outer layer at 11 lbs/cubic foot.

Insulation weights include allowances for wire, cement, canvas, bands and paint, but not special surface finishes.

To find the weight of covering on flanges, valves or fittings, multiply the weight factor by the weight per foot of covering used on straight pipe.

Valve weights are approximate. When possible, obtain weights from the manufacturer.

Cast iron valve weights are for flanged end valves; steel weights for welding end valves.

All flanged fitting, flanged valve and flange weights include the proportional weight of bolts or studs to make up all joints.

* 16 lb cu. ft. density.

3″ PIPE 3.500″ O.D. WEIGHTS OF PIPING MATERIALS

PIPE												
Schedule No.	40	80	160									
Wall Designation	Std.	XS		XXS								
Thickness—In.	.216	.300	.438	.600								
Pipe—Lbs/Ft	**7.58**	**10.25**	**14.32**	**18.58**								
Water—Lbs/Ft	**3.20**	**2.86**	**2.35**	**1.80**								

WELDING FITTINGS												
L.R. 90° Elbow	**4.6** .8	**6.1** .8	**8.4** .8	**10.7** .8								
S.R. 90° Elbow	**3** .5	**4** .5										
L.R. 45° Elbow	**2.4** .3	**3.2** .3	**4.4** .3	**5.4** .3								
Tee	**7.4** .8	**9.5** .8	**12.2** .8	**14.8** .8								
Lateral	**13** 1.8	**19** 1.8										
Reducer	**2.2** .3	**2.9** .3	**3.7** .3	**4.7** .3								
Cap	**1.4** .5	**1.8** .5	**3.5** .5	**3.7** .5								

INSULATION	Temperature Range °F	100-199	200-299	300-399	400-499	500-599	600-699	700-799	800-899	900-999	1000-1099	1100-1200
85% Magnesia Calcium Silicate	Nom. Thick., In.	1	1	1½	2	2	2½	3	3	3	3½	3½
	Lbs/Ft	**1.25**	**1.25**	**2.08**	**3.01**	**3.01**	**4.07**	**5.24**	**5.24**	**5.24**	**6.65**	**6.65**
Combination	Nom. Thick., In.						2½	3	3	3	3½	3½
	Lbs/Ft						**5.07**	**6.94**	**6.94**	**6.94**	**9.17**	**9.17**
*Asbestos Fiber—Sodium Silicate	Nom. Thick., In.	1	1	1	1½	1½	2	2	3	3	3½	3½
	Lbs/Ft	**1.61**	**1.61**	**1.61**	**2.74**	**2.74**	**3.98**	**3.98**	**6.99**	**6.99**	**8.99**	**8.99**

	Pressure Rating psi	Cast Iron		Steel						
		125	250	150	300	400	600	900	1500	2500
FLANGES	Screwed or Slip-On	**9** 1.5	**17** 1.5	**9** 1.5	**17** 1.5	**20** 1.5	**20** 1.5	**37** 1.5	**61** 1.5	**102** 1.5
	Welding Neck			**11** 1.5	**19** 1.5	**27** 1.5	**27** 1.5	**38** 1.5	**61** 1.5	**113** 1.5
	Lap Joint			**9** 1.5	**17** 1.5	**19** 1.5	**19** 1.5	**36** 1.5	**60** 1.5	**99** 1.5
	Blind	**10** 1.5	**19** 1.5	**10** 1.5	**20** 1.5	**24** 1.5	**24** 1.5	**38** 1.5	**61** 1.5	**105** 1.5
FLANGED FITTINGS	S.R. 90° Elbow	**26** 3.9	**46** 4	**32** 3.9	**53** 4		**67** 4.1	**98** 4.3	**150** 4.6	
	L.R. 90° Elbow	**30** 4.3	**50** 4.3	**40** 4.3	**63** 4.3					
	45° Elbow	**22** 3.5	**41** 3.6	**28** 3.5	**46** 3.6		**60** 3.8	**93** 3.9	**135** 4	
	Tee	**39** 5.9	**67** 6	**52** 5.9	**81** 6		**102** 6.2	**151** 6.5	**238** 6.9	
VALVES	Flanged Bonnet Gate	**66** 7	**112** 7.4	**70** 4	**125** 4.4		**155** 4.8	**260** 5	**410** 5.5	
	Flanged Bonnet Globe or Angle	**56** 7.2	**121** 7.6	**60** 4.3	**95** 4.5		**155** 4.8	**225** 5	**495** 5.5	
	Flanged Bonnet Check	**46** 7.2	**100** 7.6	**60** 4.3	**70** 4.4		**120** 4.8	**150** 4.9	**440** 5.8	
	Pressure Seal Bonnet—Gate							**208** 3	**235** 3.2	
	Pressure Seal Bonnet—Globe							**135** 2.5	**180** 3	

Boldface type is weight in pounds. Lightface type beneath weight is weight factor for insulation.

Insulation thicknesses and weights are based on average conditions and do not constitute a recommendation for specific thicknesses of materials. Insulation weights are based on 85% magnesia and hydrous calcium silicate at 11 lbs/cubic foot. The listed thicknesses and weights of combination covering are the sums of the inner layer of diatomaceous earth at 21 lbs/cubic foot and the outer layer at 11 lbs/cubic foot.

Insulation weights include allowances for wire, cement, canvas, bands and paint, but not special surface finishes.

To find the weight of covering on flanges, valves or fittings, multiply the weight factor by the weight per foot of covering used on straight pipe.

Valve weights are approximate. When possible, obtain weights from the manufacturer.

Cast iron valve weights are for flanged end valves; steel weights for welding end valves.

All flanged fitting, flanged valve and flange weights include the proportional weight of bolts or studs to make up all joints.

* 16 lb cu. ft. density.

WEIGHTS OF PIPING MATERIALS 4.000″ O.D. 3½″ PIPE

PIPE			
Schedule No.	40	80	
Wall Designation	Std.	XS	XXS
Thickness—In.	.226	.318	.636
Pipe—Lbs/Ft	**9.11**	**12.51**	**22.85**
Water—Lbs/Ft	**4.28**	**3.85**	**2.53**

WELDING FITTINGS	Std.	XS	XXS
L.R. 90° Elbow	**6.4** .9	**8.7** .9	**15.4** .9
S.R. 90° Elbow	**4.3** .6	**5.8** .6	
L.R. 45° Elbow	**3.3** .4	**4.4** .4	**7.5** .4
Tee	**9.9** .9	**12.6** .9	**20** .9
Lateral	**17** 1.8	**26** 1.8	
Reducer	**3.1** .3	**4.1** .3	**6.9** .3
Cap	**2.1** .6	**2.8** .6	**5.5** .6

INSULATION	Temperature Range °F	100-199	200-299	300-399	400-499	500-599	600-699	700-799	800-899	900-999	1000-1099	1100-1200
85% Magnesia Calcium Silicate	Nom. Thick., In.	1	1	1½	2	2½	2½	3	3	3½	3½	3½
	Lbs/Ft	**1.83**	**1.83**	**2.77**	**3.71**	**4.88**	**4.88**	**6.39**	**6.39**	**7.80**	**7.80**	**7.80**
Combination	Nom. Thick., In.						2½	3	3	3½	3½	3½
	Lbs/Ft						**6.49**	**8.71**	**8.71**	**10.8**	**10.8**	**10.8**
*Asbestos Fiber—Sodium Silicate	Nom. Thick., In.	1	1	1	1½	1½	2	2	3	3	3½	3½
	Lbs/Ft	**2.41**	**2.41**	**2.41**	**3.65**	**3.65**	**5.07**	**5.07**	**8.66**	**8.66**	**10.62**	**10.62**

	Pressure Rating psi	Cast Iron		Steel						
		125	250	150	300	400	600	900	1500	2500
FLANGES	Screwed or Slip-On	**13** 1.5	**21** 1.5	**13** 1.5	**21** 1.5	**27** 1.5	**27** 1.5			
	Welding Neck			**14** 1.5	**22** 1.5	**32** 1.5	**32** 1.5			
	Lap Joint			**13** 1.5	**21** 1.5	**26** 1.5	**26** 1.5			
	Blind	**14** 1.5	**23** 1.5	**15** 1.5	**25** 1.5	**35** 1.5	**35** 1.5			
FLANGED FITTINGS	S.R. 90° Elbow	**35** 4	**56** 4.1	**49** 4			**82** 4.3			
	L.R. 90° Elbow	**40** 4.4	**62** 4.4	**54** 4.4						
	45° Elbow	**31** 3.6	**51** 3.7	**39** 3.6			**75** 3.9			
	Tee	**54** 6	**86** 6.2	**70** 6			**133** 6.4			
VALVES	Flanged Bonnet Gate	**82** 7.1	**143** 7.5	**90** 4.1	**155** 4.5		**180** 4.8	**360** 5	**510** 5.5	
	Flanged Bonnet Globe or Angle	**74** 7.3	**137** 7.7				**160** 4.7			
	Flanged Bonnet Check	**71** 7.3	**125** 7.7				**125** 4.7			
	Pressure Seal Bonnet—Gate						**140** 2.5	**295** 2.8	**380** 3	
	Pressure Seal Bonnet—Globe									

Boldface type is weight in pounds. Lightface type beneath weight is weight factor for insulation.

Insulation thicknesses and weights are based on average conditions and do not constitute a recommendation for specific thicknesses of materials. Insulation weights are based on 85% magnesia and hydrous calcium silicate at 11 lbs/cubic foot. The listed thicknesses and weights of combination covering are the sums of the inner layer of diatomaceous earth at 21 lbs/cubic foot and the outer layer at 11 lbs/cubic foot.

Insulation weights include allowances for wire, cement, canvas, bands and paint, but not special surface finishes.

To find the weight of covering on flanges, valves or fittings, multiply the weight factor by the weight per foot of covering used on straight pipe.

Valve weights are approximate. When possible, obtain weights from the manufacturer.

Cast iron valve weights are for flanged end valves; steel weights for welding end valves.

All flanged fitting, flanged valve and flange weights include the proportional weight of bolts or studs to make up all joints.

* 16 lb cu. ft. density.

4″ PIPE 4.500″ O.D. WEIGHTS OF PIPING MATERIALS

PIPE

Schedule No.	40	80	120	160	
Wall Designation	Std.	XS			XXS
Thickness—In.	.237	.337	.438	.531	.674
Pipe—Lbs/Ft	**10.79**	**14.98**	**18.96**	**22.51**	**27.54**
Water—Lbs/Ft	**5.51**	**4.98**	**4.48**	**4.02**	**3.38**

WELDING FITTINGS

Fitting	40	80	120	160	XXS
L.R. 90° Elbow	**8.7** 1	**11.9** 1		**17.6** 1	**21** 1
S.R. 90° Elbow	**5.8** .7	**7.9** .7			
L.R. 45° Elbow	**4.3** .4	**5.9** .4		**8.5** .4	**10.1** .4
Tee	**12.6** 1	**16.4** 1		**23** 1	**27** 1
Lateral	**21** 2.1	**33** 2.1			
Reducer	**3.6** .3	**4.9** .3		**6.6** .3	**8.2** .3
Cap	**2.6** .6	**3.4** .6		**6.5** .6	**6.7** .6

INSULATION

Temperature Range °F		100-199	200-299	300-399	400-499	500-599	600-699	700-799	800-899	900-999	1000-1099	1100-1200
85% Magnesia Calcium Silicate	Nom. Thick., In.	1	1	1½	2	2½	2½	3	3	3½	3½	4
	Lbs/Ft	**1.62**	**1.62**	**2.55**	**3.61**	**4.66**	**4.66**	**6.07**	**6.07**	**7.48**	**7.48**	**9.10**
Combination	Nom. Thick., In.						2½	3	3	3½	3½	3½
	Lbs/Ft						**6.07**	**8.30**	**8.30**	**10.6**	**10.6**	**10.6**
*Asbestos Fiber—Sodium Silicate	Nom. Thick., In.	1	1	1	1½	1½	2	2	3	3	3½	3½
	Lbs/Ft	**2.04**	**2.04**	**2.04**	**3.28**	**3.28**	**4.70**	**4.70**	**8.29**	**8.29**	**10.25**	**10.25**

FLANGES, FLANGED FITTINGS, VALVES

	Pressure Rating psi	Cast Iron		Steel						
		125	250	150	300	400	600	900	1500	2500
FLANGES	Screwed or Slip-On	**16** 1.5	**26** 1.5	**15** 1.5	**26** 1.5	**32** 1.5	**43** 1.5	**66** 1.5	**90** 1.5	**158** 1.5
	Welding Neck			**17** 1.5	**29** 1.5	**41** 1.5	**48** 1.5	**64** 1.5	**90** 1.5	**177** 1.5
	Lap Joint			**15** 1.5	**26** 1.5	**31** 1.5	**42** 1.5	**64** 1.5	**92** 1.5	**153** 1.5
	Blind	**18** 1.5	**29** 1.5	**19** 1.5	**31** 1.5	**39** 1.5	**47** 1.5	**67** 1.5	**90** 1.5	**164** 1.5
FLANGED FITTINGS	S.R. 90° Elbow	**45** 4.1	**72** 4.2	**59** 4.1	**85** 4.2	**99** 4.3	**128** 4.4	**185** 4.5	**254** 4.8	
	L.R. 90° Elbow	**52** 4.5	**79** 4.5	**72** 4.5	**98** 4.5					
	45° Elbow	**40** 3.7	**65** 3.8	**51** 3.7	**78** 3.8	**82** 3.9	**119** 4	**170** 4.1	**214** 4.2	
	Tee	**70** 6.1	**109** 6.3	**86** 6.1	**121** 6.3	**153** 6.4	**187** 6.6	**262** 6.8	**386** 7.2	
VALVES	Flanged Bonnet Gate	**109** 7.2	**188** 7.5	**100** 4.2	**175** 4.5	**195** 5	**255** 5.1	**455** 5.4	**735** 6	
	Flanged Bonnet Globe or Angle	**97** 7.4	**177** 7.8	**95** 4.3	**145** 4.8	**215** 5	**230** 5.1	**415** 5.5	**800** 6	
	Flanged Bonnet Check	**80** 7.4	**146** 7.8	**80** 4.3	**105** 4.5	**160** 4.8	**195** 5	**320** 5.6	**780** 6	
	Pressure Seal Bonnet—Gate						**215** 2.8	**380** 3	**520** 4	
	Pressure Seal Bonnet—Globe							**240** 2.7	**290** 3	

Boldface type is weight in pounds. Lightface type beneath weight is weight factor for insulation.

Insulation thicknesses and weights are based on average conditions and do not constitute a recommendation for specific thicknesses of materials. Insulation weights are based on 85% magnesia and hydrous calcium silicate at 11 lbs/cubic foot. The listed thicknesses and weights of combination covering are the sums of the inner layer of diatomaceous earth at 21 lbs/cubic foot and the outer layer at 11 lbs/cubic foot.

Insulation weights include allowances for wire, cement, canvas, bands and paint, but not special surface finishes.

To find the weight of covering on flanges, valves or fittings, multiply the weight factor by the weight per foot of covering used on straight pipe.

Valve weights are approximate. When possible, obtain weights from the manufacturer.

Cast iron valve weights are for flanged end valves; steel weights for welding end valves.

All flanged fitting, flanged valve and flange weights include the proportional weight of bolts or studs to make up all joints.

* 16 lb cu. ft. density.

WEIGHTS OF PIPING MATERIALS

5.563″ O.D. 5″ PIPE

PIPE

Schedule No.	40	80	120	160	
Wall Designation	Std.	XS			XXS
Thickness—In.	.258	.375	.500	.625	.750
Pipe—Lbs/Ft	**14.62**	**20.78**	**27.04**	**32.96**	**38.55**
Water—Lbs/Ft	**8.66**	**7.89**	**7.09**	**6.33**	**5.62**

WELDING FITTINGS

	40	80	120	160	XXS
L.R. 90° Elbow	**14.7** 1.3	**21** 1.3		**32** 1.3	**37** 1.3
S.R. 90° Elbow	**9.8** .8	**13.7** .8			
L.R. 45° Elbow	**7.3** .5	**10.2** .5		**15.6** .5	**17.7** .5
Tee	**19.8** 1.2	**26** 1.2		**39** 1.2	**43** 1.2
Lateral	**31** 2.5	**50** 2.5			
Reducer	**6** .4	**8.3** .4		**12.4** .4	**14.2** .4
Cap	**4.2** .7	**5.7** .7		**11** .7	**11** .7

INSULATION

Temperature Range °F		100-199	200-299	300-399	400-499	500-599	600-699	700-799	800-899	900-999	1000-1099	1100-1200
*Asbestos Fiber—Sodium Silicate	Nom. Thick., In.	1	1½	1½	2	2½	2½	3	3½	3½	4	4
	Lbs/Ft	**1.86**	**2.92**	**2.92**	**4.08**	**5.38**	**5.38**	**6.90**	**8.41**	**8.41**	**10.4**	**10.4**
Combination	Nom. Thick., In.						2½	3	3½	3½	4	4
	Lbs/Ft						**7.01**	**9.30**	**11.8**	**11.8**	**14.9**	**14.9**
85% Magnesia Calcium Silicate	Nom. Thick., In.	1	1	1	1½	1½	2½	2½	3	3	4	4
	Lbs/Ft	**2.34**	**2.34**	**2.34**	**3.76**	**3.76**	**7.35**	**7.35**	**9.31**	**9.31**	**14.37**	**14.37**

Pressure Rating psi	Cast Iron		Steel						
	125	250	150	300	400	600	900	1500	2500
FLANGES									
Screwed or Slip-On	**20** 1.5	**32** 1.5	**18** 1.5	**32** 1.5	**37** 1.5	**73** 1.5	**100** 1.5	**162** 1.5	**259** 1.5
Welding Neck			**22** 1.5	**36** 1.5	**49** 1.5	**78** 1.5	**103** 1.5	**162** 1.5	**293** 1.5
Lap Joint			**18** 1.5	**32** 1.5	**35** 1.5	**71** 1.5	**98** 1.5	**168** 1.5	**253** 1.5
Blind	**23** 1.5	**37** 1.5	**23** 1.5	**39** 1.5	**50** 1.5	**78** 1.5	**104** 1.5	**172** 1.5	**272** 1.5
FLANGED FITTINGS									
S.R. 90° Elbow	**58** 4.3	**94** 4.3	**80** 4.3	**113** 4.3	**123** 4.5	**205** 4.7	**268** 4.8	**435** 5.2	
L.R. 90° Elbow	**68** 4.7	**105** 4.7	**91** 4.7	**128** 4.7					
45° Elbow	**51** 3.8	**83** 3.8	**66** 3.8	**98** 3.8	**123** 4	**180** 4.2	**239** 4.3	**350** 4.5	
Tee	**90** 6.4	**145** 6.5	**119** 6.4	**172** 6.4	**179** 6.8	**304** 7	**415** 7.2	**665** 7.8	
VALVES									
Flanged Bonnet Gate	**138** 7.3	**264** 7.9	**150** 4.3	**265** 4.9	**310** 5.3	**455** 5.5	**615** 6	**1340** 7	
Flanged Bonnet Globe or Angle	**138** 7.6	**247** 8	**155** 4.3	**215** 5	**355** 5.2	**515** 5.8	**555** 5.8	**950** 6	
Flanged Bonnet Check	**118** 7.6	**210** 8	**110** 4.3	**165** 5	**185** 5	**350** 5.8	**560** 6	**1150** 7	
Pressure Seal Bonnet—Gate						**350** 3.1	**520** 3.8	**865** 4.5	
Pressure Seal Bonnet—Globe							**280** 4	**450** 4.5	

Boldface type is weight in pounds. Lightface type beneath weight is weight factor for insulation.

Insulation thicknesses and weights are based on average conditions and do not constitute a recommendation for specific thicknesses of materials. Insulation weights are based on 85% magnesia and hydrous calcium silicate at 11 lbs/cubic foot. The listed thicknesses and weights of combination covering are the sums of the inner layer of diatomaceous earth at 21 lbs/cubic foot and the outer layer at 11 lbs/cubic foot.

Insulation weights include allowances for wire, cement, canvas, bands and paint, but not special surface finishes.

To find the weight of covering on flanges, valves or fittings, multiply the weight factor by the weight per foot of covering used on straight pipe.

Valve weights are approximate. When possible, obtain weights from the manufacturer.

Cast iron valve weights are for flanged end valves; steel weights for welding end valves.

All flanged fitting, flanged valve and flange weights include the proportional weight of bolts or studs to make up all joints.

* 16 lb cu. ft. density.

6″ PIPE 6.625″ O.D. WEIGHTS OF PIPING MATERIALS

PIPE	Schedule No.	40	80	120	160	
	Wall Designation	Std.	XS			XXS
	Thickness—In.	.280	.432	.562	.718	.864
	Pipe—Lbs/Ft	**18.97**	**28.57**	**36.39**	**45.3**	**53.2**
	Water—Lbs/Ft	**12.51**	**11.29**	**10.30**	**9.2**	**8.2**
WELDING FITTINGS	L.R. 90° Elbow	**23** 1.5	**34** 1.5		**53** 1.5	**62** 1.5
	S.R. 90° Elbow	**15.2** 1	**23** 1			
	L.R. 45° Elbow	**11.3** .6	**16.7** .6		**26** .6	**30** .6
	Tee	**29.3** 1.4	**42** 1.4		**60** 1.4	**68** 1.4
	Lateral	**42** 2.9	**79** 2.9			
	Reducer	**8.7** .5	**12.6** .5		**18.8** .5	**21** .5
	Cap	**6.4** .9	**9.2** .9		**17.5** .9	**17.5** .9

INSULATION	Temperature Range °F	100-199	200-299	300-399	400-499	500-599	600-699	700-799	800-899	900-999	1000-1099	1100-1200
85% Magnesia Calcium Silicate	Nom. Thick., In.	1	1½	2	2	2½	3	3	3½	3½	4	4
	Lbs/Ft	**2.11**	**3.28**	**4.57**	**4.57**	**6.09**	**7.60**	**7.60**	**9.82**	**98.2**	**11.5**	**11.4**
Combination	Nom. Thick., In.						3	3	3½	3½	4	4
	Lbs/Ft						**10.3**	**10.3**	**13.4**	**13.4**	**16.6**	**16.6**
*Asbestos Fiber—Sodium Silicate	Nom. Thick., In.	1	1	1	1½	1½	2½	2½	3½	3½	4	4
	Lbs/Ft	**2.57**	**2.57**	**2.57**	**4.18**	**4.18**	**8.10**	**8.10**	**13.31**	**13.31**	**15.85**	**15.85**

	Pressure Rating psi	Cast Iron		Steel						
		125	250	150	300	400	600	900	1500	2500
FLANGES	Screwed or Slip-On	**25** 1.5	**42** 1.5	**22** 1.5	**45** 1.5	**54** 1.5	**95** 1.5	**128** 1.5	**202** 1.5	**396** 1.5
	Welding Neck			**27** 1.5	**48** 1.5	**67** 1.5	**96** 1.5	**130** 1.5	**202** 1.5	**451** 1.5
	Lap Joint			**22** 1.5	**45** 1.5	**52** 1.5	**93** 1.5	**125** 1.5	**208** 1.5	**387** 1.5
	Blind	**28** 1.5	**51** 1.5	**29** 1.5	**56** 1.5	**71** 1.5	**101** 1.5	**133** 1.5	**197** 1.5	**418** 1.5
FLANGED FITTINGS	S.R. 90° Elbow	**74** 4.3	**125** 4.4	**90** 4.3	**147** 4.4	**184** 4.6	**275** 4.8	**375** 5	**566** 5.3	
	L.R. 90° Elbow	**91** 4.9	**145** 4.9	**126** 4.9	**182** 4.9					
	45° Elbow	**66** 3.8	**115** 3.9	**82** 3.8	**132** 3.9	**149** 4.1	**240** 4.3	**320** 4.3	**476** 4.6	
	Tee	**114** 6.5	**195** 6.6	**149** 6.5	**217** 6.6	**279** 6.9	**400** 7.2	**565** 7.5	**839** 8	
VALVES	Flanged Bonnet Gate	**172** 7.3	**359** 8	**190** 4.3	**360** 5	**435** 5.5	**620** 5.8	**835** 6	**1595** 7	
	Flanged Bonnet Globe or Angle	**184** 7.8	**345** 8.2	**185** 4.4	**275** 5	**415** 5.3	**645** 5.8	**765** 6	**1800** 7	
	Flanged Bonnet Check	**154** 7.8	**286** 8.2	**150** 4.8	**200** 5	**360** 5.4	**445** 6	**800** 6.4	**1630** 7	
	Pressure Seal Bonnet—Gate						**580** 3.5	**750** 4	**1215** 5	
	Pressure Seal Bonnet—Globe							**730** 4	**780** 5	

Boldface type is weight in pounds. Lightface type beneath weight is weight factor for insulation.

Insulation thicknesses and weights are based on average conditions and do not constitute a recommendation for specific thicknesses of materials. Insulation weights are based on 85% magnesia and hydrous calcium silicate at 11 lbs/cubic foot. The listed thicknesses and weights of combination covering are the sums of the inner layer of diatomaceous earth at 21 lbs/cubic foot and the outer layer at 11 lbs/cubic foot.

Insulation weights include allowances for wire, cement, canvas, bands and paint, but not special surface finishes.

To find the weight of covering on flanges, valves or fittings, multiply the weight factor by the weight per foot of covering used on straight pipe.

Valve weights are approximate. When possible, obtain weights from the manufacturer.

Cast iron valve weights are for flanged end valves; steel weights for welding end valves.

All flanged fitting, flanged valve and flange weights include the proportional weight of bolts or studs to make up all joints.

* 16 lb cu. ft. density.

WEIGHTS OF PIPING MATERIALS

8.625″ O.D. 8″ PIPE

PIPE	Schedule No.	20	30	40	60	80	100	120	140		160		
	Wall Designation			Std.		XS				XXS			
	Thickness—In.	.250	.277	.322	.406	.500	.593	.718	.812	.875	.906		
	Pipe—Lbs/Ft	**22.36**	**24.70**	**28.55**	**35.64**	**43.4**	**50.9**	**60.6**	**67.8**	**72.4**	**74.7**		
	Water—Lbs/Ft	**22.48**	**22.18**	**21.69**	**20.79**	**19.8**	**18.8**	**17.6**	**16.7**	**16.1**	**15.8**		
WELDING FITTINGS	L.R. 90° Elbow			**46** 2		**69** 2				**114** 2	**117** 2		
	S.R. 90° Elbow			**31** 1.3		**46** 1.3							
	L.R. 45° Elbow			**23** .8		**34** .8				**55** .8	**56** .8		
	Tee			**54** 1.8		**76** 1.8				**118** 1.8	**120** 1.8		
	Lateral			**76** 3.8		**140** 3.8							
	Reducer			**13.9** .5		**20** .5				**32** .5	**33** .5		
	Cap			**11.3** 1		**16.3** 1				**31** 1	**32** 1		

INSULATION	Temperature Range °F		100-199	200-299	300-399	400-499	500-599	600-699	700-799	800-899	900-999	1000-1099	1100-1200
	85% Magnesia Calcium Silicate	Nom. Thick., In.	1½	1½	2	2	2½	3	3½	3½	4	4	4½
		Lbs/Ft	**4.13**	**4.13**	**5.64**	**5.64**	**7.85**	**9.48**	**11.5**	**11.5**	**13.8**	**13.8**	**16.0**
	Combination	Nom. Thick., In.						3	3½	3½	4	4	4½
		Lbs/Ft						**12.9**	**16.2**	**16.2**	**20.4**	**20.4**	**23.8**
	*Asbestos Fiber—Sodium Silicate	Nom. Thick., In.	1½	1½	1½	1½	1½	2½	2½	3½	3½	4½	4½
		Lbs/Ft	**5.38**	**5.38**	**5.38**	**5.38**	**5.38**	**10.60**	**10.60**	**15.85**	**15.85**	**20.85**	**20.85**

	Pressure Rating psi	Cast Iron		Steel						
		125	250	150	300	400	600	900	1500	2500
FLANGES	Screwed or Slip-On	**34** 1.5	**64** 1.5	**33** 1.5	**67** 1.5	**82** 1.5	**135** 1.5	**207** 1.5	**319** 1.5	**601** 1.5
	Welding Neck			**42** 1.5	**76** 1.5	**104** 1.5	**137** 1.5	**222** 1.5	**334** 1.5	**692** 1.5
	Lap Joint			**33** 1.5	**67** 1.5	**79** 1.5	**132** 1.5	**223** 1.5	**347** 1.5	**587** 1.5
	Blind	**45** 1.5	**83** 1.5	**48** 1.5	**90** 1.5	**115** 1.5	**159** 1.5	**232** 1.5	**363** 1.5	**649** 1.5
FLANGED FITTINGS	S.R. 90° Elbow	**117** 4.5	**201** 4.7	**157** 4.5	**238** 4.7	**310** 5	**435** 5.2	**639** 5.4	**995** 5.7	
	L.R. 90° Elbow	**152** 5.3	**236** 5.3	**202** 5.3	**283** 5.3					
	45° Elbow	**101** 3.9	**171** 4	**127** 3.9	**203** 4	**215** 4.1	**360** 4.4	**507** 4.5	**870** 4.8	
	Tee	**175** 6.8	**304** 7.1	**230** 6.8	**337** 7.1	**445** 7.5	**610** 7.8	**978** 8.1	**1465** 8.6	
VALVES	Flanged Bonnet Gate	**251** 7.5	**583** 8.1	**305** 4.5	**505** 5.1	**730** 6	**960** 6.3	**1180** 6.6	**2740** 7	
	Flanged Bonnet Globe or Angle	**317** 8.4	**554** 8.6	**475** 5.4	**505** 5.5	**610** 5.9	**1130** 6.3	**1160** 6.3	**2865** 7	
	Flanged Bonnet Check	**302** 8.4	**454** 8.6	**235** 5.2	**310** 5.3	**475** 5.6	**725** 6	**1140** 6.4	**2075** 7	
	Pressure Seal Bonnet—Gate						**925** 4.5	**1185** 4.7	**2345** 5.5	
	Pressure Seal Bonnet—Globe							**1550** 4	**1680** 5	

Boldface type is weight in pounds. Lightface type beneath weight is weight factor for insulation.

Insulation thicknesses and weights are based on average conditions and do not constitute a recommendation for specific thicknesses of materials. Insulation weights are based on 85% magnesia and hydrous calcium silicate at 11 lbs/cubic foot. The listed thicknesses and weights of combination covering are the sums of the inner layer of diatomaceous earth at 21 lbs/cubic foot and the outer layer at 11 lbs/cubic foot.

Insulation weights include allowances for wire, cement, canvas, bands and paint, but not special surface finishes.

To find the weight of covering on flanges, valves or fittings, multiply the weight factor by the weight per foot of covering used on straight pipe.

Valve weights are approximate. When possible, obtain weights from the manufacturer.

Cast iron valve weights are for flanged end valves; steel weights for welding end valves.

All flanged fitting, flanged valve and flange weights include the proportional weight of bolts or studs to make up all joints.

* 16 lb cu. ft. density.

10″ PIPE 10.750″ O.D. WEIGHTS OF PIPING MATERIALS

PIPE	Schedule No.	20	30	40	60	80	100	120	140	160			
	Wall Designation			Std.	XS								
	Thickness—In	.250	.307	.365	.500	.593	.718	.843	1.000	1.125			
	Pipe—Lbs/Ft	**28.04**	**34.24**	**40.5**	**54.7**	**64.3**	**76.9**	**89.2**	**104.1**	**115.7**			
	Water—Lbs/Ft	**35.77**	**34.98**	**34.1**	**32.3**	**31.1**	**29.5**	**28.0**	**26.1**	**24.6**			
WELDING FITTINGS	L.R. 90° Elbow			**82** 2.5	**109** 2.5					**226** 2.5			
	S.R. 90° Elbow			**54** 1.7	**73** 1.7								
	L.R. 45° Elbow			**40** 1	**54** 1					**109** 1			
	Tee			**91** 2.1	**118** 2.1					**222** 2.1			
	Lateral			**124** 4.4	**202** 4.4								
	Reducer			**23** .6	**31** .6					**58** .6			
	Cap			**20** 1.3	**26** 1.3					**54** 1.3			

INSULATION		Temperature Range °F	100-199	200-299	300-399	400-499	500-599	600-699	700-799	800-899	900-999	1000-1099	1100-1200
	85% Magnesia Calcium Silicate	Nom. Thick., In.	1½	1½	2	2½	2½	3	3½	3½	4	4	4½
		Lbs/Ft	**5.20**	**5.20**	**7.07**	**8.93**	**8.93**	**11.0**	**13.2**	**13.2**	**15.5**	**15.5**	**18.1**
	Combination	Nom. Thick., In.						3	3½	3½	4	4	4½
		Lbs/Ft						**15.4**	**19.3**	**19.3**	**23**	**23**	**27.2**
	*Asbestos Fiber—Sodium Silicate	Nom. Thick., In.	1½	1½	1½	1½	1½	2½	2½	3½	3½	4½	4½
		Lbs/Ft	**6.77**	**6.77**	**6.77**	**6.77**	**6.77**	**12.03**	**12.03**	**18.39**	**18.39**	**25.21**	**25.21**

	Pressure Rating psi	Cast Iron		Steel						
		125	250	150	300	400	600	900	1500	2500
FLANGES	Screwed or Slip-On	**53** 1.5	**97** 1.5	**51** 1.5	**100** 1.5	**117** 1.5	**213** 1.5	**293** 1.5	**528** 1.5	**1148** 1.5
	Welding Neck			**60** 1.5	**110** 1.5	**152** 1.5	**225** 1.5	**316** 1.5	**546** 1.5	**1291** 1.5
	Lap Joint			**51** 1.5	**110** 1.5	**138** 1.5	**231** 1.5	**325** 1.5	**577** 1.5	**1120** 1.5
	Blind	**71** 1.5	**136** 1.5	**78** 1.5	**146** 1.5	**181** 1.5	**267** 1.5	**338** 1.5	**599** 1.5	**1248** 1.5
FLANGED FITTINGS	S.R. 90° Elbow	**190** 4.8	**323** 4.9	**240** 4.8	**343** 4.9	**462** 5.2	**747** 5.6	**995** 5.8		
	L.R. 90° Elbow	**245** 5.8	**383** 5.8	**290** 5.8	**438** 5.8					
	45° Elbow	**160** 4.1	**273** 4.2	**185** 4.1	**288** 4.2	**332** 4.3	**572** 4.6	**732** 4.7		
	Tee	**293** 7.2	**479** 7.4	**353** 7.2	**527** 7.4	**578** 7.8	**1007** 8.4	**1417** 8.7		
VALVES	Flanged Bonnet Gate	**471** 7.7	**899** 8.3	**455** 4.5	**750** 5	**1035** 6	**1575** 6.9	**2140** 7.1	**3690** 8	
	Flanged Bonnet Globe or Angle	**541** 9.1	**943** 9.1	**485** 4.5	**855** 5.5	**1070** 6	**1500** 6.3	**2500** 6.8	**4160** 8	
	Flanged Bonnet Check	**453** 9.1	**751** 9.1	**370** 6	**485** 6.1	**605** 6.3	**1030** 6.8	**1350** 7	**2280** 7.5	
	Pressure Seal Bonnet—Gate						**1450** 4.9	**1860** 5.5	**3150** 6	
	Pressure Seal Bonnet—Globe							**1800** 5	**1910** 6	

Boldface type is weight in pounds. Lightface type beneath weight is weight factor for insulation.

Insulation thicknesses and weights are based on average conditions and do not constitute a recommendation for specific thicknesses of materials. Insulation weights are based on 85% magnesia and hydrous calcium silicate at 11 lbs/cubic foot. The listed thicknesses and weights of combination covering are the sums of the inner layer of diatomaceous earth at 21 lbs/cubic foot and the outer layer at 11 lbs/cubic foot.

Insulation weights include allowances for wire, cement, canvas, bands and paint, but not special surface finishes.

To find the weight of covering on flanges, valves or fittings, multiply the weight factor by the weight per foot of covering used on straight pipe.

Valve weights are approximate. When possible, obtain weights from the manufacturer.

Cast iron valve weights are for flanged end valves; steel weights for welding end valves.

All flanged fitting, flanged valve and flange weights include the proportional weight of bolts or studs to make up all joints.

* 16 lb cu. ft. density.

WEIGHTS OF PIPING MATERIALS 12.750″ O.D. 12″ PIPE

PIPE	Schedule No.	20	30		40		60	80	100	120	140	160	
	Wall Designation			Std.		XS							
	Thickness—In.	.250	.330	.375	.406	.500	.562	.687	.843	1.000	1.125	1.312	
	Pipe—Lbs/Ft	**33.38**	**43.8**	**49.6**	**53.5**	**65.4**	**73.2**	**88.5**	**107.2**	**125.5**	**139.7**	**160.3**	
	Water—Lbs/Ft	**51.10**	**49.7**	**49.0**	**48.5**	**47.0**	**46.0**	**44.0**	**41.6**	**39.3**	**37.5**	**34.9**	
WELDING FITTINGS	L.R. 90° Elbow			**119** 3		**157** 3						**375** 3	
	S.R. 90° Elbow			**80** 2		**104** 2							
	L.R. 45° Elbow			**60** 1.3		**78** 1.3						**181** 1.3	
	Tee			**132** 2.5		**167** 2.5						**360** 2.5	
	Lateral			**180** 5.4		**273** 5.4							
	Reducer			**33** .7		**44** .7						**94** .7	
	Cap			**30** 1.5		**38** 1.5						**89** 1.5	

INSULATION	Temperature Range °F		100-199	200-299	300-399	400-499	500-599	600-699	700-799	800-899	900-999	1000-1099	1100-1200
	85% Magnesia Calcium Silicate	Nom. Thick., In.	1½	1½	2	2½	3	3	3½	4	4	4½	4½
		Lbs/Ft	**6.04**	**6.04**	**8.13**	**10.5**	**12.7**	**12.7**	**15.1**	**17.9**	**17.9**	**20.4**	**20.4**
	Combination	Nom. Thick., In.						3	3½	4	4	4½	4½
		Lbs/Ft						**17.7**	**21.9**	**26.7**	**26.7**	**31.1**	**31.1**
	*Asbestos Fiber—Sodium Silicate	Nom. Thick., In.	1½	1½	1½	1½	1½	2½	2½	4	4	5	5
		Lbs/Ft	**5.22**	**5.22**	**5.22**	**5.22**	**5.22**	**14.20**	**14.20**	**24.64**	**24.64**	**32.40**	**32.40**

	Pressure Rating psi	Cast Iron		Steel						
		125	250	150	300	400	600	900	1500	2500
FLANGES	Screwed or Slip-On	**71** 1.5	**137** 1.5	**72** 1.5	**140** 1.5	**164** 1.5	**261** 1.5	**388** 1.5	**820** 1.5	**1611** 1.5
	Welding Neck			**88** 1.5	**163** 1.5	**212** 1.5	**272** 1.5	**434** 1.5	**843** 1.5	**1919** 1.5
	Lap Joint			**72** 1.5	**164** 1.5	**187** 1.5	**286** 1,5	**433** 1.5	**902** 1.5	**1573** 1.5
	Blind	**96** 1.5	**177** 1.5	**118** 1.5	**209** 1.5	**261** 1.5	**341** 1.5	**475** 1.5	**928** 1.5	**1775** 1.5
FLANGED FITTINGS	S.R. 90° Elbow	**265** 5	**453** 5.2	**345** 5	**509** 5.2	**669** 5.5	**815** 5.8	**1474** 6.2		
	L.R. 90° Elbow	**375** 6.2	**553** 6.2	**485** 6.2	**624** 6.2			**1598** 6.2		
	45° Elbow	**235** 4.3	**383** 4.3	**282** 4.3	**414** 4.3	**469** 4.5	**705** 4.7	**1124** 4.8		
	Tee	**403** 7.5	**684** 7.8	**513** 7.5	**754** 7.8	**943** 8.3	**1361** 8.7	**1928** 9.3		
VALVES	Flanged Bonnet Gate	**687** 7.8	**1298** 8.5	**635** 4	**1015** 5	**1420** 5.5	**2155** 7	**2770** 7.2	**4650** 8	
	Flanged Bonnet Globe or Angle	**808** 9.4	**1200** 9.5	**710** 5	**1410** 5.5					
	Flanged Bonnet Check	**674** 9.4	**1160** 9.5	**560** 6	**720** 6.5		**1410** 7.2	**2600** 8	**3370** 8	
	Pressure Seal Bonnet—Gate						**1975** 5.5	**2560** 6	**4515** 7	
	Pressure Seal Bonnet—Globe									

* 16 lb cu. ft. density.

Boldface type is weight in pounds. Lightface type beneath weight is weight factor for insulation.

Insulation thicknesses and weights are based on average conditions and do not constitute a recommendation for specific thicknesses of materials. Insulation weights are based on 85% magnesia and hydrous calcium silicate at 11 lbs/cubic foot. The listed thicknesses and weights of combination covering are the sums of the inner layer of diatomaceous earth at 21 lbs/cubic foot and the outer layer at 11 lbs/cubic foot.

Insulation weights include allowances for wire, cement, canvas, bands and paint, but not special surface finishes.

To find the weight of covering on flanges, valves or fittings, multiply the weight factor by the weight per foot of covering used on straight pipe.

Valve weights are approximate. When possible, obtain weights from the manufacturer.

Cast iron valve weights are for flanged end valves; steel weights for welding end valves.

All flanged fitting, flanged valve and flange weights include the proportional weight of bolts or studs to make up all joints.

14″ PIPE 14″ O.D. WEIGHTS OF PIPING MATERIALS

PIPE												
Schedule No.	10	20	30	40		60	80	100	120	140	160	
Wall Designation			Std.		XS							
Thickness—In.	.250	.312	.375	.438	.500	.593	.750	.937	1.093	1.250	1.406	
Pipe—Lbs/Ft	**36.71**	**45.7**	**54.6**	**63.4**	**72.1**	**84.9**	**106.1**	**130.7**	**150.7**	**170.2**	**189.1**	
Water—Lbs/Ft	**62.06**	**60.92**	**59.7**	**58.7**	**57.5**	**55.9**	**53.2**	**50.0**	**47.5**	**45.0**	**42.6**	

WELDING FITTINGS	10	20	30	40	XS	60	80	100	120	140	160	
L.R. 90° Elbow			**154** 3.5		**202** 3.5							
S.R. 90° Elbow			**102** 2.3		**135** 2.3							
L.R. 45° Elbow			**77** 1.5		**100** 1.5							
Tee			**159** 2.8		**203** 2.8							
Lateral			**218** 5.8		**340** 5.8							
Reducer			**63** 1.1		**83** 1.1							
Cap			**35** 1.7		**46** 1.7							

INSULATION	Temperature Range °F	100-199	200-299	300-399	400-499	500-599	600-699	700-799	800-899	900-999	1000-1099	1100-1200
85% Magnesia Calcium Silicate	Nom. Thick., In.	1½	1½	2	2½	3	3	3½	4	4	4½	4½
	Lbs/Ft	**6.16**	**6.16**	**8.38**	**10.7**	**13.1**	**13.1**	**15.8**	**18.5**	**18.5**	**21.3**	**21.3**
Combination	Nom. Thick., In.						3	3½	4	4	4½	4½
	Lbs/Ft						**18.2**	**22.8**	**27.5**	**27.5**	**32.4**	**32.4**
*Asbestos Fiber—Sodium Silicate	Nom. Thick., In.	1½	1½	1½	2	2	3	3	4	4	5	5
	Lbs/Ft	**7.90**	**7.90**	**7.90**	**11.18**	**11.18**	**18.00**	**18.00**	**25.42**	**25.42**	**33.53**	**33.53**

	Pressure Rating psi	Cast Iron		Steel						
		125	250	150	300	400	600	900	1500	2500
FLANGES	Screwed or Slip-On	**93** 1.5	**184** 1.5	**96** 1.5	**195** 1.5	**235** 1.5	**318** 1.5	**460** 1.5	**1016** 1.5	
	Welding Neck			**113** 1.5	**217** 1.5	**277** 1.5	**406** 1.5	**642** 1.5	**1241** 1.5	
	Lap Joint			**110** 1.5	**220** 1.5	**254** 1.5	**349** 1.5	**477** 1.5	**1076** 1.5	
	Blind	**126** 1.5	**239** 1.5	**142** 1.5	**267** 1.5	**354** 1.5	**437** 1.5	**574** 1.5		
FLANGED FITTINGS	S.R. 90° Elbow	**372** 5.3	**617** 5.5	**497** 5.3	**632** 5.5	**664** 5.7	**918** 5.9	**1549** 6.4		
	L.R. 90° Elbow	**492** 6.6	**767** 6.6	**622** 6.6	**772** 6.6					
	45° Elbow	**292** 4.3	**497** 4.4	**377** 4.3	**587** 4.4	**638** 4.6	**883** 4.8	**1246** 4.9		
	Tee	**563** 8	**956** 8.4	**683** 8	**968** 8.3	**1131** 8.6	**1652** 8.9	**2318** 9.6		
VALVES	Flanged Bonnet Gate	**921** 7.9	**1762** 8.8	**905** 4.9	**1525** 6	**1920** 6.3	**2960** 7	**4170** 8	**6425** 8.8	
	Flanged Bonnet Globe or Angle	**1171** 9.9								
	Flanged Bonnet Check	**885** 9.9		**1010** 5	**1155** 5.2					
	Pressure Seal Bonnet—Gate						**2620** 6	**3475** 6.5	**6380** 7.5	
	Pressure Seal Bonnet—Globe									

Boldface type is weight in pounds. Lightface type beneath weight is weight factor for insulation.

Insulation thicknesses and weights are based on average conditions and do not constitute a recommendation for specific thicknesses of materials. Insulation weights are based on 85% magnesia and hydrous calcium silicate at 11 lbs/cubic foot. The listed thicknesses and weights of combination covering are the sums of the inner layer of diatomaceous earth at 21 lbs/cubic foot and the outer layer at 11 lbs/cubic foot.

Insulation weights include allowances for wire, cement, canvas, bands and paint, but not special surface finishes.

To find the weight of covering on flanges, valves or fittings, multiply the weight factor by the weight per foot of covering used on straight pipe.

Valve weights are approximate. When possible, obtain weights from the manufacturer.

Cast iron valve weights are for flanged end valves; steel weights for welding end valves.

All flanged fitting, flanged valve and flange weights include the proportional weight of bolts or studs to make up all joints.

* 16 lb cu. ft. density

WEIGHTS OF PIPING MATERIALS

16″ O.D. 16″ PIPE

PIPE	Schedule No.	10	20	30	40	60	80	100	120	140	160		
	Wall Designation			Std.	XS								
	Thickness—In.	.250	.312	.375	.500	.656	.843	1.031	1.218	1.438	1.593		
	Pipe—Lbs/Ft	**42.1**	**52.4**	**62.6**	**82.8**	**107.5**	**136.5**	**164.8**	**192.3**	**223.6**	**245.1**		
	Water—Lbs/Ft	**81.8**	**80.5**	**79.1**	**76.5**	**73.4**	**69.7**	**66.1**	**62.6**	**58.6**	**55.9**		
WELDING FITTINGS	L.R. 90° Elbow			**201** 4	**265** 4								
	S.R. 90° Elbow			**135** 2.5	**177** 2.5								
	L.R. 45° Elbow			**100** 1.7	**132** 1.7								
	Tee			**202** 3.2	**257** 3.2								
	Lateral			**275** 6.7	**433** 6.7								
	Reducer			**78** 1.2	**102** 1.2								
	Cap			**44** 1.8	**58** 1.8								

INSULATION		Temperature Range °F	100-199	200-299	300-399	400-499	500-599	600-699	700-799	800-899	900-999	1000-1099	1100-1200
	85% Magnesia Calcium Silicate	Nom. Thick., In.	1½	1½	2	2½	3	3	3½	4	4	4½	4½
		Lbs/Ft	**6.90**	**6.90**	**9.33**	**12.0**	**14.6**	**14.6**	**17.5**	**20.5**	**20.5**	**23.6**	**23.6**
	Combination	Nom. Thick., In.						3	3½	4	4	4½	4½
		Lbs/Ft						**20.3**	**25.2**	**30.7**	**30.7**	**36.0**	**36.0**
	*Asbestos Fiber—, Sodium Silicate	Nom. Thick., In.	1½	1½	1½	1½	1½	2½	2½	3½	3½	4½	4½
		Lbs/Ft	**9.26**	**9.26**	**9.26**	**9.26**	**9.26**	**16.35**	**16.35**	**24.11**	**24.11**	**32.57**	**32.57**

	Pressure Rating psi	Cast Iron		Steel						
		125	250	150	300	400	600	900	1500	2500
FLANGES	Screwed or Slip-On	**120** 1.5	**233** 1.5	**108** 1.5	**262** 1.5	**310** 1.5	**442** 1.5	**559** 1.5	**1297** 1.5	
	Welding Neck			**142** 1.5	**288** 1.5	**351** 1.5	**577** 1.5	**785** 1.5	**1597** 1.5	
	Lap Joint			**143** 1.5	**282** 1.5	**337** 1.5	**476** 1.5	**588** 1.5	**1372** 1.5	
	Blind	**175** 1.5	**308** 1.5	**185** 1.5	**349** 1.5	**455** 1.5	**603** 1.5	**719** 1.5		
FLANGED FITTINGS	S.R. 90° Elbow	**501** 5.5	**826** 5.8	**656** 5.5	**958** 5.8	**1014** 6	**1402** 6.3	**1886** 6.7		
	L.R. 90° Elbow	**701** 7	**1036** 7	**781** 7	**1058** 7					
	45° Elbow	**391** 4.3	**696** 4.6	**481** 4.3	**708** 4.6	**839** 4.7	**1212** 5	**1586** 5		
	Tee	**746** 8.3	**1263** 8.7	**961** 8.3	**1404** 8.6	**1671** 9	**2128** 9.4	**3054** 10		
VALVES	Flanged Bonnet Gate	**1254** 8	**2321** 9	**1190** 5	**2015** 7	**2300** 7.2	**3675** 7.9	**4950** 8.2	**7875** 9	
	Flanged Bonnet Globe or Angle									
	Flanged Bonnet Check	**1166** 10.5			**1225** 6					
	Pressure Seal Bonnet—Gate						**3230** 7		**8130** 8	
	Pressure Seal Bonnet—Globe									

* 16 lb cu. ft. density.

Boldface type is weight in pounds. Lightface type beneath weight is weight factor for insulation.

Insulation thicknesses and weights are based on average conditions and do not constitute a recommendation for specific thicknesses of materials. Insulation weights are based on 85% magnesia and hydrous calcium silicate at 11 lbs/cubic foot. The listed thicknesses and weights of combination covering are the sums of the inner layer of diatomaceous earth at 21 lbs/cubic foot and the outer layer at 11 lbs/cubic foot.

Insulation weights include allowances for wire, cement, canvas, bands and paint, but not special surface finishes.

To find the weight of covering on flanges, valves or fittings, multiply the weight factor by the weight per foot of covering used on straight pipe.

Valve weights are approximate. When possible, obtain weights from the manufacturer.

Cast iron valve weights are for flanged end valves; steel weights for welding end valves.

All flanged fitting, flanged valve and flange weights include the proportional weight of bolts or studs to make up all joints.

18″ PIPE 18″ O.D. WEIGHTS OF PIPING MATERIALS

PIPE Schedule No.	10	20		30		40	60	80	100	120	140	160
Wall Designation			Std.		XS							
Thickness—In.	.250	.312	.375	.438	.500	.562	.750	.937	1.156	1.375	1.562	1.781
Pipe—Lbs/Ft	**47.4**	**59.0**	**70.6**	**82.1**	**93.5**	**104.8**	**138.2**	**170.8**	**208.0**	**244.1**	**274.2**	**308.5**
Water—Lbs/Ft	**104.3**	**102.8**	**101.2**	**99.9**	**98.4**	**97**	**92.7**	**88.5**	**83.7**	**79.2**	**75.3**	**71.0**
WELDING FITTINGS L.R. 90° Elbow			**256** 4.5		**338** 4.5							
S.R. 90° Elbow			**171** 2.8		**225** 2.8							
L.R. 45° Elbow			**128** 1.9		**168** 1.9							
Tee			**258** 3.6		**328** 3.6							
Lateral			**326** 7.5		**526** 7.5							
Reducer			**94** 1.3		**123** 1.3							
Cap			**57** 2.1		**75** 2.1							

INSULATION	Temperature Range °F	100-199	200-299	300-399	400-499	500-599	600-699	700-799	800-899	900-999	1000-1099	1100-1200
85% Magnesia Calcium Silicate	Nom. Thick., In.	1½	1½	2	2½	3	3	3½	4	4	4½	4½
	Lbs/Ft	**7.73**	**7.73**	**10.4**	**13.3**	**16.3**	**16.3**	**19.3**	**22.6**	**22.6**	**25.9**	**25.9**
Combination	Nom. Thick., In.						3	3½	4	4	4½	4½
	Lbs/Ft						**22.7**	**28.0**	**33.8**	**33.8**	**39.5**	**39.5**
*Asbestos Fiber—Sodium Silicate	Nom. Thick., In.	1½	1½	1½	2	2	3	3	4	4	5	5
	Lbs/Ft	**9.93**	**9.93**	**9.93**	**13.72**	**13.72**	**21.84**	**21.84**	**31.22**	**31.22**	**40.77**	**40.77**

	Pressure Rating psi	Cast Iron		Steel						
		125	250	150	300	400	600	900	1500	2500
FLANGES	Screwed or Slip-On	**140** 1.5		**140** 1.5	**331** 1.5	**380** 1.5	**573** 1.5	**797** 1.5	**1694** 1.5	
	Welding Neck			**160** 1.5	**355** 1.5	**430** 1.5	**652** 1.5	**1074** 1.5	**2069** 1.5	
	Lap Joint			**166** 1.5	**355** 1.5	**415** 1.5	**566** 1.5	**820** 1.5	**1769** 1.5	
	Blind	**210** 1.5	**396** 1.5	**229** 1.5	**440** 1.5	**572** 1.5	**762** 1.5	**1030** 1.5		
FLANGED FITTINGS	S.R. 90° Elbow	**621** 5.8	**1060** 6	**711** 5.8	**1126** 6	**1340** 6.2	**1793** 6.6	**2817** 7		
	L.R. 90° Elbow	**881** 7.4	**1350** 7.4	**941** 7.4	**1426** 7.4					
	45° Elbow	**461** 4.4	**870** 4.7	**521** 4.4	**901** 4.7	**1040** 4.8	**1543** 5	**2252** 5.2		
	Tee	**921** 8.6	**1625** 9	**1010** 8.6	**1602** 9	**1909** 9.3	**2690** 9.9	**4327** 10.5		
VALVES	Flanged Bonnet Gate	**1629** 8.2	**2578** 9.3	**1510** 6	**2505** 6.5	**3765** 7	**4460** 7.8	**6675** 8.5		
	Flanged Bonnet Globe or Angle									
	Flanged Bonnet Check	**1371** 10.5								
	Pressure Seal Bonnet—Gate						**3100** 5.5	**3400** 5.6	**4200** 6	
	Pressure Seal Bonnet—Globe									

Boldface type is weight in pounds. Lightface type beneath weight is weight factor for insulation.

Insulation thicknesses and weights are based on average conditions and do not constitute a recommendation for specific thicknesses of materials. Insulation weights are based on 85% magnesia and hydrous calcium silicate at 11 lbs/cubic foot. The listed thicknesses and weights of combination covering are the sums of the inner layer of diatomaceous earth at 21 lbs/cubic foot and the outer layer at 11 lbs/cubic foot.

Insulation weights include allowances for wire, cement, canvas, bands and paint, but not special surface finishes.

To find the weight of covering on flanges, valves or fittings, multiply the weight factor by the weight per foot of covering used on straight pipe.

Valve weights are approximate. When possible, obtain weights from the manufacturer.

Cast iron valve weights are for flanged end valves; steel weights for welding end valves.

All flanged fitting, flanged valve and flange weights include the proportional weight of bolts or studs to make up all joints.

*** 16 lb cu. ft. density.**

WEIGHTS OF PIPING MATERIALS

20″ O.D. 20″ PIPE

PIPE												
Schedule No.	10	20	30	40	60	80	100	120	140	160		
Wall Designation		Std.	XS									
Thickness—In.	.250	.375	.500	.593	.812	1.031	1.281	1.500	1.750	1.968		
Pipe—Lbs/Ft	**52.7**	**78.6**	**104.1**	**122.9**	**166.4**	**208.9**	**256.1**	**296.4**	**341.1**	**379.0**		
Water—Lbs/Ft	**129.5**	**126.0**	**122.8**	**120.4**	**115.0**	**109.4**	**103.4**	**98.3**	**92.6**	**87.9**		

WELDING FITTINGS	10	20	30	40	60	80	100	120	140	160		
L.R. 90° Elbow		**317** 5	**419** 5									
S.R. 90° Elbow		**212** 3.4	**278** 3.4									
L.R. 45° Elbow		**158** 2.1	**208** 2.1									
Tee		**321** 4	**407** 4									
Lateral		**396** 8.3	**628** 8.3									
Reducer		**142** 1.7	**186** 1.7									
Cap		**72** 2.3	**94** 2.3									

INSULATION		Temperature Range °F	100-199	200-299	300-399	400-499	500-599	600-699	700-799	800-899	900-999	1000-1099	1100-1200
85% Magnesia Calcium Silicate		Nom. Thick., In.	1½	1½	2	2½	3	3	3½	4	4	4½	4½
		Lbs/Ft	**8.45**	**8.45**	**11.6**	**14.6**	**17.7**	**17.7**	**21.1**	**24.6**	**24.6**	**28.1**	**28.1**
Combination		Nom. Thick., In.						3	3½	4	4	4½	4½
		Lbs/Ft						**24.7**	**30.7**	**37.0**	**37.0**	**43.1**	**43.1**
*Asbestos Fiber—Sodium Silicate		Nom. Thick., In.	1½	1½	1½	2	2	3	3	4	4	5	5
		Lbs/Ft	**10.96**	**10.96**	**10.96**	**14.86**	**14.86**	**24.24**	**24.24**	**33.79**	**33.79**	**44.03**	**44.03**

	Pressure Rating psi	Cast Iron		Steel						
		125	250	150	300	400	600	900	1500	2500
FLANGES	Screwed or Slip-On	**176** 1.5		**181** 1.5	**378** 1.5	**468** 1.5	**733** 1.5	**972** 1.5	**2114** 1.5	
	Welding Neck			**196** 1.5	**431** 1.5	**535** 1.5	**811** 1.5	**1344** 1.5	**2614** 1.5	
	Lap Joint			**211** 1.5	**428** 1.5	**510** 1.5	**725** 1.5	**1048** 1.5	**2189** 1.5	
	Blind	**276** 1.5	**487** 1.5	**298** 1.5	**545** 1.5	**711** 1.5	**976** 1.5	**1287** 1.5		
FLANGED FITTINGS	S.R. 90° Elbow	**792** 6	**1315** 6.3	**922** 6	**1375** 6.3	**1680** 6.5	**2314** 6.9	**3610** 7.3		
	L.R. 90° Elbow	**1132** 7.8	**1725** 7.8	**1352** 7.8	**1705** 7.8					
	45° Elbow	**592** 4.6	**1055** 4.8	**652** 4.6	**1105** 4.8	**1330** 4.9	**1917** 5.2	**2848** 5.4		
	Tee	**1178** 9	**2022** 9.5	**1378** 9	**1908** 9.5	**2370** 9.7	**3463** 10.1	**5520** 11		
VALVES	Flanged Bonnet Gate	**1934** 8.3	**3823** 9.5	**1855** 6	**3370** 7	**5700** 8	**5755** 8			
	Flanged Bonnet Globe or Angle									
	Flanged Bonnet Check	**1772** 11								
	Pressure Seal Bonnet—Gate									
	Pressure Seal Bonnet—Globe									

Boldface type is weight in pounds. Lightface type beneath weight is weight factor for insulation.

Insulation thicknesses and weights are based on average conditions and do not constitute a recommendation for specific thicknesses of materials. Insulation weights are based on 85% magnesia and hydrous calcium silicate at 11 lbs/cubic foot. The listed thicknesses and weights of combination covering are the sums of the inner layer of diatomaceous earth at 21 lbs/cubic foot and the outer layer at 11 lbs/cubic foot.

Insulation weights include allowances for wire, cement, canvas, bands and paint, but not special surface finishes.

To find the weight of covering on flanges, valves or fittings, multiply the weight factor by the weight per foot of covering used on straight pipe.

Valve weights are approximate. When possible, obtain weights from the manufacturer.

Cast iron valve weights are for flanged end valves; steel weights for welding end valves.

All flanged fitting, flanged valve and flange weights include the proportional weight of bolts or studs to make up all joints.

* 16 lb cu. ft. density.

24" PIPE 24" O.D. WEIGHTS OF PIPING MATERIALS

PIPE	Schedule No.	10	20		30	40	60	80	100	120	140	160	
	Wall Designation		Std.	XS									
	Thickness—In.	.250	.375	.500	.562	.687	.968	1.218	1.531	1.812	2.062	2.343	
	Pipe—Lbs/Ft	**63.4**	**94.6**	**125.5**	**140.8**	**171.2**	**238.1**	**296.4**	**367.4**	**429.4**	**483.1**	**541.9**	
	Water—Lbs/Ft	**188.0**	**183.8**	**180.1**	**178.1**	**174.3**	**165.8**	**158.3**	**149.3**	**141.4**	**134.5**	**127.0**	
WELDING FITTINGS	L.R. 90° Elbow		**458** 6	**606** 6									
	S.R. 90° Elbow		**305** 3.7	**404** 3.7									
	L.R. 45° Elbow		**229** 2.5	**302** 2.5									
	Tee		**445** 4.9	**563** 4.9									
	Lateral		**544** 10	**882** 10									
	Reducer		**167** 1.7	**220** 1.7									
	Cap		**102** 2.8	**134** 2.8									

INSULATION	Temperature Range °F	100-199	200-299	300-399	400-499	500-599	600-699	700-799	800-899	900-999	1000-1099	1100-1200
85% Magnesia Calcium Silicate	Nom. Thick., In.	1½	1½	2	2½	3	3	3½	4	4	4½	4½
	Lbs/Ft	**10.0**	**10.0**	**13.4**	**17.0**	**21.0**	**21.0**	**24.8**	**28.7**	**28.7**	**32.9**	**32.9**
Combination	Nom. Thick., In.						3	3½	4	4	4½	4½
	Lbs/Ft						**29.2**	**36.0**	**43.1**	**43.1**	**50.6**	**50.6**
*Asbestos Fiber—Sodium Silicate	Nom. Thick., In.	1½	1½	1½	2	2	3	3	4½	4½	5	5
	Lbs/Ft	**13.55**	**13.55**	**13.55**	**18.44**	**18.44**	**28.38**	**28.38**	**45.06**	**45.06**	**50.97**	**50.97**

	Pressure Rating psi	Cast Iron		Steel						
		125	250	150	300	400	600	900	1500	2500
FLANGES	Screwed or Slip-On	**255** 1.5		**245** 1.5	**577** 1.5	**676** 1.5	**1056** 1.5	**1823** 1.5	**3378** 1.5	
	Welding Neck			**295** 1.5	**632** 1.5	**777** 1.5	**1157** 1.5	**2450** 1.5	**4153** 1.5	
	Lap Joint			**295** 1.5	**617** 1.5	**752** 1.5	**1046** 1.5	**2002** 1.5	**3478** 1.5	
	Blind	**405** 1.5	**757** 1.5	**446** 1.5	**841** 1.5	**1073** 1.5	**1355** 1.5	**2442** 1.5		
FLANGED FITTINGS	S.R. 90° Elbow	**1231** 6.7	**2014** 6.8	**1671** 6.7	**2174** 6.8	**2474** 7.1	**3506** 7.6	**6155** 8.1		
	L.R. 90° Elbow	**1711** 8.7	**2644** 8.7	**1821** 8.7	**2874** 8.7					
	45° Elbow	**871** 4.8	**1604** 5	**1121** 4.8	**1634** 5	**1974** 5.1	**2831** 5.5	**5124** 6		
	Tee	**1836** 10	**3061** 10.2	**2276** 10	**3161** 10.2	**3811** 10.6	**5184** 11.4	**9387** 12.1		
VALVES	Flanged Bonnet Gate	**3062** 8.5	**6484** 9.8	**2500** 5	**4675** 7	**6995** 8.7	**8020** 9.5			
	Flanged Bonnet Globe or Angle									
	Flanged Bonnet Check	**2956** 12								
	Pressure Seal Bonnet—Gate									
	Pressure Seal Bonnet—Globe									

Boldface type is weight in pounds. Lightface type beneath weight is weight factor for insulation.

Insulation thicknesses and weights are based on average conditions and do not constitute a recommendation for specific thicknesses of materials. Insulation weights are based on 85% magnesia and hydrous calcium silicate at 11 lbs/cubic foot. The listed thicknesses and weights of combination covering are the sums of the inner layer of diatomaceous earth at 21 lbs/cubic foot and the outer layer at 11 lbs/cubic foot.

Insulation weights include allowances for wire, cement, canvas, bands and paint, but not special surface finishes.

To find the weight of covering on flanges, valves or fittings, multiply the weight factor by the weight per foot of covering used on straight pipe.

Valve weights are approximate. When possible, obtain weights from the manufacturer.

Cast iron valve weights are for flanged end valves; steel weights for welding end valves.

All flanged fitting, flanged valve and flange weights include the proportional weight of bolts or studs to make up all joints.

* 16 lb cu. ft. density.

WEIGHTS OF PIPING MATERIALS

26″ O.D **26″** PIPE

PIPE	Schedule No.	10		20
	Wall Designation		Std.	XS
	Thickness—In.	.312	.375	.500
	Pipe—Lbs/Ft	**85.7**	**102.6**	**136.2**
	Water—Lbs/Ft	**219.2**	**216.8**	**212.5**
WELDING FITTINGS	L.R. 90° Elbow		**602** 8.5	**713** 8.5
	S.R. 90° Elbow		**359** 5	**474** 5
	L.R. 45° Elbow		**269** 3.5	**355** 3.5
	Tee		**634** 6.8	**794** 6.8
	Lateral			
	Reducer		**200** 2.5	**272** 2.5
	Cap		**110** 4.3	**145** 4.3

INSULATION	Temperature Range °F		100-199	200-299	300-399	400-499	500-599	600-699	700-799	800-899	900-999	1000-1099	1100-1200
	85% Magnesia Calcium Silicate	Nom. Thick., In.	1½	1½	2	2½	3	3½	4	4½	5	5	6
		Lbs/Ft	**10.4**	**10.4**	**14.1**	**18.0**	**21.9**	**26.0**	**30.2**	**34.6**	**39.1**	**39.1**	**48.4**
	Combination	Nom. Thick., In.						3½	4½	5½	6	6½	7
		Lbs/Ft						**37.0**	**51.9**	**67.8**	**76.0**	**84.5**	**93.2**
	*Asbestos Fiber—Sodium Silicate	Nom. Thick., In.	2	2	2	2	2	3	3	4½	4½	5	5
		Lbs/Ft	**18.93**	**18.93**	**18.93**	**18.93**	**18.93**	**29.87**	**29.87**	**47.60**	**47.60**	**53.85**	**53.85**

	Pressure Rating psi	Cast Iron		Steel						
		125	250	150	300	400	600	900	1500	2500
FLANGES	Screwed or Slip-On			**250** 1.5	**570** 1.5	**650** 1.5	**950** 1.5	**1525** 1.5		
	Welding Neck			**300** 1.5	**670** 1.5	**750** 1.5	**1025** 1.5	**1575** 1.5		
	Lap Joint									
	Blind			**525** 1.5	**1050** 1.5	**1125** 1.5	**1525** 1.5	**2200** 1.5		
FLANGED FITTINGS	S.R. 90° Elbow									
	L.R. 90° Elbow									
	45° Elbow									
	Tee									
VALVES	Flanged Bonnet Gate									
	Flanged Bonnet Globe or Angle									
	Flanged Bonnet Check									
	Pressure Seal Bonnet—Gate									
	Pressure Seal Bonnet—Globe									

Boldface type is weight in pounds. Lightface type beneath weight is weight factor for insulation.

Insulation thicknesses and weights are based on average conditions and do not constitute a recommendation for specific thicknesses of materials. Insulation weights are based on 85% magnesia and hydrous calcium silicate at 11 lbs/cubic foot. The listed thicknesses and weights of combination covering are the sums of the inner layer of diatomaceous earth at 21 lbs/cubic foot and the outer layer at 11 lbs/cubic foot.

Insulation weights include allowances for wire, cement, canvas, bands and paint, but not special surface finishes.

To find the weight of covering on flanges, valves or fittings, multiply the weight factor by the weight per foot of covering used on straight pipe.

Valve weights are approximate. When possible, obtain weights from manufacturer.

Cast iron valve weights are for flanged end valves; steel weights for welding end valves.

All flanged fitting, flanged valve and flange weights include the proportional weight of bolts or studs to make up all joints.

* 16 lb cu. ft. density.

28″ PIPE 28″ O.D. WEIGHTS OF PIPING MATERIALS

PIPE	Schedule No.	10		20	30							
	Wall Designation		Std.	XS								
	Thickness—In.	.312	.375	.500	.625							
	Pipe—Lbs/Ft	**92.4**	**110.6**	**146.9**	**182.7**							
	Water—Lbs/Ft	**255.0**	**252.7**	**248.1**	**243.6**							
WELDING FITTINGS	L.R. 90° Elbow		**626** 9	**829** 9								
	S.R. 90° Elbow		**415** 5.4	**551** 5.4								
	L.R. 45° Elbow		**312** 3.6	**413** 3.6								
	Tee		**729** 7	**910** 7								
	Lateral											
	Reducer		**210** 2.7	**290** 2.7								
	Cap		**120** 4.5	**160** 4.5								

Temperature Range °F		100-199	200-299	300-399	400-499	500-599	600-699	700-799	800-899	900-999	1000-1099	1100-1200
INSULATION 85% Magnesia Calcium Silicate	Nom. Thick., In.	1½	1½	2	2½	3	3½	4	4½	5	5	6
	Lbs/Ft	**11.2**	**11.2**	**15.1**	**19.2**	**23.4**	27.8	32.3	36.9	41.6	41.6	51.4
Combination	Nom. Thick., In.						3½	4½	5½	6	6½	7
	Lbs/Ft						**39.5**	**55.4**	**72.2**	**80.9**	**89.8**	**99.0**
*Asbestos Fiber—Sodium Silicate	Nom. Thick., In.	2	2	2	2	2	3	3	4½	4½	5	5
	Lbs/Ft	20.26	20.26	20.26	20.26	20.26	31.90	31.90	52.51	52.51	59.17	59.17

	Pressure Rating psi	Cast Iron		Steel						
		125	250	150	300	400	600	900	1500	2500
FLANGES	Screwed or Slip-On			**285** 1.5	**720** 1.5	**780** 1.5	**1075** 1.5	**1800** 1.5		
	Welding Neck			**315** 1.5	**810** 1.5	**880** 1.5	**1175** 1.5	**1850** 1.5		
	Lap Joint									
	Blind			**620** 1.5	**1275** 1.5	**1425** 1.5	**1750** 1.5	**2575** 1.5		
FLANGED FITTINGS	S.R. 90° Elbow									
	L.R. 90° Elbow									
	45° Elbow									
	Tee									
VALVES	Flanged Bonnet Gate									
	Flanged Bonnet Globe or Angle									
	Flanged Bonnet Check									
	Pressure Seal Bonnet—Gate									
	Pressure Seal Bonnet—Globe									

* 16 lb cu. ft. density.

Boldface type is weight in pounds. Lightface type beneath weight is weight factor for insulation.

Insulation thicknesses and weights are based on average conditions and do not constitute a recommendation for specific thicknesses of materials. Insulation weights are based on 85% magnesia and hydrous calcium silicate at 11 lbs/cubic foot. The listed thicknesses and weights of combination covering are the sums of the inner layer of diatomaceous earth at 21 lbs/cubic foot and the outer layer at 11 lbs/cubic foot.

Insulation weights include allowances for wire, cement, canvas, bands and paint, but not special surface finishes.

To find the weight of covering on flanges, valves or fittings, multiply the weight factor by the weight per foot of covering used on straight pipe.

Valve weights are approximate. When possible, obtain weights from manufacturer.

Cast iron valve weights are for flanged end valves; steel weights for welding end valves.

All flanged fitting, flanged valve and flange weights include the proportional weight of bolts or studs to make up all joints.

WEIGHTS OF PIPING MATERIALS

30″ O.D. **30″** PIPE

PIPE	Schedule No.	10			20		30					
	Wall Designation		Std.		XS							
	Thickness—In.	.312	.375	.438	.500	.562	.625					
	Pipe—Lbs/Ft	**98.9**	**118.7**	**138.0**	**157.6**	**176.8**	**196.1**					
	Water—Lbs/Ft	**293.5**	**291.0**	**288.4**	**286.0**	**283.6**	**281.1**					
WELDING FITTINGS	L.R. 90° Elbow		**775** 10		**953** 10							
	S.R. 90° Elbow		**470** 5.9		**644** 5.9							
	L.R. 45° Elbow		**358** 3.9		**475** 3.9							
	Tee		**855** 7.8		**1065** 7.8							
	Lateral											
	Reducer		**220** 3.9		**315** 3.9							
	Cap		**125** 4.8		**175** 4.8							

Insulation		100-199	200-299	300-399	400-499	500-599	600-699	700-799	800-899	900-999	1000-1099	1100-1200
Temperature Range °F		100-199	200-299	300-399	400-499	500-599	600-699	700-799	800-899	900-999	1000-1099	1100-1200
85% Magnesia Calcium Silicate	Nom. Thick., In.	1½	1½	2	2½	3	3½	4	4½	5	5	6
	Lbs/Ft	**11.9**	**11.9**	**16.1**	**20.5**	**25.0**	**29.5**	**34.3**	**39.1**	**44.1**	**44.1**	**54.4**
Combination	Nom. Thick., In.						3½	4½	5½	6	6½	7
	Lbs/Ft						**42.1**	**58.9**	**76.5**	**85.7**	**95.1**	**104.7**
*Asbestos Fiber—Sodium Silicate	Nom. Thick., In.	2½	2½	2½	2½	2½	3	3	4½	4½	5	5
	Lbs/Ft	**27.17**	**27.17**	**27.17**	**27.17**	**27.17**	**33.43**	**33.43**	**55.18**	**55.18**	**62.18**	**62.18**

	Pressure Rating psi	Cast Iron		Steel						
		125	250	150	300	400	600	900	1500	2500
FLANGES	Screwed or Slip-On			**315** 1.5	**810** 1.5	**900** 1.5	**1175** 1.5	**2075** 1.5		
	Welding Neck			**360** 1.5	**930** 1.5	**1000** 1.5	**1300** 1.5	**2150** 1.5		
	Lap Joint									
	Blind			**720** 1.5	**1500** 1.5	**1675** 1.5	**2000** 1.5	**3025** 1.5		
FLANGED FITTINGS	S.R. 90° Elbow									
	L.R. 90° Elbow									
	45° Elbow									
	Tee									
VALVES	Flanged Bonnet Gate									
	Flanged Bonnet Globe or Angle									
	Flanged Bonnet Check									
	Pressure Seal Bonnet—Gate									
	Pressure Seal Bonnet—Globe									

* 16 lb cu. ft. density.

Boldface type is weight in pounds. Lightface type beneath weight is weight factor for insulation.

Insulation thicknesses and weights are based on average conditions and do not constitute a recommendation for specific thicknesses of materials. Insulation weights are based on 85% magnesia and hydrous calcium silicate at 11 lbs/cubic foot. The listed thicknesses and weights of combination covering are the sums of the inner layer of diatomaceous earth at 21 lbs/cubic foot and the outer layer at 11 lbs/cubic foot.

Insulation weights include allowances for wire, cement, canvas, bands and paint, but not special surface finishes.

To find the weight of covering on flanges, valves or fittings, multiply the weight factor by the weight per foot of covering used on straight pipe.

Valve weights are approximate. When possible, obtain weights from manufacturer.

Cast iron valve weights are for flanged end valves; steel weights for welding end valves.

All flanged fitting, flanged valve and flange weights include the proportional weight of bolts or studs to make up all joints.

32" PIPE 32" O.D. WEIGHTS OF PIPING MATERIALS

PIPE	Schedule No.	10		20	30	40
	Wall Designation		Std.	XS		
	Thickness—In.	.312	.375	.500	.625	.688
	Pipe—Lbs/Ft	**105.8**	**126.7**	**168.2**	**209.4**	**229.9**
	Water—Lbs/Ft	**335.0**	**332.3**	**327.0**	**321.8**	**319.2**
WELDING FITTINGS	L.R. 90° Elbow		**818** 10.5	**1090** 10.5		
	S.R. 90° Elbow		**546** 6.3	**722** 6.3		
	L.R. 45° Elbow		**408** 4.2	**541** 4.2		
	Tee		**991** 8.4	**1230** 8.4		
	Lateral					
	Reducer		**255** 3.1	**335** 3.1		
	Cap		**145** 5.2	**190** 5.2		

INSULATION	Temperature Range °F	100-199	200-299	300-399	400-499	500-599	600-699	700-799	800-899	900-999	1000-1099	1100-1200
85% Magnesia Calcium Silicate	Nom. Thick., In.	1½	1½	2	2½	3	3½	4	4½	5	5	6
	Lbs/Ft	**12.7**	**12.7**	**17.1**	**21.7**	**26.5**	**31.3**	**36.3**	**41.4**	**46.6**	**46.6**	**57.5**
Combination	Nom. Thick., In.						3½	4½	5½	6	6½	7
	Lbs/Ft						**44.7**	**62.3**	**80.9**	**90.5**	**100.4**	**110.5**
*Asbestos Fiber—Sodium Silicate	Nom. Thick., In.	3	3	3	3	3	3	3	4½	4½	5	5
	Lbs/Ft	**41.50**	**41.50**	**41.50**	**41.50**	**41.50**	**41.50**	**41.50**	**57.27**	**57.27**	**64.49**	**64.49**

	Pressure Rating psi	Cast Iron		Steel						
		125	250	150	300	400	600	900	1500	2500
FLANGES	Screwed or Slip-On			**395** 1.5	**890** 1.5	**1025** 1.5	**1375** 1.5	**2500** 1.5		
	Welding Neck			**435** 1.5	**1025** 1.5	**1150** 1.5	**1500** 1.5	**2575** 1.5		
	Lap Joint									
	Blind			**870** 1.5	**1775** 1.5	**1975** 1.5	**2300** 1.5	**3650** 1.5		
FLANGED FITTINGS	S.R. 90° Elbow									
	L.R. 90° Elbow									
	45° Elbow									
	Tee									
VALVES	Flanged Bonnet Gate									
	Flanged Bonnet Globe or Angle									
	Flanged Bonnet Check									
	Pressure Seal Bonnet—Gate									
	Pressure Seal Bonnet—Globe									

* 16 lb cu. ft. density.

Boldface type is weight in pounds. Lightface type beneath weight is weight factor for insulation.

Insulation thicknesses and weights are based on average conditions and do not constitute a recommendation for specific thicknesses of materials. Insulation weights are based on 85% magnesia and hydrous calcium silicate at 11 lbs/cubic foot. The listed thicknesses and weights of combination covering are the sums of the inner layer of diatomaceous earth at 21 lbs/cubic foot and the outer layer at 11 lbs/cubic foot.

Insulation weights include allowances for wire, cement, canvas, bands and paint, but not special surface finishes.

To find the weight of covering on flanges, valves or fittings, multiply the weight factor by the weight per foot of covering used on straight pipe.

Valve weights are approximate. When possible, obtain weights from manufacturer. Cast iron valve weights are for flanged end valves; steel weights for welding end valves.

All flanged fitting, flanged valve and flange weights include the proportional weight of bolts or studs to make up all joints.

WEIGHTS OF PIPING MATERIALS — 34″ O.D. 34″ PIPE

PIPE					
Schedule No.	10		20	30	40
Wall Designation		Std.	XS		
Thickness—In.	.312	.375	.500	.625	.688
Pipe—Lbs/Ft	**112.4**	**134.7**	**178.9**	**222.8**	**244.6**
Water—Lbs/Ft	**379.1**	**376.0**	**370.3**	**365.0**	**362.2**

WELDING FITTINGS	10	Std.	XS	30	40
L.R. 90° Elbow		**926** 11	**1230** 11		
S.R. 90° Elbow		**617** 5.5	**817** 5.5		
L.R. 45° Elbow		**463** 4.4	**615** 4.4		
Tee		**1136** 8.9	**1420** 8.9		
Lateral					
Reducer		**270** 3.3	**355** 3.3		
Cap		**160** 5.6	**210** 5.6		

INSULATION	Temperature Range °F	100-199	200-299	300-399	400-499	500-599	600-699	700-799	800-899	900-999	1000-1099	1100-1200
85% Magnesia Calcium Silicate	Nom. Thick., In.	1½	1½	2	2½	3	3½	4	4½	5	5	6
	Lbs/Ft	**13.4**	**13.4**	**18.2**	**23.0**	**28.0**	**33.1**	**38.3**	**43.7**	**49.1**	**49.1**	**60.5**
Combination	Nom. Thick., In.						3½	4½	5½	6	6½	7
	Lbs/Ft						**47.2**	**65.8**	**85.3**	**95.4**	**105.7**	**116.3**
*Asbestos Fiber—Sodium Silicate	Nom. Thick., In.	3	3	3	3	3	3	3	4½	4½	5	5
	Lbs/Ft	**38.74**	**38.74**	**38.74**	**38.74**	**38.74**	**38.74**	**38.74**	**60.50**	**60.50**	**68.04**	**68.04**

	Pressure Rating psi	Cast Iron		Steel						
		125	250	150	300	400	600	900	1500	2500
FLANGES	Screwed or Slip-On			**420** 1.5	**1075** 1.5	**1150** 1.5	**1500** 1.5	**2950** 1.5		
	Welding Neck			**465** 1.5	**1200** 1.5	**1300** 1.5	**1650** 1.5	**3025** 1.5		
	Lap Joint									
	Blind			**990** 1.5	**2025** 1.5	**2250** 1.5	**2575** 1.5	**4275** 1.5		
FLANGED FITTINGS	S.R. 90° Elbow									
	L.R. 90° Elbow									
	45° Elbow									
	Tee									
VALVES	Flanged Bonnet Gate									
	Flanged Bonnet Globe or Angle									
	Flanged Bonnet Check									
	Presssure Seal Bonnet—Gate									
	Pressure Seal Bonnet—Globe									

Boldface type is weight in pounds. Lightface type beneath weight is weight factor for insulation.

Insulation thicknesses and weights are based on average conditions and do not constitute a recommendation for specific thicknesses of materials. Insulation weights are based on 85% magnesia and hydrous calcium silicate at 11 lbs/cubic foot. The listed thicknesses and weights of combination covering are the sums of the inner layer of diatomaceous earth at 21 lbs/cubic foot and the outer layer at 11 lbs/cubic foot.

Insulation weights include allowances for wire, cement, canvas, bands and paint, but not special surface finishes.

To find the weight of covering on flanges, valves or fittings, multiply the weight factor by the weight per foot of covering used on straight pipe.

Valve weights are approximate. When possible, obtain weights from manufacturer.

Cast iron valve weights are for flanged end valves; steel weights for welding end valves.

All flanged fitting, flanged valve and flange weights include the proportional weight of bolts or studs to make up all joints.

* 16 lb cu. ft. density.

36″ PIPE 36″ O.D. WEIGHTS OF PIPING MATERIALS

PIPE	Schedule No.	10		20	30	40	
	Wall Designation		Std.	XS			
	Thickness—In.	.312	.375	.500	.625	.750	1.125
	Pipe—Lbs/Ft	**119.1**	**142.7**	**189.6**	**236.1**	**282.4**	**419.03**
	Water—Lbs/Ft	**425.9**	**422.6**	**416.6**	**411.0**	**405.1**	**387.67**
WELDED FITTINGS	L.R. 90° Elbow		**1040** 12	**1380** 12			
	S.R. 90° Elbow		**692** 5	**913** 5			
	L.R. 45° Elbow		**518** 4.8	**686** 4.8			
	Tee		**1294** 9.5	**1610** 9.5			
	Lateral						
	Reducer		**340** 3.6	**360** 3.6			
	Cap		**175** 6	**235** 6			

INSULATION	Temperature Range °F		100-199	200-299	300-399	400-499	500-599	600-699	700-799	800-899	900-999	1000-1099	1100-1200
	85% Magnesia Calcium. Silicate	Nom. Thick., In.	1½	1½	2	2½	3	3½	4	4½	5	5	6
		Lbs/Ft	**14.2**	**14.2**	**19.2**	**24.2**	**29.5**	**34.8**	**40.3**	**45.9**	**51.7**	**51.7**	**63.5**
	Combination	Nom. Thick., In.						3½	4½	5½	6	6½	7
		Lbs/Ft						**49.8**	**69.3**	**89.7**	**100.2**	**111.0**	**122.0**
	*Asbestos Fiber—Sodium Silicate	Nom. Thick., In.	3	3	3	3	3	3	3	4½	4½	5	5
		Lbs/Ft	**40.84**	**40.84**	**40.84**	**40.84**	**40.84**	**40.84**	**40.84**	**55.83**	**55.83**	**71.48**	**71.48**

	Pressure Rating psi	Cast Iron		Steel						
		125	250	150	300	400	600	900	1500	2500
FLANGES	Screwed or Slip-On			**480** 1.5	**1200** 1.5	**1325** 1.5	**1600** 1.5	**3350** 1.5		
	Welding Neck			**520** 1.5	**1300** 1.5	**1475** 1.5	**1750** 1 5	**3450** 1.5		
	Lap Joint									
	Blind			**1125** 1.5	**2275** 1.5	**2525** 1.5	**2950** 1.5	**4900** 1.5		
FLANGED FITTINGS	S.R. 90° Elbow									
	L.R. 90° Elbow									
	45° Elbow									
	Tee									
VALVES	Flanged Bonnet Gate									
	Flanged Bonnet Globe or Angle									
	Flanged Bonnet Check									
	Pressure Seal Bonnet—Gate									
	Pressure Seal Bonnet—Globe									

Boldface type is weight in pounds. Lightface type beneath weight is weight factor for insulation.

Insulation thicknesses and weights are based on average conditions and do not constitute a recommendation for specific thicknesses of materials. Insulation weights are based on 85% magnesia and hydrous calcium silicate at 11 lbs/cubic foot. The listed thicknesses and weights of combination covering are the sums of the inner layer of diatomaceous earth at 21 lbs/cubic foot and the outer layer at 11 lbs/cubic foot.

Insulation weights include allowances for wire, cement, canvas, bands and paint, but not special surface finishes.

To find the weight of covering on flanges, valves or fittings, multiply the weight factor by the weight per foot of covering used on straight pipe.

Valve weights are approximate. When possible, obtain weights from manufacturer.

Cast iron valve weights are for flanged end valves; steel weights for welding end valves.

All flanged fitting, flanged valve and flange weights include the proportional weight of bolts or studs to make up all joints.

* [illegible]

Appendix D

Conversion Factors

Alphabetical Conversion Factors

TO CONVERT	INTO	MULTIPLY BY
	A	
Abcoulomb	Statcoulombs	2.998×10^{10}
Acre	Sq. chain (Gunters)	10
Acre	Rods	160
Acre	Square links (Gunters)	1×10^{5}
Acre	Hectare or sq. hectometer	.4047
acres	sq feet	43,560.0
acres	sq meters	4,047.
acres	sq miles	1.562×10^{-3}
acres	sq yards	4,840.
acre-feet	cu feet	43,560.0
acre-feet	gallons	3.259×10^{5}
amperes/sq cm	amps/sq in.	6.452
amperes/sq cm	amps/sq meter	10^{4}
amperes/sq in.	amps/sq cm	0.1550
amperes/sq in.	amps/sq meter	1,550.0
amperes/sq meter	amps/sq cm	10^{-4}
amperes/sq meter	amps/sq in.	6.452×10^{-4}
ampere-hours	coulombs	3,600.0
ampere-hours	faradays	0.03731
ampere-turns	gilberts	1.257
ampere-turns/cm	amp-turns/in.	2.540
ampere-turns/cm	amp-turns/meter	100.0
ampere-turns/cm	gilberts/cm	1.257
ampere-turns/in.	amp-turns/cm	0.3937
ampere-turns/in.	amp-turns/meter	39.37
ampere-turns/in.	gilberts/cm	0.4950
ampere-turns/meter	amp/turns/cm	0.01
ampere-turns/meter	amp-turns/in.	0.0254
ampere-turns/meter	gilberts/cm	0.01257
Angstrom unit	Inch	3937×10^{-9}
Angstrom unit	Meter	1×10^{-10}
Angstrom unit	Micron or (Mu)	1×10^{-4}
Are	Acre (US)	.02471
Ares	sq. yards	119.60
ares	acres	0.02471
ares	sq meters	100.0
Astronomical Unit	Kilometers	1.495×10^{8}
Atmospheres	Ton/sq. inch	.007348
atmospheres	cms of mercury	76.0
atmospheres	ft of water (at 4°C)	33.90
atmospheres	in. of mercury (at 0°C)	29.92
atmospheres	kgs/sq cm	1.0333
atmospheres	kgs/sq meter	10,332.
atmospheres	pounds/sq in.	14.70
atmospheres	tons/sq ft	1.058
	B	
Barrels (U.S., dry)	cu. inches	7056.
Barrels (U.S., dry)	quarts (dry)	105.0
Barrels (U.S., liquid)	gallons	31.5
barrels (oil)	gallons (oil)	42.0
bars	atmospheres	0.9869
bars	dynes/sq cm	10^{6}
bars	kgs/sq meter	1.020×10^{4}
bars	pounds/sq ft	2,089.
bars	pounds/sq in.	14.50
Baryl	Dyne/sq. cm.	1.000
Bolt (US Cloth)	Meters	36.576
BTU	Liter–Atmosphere	10.409
Btu	ergs	1.0550×10^{10}
Btu	foot-lbs	778.3
Btu	gram-calories	252.0
Btu	horsepower-hrs	3.931×10^{-4}
Btu	joules	1,054.8
Btu	kilogram-calories	0.2520
Btu	kilogram-meters	107.5
Btu	kilowatt-hrs	2.928×10^{-4}
Btu/hr	foot-pounds/sec	0.2162
Btu/hr	gram-cal/sec	0.0700
Btu/hr	horsepower-hrs	3.929×10^{-4}
Btu/hr	watts	0.2931
Btu/min	foot-lbs/sec	12.96
Btu/min	horsepower	0.02356
Btu/min	kilowatts	0.01757
Btu/min	watts	17.57
Btu/sq ft/min	watts/sq in.	0.1221
Bucket (Br. dry)	Cubic Cm.	1.818×10^{4}
bushels	cu ft	1.2445
bushels	cu in.	2,150.4
bushels	cu meters	0.03524
bushels	liters	35.24
bushels	pecks	4.0
bushels	pints (dry)	64.0
bushels	quarts (dry)	32.0
	C	
Calories, gram (mean)	B.T.U. (mean)	3.9685×10^{-3}
Candle/sq. cm	Lamberts	3.142
Candle/sq. inch	Lamberts	.4870
centares (centiares)	sq meters	1.0
Centigrade	Fahrenheit	(C°x9/5)+32
centigrams	grams	0.01
Centiliter	Ounce fluid (US)	.3382
Centiliter	Cubic inch	.6103
Centiliter	drams	2.705
centiliters	liters	0.01
centimeters	feet	3.281×10^{-2}
centimeters	inches	0.3937
centimeters	kilometers	10^{-5}
centimeters	meters	0.01
centimeters	miles	6.214×10^{-6}
centimeters	millimeters	10.0
centimeters	mils	393.7
centimeters	yards	1.094×10^{-2}
centimeter-dynes	cm-grams	1.020×10^{-3}
centimeter-dynes	meter-kgs	1.020×10^{-8}
centimeter-dynes	pound-feet	7.376×10^{-8}
centimeter-grams	cm-dynes	980.7
centimeter-grams	meter-kgs	10^{-5}
centimeter-grams	pound-feet	7.233×10^{-5}
centimeters of mercury	atmospheres	0.01316
centimeters of mercury	feet of water	0.4461
centimeters of mercury	kgs/sq meter	136.0
centimeters of mercury	pounds/sq ft	27.85
centimeters of mercury	pounds/sq in.	0.1934
centimeters/sec	feet/min	1.1969
centimeters/sec	feet/sec	0.03281
centimeters/sec	kilometers/hr	0.036
centimeters/sec	knots	0.1943
centimeters/sec	meters/min	0.6
centimeters/sec	miles/hr	0.02237
centimeters/sec	miles/min	3.728×10^{-4}
centimeters/sec/sec	feet/sec/sec	0.03281
centimeters/sec/sec	kms/hr/sec	0.036
centimeters/sec/sec	meters/sec/sec	0.01
centimeters/sec/sec	miles/hr/sec	0.02237
Chain	Inches	792.00
Chain	meters	20.12
Chains (surveyors' or Gunter's)	yards	22.00
circular mils	sq cms	5.067×10^{-6}
circular mils	sq mils	0.7854
Circumference	Radians	6.283
circular mils	sq inches	7.854×10^{-7}
Cords	cord feet	8
Cord feet	cu. feet	16
Coulomb	Statcoulombs	2.998×10^{9}
coulombs	faradays	1.036×10^{-5}

(Continued). Alphabetical Conversion Factors

TO CONVERT	INTO	MULTIPLY BY
coulombs/sq cm	coulombs/sq in.	64.52
coulombs/sq cm	coulombs/sq meter	10^{4}
coulombs/sq in.	coulombs/sq cm	0.1550
coulombs/sq in.	coulombs/sq meter	1,550.
coulombs/sq meter	coulombs/sq cm	10^{-4}
coulombs/sq meter	coulombs/sq in.	6.452×10^{-4}
cubic centimeters	cu feet	3.531×10^{-5}
cubic centimeters	cu inches	0.06102
cubic centimeters	cu meters	10^{-6}
cubic centimeters	cu yards	1.308×10^{-6}
cubic centimeters	gallons (U. S. liq.)	2.642×10^{-4}
cubic centimeters	liters	0.001
cubic centimeters	pints (U.S. liq.)	2.113×10^{-3}
cubic centimeters	quarts (U.S. liq.)	1.057×10^{-3}
cubic feet	bushels (dry)	0.8036
cubic feet	cu cms	28,320.0
cubic feet	cu inches	1,728.0
cubic feet	cu meters	0.02832
cubic feet	cu yards	0.03704
cubic feet	gallons (U.S. liq.)	7.48052
cubic feet	liters	28.32
cubic feet	pints (U.S. liq.)	59.84
cubic feet	quarts (U.S. liq.)	29.92
cubic feet/min	cu cms/sec	472.0
cubic feet/min	gallons/sec	0.1247
cubic feet/min	liters/sec	0.4720
cubic feet/min	pounds of water/min	62.43
cubic feet/sec	million gals/day	0.646317
cubic feet/sec	gallons/min	448.831
cubic inches	cu cms	16.39
cubic inches	cu feet	5.787×10^{-4}
cubic inches	cu meters	1.639×10^{-5}
cubic inches	cu yards	2.143×10^{-5}
cubic inches	gallons	4.329×10^{-3}
cubic inches	liters	0.01639
cubic inches	mil-feet	1.061×10^{5}
cubic inches	pints (U.S. liq.)	0.03463
cubic inches	quarts (U.S. liq.)	0.01732
cubic meters	bushels (dry)	28.38
cubic meters	cu cms	10^{6}
cubic meters	cu feet	35.31
cubic meters	cu inches	61,023.0
cubic meters	cu yards	1.308
cubic meters	gallons (U.S. liq.)	264.2
cubic meters	liters	1,000.0
cubic meters	pints (U.S. liq.)	2,113.0
cubic meters	quarts (U.S. liq.)	1,057.
cubic yards	cu cms	7.646×10^{5}
cubic yards	cu feet	27.0
cubic yards	cu inches	46,656.0
cubic yards	cu meters	0.7646
cubic yards	gallons (U.S. liq.)	202.0
cubic yards	liters	764.6
cubic yards	pints (U.S. liq.)	1,615.9
cubic yards	quarts (U.S. liq.)	807.9
cubic yards/min	cubic ft/sec	0.45
cubic yards/min	gallons/sec	3.367
cubic yards/min	liters/sec	12.74

D

TO CONVERT	INTO	MULTIPLY BY
Dalton	Gram	1.650×10^{-24}
days	seconds	86,400.0
decigrams	grams	0.1
deciliters	liters	0.1
decimeters	meters	0.1
degrees (angle)	quadrants	0.01111
degrees (angle)	radians	0.01745
degrees (angle)	seconds	3,600.0
degrees/sec	radians/sec	0.01745
degrees/sec	revolutions/min	0.1667
degrees/sec	revolutions/sec	2.778×10^{-3}
dekagrams	grams	10.0
dekaliters	liters	10.0
dekameters	meters	10.0
Drams (apothecaries' or troy)	ounces (avoidupois)	0.1371429
Drams (apothecaries' or troy)	ounces (troy)	0.125
Drams (U.S., fluid or apoth.)	cubic cm.	3.6967
drams	grams	1.7718
drams	grains	27.3437
drams	ounces	0.0625
Dyne/cm	Erg/sq. millimeter	.01
Dyne/sq. cm.	Atmospheres	9.869×10^{-7}
Dyne/sq. cm.	Inch of Mercury at 0°C	2.953×10^{-5}
Dyne/sq. cm.	Inch of Water at 4°C	4.015×10^{-4}
dynes	grams	1.020×10^{-3}
dynes	joules/cm	10^{-7}
dynes	joules/meter (newtons)	10^{-5}
dynes	kilograms	1.020×10^{-6}
dynes	poundals	7.233×10^{-5}
dynes	pounds	2.248×10^{-6}
dynes/sq cm	bars	10^{-6}

E

TO CONVERT	INTO	MULTIPLY BY
Ell	Cm.	114.30
Ell	Inches	45
Em, Pica	Inch	.167
Em, Pica	Cm.	.4233
Erg/sec	Dyne – cm/sec	1.000
ergs	Btu	9.480×10^{-11}
ergs	dyne-centimeters	1.0
ergs	foot-pounds	7.367×10^{-8}
ergs	gram-calories	0.2389×10^{-7}
ergs	gram-cms	1.020×10^{-3}
ergs	horsepower-hrs	3.7250×10^{-14}
ergs	joules	10^{-7}
ergs	kg-calories	2.389×10^{-11}
ergs	kg-meters	1.020×10^{-8}
ergs	kilowatt-hrs	0.2778×10^{-13}
ergs	watt-hours	0.2778×10^{-10}
ergs/sec	Btu/min	$5{,}688 \times 10^{-9}$
ergs/sec	ft-lbs/min	4.427×10^{-6}
ergs/sec	ft-lbs/sec	7.3756×10^{-8}
ergs/sec	horsepower	1.341×10^{-10}
ergs/sec	kg-calories/min	1.433×10^{-9}
ergs/sec	kilowatts	10^{-10}

F

TO CONVERT	INTO	MULTIPLY BY
farads	microfarads	10^{6}
Faraday/sec	Ampere (absolute)	9.6500×10^{4}
faradays	ampere-hours	26.80
faradays	coulombs	9.649×10^{4}
Fathom	Meter	1.828804
fathoms	feet	6.0
feet	centimeters	30.48
feet	kilometers	3.048×10^{-4}
feet	meters	0.3048
feet	miles (naut.)	1.645×10^{-4}
feet	miles (stat.)	1.894×10^{-4}
feet	millimeters	304.8
feet	mils	1.2×10^{4}
feet of water	atmospheres	0.02950
feet of water	in. of mercury	0.8826
feet of water	kgs/sq cm	0.03048

(Continued). Alphabetical Conversion Factors

TO CONVERT	INTO	MULTIPLY BY
feet of water	kgs/sq meter	304.8
feet of water	pounds/sq ft	62.43
feet of water	pounds/sq in.	0.4335
feet/min	cms/sec	0.5080
feet/min	feet/sec	0.01667
feet/min	kms/hr	0.01829
feet/min	meters/min	0.3048
feet/min	miles/hr	0.01136
feet/sec	cms/sec	30.48
feet/sec	kms/hr	1.097
feet/sec	knots	0.5921
feet/sec	meters/min	18.29
feet/sec	miles/hr	0.6818
feet/sec	miles/min	0.01136
feet/sec/sec	cms/sec/sec	30.48
feet/sec/sec	kms/hr/sec	1.097
feet/sec/sec	meters/sec/sec	0.3048
feet/sec/sec	miles/hr/sec	0.6818
feet/100 feet	per cent grade	1.0
Foot – candle	Lumen/sq. meter	10.764
foot-pounds	Btu	1.286×10^{-3}
foot-pounds	ergs	1.356×10^{7}
foot-pounds	gram-calories	0.3238
foot-pounds	hp-hrs	5.050×10^{-7}
foot-pounds	joules	1.356
foot-pounds	kg-calories	3.24×10^{-4}
foot-pounds	kg-meters	0.1383
foot-pounds	kilowatt-hrs	3.766×10^{-7}
foot-pounds/min	Btu/min	1.286×10^{-3}
foot-pounds/min	foot-pounds/sec	0.01667
foot-pounds/min	horsepower	3.030×10^{-5}
foot-pounds/min	kg-calories/min	3.24×10^{-4}
foot-pounds/min	kilowatts	2.260×10^{-5}
foot-pounds/sec	Btu/hr	4.6263
foot-pounds/sec	Btu/min	0.07717
foot-pounds/sec	horsepower	1.818×10^{-3}
foot-pounds/sec	kg-calories/min	0.01945
foot-pounds/sec	kilowatts	1.356×10^{-3}
Furlongs	miles (U.S.)	0.125
furlongs	rods	40.0
furlongs	feet	660.0
	G	
gallons	cu cms	3,785.0
gallons	cu feet	0.1337
gallons	cu inches	231.0
gallons	cu meters	3.785×10^{-3}
gallons	cu yards	4.951×10^{-3}
gallons	liters	3.785
gallons (liq. Br. Imp.)	gallons (U.S. liq.)	1.20095
gallons (U.S.)	gallons (Imp.)	0.83267
gallons of water	pounds of water	8.3453
gallons/min	cu ft/sec	2.228×10^{-3}
gallons/min	liters/sec	0.06308
gallons/min	cu ft/hr	8.0208
gausses	lines/sq in.	6.452
gausses	webers/sq cm	10^{-8}
gausses	webers/sq in.	6.452×10^{-8}
gausses	webers/sq meter	10^{-4}
gilberts	ampere-turns	0.7958
gilberts/cm	amp-turns/cm	0.7958
gilberts/cm	amp-turns/in	2.021
gilberts/cm	amp-turns/meter	79.58
Gills (British)	cubic cm.	142.07
gills	liters	0.1183
gills	pints (liq.)	0.25
Grade	Radian	.01571
Grains	drams (avoirdupois)	0.03657143

TO CONVERT	INTO	MULTIPLY BY
grains (troy)	grains (avdp)	1.0
grains (troy)	grams	0.06480
grains (troy)	ounces (avdp)	2.0833×10^{-3}
grains (troy)	pennyweight (troy)	0.04167
grains/U.S. gal	parts/million	17.118
grains/U.S. gal	pounds/million gal	142.86
grains/Imp. gal	parts/million	14.286
grams	dynes	980.7
grams	grains	15.43
grams	joules/cm	9.807×10^{-5}
grams	joules/meter (newtons)	9.807×10^{-3}
grams	kilograms	0.001
grams	milligrams	1,000.
grams	ounces (avdp)	0.03527
grams	ounces (troy)	0.03215
grams	poundals	0.07093
grams	pounds	2.205×10^{-3}
grams/cm	pounds/inch	5.600×10^{-3}
grams/cu cm	pounds/cu ft	62.43
grams/cu cm	pounds/cu in	0.03613
grams/cu cm	pounds/mil-foot	3.405×10^{-7}
grams/liter	grains/gal	58.417
grams/liter	pounds/ gal	8.345
grams/liter	pounds/cu ft	0.062427
grams/liter	parts/million	1,000.0
grams/sq cm	pounds/sq ft	2.0481
gram-calories	Btu	3.9683×10^{-3}
gram-calories	ergs	4.1868×10^{7}
gram-calories	foot-pounds	3.0880
gram-calories	horsepower-hrs	1.5596×10^{-6}
gram-calories	kilowatt-hrs	1.1630×10^{-6}
gram-calories	watt-hrs	1.1630×10^{-3}
gram-calories/sec	Btu/hr	14.286
gram-centimeters	Btu	9.297×10^{-8}
gram-centimeters	ergs	980.7
gram-centimeters	joules	9.807×10^{-5}
gram-centimeters	kg-cal	2.343×10^{-8}
gram-centimeters	kg-meters	10^{-5}
	H	
Hand	Cm.	10.16
hectares	acres	2.471
hectares	sq feet	1.076×10^{5}
hectograms	grams	100.0
hectoliters	liters	100.0
hectometers	meters	100.0
hectowatts	watts	100.0
henries	millihenries	1,000.0
Hogsheads (British)	cubic ft.	10.114
Hogsheads (U.S.)	cubic ft.	8.42184
Hogsheads (U.S.)	gallons (U.S.)	63
horsepower	Btu/min	42.44
horsepower	foot-lbs/min	33,000.
horsepower	foot-lbs/sec	550.0
horsepower (metric) (542.5 ft lb/sec)	horsepower (550 ft lb/sec)	0.9863
horsepower (550 ft lb/sec)	horsepower (metric) (542.5 ft lb/sec)	1.014
horsepower	kg-calories/min	10.68
horsepower	kilowatts	0.7457
horsepower	watts	745.7
horsepower (boiler)	Btu/hr	33.479
horsepower (boiler)	kilowatts	9.803
horsepower-hrs	Btu	2,547.
horsepower-hrs	ergs	2.6845×10^{13}
horsepower-hrs	foot-lbs	1.98×10^{6}
horsepower-hrs	gram-calories	641,190.
horsepower-hrs	joules	2.684×10^{6}

(Continued). Alphabetical Conversion Factors

TO CONVERT	INTO	MULTIPLY BY
horsepower-hrs	kg-calories	641.1
horsepower-hrs	kg-meters	2.737×10^{5}
horsepower-hrs	kilowatt-hrs	0.7457
hours	days	4.167×10^{-2}
hours	weeks	5.952×10^{-3}
Hundredweights (long)	pounds	112
Hundredweights (long)	tons (long)	0.05
Hundredweights (short)	ounces (avoirdupois)	1600
Hundredweights (short)	pounds	100
Hundredweights (short)	tons (metric)	0.0453592
Hundredweights (short)	tons (long)	0.0446429
	I	
inches	centimeters	2.540
inches	meters	2.540×10^{-2}
inches	miles	1.578×10^{-5}
inches	millimeters	25.40
inches	mils	1,000.0
inches	yards	2.778×10^{-2}
inches of mercury	atmospheres	0.03342
inches of mercury	feet of water	1.133
inches of mercury	kgs/sq cm	0.03453
inches of mercury	kgs/sq meter	345.3
inches of mercury	pounds/sq ft	70.73
inches of mercury	pounds/sq in.	0.4912
inches of water (at 4°C)	atmospheres	2.458×10^{-3}
inches of water (at 4°C)	inches of mercury	0.07355
inches of water (at 4°C)	kgs/sq cm	2.540×10^{-3}
inches of water (at 4°C)	ounces/sq in.	0.5781
inches of water (at 4°C)	pounds/sq ft	5.204
inches of water (at 4°C)	pounds/sq in.	0.03613
International Ampere	Ampere (absolute)	.9998
International Volt	Volts (absolute)	1.0003
International volt	Joules (absolute)	$1\text{-}593 \times 10^{-19}$
International volt	Joules	9.654×10^{4}
	J	
joules	Btu	9.480×10^{-4}
joules	ergs	10^{7}
joules	foot-pounds	0.7376
joules	kg-calories	2.389×10^{-4}
joules	kg-meters	0.1020
joules	watt-hrs	2.778×10^{-4}
joules/cm	grams	1.020×10^{4}
joules/cm	dynes	10^{7}
joules/cm	joules/meter (newtons)	100.0
joules/cm	poundals	723.3
joules/cm	pounds	22.48
	K	
kilograms	dynes	980,665.
kilograms	grams	1,000.0
kilograms	joules/cm	0.09807
kilograms	joules/meter (newtons)	9.807
kilograms	poundals	70.93
kilograms	pounds	2.205
kilograms	tons (long)	9.842×10^{-4}
kilograms	tons (short)	1.102×10^{-3}
kilograms/cu meter	grams/cu cm	0.001
kilograms/cu meter	pounds/cu ft	0.06243
kilograms/cu meter	pounds/cu in.	3.613×10^{-5}
kilograms/cu meter	pounds/mil-foot	3.405×10^{-10}
kilograms/meter	pounds/ft	0.6720
Kilogram/sq. cm.	Dynes	980,665
kilograms/sq cm	atmospheres	0.9678
kilograms/sq cm	feet of water	32.81

TO CONVERT	INTO	MULTIPLY BY
kilograms/sq cm	inches of mercury	28.96
kilograms/sq cm	pounds/sq ft	2,048.
kilograms/sq cm	pounds/sq in.	14.22
kilograms/sq meter	atmospheres	9.678×10^{-5}
kilograms/sq meter	bars	98.07×10^{-6}
kilograms/sq meter	feet of water	3.281×10^{-3}
kilograms/sq meter	inches of mercury	2.896×10^{-3}
kilograms/sq meter	pounds/sq ft	0.2048
kilograms/sq meter	pounds/sq in.	1.422×10^{-3}
kilograms/sq mm	kgs/sq meter	10^{6}
kilogram-calories	Btu	3.968
kilogram-calories	foot-pounds	3,088.
kilogram-calories	hp-hrs	1.560×10^{-3}
kilogram-calories	joules	4,186.
kilogram-calories	kg-meters	426.9
kilogram-calories	kilojoules	4.186
kilogram-calories	kilowatt-hrs	1.163×10^{-3}
kilogram meters	Btu	9.294×10^{-3}
kilogram meters	ergs	9.804×10^{7}
kilogram meters	foot-pounds	7.233
kilogram meters	joules	9.804
kilogram meters	kg-calories	2.342×10^{-3}
kilogram meters	kilowatt-hrs	2.723×10^{-6}
kilolines	maxwells	1,000.0
kiloliters	liters	1,000.0
kilometers	centimeters	10^{5}
kilometers	feet	3,281.
kilometers	inches	3.937×10^{4}
kilometers	meters	1,000.0
kilometers	miles	0.6214
kilometers	millimeters	10^{6}
kilometers	yards	1,094.
kilometers/hr	cms/sec	27.78
kilometers/hr	feet/min	54.68
kilometers/hr	feet/sec	0.9113
kilometers/hr	knots	0.5396
kilometers/hr	meters/min	16.67
kilometers/hr	miles/hr	0.6214
kilometers/hr/sec	cms/sec/sec	27.78
kilometers/hr/sec	ft/sec/sec	0.9113
kilometers/hr/sec	meters/sec/sec	0.2778
kilometers/hr/sec	miles/hr/sec	0.6214
kilowatts	Btu/min	56.92
kilowatts	foot-lbs/min	4.426×10^{4}
kilowatts	foot-lbs/sec	737.6
kilowatts	horsepower	1.341
kilowatts	kg-calories/min	14.34
kilowatts	watts	1,000.0
kilowatt-hrs	Btu	3,413.
kilowatt-hrs	ergs	3.600×10^{13}
kilowatt-hrs	foot-lbs	2.655×10^{6}
kilowatt-hrs	gram-calories	859,850.
kilowatt-hrs	horsepower-hrs	1.341
kilowatt-hrs	joules	3.6×10^{6}
kilowatt-hrs	kg-calories	860.5
kilowatt-hrs	kg-meters	3.671×10^{5}
kilowatt-hrs	pounds of water evaporated from and at 212° F.	3.53
kilowatt-hrs	pounds of water raised from 62° to 212° F.	22.75
knots	feet/hr	6,080.
knots	kilometers/hr	1.8532
knots	nautical miles/hr	1.0
knots	statute miles/hr	1.151

(Continued). Alphabetical Conversion Factors

TO CONVERT	INTO	MULTIPLY BY
knots	yards/hr	2,027.
knots	feet/sec	1.689
L		
league	miles (approx.)	3.0
Light year	Miles	5.9×10^{12}
Light year	Kilometers	9.46091×10^{12}
lines/sq cm	gausses	1.0
lines/sq in.	gausses	0.1550
lines/sq in.	webers/sq cm	1.550×10^{-9}
lines/sq in.	webers/sq in.	10^{-8}
lines/sq in.	webers/sq meter	1.550×10^{-5}
links (engineer's)	inches	12.0
links (surveyor's)	inches	7.92
liters	bushels (U.S. dry)	0.02838
liters	cu cm	1,000.0
liters	cu feet	0.03531
liters	cu inches	61.02
liters	cu meters	0.001
liters	cu yards	1.308×10^{-3}
liters	gallons (U.S. liq.)	0.2642
liters	pints (U.S. liq.)	2.113
liters	quarts (U.S. liq.)	1.057
liters/min	cu ft/sec	5.886×10^{-4}
liters/min	gals/sec	4.403×10^{-3}
lumens/sq ft	foot-candles	1.0
Lumen	Spherical candle power	.07958
Lumen	Watt	.001496
Lumen/sq. ft.	Lumen/sq. meter	10.76
lux	foot-candles	0.0929
M		
maxwells	kilolines	0.001
maxwells	webers	10^{-8}
megalines	maxwells	10^{6}
megohms	microhms	10^{12}
megohms	ohms	10^{6}
meters	centimeters	100.0
meters	feet	3.281
meters	inches	39.37
meters	kilometers	0.001
meters	miles (naut.)	5.396×10^{-4}
meters	miles (stat.)	6.214×10^{-4}
meters	millimeters	1,000.0
meters	yards	1.094
meters	varas	1.179
meters/min	cms/sec	1.667
meters/min	feet/min	3.281
meters/min	feet/sec	0.05468
meters/min	kms/hr	0.06
meters/min	knots	0.03238
meters/min	miles/hr	0.03728
meters/sec	feet/min	196.8
meters/sec	feet/sec	3.281
meters/sec	kilometers/hr	3.6
meters/sec	kilometers/min	0.06
meters/sec	miles/hr	2.237
meters/sec	miles/min	0.03728
meters/sec/sec	cms/sec/sec	100.0
meters/sec/sec	ft/sec/sec	3.281
meters/sec/sec	kms/hr/sec	3.6
meters/sec/sec	miles/hr/sec	2.237
meter-kilograms	cm-dynes	9.807×10^{7}
meter-kilograms	cm-grams	10^{5}
meter-kilograms	pound-feet	7.233
microfarad	farads	10^{-6}
micrograms	grams	10^{-6}
microhms	megohms	10^{-12}
microhms	ohms	10^{-6}
microliters	liters	10^{-6}
Microns	meters	1×10^{-6}
miles (naut.)	feet	6,080.27
miles (naut.)	kilometers	1.853
miles (naut.)	meters	1,853.
miles (naut.)	miles (statute)	1.1516
miles (naut.)	yards	2,027.
miles (statute)	centimeters	1.609×10^{5}
miles (statute)	feet	5,280.
miles (statute)	inches	6.336×10^{4}
miles (statute)	kilometers	1.609
miles (statute)	meters	1,609.
miles (statute)	miles (naut.)	0.8684
miles (statute)	yards	1,760.
miles/hr	cms/sec	44.70
miles/hr	feet/min	88.
miles/hr	feet/sec	1.467
miles/hr	kms/hr	1.609
miles/hr	kms/min	0.02682
miles/hr	knots	0.8684
miles/hr	meters/min	26.82
miles/hr	miles/min	0.1667
miles/hr/sec	cms/sec/sec	44.70
miles/hr/sec	feet/sec/sec	1.467
miles/hr/sec	kms/hr/sec	1.609
miles/hr/sec	meters/sec/sec	0.4470
miles/min	cms/sec	2,682.
miles/min	feet/sec	88.
miles/min	kms/min	1.609
miles/min	knots/min	0.8684
miles/min	miles/hr	60.0
mil-feet	cu inches	9.425×10^{-6}
milliers	kilograms	1,000.
Millimicrons	meters	1×10^{-9}
Milligrams	grains	0.01543236
milligrams	grams	0.001
milligrams/liter	parts/million	1.0
millihenries	henries	0.001
milliliters	liters	0.001
millimeters	centimeters	0.1
millimeters	feet	3.281×10^{-3}
millimeters	inches	0.03937
millimeters	kilometers	10^{-6}
millimeters	meters	0.001
millimeters	miles	6.214×10^{-7}
millimeters	mils	39.37
millimeters	yards	1.094×10^{-3}
million gals/day	cu ft/sec	1.54723
mils	centimeters	2.540×10^{-3}
mils	feet	8.333×10^{-5}
mils	inches	0.001
mils	kilometers	2.540×10^{-8}
mils	yards	2.778×10^{-5}
miner's inches	cu ft/min	1.5
Minims (British)	cubic cm.	0.059192
Minims (U.S., fluid)	cubic cm.	0.061612
minutes (angles)	degrees	0.01667
minutes (angles)	quadrants	1.852×10^{-4}
minutes (angles)	radians	2.909×10^{-4}
minutes (angles)	seconds	60.0
myriagrams	kilograms	10.0
myriameters	kilometers	10.0
myriawatts	kilowatts	10.0
N		
nepers	decibels	8.686
Newton	Dynes	1 x 105

(Continued). Alphabetical Conversion Factors

O

TO CONVERT	INTO	MULTIPLY BY
OHM (International)	OHM (absolute)	1.0005
ohms	megohms	10^{-6}
ohms	microhms	10^{6}
ounces	drams	16.0
ounces	grains	437.5
ounces	grams	28.349527
ounces	pounds	0.0625
ounces	ounces (troy)	0.9115
ounces	tons (long)	2.790×10^{-5}
ounces	tons (metric)	2.835×10^{-5}
ounces (fluid)	cu inches	1.805
ounces (fluid)	liters	0.02957
ounces (troy)	grains	480.0
ounces (troy)	grams	31.103481
ounces (troy)	ounces (avdp.)	1.09714
ounces (troy)	pennyweights (troy)	20.0
ounces (troy)	pounds (troy)	0.08333
Ounce/sq. inch	Dynes/sq. cm.	4309
ounces/sq in.	pounds/sq in.	0.0625

P

TO CONVERT	INTO	MULTIPLY BY
Parsec	Miles	19×10^{12}
Parsec	Kilometers	3.084×10^{13}
parts/million	grains/U.S. gal	0.0584
parts/million	grains/Imp. gal	0.07016
parts/million	pounds/million gal	8.345
Pecks (British)	cubic inches	554.6
Pecks (British)	liters	9.091901
Pecks (U.S.)	bushels	0.25
Pecks (U.S.)	cubic inches	537.605
Pecks (U.S.)	liters	8.809582
Pecks (U.S.)	quarts (dry)	8
pennyweights (troy)	grains	24.0
pennyweights (troy)	ounces (troy)	0.05
pennyweights (troy)	grams	1.55517
pennyweights (troy)	pounds (troy)	4.1667×10^{-3}
pints (dry)	cu inches	33.60
pints (liq.)	cu cms.	473.2
pints (liq.)	cu feet	0.01671
pints (liq.)	cu inches	28.87
pints (liq.)	cu meters	4.732×10^{-4}
pints (liq.)	cu yards	6.189×10^{-4}
pints (liq.)	gallons	0.125
pints (liq.)	liters	0.4732
pints (liq.)	quarts (liq.)	0.5
Planck's quantum	Erg – second	6.624×10^{-27}
Poise	Gram/cm. sec.	1.00
Pounds (avoirdupois)	ounces (troy)	14.5833
poundals	dynes	13,826.
poundals	grams	14.10
poundals	joules/cm	1.383×10^{-3}
poundals	joules/meter (newtons)	0.1383
poundals	kilograms	0.01410
poundals	pounds	0.03108
pounds	drams	256.
pounds	dynes	44.4823×10^{4}
pounds	grains	7,000.
pounds	grams	453.5924
pounds	joules/cm	0.04448
pounds	joules/meter (newtons)	4.448
pounds	kilograms	0.4536
pounds	ounces	16.0
pounds	ounces (troy)	14.5833
pounds	poundals	32.17
pounds	pounds (troy)	1.21528
pounds	tons (short)	0.0005
pounds (troy)	grains	5,760.
pounds (troy)	grams	373.24177
pounds (troy)	ounces (avdp.)	13.1657
pounds (troy)	ounces (troy)	12.0
pounds (troy)	pennyweights (troy)	240.0
pounds (troy)	pounds (avdp.)	0.822857
pounds (troy)	tons (long)	3.6735×10^{-4}
pounds (troy)	tons (metric)	3.7324×10^{-4}
pounds (troy)	tons (short)	4.1143×10^{-4}
pounds of water	cu feet	0.01602
pounds of water	cu inches	27.68
pounds of water	gallons	0.1198
pounds of water/min	cu ft/sec	2.670×10^{-4}
pound-feet	cm-dynes	1.356×10^{7}
pound-feet	cm-grams	13,825.
pound-feet	meter-kgs	0.1383
pounds/cu ft	grams/cu cm	0.01602
pounds/cu ft	kgs/cu meter	16.02
pounds/cu ft	pounds/cu in.	5.787×10^{-4}
pounds/cu ft	pounds/mil-foot	5.456×10^{-9}
pounds/cu in.	gms/cu cm	27.68
pounds/cu in.	kgs/cu meter	2.768×10^{4}
pounds/cu in.	pounds/cu ft	1,728.
pounds/cu in.	pounds/mil-foot	9.425×10^{-6}
pounds/ft	kgs/meter	1.488
pounds/in.	gms/cm	178.6
pounds/mil-foot	gms/cu cm	2.306×10^{6}
pounds/sq ft	atmospheres	4.725×10^{-4}
pounds/sq ft	feet of water	0.01602
pounds/sq ft	inches of mercury	0.01414
pounds/sq ft	kgs/sq meter	4.882
pounds/sq ft	pounds/sq in.	6.944×10^{-3}
pounds/sq in.	atmospheres	0.06804
pounds/sq in.	feet of water	2.307
pounds/sq in.	inches of mercury	2.036
pounds/sq in.	kgs/sq meter	703.1
pounds/sq in.	pounds/sq ft	144.0

Q

TO CONVERT	INTO	MULTIPLY BY
quadrants (angle)	degrees	90.0
quadrants (angle)	minutes	5,400.0
quadrants (angle)	radians	1.571
quadrants (angle)	seconds	3.24×10^{5}
quarts (dry)	cu inches	67.20
quarts (liq.)	cu cms	946.4
quarts (liq.)	cu feet	0.03342
quarts (liq.)	cu inches	57.75
quarts (liq.)	cu meters	9.464×10^{-4}
quarts (liq.)	cu yards	1.238×10^{-3}
quarts (liq.)	gallons	0.25
quarts (liq.)	liters	0.9463

R

TO CONVERT	INTO	MULTIPLY BY
radians	degrees	57.30
radians	minutes	3,438.
radians	quadrants	0.6366
radians	seconds	2.063×10^{5}
radians/sec	degrees/sec	57.30
radians/sec	revolutions/min	9.549
radians/sec	revolutions/sec	0.1592
radians/sec/sec	revs/min/min	573.0
radians/sec/sec	revs/min/sec	9.549
radians/sec/sec	revs/sec/sec	0.1592
revolutions	degrees	360.0
revolutions	quadrants	4.0
revolutions	radians	6.283
revolutions/min	degrees/sec	6.0
revolutions/min	radians/sec	0.1047
revolutions/min	revs/sec	0.01667

(Continued). Alphabetical Conversion Factors

TO CONVERT	INTO	MULTIPLY BY
revolutions/min/min	radians/sec/sec	1.745×10^{-3}
revolutions/min/min	revs/min/sec	0.01667
revolutions/min/min	revs/sec/sec	2.778×10^{-4}
revolutions/sec	degrees/sec	360.0
revolutions/sec	radians/sec	6.283
revolutions/sec	revs/min	60.0
revolutions/sec/sec	radians/sec/sec	6.283
revolutions/sec/sec	revs/min/min	3,600.0
revolutions/sec/sec	revs/min/sec	60.0
Rod	Chain (Gunters)	.25
Rod	Meters	5.029
Rods (Surveyors' meas.)	yards	5.5
rods	feet	16.5
S		
Scruples	grains	20
seconds (angle)	degrees	2.778×10^{-4}
seconds (angle)	minutes	0.01667
seconds (angle)	quadrants	3.087×10^{-6}
seconds (angle)	radians	4.848×10^{-6}
Slug	Kilogram	14.59
Slug	Pounds	32.17
Sphere	Steradians	12.57
square centimeters	circular mils	1.973×10^{5}
square centimeters	sq feet	1.076×10^{-3}
square centimeters	sq inches	0.1550
square centimeters	sq meters	0.0001
square centimeters	sq miles	3.861×10^{-11}
square centimeters	sq millimeters	100.0
square centimeters	sq yards	1.196×10^{-4}
square feet	acres	2.296×10^{-5}
square feet	circular mils	1.833×10^{8}
square feet	sq cms	929.0
square feet	sq inches	144.0
square feet	sq meters	0.09290
square feet	sq miles	3.587×10^{-8}
square feet	sq millimeters	9.290×10^{4}
square feet	sq yards	0.1111
square inches	circular mils	1.273×10^{6}
square inches	sq cms	6.452
square inches	sq feet	6.944×10^{-3}
square inches	sq millimeters	645.2
square inches	sq mils	10^{6}
square inches	sq yards	7.716×10^{-4}
square kilometers	acres	247.1
square kilometers	sq cms	10^{10}
square kilometers	sq ft	10.76×10^{6}
square kilometers	sq inches	1.550×10^{9}
square kilometers	sq meters	10^{6}
square kilometers	sq miles	0.3861
square kilometers	sq yards	1.196×10^{6}
square meters	acres	2.471×10^{-4}
square meters	sq cms	10^{4}
square meters	sq feet	10.76
square meters	sq inches	1,550.
square meters	sq miles	3.861×10^{-7}
square meters	sq millimeters	10^{6}
square meters	sq yards	1.196
square miles	acres	640.0
square miles	sq feet	27.88×10^{6}
square miles	sq kms	2.590
square miles	sq meters	2.590×10^{6}
square miles	sq yards	3.098×10^{6}
square millimeters	circular mils	1,973.
square millimeters	sq cms	0.01
square millimeters	sq feet	1.076×10^{-5}
square millimeters	sq inches	1.550×10^{-3}
square mils	circular mils	1.273
square mils	sq cms	6.452×10^{-6}
square mils	sq inches	10^{-6}
square yards	acres	2.066×10^{-4}
square yards	sq cms	8,361.
square yards	sq feet	9.0
square yards	sq inches	1,296.
square yards	sq meters	0.8361
square yards	sq miles	3.228×10^{-7}
square yards	sq millimeters	8.361×10^{5}
T		
temperature (°C) +273	absolute temperature (°C)	1.0
temperature (°C) +17.78	temperature (°F)	1.8
temperature (°F) +460	absolute temperature (°F)	1.0
temperature (°F) −32	temperature (°C)	5/9
tons (long)	kilograms	1,016.
tons (long)	pounds	2,240.
tons (long)	tons (short)	1.120
tons (metric)	kilograms	1,000.
tons (metric)	pounds	2,205.
tons (short)	kilograms	907.1848
tons (short)	ounces	32,000.
tons (short)	ounces (troy)	29,166.66
tons (short)	pounds	2,000.
tons (short)	pounds (troy)	2,430.56
tons (short)	tons (long)	0.89287
tons (short)	tons (metric)	0.9078
tons (short)/sq ft	kgs/sq meter	9,765.
tons (short)/sq ft	pounds/sq in.	2,000.
tons of water/24 hrs	pounds of water/hr	83.333
tons of water/24 hrs	gallons/min	0.16643
tons of water/24 hrs	cu ft/hr	1.3349
V		
Volt/inch	Volt/cm.	.39370
Volt (absolute)	Statvolts	.003336
W		
watts	Btu/hr	3.4129
watts	Btu/min	0.05688
watts	ergs/sec	107.
watts	foot-lbs/min	44.27
watts	foot-lbs/sec	0.7378
watts	horsepower	1.341×10^{-3}
watts	horsepower (metric)	1.360×10^{-3}
watts	kg-calories/min	0.01433
watts	kilowatts	0.001
Watts (Abs.)	B.T.U. (mean)/min.	0.056884
Watts (Abs.)	joules/sec.	1
watt-hours	Btu	3.413
watt-hours	ergs	3.60×10^{10}
watt-hours	foot-pounds	2,656.
watt-hours	gram-calories	859.85
watt-hours	horsepower-hrs	1.341×10^{-3}
watt-hours	kilogram-calories	0.8605
watt-hours	kilogram-meters	367.2
watt-hours	kilowatt-hrs	0.001
Watt (International)	Watt (absolute)	1.0002
webers	maxwells	10^{8}
webers	kilolines	10^{5}

Synchronous Speeds

$$\text{Synchronous Speed} = \frac{\text{Frequency x 120}}{\text{No. of Poles}}$$

	FREQUENCY				FREQUENCY	
Poles	60 cycle	50 cycle	25 cycle	Poles	60 cycle	50 cycle
2	3600	3000	1500	42	171.4	142.9
4	1800	1500	750	44	163.6	136.4
6	1200	1000	500	46	156.5	130.4
8	900	750	375	48	150	125
10	720	600	300	50	144	120
12	600	500	250	52	138.5	115.4
14	514.3	428.6	214.3	54	133.3	111.1
16	450	375	187.5	56	128.6	107.1
18	400	333.3	166.7	58	124.1	103.5
20	360	300	150	60	120	100
22	327.2	272.7	136.4	62	116.1	96.8
24	300	250	125	64	112.5	93.7
26	276.9	230.8	115.4	66	109.1	90.9
28	257.1	214.3	107.1	68	105.9	88.2
30	240	200	100	70	102.9	85.7
32	225	187.5	93.7	72	100	83.3
34	211.8	176.5	88.2	74	97.3	81.1
36	200	166.7	83.3	76	94.7	78.9
38	189.5	157.9	78.9	78	92.3	76.9
40	180	150	75	80	90	75

Courtesy Ingersoll-Rand Co.

Temperature Conversion

NOTE: The center column of numbers in boldface refers to the temperature in degrees, either Centigrade or Fahrenheit, which it is desired to convert into the other scale. If converting from Fahrenheit to Centigrade degrees, the equivalent temperature will be found in the left column, while if converting from degrees Centigrade to degrees Fahrenhiet, the answer will be found in the column on the right.

Centigrade		Fahrenheit
-273.17	**-459.7**	
-268	**-450**	
-262	**-440**	
-257	**-430**	
-251	**-420**	
-246	**-410**	
-240	**-400**	
-234	**-390**	
-229	**-380**	
-223	**-370**	
-218	**-360**	
-212	**-350**	
-207	**-340**	
-201	**-330**	
-196	**-320**	
-190	**-310**	
-184	**-300**	
-179	**-290**	
-173	**-280**	
-169	**-273**	-459.4
-168	**-270**	-454
-162	**-260**	-436
-157	**-250**	-418
-151	**-240**	-400
-146	**-230**	-382
-140	**-220**	-364
-134	**-210**	-346
-129	**-200**	-328
-123	**-190**	-310
-118	**-180**	-292
-112	**-170**	-274
-107	**-160**	-256
-101	**-150**	-238
-96	**-140**	-220
-90	**-130**	-202
-84	**-120**	-184
-79	**-110**	-166
-73.3	**-100**	-148.0
-67.8	**-90**	-130.0
-62.2	**-80**	-112.0
-59.4	**-75**	-103.0
-56.7	**-70**	-94.0
-53.9	**-65**	-85.0
-51.1	**-60**	-76.0
-48.3	**-55**	-67.0
-45.6	**-50**	-58.0
-42.8	**-45**	-49.0
-40.0	**-40**	-40.0
-37.2	**-35**	-31.0
-34.4	**-30**	-22.0
-31.7	**-25**	-13.0
-28.9	**-20**	-4.0
-26.1	**-15**	5.0
-23.3	**-10**	.4.0

Centigrade		Fahrenheit
-20.6	**-5**	23.0
-17.8	**0**	32.0
-17.2	**1**	33.8
-16.7	**2**	35.6
-16.1	**3**	37.4
-15.6	**4**	39.2
-15.0	**5**	41.0
-14.4	**6**	42.8
-13.9	**7**	44.6
-13.3	**8**	46.4
-12.8	**9**	48.2
-12.2	**10**	50.0
-11.7	**11**	51.8
-11.1	**12**	53.6
-10.6	**13**	55.4
-10.0	**14**	57.2
-9.4	**15**	59.0
-8.9	**16**	60.8
-8.3	**17**	62.6
-7.8	**18**	64.4
-7.2	**19**	66.2
-6.7	**20**	68.0
-6.1	**21**	69.8
-5.6	**22**	71.6
-5.0	**23**	73.4
-4.4	**24**	75.2
-3.9	**25**	77.0
-3.3	**26**	78.8
-2.8	**27**	80.6
-2.2	**28**	82.4
-1.7	**29**	84.2
-1.1	**30**	86.0
-0.6	**31**	87.8
0.0	**32**	89.6
0.6	**33**	91.4
1.1	**34**	93.2
1.7	**35**	95.0
2.2	**36**	96.8
2.8	**37**	98.6
3.3	**38**	100.4
3.9	**39**	102.2
4.4	**40**	104.0
5.0	**41**	105.8
5.6	**42**	107.6
6.1	**43**	109.4
6.7	**44**	111.2
7.2	**45**	113.0
7.8	**46**	114.8
8.3	**47**	116.6
8.9	**48**	118.4
9.4	**49**	120.2
10.0	**50**	122.0
10.6	**51**	123.8

Centigrade		Fahrenheit
11.1	**52**	125.6
11.7	**53**	127.4
12.2	**54**	129.2
12.8	**55**	131.0
13.3	**56**	132.8
13.9	**57**	134.6
14.4	**58**	136.4
15.0	**59**	138.2
15.6	**60**	140.0
16.1	**61**	141.8
16.7	**62**	143.6
17.2	**63**	145.4
17.8	**64**	147.2
18.3	**65**	149.0
18.9	**66**	150.8
19.4	**67**	152.6
20.0	**68**	154.4
20.6	**69**	156.2
21.1	**70**	158.0
21.7	**71**	159.8
22.2	**72**	161.6
22.8	**73**	163.4
23.3	**74**	165.2
23.9	**75**	167.0
24.4	**76**	168.8
25.0	**77**	170.6
25.6	**78**	172.4
26.1	**79**	174.2
26.7	**80**	176.0
27.2	**81**	177.8
27.8	**82**	179.6
28.3	**83**	181.4
28.9	**84**	183.2
29.4	**85**	185.0
30.0	**86**	186.8
30.6	**87**	188.6
31.1	**88**	190.4
31.7	**89**	192.2
32.2	**90**	194.0
32.8	**91**	195.8
33.3	**92**	197.6
33.9	**93**	199.4
34.4	**94**	201.2
35.0	**95**	203.0
35.6	**96**	204.8
36.1	**97**	206.6
36.7	**98**	208.4
37.2	**99**	210.2
37.8	**100**	212.0
40.6	**105**	221
43.3	**110**	230
46.1	**115**	239
48.9	**120**	248
51.7	**125**	257

Centigrade		Fahrenheit
54.4	**130**	266
57.2	**135**	275
60.0	**140**	284
62.8	**145**	293
65.6	**150**	302
68.3	**155**	311
71.1	**160**	320
73.9	**165**	329
76.7	**170**	338
79.4	**175**	347
82.2	**180**	356
85.0	**185**	365
87.8	**190**	374
90.6	**195**	383
93.3	**200**	392
96.1	**205**	401
98.9	**210**	410
100.0	**212**	414
102	**215**	419
104	**220**	428
107	**225**	437
110	**230**	446
113	**235**	455
116	**240**	464
118	**245**	473
121	**250**	482
124	**255**	491
127	**260**	500
129	**265**	509
132	**270**	518
135	**275**	527
138	**280**	536
141	**285**	545
143	**290**	554
146	**295**	563
149	**300**	572
154	**310**	590
160	**320**	608
166	**330**	626
171	**340**	644
177	**350**	662
182	**360**	680
188	**370**	698
193	**380**	716
199	**390**	734
204	**400**	752
210	**410**	770
216	**420**	788
221	**430**	806
227	**440**	824
232	**450**	842
238	**460**	860
243	**470**	878
249	**480**	896
254	**490**	914
260	**500**	932

The formulas at the right may also be used for converting Centigrade or Fahrenheit degrees into the other scales.

Degrees Cent., $°C = \frac{5}{9}(°F + 40) - 40$

$= \frac{5}{9}(°F - 32)$

Degrees Fahr., $°F = \frac{9}{5}(°C + 40) - 40$

$= \frac{9}{5}\ C + 32$

Degrees Kelvin, $°K = °C + 273.2$

Degrees Rankine, $°R = °F + 459.7$

Courtesy Ingersoll-Rand Co.

Altitude and Atmospheric Pressures

Altitude above Sea Level			Temperature**		Barometer*		Atmospheric Pressure	
Feet*	Miles	Meters*	°F	°C	Inches Hg Abs.	mm Hg Abs.	PSIA	Kg/sq cm Abs.
-5000	——	-1526	77	25	35.58	903.7	17.48	1.229
-4500	——	-1373	75	24	35.00	889.0	17.19	1.209
-4000	——	-1220	73	23	34.42	874.3	16.90	1.188
-3500	——	-1068	71	22	33.84	859.5	16.62	1.169
-3000	——	-915	70	21	33.27	845.1	16.34	1.149
-2500	——	-763	68	20	32.70	830.6	16.06	1.129
-2000	——	-610	66	19	32.14	816.4	15.78	1.109
-1500	——	-458	64	18	31.58	802.1	15.51	1.091
-1000	——	-305	63	17	31.02	787.9	15.23	1.071
-500	——	-153	61	16	30.47	773.9	14.96	1.052
0	——	0	59	15	29.92	760.0	14.696	1.0333
500	——	153	57	14	29.38	746.3	14.43	1.015
1000	——	305	55	13	28.86	733.0	14.16	.956
1500	——	458	54	12	28.33	719.6	13.91	.978
2000	——	610	52	11	27.82	706.6	13.66	.960
2500	——	763	50	10	27.32	693.9	13.41	.943
3000	——	915	48	9	26.82	681.2	13.17	.926
3500	——	1068	47	8	26.33	668.8	12.93	.909
4000	——	1220	45	7	25.84	656.3	12.69	.892
4500	——	1373	43	6	25.37	644.4	12.46	.876
5000	0.95	1526	41	5	24.90	632.5	12.23	.860
6000	1.1	1831	38	3	23.99	609.3	11.78	.828
7000	1.3	2136	34	1	23.10	586.7	11.34	.797
8000	1.5	2441	31	-1	22.23	564.6	10.91	.767
9000	1.7	2746	27	-3	21.39	543.3	10.50	.738
10,000	1.9	3050	23	-5	20.58	522.7	10.10	.710
15,000	2.8	4577	6	-14	16.89	429.0	8.29	.583
20,000	3.8	6102	-12	-24	13.76	349.5	6.76	.475
25,000	4.7	7628	-30	-34	11.12	282.4	5.46	.384
30,000	5.7	9153	-48	-44	8.903	226.1	4.37	.307
35,000	6.6	10,679	-66		7.060	179.3	3.47	.244
40,000	7.6	12,204	-70	-57	5.558	141.2	2.73	.192
45,000	8.5	13,730	-70	-57	4.375	111.1	2.15	.151
50,000	9.5	15,255	-70	-57	3.444	87.5	1.69	.119
55,000	10.4	16,781	-70	-57	2.712	68.9	1.33	.0935
60,000	11.4	18,306	-70	-57	2.135	54.2	1.05	.0738
70,000	13.3	21,357	-67	-55	1.325	33.7	.651	.0458
80,000	15.2	24,408	-62	-52	†8.273^{-1}	21.0	.406	.0285
90,000	17.1	27,459	-57	-59	5.200^{-1}	13.2	.255	.0179
100,000	18.9	30,510	-51	-46	3.290^{-1}	8.36	.162	.0114
120,000	22.8	36,612	-26	-48	1.358^{-1}	3.45	——	——
140,000	26.6	42,714	4	-16	5.947^{-2}	1.51	——	——
160,000	30.4	48,816	28	-2	2.746^{-2}	†6.97^{-1}	——	——
180,000	34.2	54,918	19	-7	1.284^{-2}	3.26^{-1}	——	——
200,000	37.9	61,020	-3	-19	5.846^{-3}	1.48^{-1}	——	——
220,000	41.7	67,122	-44	-42	2.523^{-3}	6.41^{-2}	——	——
240,000	45.5	73,224	-86	-66	9.955^{-4}	2.53^{-2}	——	——
260,000	49.3	79,326	-129	-90	3.513^{-4}	8.92^{-3}	——	——
280,000	53.1	85,428	-135	-93	1.143^{-4}	3.67^{-3}	——	——
300,000	56.9	91,530	-127	-88	3.737^{-5}	9.49^{-4}	——	——
400,000	75.9	122,040	——	——	6.3^{-7}	1.60^{-5}	——	——
500,000	94.8	152,550	——	——	1.4^{-7}	3.56^{-6}	——	——
600,000	114	183,060	——	——	5.9^{-8}	1.50^{-6}	——	——
800,000	152	244,080	——	——	1.6^{-8}	4.06^{-7}	——	——
1,000,000	189	305,100	——	——	5.1^{-9}	1.30^{-7}	——	——
1,200,000	228	366,120	——	——	2.0^{-9}	5.08^{-8}	——	——
1,400,000	266	427,140	——	——	8.2^{-10}	2.08^{-8}	——	——
1,600,000	304	488,160	——	——	3.8^{-10}	9.65^{-9}	——	——
1,800,000	342	549,180	——	——	1.8^{-10}	4.57^{-9}	——	——
2,000,000	379	610,200	——	——	9.2^{-11}	2.34^{-9}	——	——

Data from NASA Standard Atmosphere (1962).

*Temperature and barometer are approximate for negative altitudes.

**Temperatures are average existing at 40° latitude and are rounded to even numbers.

†Negative exponent shows number of spaces the decimal point must be moved to the left.

Courtesy Ingersoll-Rand Co.

Pressure Conversion Chart

BY FACTOR TO OBTAIN →

MULTIPLY GIVEN NUMBER OF ↓

GIVEN	lb/in²	in H_2O (at +39.2°F)	cm H_2O (at +4°C)	in Hg (at +32°F)	mm Hg (Torr) (at 0°C)	dyne/cm² (1μ bar)	newton/m² (PASCAL)	kgm/cm²	bar	atm. (A_n)	lb/ft²	ft H_2O (at +39.2°F)
lb/in²	1.0000	2.7680×10^{1}	7.0308×10^{1}	2.0360	5.1715×10^{1}	6.8948×10^{4}	6.8948×10^{3}	7.0306×10^{-2}	6.8947×10^{-2}	6.8045×10^{-2}	1.4400×10^{2}	2.3067
in H_2O (at +39.2°F)	3.6127×10^{-2}	1.0000	2.5400	7.3554×10^{-2}	1.8683	2.4908×10^{3}	2.4908×10^{2}	2.5399×10^{-3}	2.4908×10^{-3}	2.4582×10^{-3}	5.2022	8.3333×10^{-2}
cm H_2O (at +4°C)	1.4223×10^{-2}	0.3937	1.0000	2.8958×10^{-2}	0.7355	9.8064×10^{2}	9.8064×10^{1}	9.9997×10^{-4}	9.8064×10^{-4}	9.6781×10^{-4}	2.0481	3.2808×10^{-2}
in Hg (at +32°F)	4.9116×10^{-1}	1.3596×10^{1}	3.4532×10^{1}	1.0000	2.5400×10^{1}	3.3864×10^{4}	3.3864×10^{3}	3.4532×10^{-2}	3.3864×10^{-2}	3.3421×10^{-2}	7.0727×10^{1}	1.1330
mm Hg (Torr) (at 0°C)	1.9337×10^{-2}	5.3525×10^{-1}	1.3595	3.9370×10^{-2}	1.0000	1.3332×10^{3}	1.3332×10^{2}	1.3595×10^{-3}	1.3332×10^{-3}	1.3158×10^{-3}	2.7845	4.4605×10^{-2}
dyne/cm² (1μ bar)	1.4504×10^{-5}	4.0147×10^{-4}	1.0197×10^{-3}	2.9530×10^{-5}	7.5006×10^{-4}	1.0000	1.0000×10^{-1}	1.0197×10^{-6}	1.0000×10^{-6}	9.8692×10^{-7}	2.0886×10^{-3}	3.3456×10^{-5}
newton/m² (PASCAL)	1.4504×10^{-4}	4.0147×10^{-3}	1.0197×10^{-2}	2.9530×10^{-4}	7.5006×10^{-3}	1.0000×10^{1}	1.0000	1.0197×10^{-5}	1.0000×10^{-5}	9.8692×10^{-6}	2.0885×10^{-2}	3.3456×10^{-4}
kgm/cm²	1.4224×10^{1}	3.9371×10^{2}	1.00003×10^{3}	2.8959×10^{1}	7.3556×10^{2}	9.8060×10^{5}	9.8060×10^{4}	1.0000	9.8060×10^{-1}	9.678×10^{-1}	2.0482×10^{3}	3.2809×10^{1}
bar	1.4504×10^{1}	4.0147×10^{2}	1.0197×10^{3}	2.9530×10^{1}	7.5006×10^{2}	1.0000×10^{6}	1.0000×10^{5}	1.0197	1.0000	9.8692×10^{-1}	2.0885×10^{3}	3.3456×10^{1}
atm. (A_n)	1.4696×10^{1}	4.0679×10^{2}	1.0333×10^{3}	2.9921×10^{1}	7.6000×10^{2}	1.0133×10^{6}	1.0133×10^{5}	1.0332	1.0133	1.0000	2.1162×10^{3}	3.3900×10^{1}
lb/ft²	6.9445×10^{-3}	1.9223×10^{-1}	4.882×10^{-1}	1.4139×10^{-2}	3.591×10^{-1}	4.7880×10^{2}	4.7880×10^{1}	4.8824×10^{-4}	4.7880×10^{-4}	4.7254×10^{-4}	1.0000	1.6019×10^{-2}
ft H_2O (at +39.2°F)	4.3352×10^{-1}	1.2000×10^{1}	3.0480×10^{1}	8.826×10^{-1}	2.2419×10^{1}	2.9890×10^{4}	2.9890×10^{3}	3.0479×10^{-2}	2.9890×10^{-2}	2.9499×10^{-2}	6.2427×10^{1}	1.0000

Index

ACI bearing strengths, 180
American Institute of Steel Construction. *See* AISC.
Anchor bolts
  analysis, preloaded bolt, 184, 186
  bolt area, required, 184
  bolt loads, allowable, 187
  bolt load, minimum required, 184
  bolt spacing, 186
  common types of, 190
  large bolts, undesirability of, 184
  loading force, distribution of, 186
  loadings induced on, 184
  lubricant, 190
  philosophy, design, 184
  size and number, 228
  stress in, 184, 186
  tension on gross area, 187
  torque, anchor bolt, 189–190, 229
ASME Piping Codes
  ASME B31.1, 48
  ASME B31.3, 48
  ASME B31.4, 48
  ASME B31.5, 48
  ASME B31.8, 48
ASME Section III, 48. *Also see* Pressure vessels.
  for piping, 48
  for pressure vessels, 48
ASME Section VIII, Division II
  for piping, 48
Aspect ratio, 85

Baseplate design, 186–189
  anchor bolt size range, 186
  bearing pressure on, 189
  concrete foundation for, 186
  concrete mixes, 186, 187
  concrete modulus of elasticity of, 186
  concrete and steel, relative strength of, 186
  gusset plates, 188–189
  k-factor, offset, 188
  steel, modulus of elasticity, 187
  steel-concrete moduli ratio, 186
  tension on gross area, 187
  torque, anchor bolt, 189–190, 229
Bernoulli equation, 2
Bingham, 6–7
Boundary conditions for saddle plate design, 178
Buckling coefficients for saddle plate design, 175–178

Centroid, section, 212
Circumferential stress, moment, 170
Codes, vessel
  differences in, 159
  foreign, 159
Cold-spring, 49
Colebrook equation, 4. *Also see* Friction factors.
Compressible flow
  adiabatic flow, 2
  compressibility effects, 24
  introduction to, 1–2, 24
  isothermal flow, 1
  modulus, bulk compressibility, 24
  non-steady flow, 24
  sound, velocity of, 24
  steady flow, 24
Concrete mixes for baseplate design, 186–187
Concrete modulus of elasticity, 186
Conical sections, 199, 224
Cost-plus contractor, 183
Creep, 49
Critical damping factor, 202, 204
Critical pressure, 83

Critical temperature, 83
Critical wind velocity, 236

Damping coefficient, 202–204
Deflections, wind, 199–201, 242
Degree of freedom, 201
Discontinuity, 236
Drag, 195, 203
Ductile materials, 50, 52
Dynamic magnification factor, 201–204
Dynamic response, 200

EJMA. *See* Expansion joints, bellows.
Electrical tracing, 103
Equivalent length, 2
Expansion joints
  bellows, corrugated, 77
  gimbal joint, 79
  hinged joint, 78–79
  in-line pressure balanced, 79
  multi-ply, 80
  pipe span, allowable, 78
  pressure thrust, 78–79
  single ply, 80
  standards of the Expansion Joint Manufacturers Association (EJMA), 80
  stiffness, rotational, 78
  stiffness, translational, 78
  tie rods, 78–79
    reasons for, 78
  universal joint, pressure-balanced, 78

Fanning equation, 3
Fluid Mechanics, piping. *See* Hydraulics.
Fourier number, 151
Friction factors, 4
  Colebrook equation, 4
  laminar flow, 4
  Moody friction factors, 4
  Prandtl solution, 5
  turbulent flow, 4
  von Karman solution, 5

Gimbal joint, 79
Grashof number, 132, 134, 153
Gusset plates, 188–189
Gust (wind) effects, 194–196, 236–237
Guy wires, 249

Head
  foot of, 2
  pressure, 1
  static, 1
  velocity. *See* Velocity head.

Heads
  manufacture of, 160
  thickness of, 160
Heat transfer
  control mass, 115, 131
  control volume, 115, 131
  electrical tracing, 103
  Fourier number, 151
  Grashof number, 132, 134, 153
  in jacketed pipe, 112–115
  LMTD (log mean)
    chart for, 114
    definition of, 114
  Nusselt number, 132, 134, 153
  in pipe shoes, 135–136
    application of, 156
    heat balance for, 136
    temperature distribution in, 136
  in pipe supports, 133
  in piping
    temperature distribution in, 134
    typical applications of, 133–134
  Prandtl number, 112, 139–140
  in process systems, 103
  in residual systems
    applications of, 132
    deflections, thermal, 134–135
    overall heat transfer coefficient, 134
    in piping, 134, 154–155
  tubular tracers. *See* Tracing.
  in vessel skirts
    application of, 152–154
    coefficients of, 132
    convection, significance of, 133
    free convection, 133
    rate of, 133
    temperature, distribution of, 132–133
Heat transfer design example, 148–150
  static analysis, 148–150
  transient analysis, 150–152
Heisler's chart, 151
Hesse formula, 82
Horizontal pressure vessels
  saddle bearing plate design, 180
    ACI bearing strengths, 180
    bearing plate thickness, 180
    factor of safety for, 180
  saddle plate buckling analysis, 251–252
  saddle plates
    application of, 249–252
    boundary conditions for, 178
    buckling coefficients for, 175–178
    design of, 174–179
    effective area, 174, 178

effective width, 173, 178, 179
horizontal reaction, 179, 252
stiffener plates, 174, 179
stress
criterion for residual, 178
elastic buckling, 179
inelastic buckling, 179
U.S. Steel design method, 174–179
web plates, 174
wear plate requirements, 215
Zick analysis, 166, 215
bending moment diagram, 167
constant, circumferential bending moment, 170
introduction to, 166
saddle supports, location, criteria for, 172
shear stress, 171
shell
stiffened by head, 171
unstiffened, saddles away from head, 171
stiffening rings, 172, 174
stress
allowable compressive, 166
circumferential compressive, 171
circumferential at horn of saddle, 171
head used as a stiffener, 171
"Hot-spring," 49
Hydraulic radius,
definition of, 21
tabulated values, 24
Hydraulics
basic equations, 1
Bernoulli equation, 2
modified form of, 3
compressible flow
adiabatic flow, 2
compressibility effects, 24
introduction to, 1–2, 24
isothermal flow, 1
modulus, bulk compressibility, 24
non-steady flow, 24
sound, velocity of, 24
steady flow, 24
incompressible flow, 1
non-Newtonian fluids
Bingham, 6–7
introduction to, 5–7
Metzer and Reed, 7
pseudoplastic, 6–7
rheological constants, 8
rheopectic, 6–7
thixotropic, 6–7
time-dependent, 6–7
time-independent, 6–7
viscoelastic, 6–7
yield-pseudoplastic, 6–7
piping, reasonable velocities in, 25
problem formulation, 24
two-K method, 8, 21
viscosity, 24–26

Incompressible flow. *See* Hydraulics.
Internal pressure, 159–160

Jacketed pipe
annulus, hydraulic radius for, 112
applications of, 112–115, 139–140
details of, 104–106, 112–113
expansion joints for, 105–106
heat transfer, 112–115
coefficient, film, 112
coefficient, overall, 112
rates of, 112–115
pressure drop in, 115–117
rules of thumb for, 103
versus traced pipe, 103–106
Joints, expansion. *See* Expansion joints.
Laminar flow, 4. *Also see* Friction factors.
Lumped-mass approach, 204–205
Lump-sum contractor, 183

Maximum allowable working pressure, 160
Mitchell equation, 210, 212
Moments
equations for, 198
of inertia, for tube bundle, 222–223
wind-induced, 198
Moody friction factors. *See* Friction factors.
Myklestad method, 200–201

Non-Newtonian fluids. *See* Hydraulics.
Nusselt number, 132, 134, 153

Ovaling, 205, 208

Pipe loops, 59–68
Pipe lug supports, 70–72, 98–99
Pipe materials
ductile materials, 50, 52
non-ductile materials, 50
plastic deformation, 50–52
stress-strain curves, 50–51
Pipe shoes, heat transfer in, 135–136
Pipe supports, heat transfer in, 133
Piping codes. *See* ASME.
Piping expansion joints. *See* Expansion joints.
Piping mechanics
anchor, pipe, definition, 58
API, 47

equipment nozzle loads, 94
extraneous piping loads
"cold spring" for, 80
vibration
applications for, 100–101
natural frequency of beam elements, 86
vortex shedding, 83–87
resonance, 83
Reynolds number, 195, 200, 201, 236
Strouhal number, 84–85
vortex force, 83
vortex streets, 83
flexibility (compliance) matrix, 53
flexibility method, 59–68, 81
advantages of, 53, 68
application of, 95–98
"hot-spring," 49
nozzle flexibility factors,
angle of twist, 70
circumferential, 70
longitudinal, 70
Oak Ridge Phase 3 Report, 70
rotation deformation of, 70
rotational spring rate, 70
pipe loops, 59–68
pipe lug supports, 70–72, 98–99
pipe restraints
moment restraints (MRS), 57–59, 77, 88–94
rotational 58, 68
translational, 58, 68
pipe roughness, 5
pipe stress
circumferential bending/membrane, 71
"cold-spring," 49
creep, 49
"hot-spring," 49
internal pressure, circumferential stress, 49
longitudinal stress, 49
pipe weight, bending stress, 49
pressure, 72
prestressed piping, 80
primary stress, 49–50, 72
range, allowable, 42
residual stress, 51
secondary stress, 49–52, 72
self-spring, 49
"shakedown," 52
thermal expansion, 49
torsional or shear stress, 49
self-spring, 49
shear flow, 58–59
spring supports, 72, 75, 76
guided load column, 72
jamming of, 77
stiffness
beam element, 54
concrete, 69
matrix, 53–54
method, 81
advantages, 53, 68
applications of, 88–94
piping elements, 55–56, 69
translational, 54
Pipe Stress. *See* Piping mechanics.
Piping systems
adiabatic process, 83
API 520 Part 2, 82
ASME 31.1, 82
critical pressure, 83
critical pressure ratio, 83
critical temperature, 83
Hesse formula, 82
impulse-momentum principle, as applied to a pipe elbow, 81
nozzle correction factor, 82
nozzle discharge coefficient, 82
nozzles, 83
Prandtl number, 112, 139–140
Pressure vessels
ASME Section VIII Division I, 160
components, 159–160
design, philosophy of, 159
external pressure, 160
heads, 160
horizontal
saddle bearing plate design, 180
saddle plate buckling analysis, 251–252
saddle plate design, 174–179
application of, 249–252
boundary conditions for, 178
buckling coefficients for, 175–178
effective area, 174, 178
effective width, 173, 178, 179
horizontal reaction, 179, 252
stiffener plates, 174, 179
stress, criterion for residual, 178
stress, elastic buckling, 179
stress, inelastic buckling, 179
U.S. Steel design method, 174–179
wear plate requirements, 215
web plates, 174
Zick analysis, 166, 215
bending moment diagram, 167
compressive B-factor, 174
constant, circumferential bending moment, 170
head used as stiffener, 171
saddle support location, 172

shear stress in head/shell, 171
shell
stiffened by head, 171
unstiffened, saddles away from head, 171
stiffening rings, 172, 174
stress, allowable compressive, 166
stress, circumferential compressive, 171
stress, location of, 168–169
tangential shear, 167–171
wear plates, 171–172
internal pressure
component thickness, 159
maximum allowable working pressure, 160
quality of welds, 159
upset conditions, 160
vertical
anchor bolts
analysis, preloaded bolt, 184, 186
bolt area, required, 184
bolt loads, allowable, 187
bolt load, minimum required, 184
bolt spacing, 186
common types of, 190
large bolts, undesirability of, 184
loading force, distribution of, 186
loadings induced on, 184
lubricant, 190
philosophy, design, 184
size and number, 228
stress in, 184, 186
tension on gross area, 187
torque, anchor bolt, 189–190, 229
ANSI-1982, 215
baseplate design, 186–189
anchor bolt size range, 186
bearing pressure on, 189
concrete foundation for, 186
concrete mixes, 186, 187
concrete modulus of elasticity of, 186
concrete and steel, relative strength of, 186
gusset plates, 188–189
k-factor, offset, 188
steel, modulus of elasticity, 187
steel-concrete moduli ratio, 186
stress, compressive, on concrete, 188
thickness, baseplate, 188
centroid, section, 212
combined loads on, 181
compression plate, 189
cone, truncated, equivalent radius for, 214
conical head, equivalent radius for, 214
conical sections, equivalent radii for, 224
earthquake, *See* Seismic design.
loads, wind and seismic, 190–191
moments
equations for, 198
of inertia, for tube bundle, 222–223
pressure sections, centroids of, 198
vectors, section force, 198
wind-induced, 198
wind pressure, distribution of, 198
section properties of, 181
seismic analysis of, loads, combined, 190–191
seismic design
baseplate design, 238
coefficients, Mitchell, 210, 213
coefficients, structure type, 210
criteria, quasi-static, 210
criteria, 238
Mitchell equation, 210, 212
compared to Rayleigh equation, 237–238
occupancy importance factor, 210
period
characteristic site, 238
numeric integration of vibration, 238–239
of tower, 210, 212
Rayleigh equation, 212
compared to Mitchell equation, 237–238
seismic zone factor/map, 210–211
site structure interaction factor, 210, 212
equation for, 212
shear forces
earthquake force, total, 212
lateral force, equation for, 212
vertical distribution of, 212
seismic moments, equation for, 212
skirt design, 238
structural period response factor, 210
Uniform Building Code, 209–210
self-supporting, 180
skirts
controlling criteria for, 184
design of, 183, 185
cost-plus contractor, 183
lump-sum contractor, 183
stress equation, 183
supports, 183, 185
thickness, 183–184
stress, bending, 181
combined loading, 181
compressive B factor, 191
compressive, leeward side, 181
discontinuity, 236
elements in, 182
tensile, windward side, 181
vacuum, 183
towers
centroids, section, 230–231

definition of, 181
equivalent circle method, 214
section moment of inertia, 241–243
skirt and baseplate design, 228–229
anchor bolts, 228
anchor bolt torque, 229
compression ring thickness, 229
skirt thickness, 229
weld size, minimum for skirt-to-base plate, 229
skirt detail, 230
stress, discontinuity
criteria for, 214
for conical sections, 214
stresses, wind section, 226–228
transition piece, 241, 243–244
vibration ensemble, 216
of lumped masses, 232, 246
wind deflections
modes of, 199
schematic diagram of, 201
superposition, method of, 199
wind ensemble, 242
vibration, wind-induced
angular natural undamped frequency, 205
applications of, 232–236, 241–249
area-moment method, 205–207
conjugate beam. *See* Area moment.
controlling length, 203
critical damping factor, 202, 204
critical wind velocity, 208–209, 236, 248–249
total wind force, 209
Zorilla criteria, 209
damping coefficient, 203
damping ratio, 202–203
degree of freedom, single, 201
differential equations for, 201–202
dynamic magnification factor, 201–202, 203, 204
dynamic response, 200
example of, 232–236
first period of, 204
force amplitude, 235
force amplitude, dynamic, 200
forced vibration theory, 200
frequency
natural, 248
ratio, 202
vortex shedding, 208, 248
guy wires, disadvantages of, 249
Holzer procedure, 200
lock-in effect, 200
logarithmic decrement, 203–204
lumped mass approach, 204–205
mode shapes, 200
Myklestad method, 200, 201
ovaling, 205
natural frequency of, 205
vibration due to, 208
wind velocity, resonance, 208
period of vibration, 234–235, 248
phase angle, 202
Rayleigh equation, 200, 201, 204, 205
resonance, 236
Reynolds number, 195, 200, 201, 236
soil types, 204
stresses, dynamic, 236
tower
fluid forces on, 203
model for, 201–202
moment distribution in, 205
stiffness, 205
vibration ensemble, 209
of lumped masses, 232
vibration, first peak amplitude, 200
vortex shedding, 199
vortex strakes, 249
wind tunnel tests, 236
wind analysis of, loads, combined, 190–191
wind design speed
ASA 58.1-1955, 194
ANSI-A58.1-1972, 192
basic wind pressure, 192
effective velocity pressure, 192
gust response factor, dynamic, 192
ANSI A58.1-1982, 196, 236–237
effective velocity pressure, 192
gust response factor, 192
importance coefficient, 192
velocity pressure coefficient, 192
wind speed, variation of, 192
wind tunnel tests, 192
centroid of spandrel segment, for wind section, 218
coefficient, drag, 195
structural damping, 217
conical sections, 199
constant exposure category, 195
cross-sectional area, effective, 217
cylinder, pressure fields around, 196
equivalent diameter method, 236–237
vs. ANSI-A58.1-1982, 236–237
exposure factor, 196
fatigue failure, 198
flexible structures, defined, 197
gust duration, 196
vs. gust diameter, 197
gust frontal area, 196

gust response, dynamic, 194
gust response factor, 195, 196, 217, 236–237
gust size, 196
isopleths, 192–193
Kutta-Joukowski theorem, 195
loading analysis, quasi-static, 196
logarithmic law, 192
parabolic area, centroid of, 219
parabolic function, 194
peak values, types of, 196
power law, 192
probability of exceeding, 196
response spectra, 198
return period, 192
similarity parameters, 195
structure size factor, 196, 197
surface roughness, 195
tower
cross-sectional area of, 198
fluid force exerted on, 194–195
gust velocity vs. structural response, 197
natural frequency of, 197
wind area section properties, 219
wind force distribution, 218
wind distribution
parabolic, 194, 218–219
triangular, 194
wind load
applications of, 215–231, 241–245
equivalent static, 195
mean, 195
weld size, skirt-to-base plate, 189
welding, joint efficiencies for, 161–165, 172
Zick analysis, 166, 215
bending moment diagram, 167
compressive B-factor, 174
constant, circumferential bending moment, 170
head used as stiffener, 171
saddle support location, 172
shear stress in head/shell, 171
shell
stiffened by head, 171
unstiffened, saddles away from head, 171
stiffening rings, 172, 174
stress, allowable compressive, 166
stress, circumferential compressive, 171
stress, location of, 168–169
tangential shear, 167–171
wear plates, 171–172

Residual systems, heat transfer in, 132–135
in piping, 154–155
Reynolds number, 195, 200, 201, 236
drag coefficient vs., 203
Newtonian fluids, 21, 30, 32–41, 139–140, 142, 145, 147
non-Newtonian fluids. *See* Hydraulics, Non-Newtonian fluids.
Strouhal coefficient vs., 85
vortex shedding, for, 83–85

Saddle plate design, 174–179
application of, 249–252
boundary conditions for, 178
buckling coefficients for, 175–178
effective area, 174, 178
effective width, 173, 178, 179
horizontal reaction, 179, 252
stiffener plates, 174, 179
stress, criterion for residual, 178
stress, (in-) elastic buckling, 179
U.S. Steel design method, 174–179
wear plate requirements, 215
web plates, 174
Seismic design
baseplate design, 238
coefficients, Mitchell, 210, 213
coefficients, structure type, 210
criteria, quasi-static, 210
compared to wind, 238
Mitchell equation, 210, 212
compared to Rayleigh equation, 237–238
moments, equation for, 212
occupancy importance factor, 210
period, characteristic site, 238
period, vibration
numeric integration of, 238–239
tower, 210, 212
Rayleigh equation, 212
compared to Mitchell equation, 237–238
seismic zone factor/map, 210, 211
shear forces
earthquake force, total, 212
lateral force, equation for, 212
vertical distribution of, 212
site structure interaction factor, 210, 212
equation for, 212
skirt design, 238
structural period response factor, 210
Uniform Building Code, 209–210
Skirts, 185
controlling criteria for, 184
cost-plus contractor, 183
design of, 183
lump-sum contractor, 183
stress equation, 183
supports, 185
thickness, 183–184

Strouhal number, 84
Reynolds number vs., 85
vibration, vortex shedding, 84–85, 200, 208
Supports, 72, 75, 76. *Also see* Piping mechanics.

Thermal design. *See* Heat transfer tie rods, 78–79
Towers
centroids, section, 230–231
definition of, 181
equivalent circle method, 214
section moment of inertia, 241–243
skirt and baseplate design, 228–229
anchor bolts, 228
anchor bolt torque, 229
compression ring thickness, 229
skirt thickness, 229
weld size, minimum for skirt-to-base plate, 229
skirt detail, 230
stress, discontinuity
criteria for, 214
for conical sections, 214
stresses, wind section, 226–228
transition piece, 241, 243–244
vibration ensemble, 216
of lumped masses, 232, 246
wind deflections of
modes of, 199
schematic diagram of, 201
superposition, method of, 199
wind ensemble, 242
Tracing
of pipes
applications of, 136–139
condensate return for, 110
condensate load, determining, 111
guidelines for, 110–111
spargers, 111
separation keys, 111
typical layout, 111
water hammer, 111
hot oil, application of, 137–139
steam, application of, 136–137
versus jacketed pipe, 103–106
with heat transfer cement, 106, 109–110
advantages, 106
procedure for, 109
film coefficient, natural convection, 108–109
heat balance for, 110
heat transfer rates of, 110
without heat transfer cement, 106–109
advantages of, 106
disadvantages of, 106
equivalent insulation thickness, 107
heat balance for, 107
heat transfer, rules of, 107
modes of heat transfer, 107
outside film coefficient, 107
overall heat transfer coefficient, 107
procedure for design, 107
of vessels and equipment
agitators
film coefficients for, 143
use of, 115
applications of, 130, 140–148
film coefficient, vessel-side, 147
heat duty of, jacketed heads, 146
heat transfer coefficients, reasonable values of, 130
transient, 115
criteria for, 115
importance of, 130
internal baffle plates, heat duty of, 144
jacketed walls, heat transfer film coefficient, 145
jackets, types of, 115, 128–131
non-Newtonians, use of, 146
plate channels, equivalent velocity of, 147
reasons for, 115
Turbulent flow, 4. *Also see* Friction factors.

Velocity head
introduction, 3, 8
method, 3
two-K method, 8, 21
values of, 9–20, 21, 22–23, 30–32
Vessels. *See* Pressure vessels.
Vibration, wind-induced
angular natural umdamped frequency, 205
applications of, 232–236, 241–249
area-moment method, 205–207
conjugate beam. *See* Area moment.
controlling length, 203
critical damping factor, 202, 204
critical wind velocity, 208–209, 236, 248–249
total wind force, 209
Zorilla criteria, 209
damping coefficient, 203
damping ratio, 202–203
degree of freedom, single, 201
differential equations for, 201–202
dynamic magnification factor, 201–202, 203, 204
dynamic response, 200
example of, 232–236
first period of, 204
force amplitude, 235
force amplitude, dynamic, 200
forced vibration theory, 200
frequency
natural, 248

    ratio, 202
    vortex shedding, 208, 248
  guy wires, disadvantages of, 249
    Holzer procedure, 200
    lock-in effect, 200
    logarithmic decrement, 203–204
  lumped mass approach, 204–205
  mode shapes, 200
  Myklestad method, 200, 201
  ovaling, 205
    natural frequency of, 205
    vibration due to, 208
    wind velocity, resonance, 208
  period of vibration, 234–235, 248
  phase angle, 202
  Rayleigh equation, 200, 201, 204, 205
  resonance, 236
  Reynolds number, 195, 200, 201, 236
  soil types, 204
  stresses, dynamic, 236
  tower
    fluid forces on, 203
    model for, 201–202
    moment distribution in, 205
      equations for, 205
      stiffness, 205
    vibration ensemble, 209
      of lumped masses, 232
    vibration, first peak amplitude, 200
    vortex shedding, 83–87, 199
    vortex strakes, 249
    wind tunnel tests, 236
Viscosity, 24–25
von Karman solution, 5
Vortex shedding, 83–87
  aspect ratio, 85
  cylinders, 83
  damping vs. amplitude, 87
  guidelines for, 85
  mode shapes, 85
  reduced damping, 85

Weld sizes
  recommended values, for plates, 71
  skirt to baseplate, 189
Welding, joint efficiencies for, 161–165, 172
Wind design speed
  ASA 58.1-1955, 194
  ANSI A58.1-1972
    basic wind pressure, 192
    effective velocity pressure, 192
    gust response factor, dynamic, 192
  ANSI A58.1-1982, 196, 236–237
    effective velocity pressure, 192
    gust response factor, 192
    importance coefficient, 192
    velocity pressure coefficient, 192
    wind speed, variation of, 192
    wind tunnel tests, 192
  centroid of spandrel segment, for wind section, 218
  coefficient, drag, 195
    structural damping, 217
  conical sections, 199
  constant exposure category, 195
  cross-sectional area, effective, 217
  cylinder, pressure fields around, 196
  equivalent diameter method, 236–237
    vs. ANSI-A58.1-1982, 236–237
  exposure factor, 196
  fatigue failure, 198
  flexible structures, defined, 197
  gust duration, 196
    vs. gust diameter, 197
  gust frontal area, 196
  gust response, dynamic, 194
  gust response factor, 195, 196, 217, 236–237
  gust size, 196
  isopleths, 192–193
  Kutta-Joukowski Theorem, 195
  loading analysis, quasi-static, 196
  logarithmic law, 192
  parabolic area, centroid of, 219
  parabolic function, 194
  peak values, types of, 196
  power law, 192
  probability of exceeding, 196
  response spectra, 198
  return period, 192
  similarity parameters, 195
  structure size factor, 196, 197
  surface roughness, 195
  tower
    cross-sectional area of, 198
    fluid force exerted on, 194–195
    gust velocity vs. structural response, 197
    natural frequency of, 197
  wind area section properties, 219
  wind force distribution, 218
  wind distribution
    parabolic, 194, 218–219
    triangular, 194
  wind load
    applications of, 215–231, 241–245
    equivalent static, 195
    mean, 195

Yield, 159
  octahedral shear stress theory, 236

Zick analysis, 166, 215
  bending moment diagram, 167
  compressive B-factor, 174
  constant, circumferential bending moment, 170
  head used as stiffener, 171
  saddle support location, 172
  shear stress in head/shell, 171
  shell
    stiffened by head, 171
    unstiffened, saddles away from head, 171
  stiffening rings, 172, 174
  stress, allowable compressive, 166
  stress, circumferential compressive, 171
  stress, location of, 168–169
  tangential shear, 167–171
  wear plates, 171–172